Lecture Notes in Computer Science 16923

The series Lecture Notes in Computer Science (LNCS), including its subseries Lecture Notes in Artificial Intelligence (LNAI) and Lecture Notes in Bioinformatics (LNBI), has established itself as a medium for the publication of new developments in computer science and information technology research, teaching, and education.

LNCS enjoys close cooperation with the computer science R & D community, the series counts many renowned academics among its volume editors and paper authors, and collaborates with prestigious societies. Its mission is to serve this international community by providing an invaluable service, mainly focused on the publication of conference and workshop proceedings and postproceedings. LNCS commenced publication in 1973.

David Duenas-Cid · Gregor Eibl ·
Lieselot Danneels · Michael Koddebusch ·
Edimara Mezzomo Luciano ·
Gianluca Carlo Misuraca · Gerhard Schwabe ·
Teta Stamati · Leif Sundberg · Peter Parycek
Editors

Electronic Participation

18th IFIP WG 8.5 International Conference, ePart 2026
Athens, Greece, August 30 – September 3, 2026
Proceedings

Springer

Editors
David Duenas-Cid
Kozminski University
Warsaw, Poland

Lieselot Danneels
Ghent University
Gent, Belgium

Edimara Mezzomo Luciano
Pontifícia Universidade Católica do Rio
Grande do Sul
Porto Alegre, Brazil

Gerhard Schwabe
University of Zurich
Zurich, Switzerland

Leif Sundberg
Mid Sweden University/ Umeå University
Sundsvall, Sweden

Gregor Eibl
University for Continuing Education Krems
Krems an der Donau, Austria

Michael Koddebusch
Muenster University
Münster, Germany

Gianluca Carlo Misuraca
Politecnico di Milano
Milan, Italy

Teta Stamati
Harokopio University of Athens
Kallithea, Greece

Peter Parycek
University for Continuing Education Krems
Krems an der Donau, Austria

ISSN 0302-9743 ISSN 1611-3349 (electronic)
Lecture Notes in Computer Science
ISBN 978-3-032-36675-7 ISBN 978-3-032-36676-4 (eBook)
https://doi.org/10.1007/978-3-032-36676-4

This Springer imprint is published by the registered company Springer Nature Switzerland AG
The registered company address is: Gewerbestrasse 11, 6330 Cham, Switzerland

Preface

The EGOV-CeDEM-ePart 2026 conference, for short EGOV 2026, was organized by the IFIP WG 8.5 Information Systems in Public Administration. The conference was dedicated to digital or electronic government, open government, local government (smart cities), smart governance, artificial intelligence (AI), e-democracy, policy informatics, and electronic participation. With a long tradition along its various branches, the EGOV-CeDEM-ePart conference has earned its reputation of being the key and leading conference worldwide in the research domains of digital/electronic, open, and smart government, as well as electronic participation.

The call for papers attracted completed research papers, work-in-progress papers on ongoing research (including doctoral papers), posters, as well as workshops and practitioner papers. This volume contains only full research papers. All submissions were assessed through a double-blind peer-review process, with at least three reviewers per submission, and 27 papers were accepted from 84 submissions, giving an acceptance rate of 32%. The submitted papers were of high quality, and many reviewers indicated that the topic of the papers they reviewed were novel.

The review process was based on a double-blind reviewing process and conflicts of interest were actively avoided. Authors of papers submitted their papers to a track. The track chairs had access only to the papers within their own track and could assign reviewers and propose acceptance decisions. The lead track chairs were editors of the proceedings, in addition to the general chairs. Track chairs were not allowed to submit to their own track, nor were their close collaborators, to avoid any conflict of interest. Track chairs could either submit to another track or to the 'track chairs' track. The latter was handled by one of the general chairs, and these submissions were reviewed in the same way as all other submissions. The general chairs checked after the submission deadline whether there was any conflict of interest among the papers submitted to tracks, and if so, the respective paper was moved to another track. The track chairs checked that all papers were submitted anonymously. If not, the authors were asked to resubmit within a couple of days. Track chairs assigned the reviewers and selected the program committee members in such a way that experts in each area reviewed the papers and there were no conflicts of interest. After at least three reviews were received, the track chairs made a proposal for a decision per paper. The decisions were discussed in a meeting with the general and track chairs to ensure that the decisions were made in a consistent manner across track.

Among the full research paper submissions, 27 papers (empirical and conceptual) were accepted for this year's Springer LNCS Electronic Participation proceedings (vol. 16923). This volume includes papers distributed over the following tracks:

- General E-Democracy and e-Participation
- Digital Technologies and Sustainable Development
- Digital Society
- Artificial Intelligence Policy, Governance, and Impact

- Governance, Digital Legislation, and Policy
- Managing E-Government

As in previous years and per the recommendation of the Paper Awards Committee, the IFIP EGOV-CeDEM-ePart 2026 conference granted outstanding paper awards in three distinct categories:

- The most innovative research contribution
- The most compelling critical research reflection
- The most promising practical concept

The winners in each category were announced during the awards ceremony at the conference.

The IFIP EGOV 2026 conference was hosted by the Harokopio University of Athens and the University of the Aegean, Greece. The conference sessions took place at the Harokopio University.

As part of the conference, the PhD colloquium and the Junior Faculty School were hosted at Harokopio University of Athens. The PhD Colloquium was a full-day event on 30 August, open only to those who were admitted, and offered a limited number of PhD bursaries. The Junior Faculty School was a full-day event on 31 August, for sharing experiences to advance the career of young researchers interested in digitalization and government.

Many people behind the scenes make large events like this conference happen. We would like to thank the members of the Program Committee, the reviewers, and the track chairs for their great efforts in reviewing the submitted papers. We would also like to express our deep gratitude to Teta Stamati, Harokopio University of Athens, and Euripidis Loukis and Yannis Charalabidis, University of the Aegean, for hosting the conference. Last, but not least, we wish to thank Efthimios Tambouris, University of Macedonia, for supporting the local organization.

The papers in these proceedings range from societal to scientific, and from rigorous to innovative. We hope you enjoy reading them!

September 2026

David Duenas-Cid
Gregor Eibl
Lieselot Danneels
Edimara Mezzomo Luciano
Gianluca Carlo Misuraca
Gerhard Schwabe
Teta Stamati
Leif Sundberg
Peter Parycek

Organization

Conference Chairs

Yannis Charalabidis	University of the Aegean, Greece
Lieselot Danneels	Ghent University, Belgium
David Duenas-Cid	Kozminski University, Poland
Marijn Janssen	Delft University of Technology, The Netherlands
Ida Lindgren	Linköping University, Sweden
Euripidis Loukis	University of the Aegean, Greece
Peter Parycek	Fraunhofer FOKUS, Germany/University for Continuing Education, Austria
Gabriela Viale Pereira	University for Continuing Education Krems, Austria
Gerhard Schwabe	University of Zurich, Switzerland
Teta Stamati	Harokopio University of Athens, Greece
Efthimios Tambouris	University of Macedonia, Greece

Program Committee Chairs

General eGovernment and E-Governance Track

Gabriela Viale Pereira	University for Continuing Education Krems, Austria
Marijn Janssen	Delft University of Technology, The Netherlands
Ida Lindgren	Linköping University, Sweden

General eDemocracy and eParticipation Track

David Duenas-Cid	Kozminski University, Poland
Noella Edelmann	University for Continuing Education Krems, Austria
Marius Rohde Johannessen	University of South-Eastern, Norway
Sara Hofmann	University of Agder, Norway

Digital Technologies and Sustainability Development Track

Gianluca Misuraca	Universidad Politécnica de Madrid, Spain
Iryna Susha	Utrecht University, The Netherlands
René Reich	KU Leuven, Belgium
Jolien Ubacht	Delft University of Technology, The Netherlands

Digital Society Track

Leif Sundberg	Mid Sweden University, Sweden
Thomas Lampoltshammer	University for Continuing Education Krems, Austria
Lieselot Danneels	Ghent University, Belgium

Emerging Issues and Innovations Track

A. Paula Rodriguez Müller	Joint Research Centre, European Commission, Spain
Anastasija Nikiforova	Tartu University, Estonia
Francesco Mureddu	Lisbon Council, Belgium

AI, Data Analytics, and Automated Decision-Making Track

Evangelos Kalampokis	University of Macedonia, Greece
Habin Lee	Brunel University of London, UK
Euripidis Loukis	University of the Aegean, Greece

AI Governance and its Societal Challenges Track

Csaba Csáki	Corvinus University of Budapest, Hungary
Gregor Eibl	University for Continuing Education Krems, Austria
Aya Rizk	Linköping University, Sweden
Roel Dobbe	Technical University of Delft, The Netherlands

Smart Cities (Government, Districts, Communities and Regions) Track

Shefali Virkar	WU Vienna University of Economics and Business, Austria
Joep Crompvoets	KU Leuven, Belgium
Manuel Pedro Rodríguez Bolívar	University of Granada, Spain

Open Data: Social and Technical Aspects Track

Anneke Zuiderwijk	Delft University of Technology, The Netherlands
J. Ramon Gil-Garcia	University at Albany, State University of New York, USA
Efthimios Tambouris	University of Macedonia, Greece

Governance, Digital Legislation, and Policy Track

Edimara Mezzomo Luciano	Pontifical Catholic University of Rio Grande do Sul, Brazil
Peter Parycek	Fraunhofer FOKUS, Germany
Anna-Sophie Novak	University for Continuing Education Krems, Austria

Managing E-Government Track

Michael Koddebusch	ERCIS, University of Münster, Germany
Ulrik Bisgaard Ulsrod Røhl	Copenhagen Business School, Denmark
Maria A. Wimmer	University of Koblenz, Germany
Vera Spitzer	University of Koblenz, Germany

Practitioners — Researchers Collaboration Track

Lucy Temple	University for Continuing Education Krems, Austria
Ulf Melin	Linköping University, Sweden
Anthony Simonofski	University of Namur, Belgium

PhD Colloquium Chairs

Lieselot Danneels	Ghent University, Belgium
Michael Koddebusch	University of Münster, Germany
Gabriela Viale Pereira	University for Continuing Education Krems, Austria
Ramon Gil-Garcia	University at Albany, USA

Junior Faculty School Chairs

Lieselot Danneels	Ghent University, Belgium
Ramon Gil-Garcia	University at Albany, USA
Ida Lindgren	Linköping University, Sweden
Gabriela Viale Pereira	University for Continuing Education Krems, Austria

Workshops and Poster Chairs

Panos Panagiotopoulos	Queen Mary University of London, UK
Ida Lindgren	Linköping University, Sweden

Chair of Outstanding Papers Awards

Manuel Pedro Rodríguez Bolívar	University of Granada, Spain
Noella Edelmann	University for Continuing Education Krems, Austria

Program Committee

Valerie Albrecht	Danube-University Krems, Austria
Cristina Alcaide Muñoz	University of Granada, Spain
Laura Alcaide-Muñoz	University of Granada, Spain
Charalampos Alexopoulos	University of the Aegean, Greece
Eric Andersson	Mid Sweden University, Sweden
Gunnar Auth	HSF Meissen University of Applied Sciences, Germany
Thiago Ávila	FGV's São Paulo School of Business Administration, Brazil

Cilia Baanstra	Delft University of Technology, Netherlands
Frank Bannister	Trinity College Dublin, Ireland
Ana Alice Baptista	University of Minho, Portugal
Lais Barbudo Carrasco	Mid Sweden University, Sweden
Luca Tom Bauer	Universität Bremen, Germany
Till Becker	University of Applied Sciences Emden/Leer, Germany
Iris Beerepoot	Utrecht University, Netherlands
Ramin Behbehani	Brunel University of London, UK
Peter Bellström	Karlstad University, Sweden
Flavia Bernardini	Universidade Federal Fluminense , Brazil
Lasse Berntzen	University of South-Eastern Norway, Norway
Nitesh Bharosa	Delft University of Technology, Netherlands
Ellinor Blom Lussi	Lund University, Sweden
Nicolas Bono Rossello	University of Namur, Belgium
Maria Booth	Linköping University, Sweden
Luzia Borgmann	RWTH Aachen University, Germany
Stylianos Bourmpoulias	University of Macedonia, Greece
Ida Bring Løberg	Norwegian Directorate of Labor and Welfare, Norway
Karin Brodén	Umeå University, Sweden
Peter André Busch	University of Agder, Norway
Eric Börner	TU Dresden, Germany
Luiz Paulo Carvalho	Federal University of Rio de Janeiro, Brazil
Cesar Casiano Flores	University of Twente, Netherlands
Seong Wook Chae	Hoseo University, South Korea
Letizia Chiappini	University of Twente, Netherlands
Antoine Clarinval	Umeå University, Sweden
Andreiwid Correa	Federal Institute of Sao Paulo, Brazil
Ilaria Crivellar	Norwegian University of Science and Technology, Norway
Joep Crompvoets	KU Leuven, Belgium
Matheus Cruz	Universidade Federal Fluminense, Brazil
Csaba Csaki	Corvinus University of Budapest, Hungary
Frank Danielsen	University of Agder, Norway
Lieselot Danneels	Ghent University/Vlerick Business School, Belgium
Lukas Daßler	Technische Universität München, Germany
Rosanna De Rosa	University of Naples Federico II, Italy
Simon Dechamps	University of Namur, Belgium
Roel Dobbe	Delft University of Technology, Netherlands
David Duenas-Cid	Kozminski University, Poland

Matthias Döring	University of Southern Denmark, Denmark
Noella Edelmann	Danube University Krems, Austria
Gregor Eibl	Danube University Krems, Austria
Eiri Elvestad	University of South-Eastern Norway, Norway
Maurus Enbergs	Delft University of Technology, Netherlands
Marie Eneman	University of Gothenburg, Sweden
Tove Engvall	Mid Sweden University, Sweden
Melanie Fink	University of Leiden, Netherlands
Fotios Fitsilis	Hellenic Parliament, Greece
Ana Mônica Fitz De Oliveira	Federal University of Rio Grande do Sul, Brazil
Steffen Fock	University of Bremen, Germany
Mary Francoli	Carleton University, Canada
Robin Fritzsche	Universität Bremen, Germany
Asbjørn Følstad	SINTEF, Norway
Alexandros Gerontas	University of Macedonia, Greece
J. Ramon Gil-Garcia	University at Albany, USA
Dimitris Gouscos	University of Athens, Greece
Kathleen Gregory	Leiden University, The Netherlands
Anushri Gupta	Indian Institute of Management, Ahmedabad, India
Stephan Haller	Bern University of Applied Sciences, Germany
Sofia Hamberg	Linköping University, Sweden
Shengnan Han	Stockholm University, Sweden
Ida Heggertveit	University of Agder, Norway
Andreas Hein	Technical University of Munich, Germany
Erica Hellmer	Mid Sweden University, Sweden
Sara Hofmann	University of Agder, Norway
Syed Sajid Hussain	Norwegian University of Science and Technology, Norway
Quan Huynh	Eindhoven University of Technology, Netherlands
Charlotte Högberg	Lund University, Sweden
Helena Iacobaeus	Linköping University, Sweden
Tupokigwe Isagah	UNU-EGOV, Portugal
Tomasz Janowski	Gdańsk University of Technology, Poland
Marijn Janssen	Delft University of Technology, Netherlands
Carlos E. Jimenez-Gomez	National Center for State Courts, USA
Marius Rohde Johannessen	University of South-Eastern Norway, Norway
Gideon Mekonnen Jonathan	Stockholm University, Sweden
Ina Jäger	Technical University of Munich, Germany
Evangelos Kalampokis	University of Macedonia, Greece
Niyazi Karabulut	Middle East Technical University, Türkiye
Nikos Karacapilidis	University of Patras, Greece

Evika Karamagioli	Paris 8 University, France
Areti Karamanou	University of Macedonia, Greece
Martin Karlsson	Mälardalen University, Sweden
Singara Rao Karna	Centre for Electronic Government and Open Democracy, Australia
Muhammad Irfan Khalid	University of Agder, Norway
Junchul Kim	Brunel University of London, UK
Sana Kiran	Maynooth University, Ireland
Fabian Kirstein	Fraunhofer FOKUS, Germany
Michael Koddebusch	University of Münster, Germany
Niklas Korte	University of Münster, Germany
Hun-Yeong Kwon	Korea University, South Korea
Thomas Lampoltshammer	University for Continuing Education Krems, Austria
Habin Lee	Brunel University of London, UK
Hong Joo Lee	Catholic University of Korea, South Korea
Kristina Lemmer	University of Siegen, Germany
Azi Lev-On	Ariel University, Israel
Yanliu Lin	Utrecht University, Netherlands
Ida Lindgren	Linköping University, Sweden
Johan Linåker	RISE Research Institutes of Sweden, Sweden
Martin Lnenicka	University of Hradec Králové, Czech Republic
Carolina Lopes	Pontifical Catholic University of Rio Grande do Sul, Brazil
Francisco J. Lopez-Pellicer	University of Zaragoza, Spain
Euripidis Loukis	University of the Aegean, Greece
Durval Lucas Jr.	Federal University of São Paulo, Brazil
Edimara Luciano	Pontifical Catholic University of Rio Grande do Sul, Brazil
Luis F. Luna-Reyes	University at Albany, SUNY, USA
Marie Anne Macadar	Federal University of Rio de Janeiro, Brazil
Christian Madsen	IT University of Copenhagen, Denmark
Maria Ioanna Maratsi	University of the Aegean, Greece/Stockholm University, Sweden
Amal Marzouki	Université du Québec à Rimouski, Canada
John McNutt	University of Delaware, USA
Ulf Melin	Linkoping University, Sweden
Nimród Mike	Corvinus University of Budapest, Hungary
Petar Milić	University in Priština — Kosovska Mitrovica, Kosovo
Gianluca Carlo Misuraca	Politecnico di Milano, Italy
Catherine Mkude	Open University of Tanzania, Tanzania

Igor Sampaio	Universidade Federal Fluminense, Brazil
Caterina Santoro	KU Leuven, Belgium
Dwi Sari	SUNY Albany, USA
Diogo Sasdelli	University for Continuing Education Krems, Austria
Antonia Sattlegger	Delft University of Technology, The Netherlands
Sven Schade	Joint Research Centre - European Commission, Italy
Michaela Schmidt	Norwegian University of Science and Technology, Norway
Johannes Schneider	University of Liechtenstein, Liechtenstein
Hendrik Scholta	Universität Speyer, Germany
Johannes Scholz	University of Salzburg, Austria
Luiza Schuch de Azambuja	Tallinn University of Technology, Estonia
Gerhard Schwabe	Zurich University, Switzerland
Timon Sengewald	FAU Erlangen-Nürnberg, Germany
Uwe Serdült	Ritsumeikan University, Japan
Ashraf Shaharudin	Delft University of Technology, The Netherlands
Tobias Siebenlist	Rhine-Waal University of Applied Sciences, Germany
Anthony Simonofski	Université de Namur, Belgium
Emyana Sirait	Delft University of Technology, The Netherlands
Søren Skaarup	IT University of Copenhagen, Denmark
Erik Slingerland	HU University of Applied Sciences Utrecht, The Netherlands
Kari Smolander	Lappeenranta-Lahti University of Technology, Finland
Juliano Soares	Universidade Estadual de Campinas, Brazil
Carl-Johan Sommar	Linköping University, Sweden
Vera Spitzer	Universität Koblenz, Germany
Janick Spycher	University of Fribourg, Switzerland
Per Steiner	Mid Sweden University, Sweden
Stefan Stepanovic	Université de Lausanne, Switzerland
Leif Sundberg	Mid Sweden University, Sweden
Anneli Sundqvist	Oslo Metropolitan University, Norway
Iryna Susha	Örebro University, Sweden
Dimitris Symeonidis	University of Tartu, Estonia
Iraklis Symeonidis	RISE Research Institutes of Sweden, Sweden
Øystein Sæbø	University of Agder, Norway
Efthimios Tambouris	University of Macedonia, Greece
Evrim Tan	KU Leuven, Belgium
Luca Tangi	Joint Research Centre - European Commission, Italy

Lucy Temple	University for Continuing Education Krems, Austria
Sara Thabit-Gonzalez	Joint Research Centre - European Commission, Estonia
Malin Tinjan	University of Gothenburg, Sweden
Hsien-Lee Tseng	National University of Tainan, Taiwan
Bernát Török	University of Public Service - Ludovika, Hungary
Jolien Ubacht	Delft University of Technology, The Netherlands
Louise van der Peet	Delft University of Technology, The Netherlands
Colin van Noordt	KU Leuven, Belgium
Gabriela Viale Pereira	Danube University Krems, Austria
Dulce Villegas	Universität Bremen, Germany
Shefali Virkar	WU Vienna University of Economics and Business, Austria
Gianluigi Viscusi	Linköping University, Sweden
Jose Viterbo	Universidade Federal Fluminense, Brazil
Jörn von Lucke	Zeppelin Universität Friedrichshafen, Germany
Craig Waltz	State University of New York at Albany, USA
Jan Westermann	University of Bremen, Germany
Guilherme Wiedenhöft	Federal University of Rio Grande, Brazil
Elin Wihlborg	Linköping University, Sweden
Moritz Wild	University for Continuing Education Krems, Austria
Maria A. Wimmer	University of Koblenz, Germany
Marc Wyszynski	Universität Bremen, Germany
Alexandros Xenakis	University of the Peloponnese, Greece
Xuan Xia	University of Lausanne, Switzerland
Dujuan Yang	Eindhoven University of Technology, The Netherlands
Liesanth Yde Nirmalarajan	Aalborg University, Denmark
Thomas Zefferer	A-SIT Plus GmbH, Austria
Dimitris Zeginis	University of Macedonia, Greece
Ying Zhang	Delft University of Technology, The Netherlands
Qinfeng Zhu	University of Groningen, The Netherlands
Anneke Zuiderwijk	Delft University of Technology, The Netherlands

Contents

Tenant in Your Own House the Wickedness of Regaining Digital Sovereignty over Digital Public Infrastructures

Nitesh Bharosa[1(✉)], Wouter Welling[2], and Mark de Reuver[1]

[1] Delft University of Technology, Delft, The Netherlands
{n.bharosa,g.a.dereuver}@tudelft.nl
[2] Digicampus, Ministry of the Interior and Kingdom Relations, The Hague, The Netherlands
wouter.welling@minbzk.nl

Abstract. Escalating geopolitical tensions and technological dependencies have rendered the pursuit of digital sovereignty a paramount challenge for governments. Sovereignty is especially desirable for digital public infrastructures given their foundational role in authenticating, transacting, and exchanging information. However, the literature provides few policy strategies for sovereignty and few in-depth analyses of sovereignty challenges. Moreover, pathways towards higher levels of digital sovereignty for a digital public infrastructure are yet to be theorised and tested. This paper provides a deeper analysis of the challenges to sovereignty over digital public infrastructures, enabling governments to unearth pathways. Drawing on an empirical case study in the Netherlands and the PESTEL framework, this paper derives a set of challenges that must be addressed. The analysis shows that regaining digital sovereignty is a wicked problem with low normative consensus and many dependencies and uncertainties. This paper develops two pathways to address the wicked problem: fire-fighting measures that aim to address the immediate loss of digital sovereignty, vs fire prevention that aims to orchestrate an ecosystem of more sovereign alternatives for digital public infrastructure components. The analysis and pathways can inform policymakers looking to stimulate digital sovereignty.

Keywords: Digital sovereignty · GovTech · digital autonomy · digital public infrastructure

1 Introduction

In the pre-digital era, the state maintained sovereign control over its vital assets—roads, bridges, energy systems, and water infrastructures. Debates about who exercises sovereign control, how, on what, and why have shaped today's public agencies and institutions. As societies digitalise, these foundational functions

D. Duenas-Cid et al. (Eds.): ePart 2026, LNCS 16923, pp. 1–17, 2027.
https://doi.org/10.1007/978-3-032-36676-4_1

are increasingly mediated by digital systems. This shift brings digital public infrastructures to the forefront as a critical locus of sovereignty [1]. A digital public infrastructure is an open system with fluid boundaries, dynamics, and interactions subject to external influences. In contrast to physical infrastructure components, digital infrastructure components can be designed, operated, and delivered from anywhere in the world, as a service [2,3]. Digital public infrastructures have become the battleground for competition for global leadership, leading to ever-increasing geopolitical tensions [1]. Across Europe, politicians, professionals, and scholars debate who should control the digital public infrastructure. According to Bria et al. [12], over 80% of Europe's digital infrastructure and technologies are imported and consumed as services, creating systemic vulnerabilities and hampering the region's capacity for innovation and self-reliance. Europe has no notable operating system developers, no relevant search engines, no global social network and no competitive cloud infrastructure [1]. The restriction of digital services for the International Criminal Court in The Hague proved a breaking point for European politicians. This breaking point is visible in the Netherlands. In 2025, several cases of choosing foreign tech suppliers in the public sector have been scrutinised both by politicians and academics. Examples include the Dutch Tax Office's decision to migrate to Microsoft Office 365, leading to concerns that the US government could collect sensitive financial data or digitally shut down the Dutch Tax office using a 'kill-switch', a term for the unilateral operational-disruption capability of a foreign jurisdiction. Beyond confidentiality and availability, a recurring concern is the potential for blackmailing governments that depend on foreign tech in geopolitical negotiations. Accordingly, it is no surprise that the newly formed Dutch cabinet includes the nation's first 'State Secretary of Digital Economy and Digital Sovereignty'. The new Dutch coalition agreement declares digital sovereignty as a starting point for new policies on digitalisation. Just days after the agreement, the Dutch Railway Agency outsourced some digital infrastructure to an American company, fuelling the debate. Politicians call for boosting digital sovereignty and reducing dependence on foreign tech. Often, such calls are bolstered by arguments such as keeping jobs of the future in the Netherlands, boosting Dutch competitiveness, and improving earning potential.

Regardless of the drivers, the concept of digital sovereignty over digital public infrastructures is not explicitly defined in the Dutch coalition agreement or policy documents. Depending on political position, measures range from nationalisation and state-owned infrastructure to market protectionism and restrictions on market access. This paper provides conceptual clarity and identifies challenges and pathways for digital sovereignty in the public sector. Accordingly, the central question is: What are the challenges and pathways to boosting digital sovereignty over digital public infrastructures?

This paper proceeds as follows. Section 2 provides conceptual clarification on digital sovereignty. Section 3 presents the research approach and introduces the PESTEL framework as an analytical framework for categorising empirical challenges for digital sovereignty. Drawing on the analytical framework, Sect. 4 analyses the Dutch Digital Identity case study, specifically analysing the proposed acquisition of the Dutch managed service provider Solvinity by the US-based infrastructure giant Kyndryl. Section 5 presents the results of the PESTEL analysis. Section 6 develops two pathways for boosting digital sovereignty. Finally, Sect. 7 reflects on the research findings and concludes the paper by outlining avenues for future research.

2 Digital Sovereignty in the Public Sector

2.1 Digital Sovereignty

The origins of the digital sovereignty concept lie in academic papers and digital activists' manifestos, focusing on areas of internet access, data ownership and artificial intelligence (AI). Over the last decade, digital sovereignty has become a central element in policy discourses on digital issues [4]. Couture & Toupin stress independence, autonomy and control dimensions of digital sovereignty [5] (p. 2317). Moerel and Timmers [1] argue that while digital sovereignty is currently the common terminology in debates and literature, digital strategic autonomy describes the means (the how) to achieve sovereignty (the what) and operationalise it. They define digital strategic autonomy as: "the capabilities, capacities, and control to decide and act autonomously on essential [digital] aspects of our economy, society and democracy" [6, p. 577]. Here, capabilities are knowledge, skills and tools; capacity is the resources available; and control is the influence over capabilities, capacities, decisions and actions. Building capacity and capabilities for digital autonomy is not easy, even for governments. The review of definitions reveals that digital sovereignty is hard to capture in a single definition, and even harder to operationalise in clear-cut metrics. The lack of clarity creates opportunities for new concepts such as 'Sovereignty-as-a-service', promoted by BigTech companies to redefine digital sovereignty [7].

2.2 Digital Public Infrastructures

Although digital sovereignty can refer to any digital system, digital public infrastructures are the most urgent objects that demand greater sovereignty in the public sector. Digital public infrastructures are complex and dynamic configurations of socio-technical components. Examples of socio-technical components include digital identities, data registers, portals, web apps, data processing and analysis applications, data exchange systems, AI/algorithmic systems, storage devices, cryptography management, workflow and office software [3]. Figure 1 provides an overview of socio-technical components in a digital public infrastructure.

Each component can have its own policies, rules, social structures, and stacks of software, hardware, and networking systems, and can be offered via different service models, such as software-as-a-service, platform-as-a-service, or infrastructure-as-a-service. Components can be provided by multiple public or private organisations, headquartered in multiple geographic regions. In many

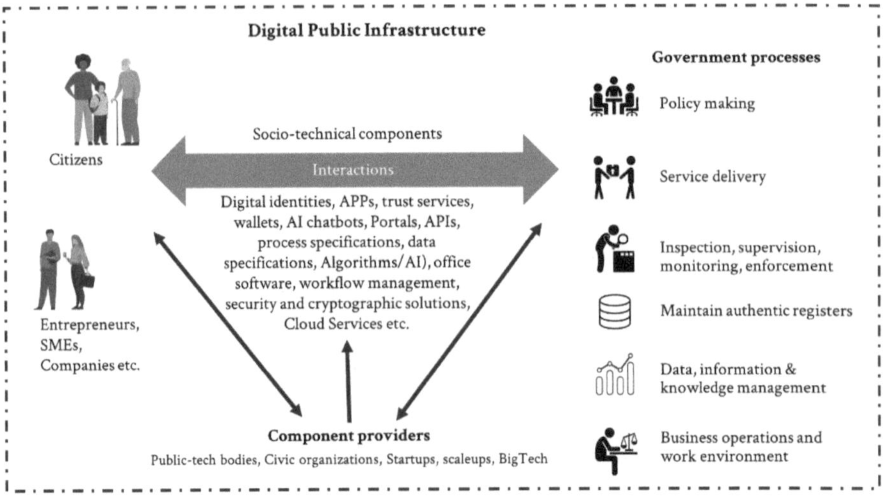

Fig. 1. Socio-technical components in a digital public infrastructure.

cases, the architecture design and source code are not open, raising concerns about public values such as privacy, transparency, democratic control, and portability (the ability to switch to other suppliers).

2.3 The Sovereignty Gap

The dependencies on various digital public infrastructure components are subject to the 'sovereignty gap'. The sovereignty gap, as defined by Moerel and Timmers [1], is the disparity between a nation-state's aspirations for control over its own affairs and the concrete reality of its digital dependencies. Four pressures widen the gap:

- BigTech dominance: there is an increasing dependence on digital technology that is largely in the hands of a limited number of foreign hyperscalers (e.g., Amazon, Microsoft, Google). This market concentration creates a vendor lock-in and reduces the market potential for local competition.
- Geopolitical tensions: rising rivalries, particularly between the U.S. and China, lead to extraterritorial claims. Examples include the U.S. CLOUD Act, which allows foreign authorities to access data stored on providers' servers abroad, and export restrictions on critical technologies such as chips.
- Cyber threats: the scale and sophistication of cyberattacks from both state and non-state actors can undermine national security and the international order without the perpetrator ever entering the state's physical territory.
- Declining internal knowledge and capacity: when agencies stop engineering and operating their own systems and transform into consumers that outsource knowledge building, they cannot critically examine the dependencies they grow into.

To close the sovereignty gap, Moerel and Timmers [1] argue that the EU and its Member States must move toward sovereignty properly understood. This involves developing sovereign assets and a unified European digital identity. We return to one such effort, the Eurostack, in Sect. 5.

3 Research Approach

This explorative paper adopts a qualitative case study to examine digital sovereignty in the public sector. Given that public digital infrastructures cannot be separated from their institutional, operational and technical context, a case study approach is suitable. Further, a case study approach is especially suitable given the large number of challenges at play, which a quantitative approach cannot do justice to.

3.1 Case Study Selection

The unit of analysis of our single case study is the proposed 2025 acquisition of the Dutch-British service provider Solvinity by the U.S.-based infrastructure giant Kyndryl. Solvinity provides secure cloud hosting and data services for a range of government services, including the digital identification system (DigiD). This case was selected because it illustrates the sovereignty gap—the disparity between a state's aspiration for autonomy and its digital dependencies. It serves as a critical stress test for European digital autonomy, highlighting how shifting ultimate beneficial ownership to a foreign company (Kyndryl) subjects vital Dutch infrastructure to extraterritorial claims, such as the U.S. CLOUD Act and FISA Section 702.

Our case is situated in the Netherlands, which is especially well-suited for studying questions of digital sovereignty. Historically, the Dutch public sector has operated under a highly decentralised IT model, in which individual ministries, executive agencies, and municipalities independently develop and manage their own digital systems. To counter the fragmentation and inefficiencies of this siloed approach, the Dutch government is gradually moving toward a more centralised approach by providing generic components for its digital public infrastructure. A central actor in this ecosystem is Logius, the digital government service operating under the Ministry of the Interior and Kingdom Relations' mandate. While the ministry is politically responsible for the overarching policy, strategy, and safeguarding of the national digital infrastructure, Logius is tasked with managing and procuring essential shared services, such as citizen authentication (i.e., the DigiD system).

3.2 Data Sources

For the case study, a Document Corpus was collected (Appendix A) comprising formal submissions to the Dutch parliamentary committee on Digital Affairs. The inclusion criteria are all documents made public by the Dutch parliament regarding the Solvinity takeover. We analysed technical briefings from Logius and formal inquiries from the Dutch House of Representatives. Factual claims in the corpus were verified by searching for news items and other public sources.

3.3 Analytical Framework: PESTEL

Data was analysed in a two-step sequence. First, the empirical challenges of the Solvinity-Kyndryl case were extracted from the document corpus. The longlist of challenges was then categorised through deductive coding, using the six elements of PESTEL as core categories, referring to the Political, Economic, Social, Technological, Environmental, and Legal perspectives on a topic [8]. This framework is instrumental in mapping the challenges underlying digital sovereignty concerns—a system of interrelated problems that cannot be solved in isolation.

Second, the categorised challenges were mapped against Head and Alford's [9] three dimensions to evaluate the underlying nature of the policy challenges. In this way, this study utilises Head and Alford's conceptualisation of 'wicked problems' as a secondary analytical lens. According to Head and Alford [9], wicked problems are notoriously difficult to resolve because they are characterised by three core dimensions: (1) social pluralism (multiple stakeholders with diverging values and interests), (2) institutional complexity (fragmented, multi-level interdependencies across organisations), and (3) scientific uncertainty (a lack of definitive, testable technical solutions). This lens identifies the specific barriers to digital sovereignty and formally diagnoses the phenomenon as a wicked problem. This diagnosis directly informs the discussion, justifying why linear policy interventions are insufficient and why adaptive strategies such as ecosystem orchestration are required.

4 Case Study

Solvinity is a mostly British-owned managed cloud provider with a pivotal, often invisible, role in the Netherlands' digital ecosystem. Commissioned by Logius, Solvinity has operated the secure digital platform upon which DigiD runs since 2020. DigiD is the cornerstone of Dutch e-government. It serves as the digital identity for over 16.5 million citizens and processes approximately 550 million authentications annually. It grants access to vital services: filing tax returns, managing student loans, accessing healthcare records, and obtaining municipal permits and civil registry documents. Additionally, Solvinity manages infrastructure for MijnOverheid (the government's digital message box for citizens) and other services (we leave the other components out of scope for this paper). Consequently, Solvinity does not merely host data; it operationalises the interface between the Dutch state and its constituents. This includes critical digital public infrastructure components such as servers, storage arrays, virtualisation layers, and security perimeters—ensuring high availability, security, and compliance for services designated as vital processes for the public sector.

In November 2025, Kyndryl, a US-based IT infrastructure services provider, announced its intention to acquire Solvinity. However, the acquisition sparked immediate and intense concern among privacy advocates, the Dutch House of Representatives, and the cybersecurity community. The core anxiety centres on the shift of ultimate beneficial ownership from a Dutch-British entity to a US-based multinational. This change fundamentally alters the jurisdictional exposure of the infrastructure supporting DigiD, moving it from the protection of Dutch and EU law into the sphere of influence of US federal law.

Logius, the digital government service of the Ministry of the Interior and Kingdom Relations, is the formal 'owner' of DigiD. Following the announcement of the acquisition, Logius Director emphasised in a technical briefing to parliament that DigiD is Dutch and remains Dutch, drawing a sharp distinction between the application (software, logic, and data governance) and the infrastructure (servers, cooling, and platform management). This rationale highlights a critical 'ownership paradox' in modern digital public infrastructure. Logius contends that sovereignty is preserved because it retains the application logic and cryptographic keys, and the data resides in a government data centre. In this view, Solvinity is merely a supplier with no authority over the data.

5 Analysis: Structural Drivers of the Sovereignty Gap

The Solvinity takeover exposes the structural challenges to sovereignty over public digital infrastructure. To understand why vendor lock-in is the default, we analyse the macro-environmental factors using the PESTEL framework.

5.1 Political Factors

When zooming in on the political level, the following factors emerge:

- Fragmentation of consensus: Position papers reveal a fundamental tension between a historical reliance on an open private market and the urgent need for digital sovereignty. On the one hand, a broad parliamentary majority demands the cabinet 'draw a line' to protect vital assets like the digital identity system (DigiD). On the other hand the Ministry of Economic Affairs and the cabinet prioritize an open investment climate and transatlantic trade, ultimately resulting in an internal division that complicates decisive action.
- Tensions between national directives and local autonomy. For example, the City of Amsterdam's push for local digital autonomy as soon as possible has a different timeline and scope than the ongoing planning efforts from the National Government in The Hague.
- Dependence on US Intelligence: The political willingness to confront US tech dominance is tempered by the reality that European defence and intelligence capabilities are heavily integrated with US systems. A strict 'sovereignty first' policy could have diplomatic repercussions in other security domains, for instance, by leveraging global networks of informational and financial exchange for strategic advantage [10].

5.2 Economic Factors: the Efficiency Trap and Hyperscaler Dominance

When zooming in on the economic level, the following factors emerge:

- Oligopolistic market structure: The European cloud market is dominated by US hyperscalers (AWS, Microsoft Azure, Google Cloud), who control over 80% of the market share [12]. High market concentration, barriers to entry, and switching barriers (such as complex migration processes and interoperability limitations) are at play [11]. This concentration creates an economic reality where sovereign alternatives often lack scale, feature set, and cost-efficiency of their US counterparts.
- Procurement economics: Dutch public procurement law (Aanbestedingswet 2012) mandates the selection of the 'most economically advantageous tender'. US hyperscalers, leveraging massive global economies of scale, consistently outbid local European providers on price and feature velocity. The hyperscaler's offering aligns well with the New Public Management paradigm that prioritises cost efficiency in the public sector. This economic logic entrenches vendor lock-in, as switching costs to a less mature domestic provider are perceived as fiscally irresponsible in the short term. Nonetheless, public organisations can always cite risks such as dependency or privacy as grounds for choosing local or European alternatives.
- Market consolidation: The Solvinity acquisition is part of a broader trend of consolidation where mid-sized European tech firms are acquired by larger US or Asian capital (e.g., the acquisition of High Tech Campus Eindhoven by a Singaporean fund). This effectively hollows out the European industrial base for digital infrastructure components, reducing the pool of sovereign vendors available to the government.

5.3 Social Factors: Path Dependency and Convenience Bias

When zooming in on the social level, the following factors emerge:

- Convenience bias: Dutch society is highly digitised and accustomed to seamless digital services, which shape citizens' expectations of how government services should function. Citizens demand frictionless, highly reliable, app-like experiences (convenience bias), which Big Tech delivers to a great extent. A sovereign alternative that suffers downtime or poor UX could damage state legitimacy just as much as a theoretical foreign data breach.
- Path dependency: Once a government agency adopts a one-stop shop ecosystem (e.g., Microsoft 365 and Azure), the integration of workflows, data formats, and employee training creates a lock-in that is social as well as bureaucratic. Moving away from the one-stop shop requires not just technical migration but a cultural shift in how civil servants work.

- Trust fragility: The reliability of DigiD is essential for social trust. Any disruption or data breach resulting from a sovereignty dispute—or a botched migration to a sovereign provider—would severely damage the legitimacy of the state. This fear of disruption often paralyses steps towards sovereignty.

5.4 Technological Factors: Integration and Legacy Debt

When zooming in on the technological level, the following factors emerge:

- Vertical vs horizontal integration: The Dutch government historically operated vertically integrated silos, with each ministry running its own IT. The move to centralised agencies like Logius attempted to horizontalise this by creating shared services. However, this horizontal layer was outsourced to private service providers. This created single points of failure. If the horizontal infrastructure provider (Solvinity) is compromised, the failure cascades vertically across all dependent services (Tax, Education, Health).
- Legacy debt: The government IT landscape consists of fragmented legacy systems. Migrating these systems to a modern, containerised sovereign cloud environment is technically complex and resource-intensive. Seeking sovereignty through highly customised, localised solutions can lead to technical isolation, increase legacy debt, and make future migrations even harder.
- AI and compute dependency: The shift to AI-driven governance (e.g., fraud detection algorithms) requires massive compute power (GPUs). In this area, Europe has almost zero capacity compared to US firms. This creates a technological dependency that goes beyond storage; it is a dependency on processing capability that European sovereign clouds cannot currently match.

On the technical level, it is imperative to note the concept of a 'European stack' (Eurostack). Eurostack is a strategic initiative to build a native, interoperable digital infrastructure for Europe [12]. This stack is designed to reduce one-sided dependencies on non-European hyperscalers. By setting common technical standards and open APIs, it prevents the monopolistic behaviour seen in Big Tech contracts, where proprietary standards make switching suppliers nearly impossible. The development of a European digital stack is still ongoing.

5.5 Environmental Factors: Grid Congestion and Locations for Data Centres

When zooming in on the environmental level, the following factors emerge:

- The energy bottleneck: a critical, often overlooked factor in the Netherlands is grid congestion (energy network congestion). This environmental constraint has become a hard limit on digital sovereignty. The Dutch electricity grid is effectively full. In major digital hubs such as Amsterdam and Flevoland, grid operators have imposed moratoriums on new large-scale connections due to capacity constraints.

- Timeline: even if the Dutch government decided today to build a Sovereign Cloud Data Centre to replace Solvinity or AWS, it could face a 5-to-10-year wait for a power connection, competing with other projects (e.g., houses, schools, hospitals) on the same list. Getting ahead of the queue would require a broad political consensus.
- Incumbent advantage: Existing commercial data centres already possess the large electricity connections required to operate energy-intensive digital infrastructure. Through colocation models, hyperscalers can lease space and power capacity within these already connected facilities rather than building entirely new infrastructure. Where the grid is congested, as in parts of the Netherlands, new large-scale connections are hard to obtain. As a result, new sovereign infrastructure is constrained not only by technology or policy but by energy availability.

5.6 Legal Factors: Procurement and Screening

When zooming in on the legal level, we see multiple laws that, at first sight, could be used to block the takeover in the name of fostering digital sovereignty.

- Non-discrimination in EU procurement law: EU public procurement directives generally forbid discrimination based on nationality ('Buy European' is difficult to mandate explicitly). While exceptions for 'national security' exist (Article 346 TFEU), they are rarely invoked for general IT infrastructure, which is often treated as a commodity rather than a strategic asset.
- Competition Act (Mededingingswet): focusing on mergers and acquisitions that can reduce competition. While finalising this paper, the authority responsible for this screening (ACM) has published its findings, suggesting that there is no basis in this act to block the takeover of Solvinity.
- Dutch Vifo Act: The Investments, Mergers and Acquisitions Security Screening Act is another legal instrument available to block the takeover. The Bureau Toetsing Investeringen is currently investigating the takeover.
- Dutch WOZT Act: the undesirable control of telecommunications act protects the telecom sector. There is an ongoing legal debate over whether Solvinity, as a cloud/hosting provider, falls under the definitions of this act. Advocates argue for a broad interpretation given the criticality of the services to the national communication infrastructure.

Looking to create legal instruments for boosting digital autonomy, the European Commission recently published a Cloud Sovereignty Framework [13]. This framework assigns Sovereignty Effective Assurance Levels (SEALs) based on eight sovereignty objectives (SOV-1 to SOV-8). The SEAL framework provides a graded scale to measure the sovereignty of a cloud service. Given its novelty, this instrument has not yet been deployed in the Netherlands. Achieving SEAL-4 (Full Sovereignty) is expensive. It requires dedicated hardware, vetted EU-only personnel, and potentially bespoke software development.

5.7 Diagnosis: Wickedness of the Challenges

Rittel and Webber [14] used the distinction between 'tame' and 'wicked' problems, along with other labels such as 'messy,' 'ill-defined,' 'untamed,' and 'ambiguous.' Characteristic features of these wicked problems are the large number of players involved, the conflicts of values, and the unlimited number of potential policy alternatives. Head & Alford [9] state that wicked problems are generally associated with social pluralism (multiple stakeholder interests and values), institutional complexity (the context of interorganizational cooperation and multilevel governance), and scientific uncertainty (fragmentation and gaps in reliable knowledge). The PESTEL analysis reveals varying levels of complexity, uncertainty, and disagreement regarding digital sovereignty. We look at three key characteristics of a wicked problem, in relation to digital sovereignty:

- There is no stopping rule: when is a state 'sovereign enough'? Since complete autarky is impossible, sovereignty should be approached as an ongoing negotiation rather than a final destination.
- Any solution (build local data centres) throws up more problems that need to be solved in a different policy arena (e.g. access to energy grids).
- Opposing values: the conflict is not just Europe vs the US, but also internal. For instance, the Ministry of Economic Affairs can define the problem as 'maintaining an open economy', while Logius/Ministry of Interior defines it as 'safeguarding national security'. Wicked problems impede solutions because stakeholders cannot even agree on the problem's definition.

Head and Alford [9] have proposed some generic strategies for dealing with wicked problems under governmental and administrative constraints. The next section captures such strategies under the umbrella concept of ecosystem orchestration.

6 Pathways to Address the Sovereignty Gap

As demonstrated in Sect. 5, the sovereignty gap is not merely a technical vulnerability but a systemic issue driven by entrenched structural factors. The convergence of economic lock-in to foreign hyperscalers, societal demands for frictionless digital services, hard environmental constraints such as grid congestion, and a fragmented political landscape render digital sovereignty a truly wicked problem. Because these interlinked dependencies cannot be resolved through a single policy intervention, governments face a dual imperative: they must act decisively to prevent the acute loss of sovereign assets while simultaneously unwinding long-term market dependencies.

This section develops two complementary action pathways. The first pathway—fire-fighting—focuses on reactive legal and institutional interventions (see Sect. 5.6) to block imminent threats to digital control, as illustrated by the Solvinity takeover. The second pathway—fire prevention—demands a proactive shift toward ecosystem orchestration to overcome technical legacy debt (see

Sect. 5.4) and cultivate a viable, sovereign European tech stack over time. Both pathways must be pursued simultaneously; the former addresses today's crisis, while the latter builds the structural resilience of tomorrow.

6.1 Action Pathway I: Fire-Fighting

The Dutch government has legal instruments to address the immediate threat posed by the takeover. These measures are reactive but necessary to prevent an immediate loss of control.

- The first line of defence is blocking the concentration/merger based on Article 27, paragraph 1(b) of the Competition Act (Mededingingswet). However, the Netherlands Authority for Consumers and Markets (ACM) recently ruled that this act does not block the takeover.
- The second line of defence is the Vifo Act. This act allows the Minister of Economic Affairs to block a transaction if it poses a risk to national security. Given that DigiD facilitates tax, justice, and health services, a disruption or compromise constitutes a national security risk.
- A carve-out. A less drastic measure than the ones above would be to encourage a carve-out, in which only the specific unit at Solvinity that manages DigiD—including the associated back-end infrastructure and expertise—is purchased by the state or a public-private consortium. The state can then place the management of the carve-out with a public service provider or a steward-owned organisation. A similar approach is taken in the carve-out of iDEAL (a European Payment Service Provider). iDEAL was a purely Dutch sovereign payment standard. To survive global competition and maintain sovereignty, Dutch banks carved out iDEAL. They merged it into the EPI to build Wero, a pan-European payment system.

6.2 Action Pathway II: Fire Prevention

Blocking one acquisition is a temporary fix. A digital public infrastructure is full of components that are prone to takeover. A more systemic and robust solution requires shifting the government's role from a passive buyer of monolithic IT to an active ecosystem orchestrator that takes a strategic approach to managing digital dependencies. Moerel and Timmers [1] identify four high-level approaches to digital strategic autonomy (as means for digital sovereignty), which further describe the 'strategic' aspect: (i) risk management, (ii) strategic partnerships with like-minded actors, (iii) global/digital commons such as Eurostack, and (iv) autarky, where a single state seeks full control and all capabilities and capacities, and thus full digital autonomy. The approaches are not mutually exclusive, but in analysing and designing policies, one approach can often serve as the basic premise. The autarky approach would, however, economically and practically be undesirable and hardly feasible. States would forgo economies of scale, specialisation, and global supply chains, and remain highly interdependent—directly or indirectly—with foreign providers. Moreover, in times of budget cuts at public agencies, the financial demands of this approach lack political commitment.

In the Solvinity case, the failure lies in the lack of orchestration over the digital public infrastructure. Orchestration involves managing a complex ecosystem of vital components (which must be sovereign) and service components (which can be commercial). The government allowed a vital component (the digital identity hosting infrastructure) to be treated as a commercial commodity subject to market forces. A true ecosystem orchestrator could approach such a critical component more strategically, acknowledging the risks, steering toward either full state-ownership or cultivating a competitive market of alternative (European) suppliers, rather than a winner-takes-all tender, to prevent exactly this type of vendor lock-in. Orchestration of a digital public infrastructure is by no means a simple task. And we cannot automatically assume the state can smoothly pivot to an ecosystem orchestrator, ignoring the very bureaucratic inertia and path dependency that led to the current situation. Orchestration will demand interventions on each of the PESTEL levels. For instance, there needs to be political consensus and coordination. Yet, political will is irrelevant if there is no access to energy. The grid congestion crisis in the Netherlands means that building new, sovereign government data centres is practically impossible in the short term. This physical constraint forces reliance on existing (commercial) facilities. This is where a Eurostack could make all the difference, for instance, by building European data centres in countries with sufficient energy. However, while a Eurostack offers a path toward sovereignty, its development is progressing slowly.

7 Discussion

This research asked: what are the challenges and pathways for boosting digital sovereignty in the public sector? The variety, veracity, and interwovenness of components in digital public infrastructures show that digital sovereignty includes both tame and knowledge problems, as well as political/economic problems and untamed problems. Tackling these problems in a multi-actor arena with shifting perceptions of state and market ownership makes digital sovereignty a wicked concept. Theoretically, this paper bridges public policy literature on wicked problems with the emerging field of digital sovereignty. By applying Head and Alford's framework, we conclude that digital sovereignty is a mess—a system of interrelated problems where solving one (e.g., technical security) often aggravates another (e.g., economic competitiveness).

7.1 Implications for Policy and Practice

This study concludes that the Netherlands cannot navigate the digital sovereignty gap through national isolationism or passive procurement. These structural challenges necessitate a fundamental shift in how governments approach digital public infrastructure. Drawing on the case study, we formulate two implications for policymakers and public administrators looking to boost digital sovereignty. First, public agencies must abandon the role of passive consumers of monolithic IT services. Instead, they must orchestrate an ecosystem of sovereign and commercial components, treating vital components (e.g. digital identity systems, algorithms, authentic data registries, storage) as strategic state assets rather than market commodities. The transition to ecosystem orchestration demands the further development of GovTech market-shaping capabilities. Second, European policymakers should begin operationalising the European Commission's Cloud Sovereignty Framework (SEAL). For critical digital public infrastructure components, a SEAL 3 or 4 designation could be mandated, ensuring that infrastructure is governed by vetted personnel and immune to extraterritorial legislation. This requires the ability to collectively define the critical components of digital public infrastructure. Moreover, short-term cost efficiency must no longer override long-term strategic autonomy.

7.2 Theoretical Implications

At a high level, the firefighting-versus-prevention metaphor resonates with Mazzucato's ideas on market correction (ex-post) and market shaping (ex-ante). However, our fire-preventing pathway specifies these ideas to a particular category of market-shaping actions: ecosystem-oriented market shaping, aimed at the deliberate formation of a diversified supply side. Digital sovereignty requires viable alternatives, which typically do not emerge spontaneously, as observed in our case analysis. This implies a more proactive role for public actors in orchestrating the conditions under which alternative ecosystem actors can emerge and scale. These conditions include organising steady, predictable demand for, for instance, open-source solutions that meet prescribed standards, in government procurement or public-private living labs. Such targeted and steady actions reflect GovTech's ideas of public-private cocreation, which go beyond reactive moves. Our two pathways should not be understood as mutually exclusive, but rather as a case of organisational ambidexterity. The DigiD case represents an acute instance with high media coverage and direct visibility to the general public, which risks focusing solely on corrective measures. While such fire-fighting measures are important, they should be combined with long-term focused efforts of market shaping. Government bodies can learn from the ambidexterity literature how to combine short- and long-term goals, for instance through structural separation (e.g., allocating short- and long-term responsibilities to different departments), temporal cycling (e.g., responding in times of acute crisis and building capacity later), or contextual ambidexterity. This shifts the focus of digital sovereignty from dependency mitigation toward ecosystem-based capability building. Ultimately, the efficacy of both pathways must be evaluated against their ability to address the sovereignty gap. Pathway I (fire-fighting) does not close the sovereignty gap; rather, it serves as a critical mitigation strategy to prevent it from widening abruptly. By utilising legal instruments such as the Vifo Act or establishing proxy agreements, the state mitigates acute jurisdictional threats and halts the immediate transfer of vital infrastructure to foreign control. Conversely, Pathway II (fire prevention) serves as the structural mechanism for actively closing the sovereignty gap over time. By transitioning toward proactive ecosystem orchestration and adopting a European digital stack, the state progressively reduces its structural reliance on non-European hyperscalers. Together, Pathway I holds the line, while Pathway II builds the foundation necessary to align the state's operational reality with its aspirations for digital sovereignty.

7.3 Limitations and Future Research Avenues

This paper is limited in its use of theories, methods and depth. The single-case study approach and data analysis method are insufficient for generalisation or theory-building. The PESTEL analysis provides an initial first-order analysis and refrains from analysing second-order causal relationships between factors. These limitations can be taken as starting points for further research. Moreover, to support policy development, this paper uncovers the following avenues for future work:

Testing adaptive governance tools: because wicked problems require provisional courses of action rather than searching for permanent solutions, researchers could empirically test the efficacy of tools like the SEAL framework in diverse public sector contexts to determine if they successfully mandate sovereignty without stifling innovation.

Institutional design for GovTech orchestration: further investigation is needed to define the structural requirements of a governing institution capable of ecosystem orchestration. Research could explore how such an entity can overcome traditional bureaucratic silos to successfully integrate industrial policy, IT procurement, and national security.

Theorising relational sovereignty: traditional sovereignty implies exclusive territorial control, which is impossible in the digital realm. Future theoretical work must shift toward relational sovereignty—the capacity of a state to manage and negotiate risks across highly interdependent global (tech) supply chains.

Finally, more academic work is needed on the concept of a 'sovereignty premium'. If digital sovereignty requires government-owned hardware, software and localised talent, it contradicts the cost-saving mandates of new public management. If sovereignty and cost efficiency are both public values, how can policies encompass both? Hence, future studies spanning political science and information systems must theorise new procurement models that account for the economic and societal premium of maintaining digital autonomy.

Appendix: Document Corpus of Solvinity-Kyndryl Case Study

This corpus comprises the primary source materials submitted to the Parliamentary Committee on Digital Affairs (DiZa) for the hearing on the consequences of the Solvinity takeover.

- Hearing Convocation: Tweede herziene convocatie rondetafelgesprek Solvinity voor DigiD (27-01-2026), https://www.tweedekamer.nl/kamerstukken/detail?id=2026D00898&did=2026D00898.
- Technical Briefing: Presentation by Logius & Landsadvocaat on DigiD Infrastructure Security, https://debatdirect.tweedekamer.nl/2026-01-27/digitale-zaken/thorbeckezaal/rondetafelgesprek-inzake-de-consequenties-van-de-beoogde-overname-van-solvinity-voor-digid-18-30/onderwerp.
- Government Briefing: Kamerbrief over de stand van zaken en aanpak van de casus Solvinity (Feb 2026), https://www.rijksoverheid.nl/documenten/kamerstukken/2026/02/10/kamerbrief-over-stand-van-zaken-en-aanpak-casus-solvinity.
- Legal Opinion: Advice of the Government Attorney regarding CLOUD Act and FISA exposure, https://debatdirect.tweedekamer.nl/2026-01-21/digitale-zaken/thorbeckezaal/naar-aanleiding-van-de-voorgenomen-overname-van-het-bedrijf-solvinity-deels-vertrouwelijk-16-00/onderwerp.
- Bert Hubert: Government & Operational Dependencies: The Toll Road Paradox, https://berthub.eu/2026Z01347.
- Roundtable position papers—Prof. Paul Timmers (*Geopolitics of Technology*), Instituut Clingendael (*Strategic Autonomy and the EU Cloud Landscape*), Bits of Freedom and DINL/NLdigital (*Privacy and Citizen Trust in Digital Identity*)—all at https://debatdirect.tweedekamer.nl/2026-01-27/bestuur/thorbeckezaal/rondetafelgesprek-inzake-de-consequenties-van-de-beoogde-overname-van-solvinity-voor-digid-18-30/stukken.
- Prof. Lokke Moerel: Legal Analysis: Extraterritoriality and the Vifo Act, https://ibestuur.nl/markt-en-overheid/digitale-infrastructuur/digid-en-solvinity-over-mitigatie-en-zeggenschap.

References

1. Moerel, L., Timmers, P.: RESEARCH IN FOCUS Reflections on Digital Sovereignty (2021)
2. Eaves, D., Rao, K.: Digital Public Infrastructure: a framework for conceptualisation and measurement (2025)
3. Bharosa, N.: The rise of GovTech: Trojan horse or blessing in disguise? A research agenda. Gov. Inf. Q. **39**, 101692 (2022). https://doi.org/10.1016/j.giq.2022.101692
4. Pohle, J., Thiel, T.: Digital sovereignty. Internet Policy Rev. **9**(4) (2020). https://doi.org/10.14763/2020.4.1532
5. Couture, S., Toupin, S.: What does the notion of "sovereignty" mean when referring to the digital? New Media Soc. **21**, 2305–2322 (2019). https://doi.org/10.1177/1461444819865984
6. Timmers, P.: Sovereignty in the digital age. In: Introduction to Digital Humanism: A Textbook, pp. 571–592 (2023). https://doi.org/10.1007/978-3-031-45304-5_36
7. Grohmann, R., Costa Barbosa, A.: Sovereignty-as-a-service: how big tech companies co-opt and redefine digital sovereignty. Media Cult. Soc. (2025). https://doi.org/10.1177/01634437251395003

8. Andersen, P.D.: The PESTEL framework and its variants for analysing the strategic environment: evolution. Limitat. Adopt. Energy Policy Res. (2025). https://doi.org/10.2139/ssrn.5266562
9. Head, B.W., Alford, J.: Wicked problems: implications for public policy and management. Adm. Soc. **47**, 711–739 (2015). https://doi.org/10.1177/0095399713481601
10. Farrell, H., Newman, A.L.: Weaponized interdependence: how global economic networks shape state coercion. Int. Secur. **44**, 42–79 (2019). https://doi.org/10.1162/ISEC_a_00351
11. OECD: Competition in the Provision of Cloud Computing Services. OECD Roundtables on Competition Policy Papers, No. 323. (2025). https://doi.org/10.1787/595859c5-en
12. Bria, F., Timmers, P., Gernone, F.: EuroStack - A European Alternative for Digital Sovereignty (2025)
13. EUROPEAN COMMISSION: Cloud Sovereignty Framework - Version 1.2.1. , Brussels (2025)
14. Rittel, H.W.J., Webber, M.M.: Dilemmas in a general theory of planning. Policy Sci. **4**, 155–169 (1973)

From the Modern Citizen to Citizens as Multiplicities: Citizen-Centricity as a Sign of a New Episteme?

Simon Dechamps(✉)

Université de Namur, Rue de Bruxelles 61, 5000 Namur, Belgique
simon.dechamps@unamur.be

Abstract. Citizen-centricity has become a dominant paradigm in digital government, often framed either as a novel transformation of public administration or as an incremental improvement in service delivery. Yet little research has examined how this paradigm reshapes the production of knowledge about citizens within digital governance. This study addresses this gap by analyzing citizen-centricity as an epistemic transformation rather than solely a managerial or technological innovation.

Drawing on fourteen semi-structured interviews with officials involved in a digital government initiative in Belgian public administrations, this article mobilizes Michel Foucault's archaeology of knowledge to examine how contemporary governance constructs knowledge about citizens. The findings reveal three interrelated dynamics: (i) the proliferation of user multiplicities as fragmented and context-dependent categories, (ii) the conceptualization of users as variable configurations of attributes, and (iii) the co-production of the user through interactions between citizens and public administrations. These dynamics indicate a shift away from the modern episteme described by Foucault, which centers knowledge on the human subject as an abstract, stable, and universal unit. Taken together, the findings contribute to eGovernment research by reframing citizen-centricity as an indicator of epistemic transformation in the production of knowledge about citizens, while also highlighting implications for practitioners and suggesting research directions to address them.

Keywords: Citizens · eGov · Citizen-Centricity · episteme · Discourses · Foucault

1 Introduction

In recent decades, the concept of citizen-centricity has emerged as a central framing for digital transformation in public administration. Often presented as a radical innovation, citizen-centric approaches emphasize tailoring services, processes, and technologies to the needs, preferences, and experiences of citizens [1]. Governments are encouraged to adopt agile practices, leverage data analytics, and redesign digital platforms to anticipate and respond to citizen demands, creating a discourse in which public organizations appear compelled to reinvent themselves to maintain legitimacy and effectiveness [2].

D. Duenas-Cid et al. (Eds.): ePart 2026, LNCS 16923, pp. 18–33, 2027.
https://doi.org/10.1007/978-3-032-36676-4_2

While this framing portrays citizen-centricity as novel and urgent, historical and comparative perspectives suggest a more nuanced view. Centering the citizen is not a uniquely contemporary phenomenon; administrative reforms from early modern Europe to the late twentieth century have repeatedly sought to align public services with societal expectations [3–5]. What distinguishes current practice is not the principle of citizen-orientation itself, but the ways knowledge about citizens is generated, structured, and mobilized, increasingly through digital technologies, data analytics, and participatory design practices.

This study approaches the way citizen-centricity reframes the "citizen" through the lens of epistemic analysis, drawing on Foucault's notion of regimes of knowledge [6–8]. Traditional public administration reflects two prior epistemes (an episteme is a particular regime of knowledge): the classical episteme, in which knowledge unfolds toward a transcendent unity, and the modern episteme, centered on the human as a stable, universal unit whose actions and experiences are systematized [8, 9]. In this study, we argue that citizen-centric practices suggest the emergence of a contemporary epistemic configuration in which the multiplicities of citizens appear as co-produced multiplicities. These observations are presented in the results section and serve as partial evidence of this epistemic shift, rather than definitive conclusions.

Empirically, the study draws on 14 interviews conducted within Belgian public administrations. Belgium provides a suitable context due to its long-standing administrative tradition, central position in the European digital context, and diverse digital governance initiatives. Rather than assessing citizen-centricity as an innovation per se, the study investigates how current practices reflect transformations in the production and organization of knowledge about citizens, highlighting patterns that depart from the modern episteme.

The contribution of this article is threefold. First, it repositions citizen-centricity within a discursive and epistemic framework, highlighting the distinction between procedural novelty and deeper knowledge transformations. Second, it provides empirical insights suggesting that the conceptualization of citizens is indicative of an emerging epistemic shift. Third, it situates how citizen-centricity both extends and challenges current understandings of digital government, administration, and the citizens.

The article is structured as follows. Section 2 reviews the historical and contemporary conceptualizations of citizen-centricity. Section 3 situates them within broader administrative and epistemic frameworks. Section 4 describes the methodology, followed by Sect. 5 presenting the results. Section 6 discusses how these results illuminate emerging epistemic patterns, and Sect. 7 concludes with implications and limitations.

2 Background

The notion of citizen-centricity has gained prominence in public administration and digital governance research over the past two decades. Broadly defined, citizen-centricity refers to the design and delivery of public services with a primary focus on citizens' needs, experiences, and expectations [1, 10]. In practice, citizen-centric approaches involve tailoring administrative processes, digital platforms, and policy instruments to enhance responsiveness, accessibility, and user satisfaction [1, 2]. In the context of digital government, citizen-centricity has become closely associated with transparency, co-production, and the adoption of agile and participatory technologies [1, 10].

A substantial body of literature frames citizen-centricity as an urgent and novel challenge for contemporary public organizations. For instance, [1] argue that digital transformation initiatives must reorient administrative structures around citizen needs, framing the shift as both unprecedented and imperative. Similarly, [11] depict citizen-centricity as a radical departure from traditional, bureaucratic governance models, suggesting that failing to adopt such approaches may threaten organizational legitimacy and effectiveness. This framing is reinforced by studies emphasizing the role of new technologies, such as open government data platforms and digital co-creation tools, in facilitating citizen engagement [12, 13].

Despite this widespread framing of citizen-centricity as a novel phenomenon, critical voices question its supposed originality. Several authors note that citizen-focused governance has long been a concern for public administrations, albeit articulated through different practices and discourses [3–5]. For example, early open data initiatives, participatory budgeting experiments, and service delivery reforms illustrate that attention to citizen needs is not inherently new, but rather a longstanding principle, now amplified through digital technologies [14]. In this respect, citizen-centricity can be understood less as a radical innovation and more as a discursive and performative construct, legitimizing ongoing administrative transformations and emphasizing the "urgency" of digital adoption [15, 16].

The contemporary discourse of citizen-centricity often presents responsiveness to citizens' needs as a novel feature of digital governance. Yet, the principle of aligning administrative action with the expectations of the governed has a much longer history, dating back at least to the emergence of modern public administration after the collapse of the *Ancien Régime* in Europe. In the late eighteenth and early nineteenth centuries, the creation of professionalized bureaucracies in France, Britain, and other European states was explicitly aimed at improving the delivery of services to the public, rationalizing state functions, and ensuring predictable access to justice, taxation, and public goods [17].

Scholars of administrative history have shown that early modern reforms frequently emphasized user-oriented practices. For instance, the Napoleonic administrative codifications sought to standardize procedures while also facilitating citizen access and clarity in interactions with the state [4, 18]. Similarly, the rise of meritocratic civil services in Britain and Prussia was motivated not only by efficiency, but also by the expectation that civil servants should serve the public impartially and responsively, balancing bureaucratic control with the needs of citizens [5, 19]. These reforms illustrate that citizen orientation, whether framed as service quality, impartiality, or responsiveness, was embedded in administrative design from the earliest stages of modern public governance.

From a historiographical perspective, these developments challenge the notion that citizen-centricity is inherently tied to digital technologies or contemporary management paradigms. Rather, attention to citizens has long been a structuring concern, evolving in its forms and technologies, but not in its underlying principle. What has changed is the discourse surrounding urgency and novelty: modern digital governance literature frames citizen-centricity as an innovation required by a VUCA environment [11, 14] (Volatile,

Uncertain, Complex, and Ambiguous), while historically, responsiveness to citizens was perceived as an essential, ongoing function of public administration, codified and institutionalized from the early modern period onward.

In sum, understanding citizen-centricity in light of its historical antecedents allows for a more nuanced interpretation of its purported novelty, situating contemporary initiatives within a continuum of public sector efforts to balance administrative control with responsiveness to the governed [16]. That is the aim of the following sub-section.

2.1 Recent History of a Contested Concept

The term 'citizen-centricity' itself did not exist several centuries ago. However, it is important to analyze the broader tradition of public service within which it is embedded.

The notion of public service in Belgium did not appear at the time of the Kingdom's founding in 1830. Instead, references were sometimes made to a service rendered to citizens, not based solely on the sovereign's will, but on the citizens themselves: the idea of a service "to the public" advanced under the *Ancien Régime* found a new justification: the common good of citizens according to their own will, rather than the sovereign's good will [20]. Gradually, the term 'public service' entered the literature, although its meaning remained vague, broadly referring to the actions of public institutions [20].

In 1901, a first formal definition of public service was provided in France, a legal context similar to Belgium, and very influential over the Kingdom: "An organization managed by a public administration or on behalf of a public administration in order to satisfy a collective need" [21]. At the same time, some authors emphasized that this new conception of public services oriented toward collective needs corresponded to a disciplinary diagrammatic structure [22], rather than a sovereign one: "The State is not, as has long been believed, a power that commands, a sovereignty: it is a cooperation of public services organized and controlled by the government. This notion of public service is central: around it revolves all modern public law" [23].

In Belgium, these definitions were largely adopted, always emphasizing the pursuit of collective needs as the ultimate goal [24].

About twenty years ago, the European Commission integrated into notions of public services elements that went beyond the universal satisfaction of collective needs. Explicit attention was given to service continuity, quality, and price accessibility, as well as to the protection of users and consumers. While this shift may seem minor, it marked a significant change: the focus moved from the satisfaction of collective needs of an indeterminate unity of citizens to the protection of users and consumers as an undefined multiplicity. The European Commission provided little justification for this largely unnoticed modification [25].

It was around this time that the first mentions of "citizen-centricity" appeared in law and the literature. A quick search in one of the leading journals in eGov, *Government Information Quarterly*, shows that in 1992, Durrance used the term "citizen-centered," probably to emphasize a democratic conception of information systems that he mentioned just before [26].

It was only in the early 2000s that researchers began to use this term more consistently. To our knowledge, it was in 2004 that authors first attempted to approach the term from a functional perspective: "With a focus on customers' requirements, these sites broke the long-standing tradition of distributing government information along functional lines."

In this perspective, public service is no longer emphasized as a response to a collective need but rather as a tailored response to various individual needs. Since then, a growing body of literature has documented this movement. Recently, [1] situate citizen-centricity as a desired evolution within public service governance models. [10] built a comprehensive typology of the concept, depicting four coexisting understandings: user identification, participative design of the service, governance style, and finally, service overall quality once launched.

3 Analytical Lens

To analyze the historical foundations and transformations of public services in Belgium, we adopt a Foucauldian analytical lens called the Archeology of Knowledge [7]. One of the many advantages of using Foucault lies in the fact that, although his work makes no universalist claims, it nevertheless concerns historical administrative territories and practices that were, in part, shared by France and what is today Belgium. Following the Archeology of Knowledge, we conceptualize the shift towards citizen-centricity in public services historically.

Foucault is often used to analyze shifts in historical paradigms. He famously analyzed the shift from a sovereign to a disciplinary regime of power [22]. His two main contributions in this regard consist of conceiving power as a relation of forces, an informal and relational element. Power is therefore not a thing located on the side of a ruler, a dominant group, or even the State, but rather a set of force relations that follow general patterns characteristic of a given epoch: diagrams of force.

Before developing this analysis of power, Foucault had long worked on forms of knowledge. According to an approach often described as post-Kantian, Foucault considers knowledge to be composed of two disjunctive forms: what is seen and what is said. These two forms correspond to specific conditions of seeing and speaking: visibilities and statements. These conditions remain relatively stable within a given historical period, which is therefore referred to as an *episteme* [8]. Foucault situates changes within a relatively short period, often just a few years.

Deleuze's [9] analysis of Foucault further clarifies the subtlety of these changes, highlighting that while the general characteristics of institutions may persist, the underlying conditions of possibility (diagrams of power and episteme of knowledge) diverge between periods.

The Anglophone reception of Foucault, particularly within the field of Information Systems, has largely focused on his analysis of power. We, however, seek to show how his analyses of knowledge also enable a deeper understanding of so-called "citizen-centric" practices: to which conditions of seeing and saying, that is, to which episteme do they correspond?

3.1 Knowledge: From Classical to Modern Episteme

Foucault's notion of episteme provides an analytical lens to understand how knowledge is structured and how conditions govern what can be seen and said in a given historical period [8]. Each episteme establishes a set of conditions of possibility: rules and constraints that determine both what can be observed and what can be enunciated. While these dimensions are formally distinct (the visible versus the enunciable), they are mutually constitutive: what is sayable is informed by what is observable, and what is observable gains significance through discursive frameworks.

According to Deleuze [9], in Foucault's analysis, the classical episteme unfolds things and beings according to continuous series, reflecting an order of infinity inherent to creatures. Everything that can be elevated to infinity is ultimately related to God or the Nature, constituting a thought of representation and elevation. The conditions of seeing are thus structured by analogical and hierarchical relations, while the conditions of saying are limited to statements that correspond to these visualizable orders. The classical episteme is therefore one of the "constitutive infinites": knowledge expands outward, disclosing continuous series and hierarchical relations. It reaches a closure with what Nietzsche called the "death of God", marking the end of this transcendental referential framework [8, 9]. Public administration, at that time, was embodied in the person of the sovereign, who, in the name of God, unfolded public offices through a sprawling hierarchy [9, 22]. From this perspective, things were elevated toward the infinity represented by the sovereign, in a movement of unfolding of his offices. The administration functions primarily as an operational extension of the sovereign. From the user's perspective, public service does not yet aim at citizen satisfaction, but at the fulfillment of a duty toward authority, which itself guarantees divine or naturalized order. This aspect of Foucault's work is particularly evident in [22], where he shows how punishments were administered for offenses against the king, in his name, as a way of publicly demonstrating and reinforcing the sovereign's authority.

In contrast, the modern episteme is characterized by forces that fold around finitude rather than unfold toward infinity. Deleuze [9] emphasizes that these forces are never purely internal to man; they originate from external systems of power and knowledge, structuring what can be observed and said. The conditions of seeing are now organized around relations, functions, and systemic determination, focusing on life, labor, and language as the triple roots of finitude. Correspondingly, the conditions of saying expand to articulate these functional relationships, causal regularities, and population-level phenomena. This constitutes a "constitutive finitude", where knowledge is organized around what is finite yet relationally structured. The modern episteme thus reorients attention from the infinite to the finite, culminating conceptually in the "death of man", the closure of classical humanist referentiality announced by Foucault in 1968 [8]. Within this organization of knowledge, it becomes clear how the conditions of seeing and speaking about public administration consistently refer back to grouped populations: forms of folded unity centered on Man. It is no longer the sovereign, as the incarnation of infinity, who unfolds his offices; rather, public administration folds back upon a foundational unit: the human being, the citizen, the user it serves. Yet does citizen-centricity not at least call into question this unitary foundation of folding, that of Man, or the citizen conceived as an aggregated mass? The administration undergoes a profound transformation. It is no longer a mere extension of the sovereign, but a specialized apparatus serving the population. This administration form is not out of concern for individuals, but acts rather for a statistical population, a central category of the governmentality sciences [8, 22].

Accordingly, the administration provides services addressing collective needs (hygiene, education, infrastructure, social security). The citizen is not yet "at the center"; they are primarily a member of a population, an object of management, regulation, and normalization. Public service becomes a task imposed on civil servants, themselves embedded in disciplinary devices for organizing work (schedules, procedures, hierarchies, files, inspections). Administrative procedures align with a bureaucratic ideal of rationality, regularity, and predictability [17]. The primary objective is not user satisfaction, but the production of a stable social order, coherent behaviors, and a population homogenized by rule.

3.2 Continuity and Subtle Variation: Where Are We Now?

As Deleuze [9] notes, Foucault's historical method reveals that the surface characteristics of institutions may remain constant, while the underlying diagrams and epistemes shift. Applied to public services, this implies that although principles such as universality, neutrality, and procedural regularity have endured, the conditions of administration (how citizens are seen, counted, and interacted with) have changed. These changes, often subtle, provide a foundation to understand the continuity and transformation of public service rationality across historical periods (Table 1).

Table 1. The epistemes and diagrams conditioning knowledge and power according to Foucault.

Epoche	Knowledge
Approximately XVVth – XVIIIth	**Classical episteme: constitutive infinity that unfolds.** In the classical episteme, knowledge unfolds beings along continuous series, organized according to the order of constitutive infinitudes. Things can be elevated to infinity because they ultimately refer to God as the transcendent principle of unity and totalization [8, 9]. The conditions of seeing and saying are structured by representation, ordering, and resemblance
Approximately XIXth – XXth	**Modern episteme: constitutive finitude that folds.** Knowledge is reorganized around constitutive finitude: beings no longer unfold toward a divine infinity, but fold into the triad of life–labor–language, which constitutes the very condition of possibility for the human sciences [8, 9]. The conditions of seeing and saying are structured according to a logic of the finite, historicity, and the production of norms internal to human life
Approximately XIXst	**Contemporary episteme: ?** Foucault anticipates the end of the modern episteme with what he calls "the death of man" [8], but he does not explore it in further detail in his subsequent works

Regarding the current period, it is interesting to observe that Foucault and Deleuze merely delineate the diagram of forces while sidestepping the question of the emergence of a new episteme. It should be noted that, at the time of its publication, [8] which announces the "death of man" as the advent of a new episteme, provoked controversy due to widespread misunderstanding. Yet the philosopher announces neither a metaphysical drama nor a negation of the human being, as has sometimes been alleged. Rather, he articulates the logical conclusion of his archaeology of knowledge: if "Man" is indeed a recent construct insofar as it grounds the conditions of knowledge, then, being historically situated, this construct will necessarily be replaced by another foundation for the conditions of knowledge [9]. The objective of this article is to examine the extent to which we are shifting from one episteme to another, arguing for the heuristic power of the concept of citizen-centricity to do so.

4 Methodology

This study adopts a qualitative research design. Specifically, it uses semi-structured interviews to investigate how the notion and practice of "citizen-centricity" in digital public services is interpreted, negotiated, and enacted within Belgian public administrations. Qualitative methods are especially well-suited for exploring such complex, context-bound, and historically embedded phenomena, as they allow researchers to access meanings, perceptions, and institutional practices in depth [27–29].

4.1 Data Collection

Data were collected through 14 semi-structured interviews with civil servants and public officials from a variety of Belgian federal, regional, and local administrations. Participants were selected based on their involvement in a project that was required by a regulation to be citizen-centric. This implementation project aims to deploy a platform that consolidates several hundred administrative procedures across all levels of government, along with administrative information and support services. Due to its scope, the project can be understood as the coordination of the implementation of a European regulation intended to modernize public services and make them citizen-centric, rather than as the digital transformation of one or several individual public services in the strict sense. This is important insofar as our aim is precisely to understand how citizen-centricity reshapes the way public services are approached. Participants were selected to best represent the diversity of institutional levels involved, as well as variation in seniority and job roles (Table 2).

An interview guide (https://doi.org/10.5281/zenodo.20289634) was developed, drawing on the historical and theoretical framework of the study (Sects. 2–3). The guide included open-ended questions on the evolution of service delivery in their institution; the meaning and use of "citizen-centricity" (or equivalent terms) in their unit; practices related to user-orientation or user-involvement; challenges and constraints in implementing user-centered design, … As this study was based on an embedded research program, the formulation of the research questions was informed by practice, which made it possible to reach saturation quickly, as no significantly new findings emerged after the 11th interview. Interviews lasted between 45 and 90 min. They were conducted either in person or via video calls, depending on participants' availability. Before conducting each interview, participants provided informed consent and were assured of confidentiality and anonymization of their responses.

Table 2. Interviewees and their position.

Interviewee	Position	Experience
I1	Managing Director	Senior
I2	Project Manager Digital Transformation	Medior
I3	Communication Specific Attaché	Medior
I4	Webmaster Attaché	Medior
I5	Consumer Experience Strategic Advisor	Senior
I6	Project Manager	Medior
I7	Customer Relations and Service Project Manager	Medior
I8	Advisor, Project Manager	Senior
I9	Digital Content Creator	Junior
I10	Communication Quality Service Copywriter	Junior
I11	Head of Communication Quality	Senior
I12	Consultant on a Federal Digital Transformation Project	Junior
I13	Consultant on a Federal Digital Transformation Project	Senior
I14	Webmaster Attaché	Medior

4.2 Data Analysis

The analysis followed an iterative thematic coding process, using inductive codes emergent from interview and document data. All interview recordings were transcribed verbatim, and the resulting transcripts, along with collected documents, were imported into qualitative data analysis software (NVivo) to facilitate coding, retrieval, and thematic comparison. The collected data were analyzed using the grounded theory-inspired approach developed by [28] and refined by [27, 29]. This involved three coding phases:

- Open Coding: In this initial phase, transcripts were read line-by-line to identify conceptual indicators related to how participants spoke about novelty, past practices, and change. Codes such as "average citizen", "addressing needs", "satisfying the citizen", "citizen specificities", and "targeting" emerged consistently.
- Axial Coding: We then reassembled the data by identifying relationships between codes, clustering them into higher-order themes. This allowed us to detect discursive patterns, such as "multiple citizens", "average citizen", …
- Selective Coding: Finally, we identified overarching narrative structures and discursive mechanisms that aligned with the conceptual dimensions outlined in our framework: knowledge epistemes (classical, modern, or contemporary).

Following established principles of thematic saturation [29, 30], our analysis indicates that saturation was reached after eleven interviews. This relatively low number can likely be explained by the fact that the interview questions were developed based on observations from prior studies conducted within the same institution, on similar topics.

The resulting analysis provides a basis for theorizing how Citizen-centricity represents an element that no longer aligns, for the most part, with classical and modern epistemes (selective coding).

5 Results

Across interviews, at least three characteristics were diverging from previous epistemes. First, users are not regarded as a homogeneous group, but rather as multiple groups: multiplicities of users (Sect. 5.1). Second, even within these groups, citizens are decomposed into multiple variables and dimensions. Consequently, each individual citizen becomes multiplicities in themselves, varying according to the dimensions considered (Sect. 5.2). Finally, the data constituting these multiple multiplicities are derived from real-world, concrete situations, collected directly at the level of citizens, rather than from aggregates or averages. Thus, knowledge about citizens is no longer to be conceived as a top-down, encompassing view, but rather as information flowing upward toward the government (Sect. 5.3).

It should be noted that, within the limited scope of this article, we highlight inconsistencies with the modern episteme. However, as Foucault emphasized, during periods of change, epistemes can coexist. Accordingly, our analyses point toward such coexistence, such as this excerpt echoing how Deleuze described the modern episteme as a constitutive finitude that folds (Table 1): "*Sometimes, what's interesting is to aggregate feedback*" (I10).

5.1 The Citizen as a Non-universal Subject: Towards Multiplicities of Users

Interviews first reveal a weakening of the idea of the citizen as a universal and homogeneous category. Instead, administrative actors describe citizens through multiple and context-dependent classifications, such as clients, users, or targeted populations. This reflects a shift from abstract categorization to situated understandings of publics grounded in specific administrative problems and service interactions.

For example, one respondent emphasizes the need to rely on "*experts of lived experience*" to reach differentiated publics that standard administrative tools fail to capture (I2). This indicates that existing categorical schemes are perceived as insufficient to account for the heterogeneity of citizen experiences.

Similarly, rather than assuming a unified public, organizations increasingly operate through differentiated targeting logics:" *We have as many targets as there are people likely to visit our site*" (I13). However, this multiplicity also generates operational tensions, as organizations struggle to reconcile inclusiveness with manageability:" *It is easier to address well-defined target audiences than to address a wide range of profiles*" (I6).

Overall, these empirical elements indicate a progressive erosion of the citizen as a singular administrative reference point, replaced by a plurality of situational categories that vary across contexts and institutional needs: multiplicities of users.

5.2 The Citizen as Data-Constructed Entity: Towards Users as Multiplicities

A second set of findings concerns the increasing reliance on multiple dimensions to construct representations of citizens. Rather than being treated as pre-given subjects, citizens are described through multiple data traces, behavioral indicators, and segmented profiles.

One interviewee explains: "*the idea is to really manage all the data we receive [...] and then move toward solicited data to complete certain information*" (I6). This illustrates a shift towards continuous accumulation and refinement of citizen-related data.

In this configuration, the citizen is not a stable entity but an analytically constructed object that is continuously updated through digital interactions. As such, representations of citizens become modular, combining different variables depending on administrative purposes. This leads to iterative forms of knowledge production in which citizens are constantly redefined through system-citizen interactions.

As another participant notes, adaptation becomes a central operational principle: "*we try to keep an eye on how people are using our digital tools to always try to adapt accordingly*" (I9). Yet this adaptive process remains partial and imperfect, highlighting the contingent nature of data-driven understandings of citizens.

Overall, these findings suggest that citizens are not only differentiated into multiple categories, but are also internally decomposed into variable dimensions, such that each user emerges as a set of dynamic and context-dependent multiplicities.

5.3 The Citizen as Co-producer: Towards Co-produced Multiplicities

A third dynamic concerns the active participation of citizens in the production of knowledge about themselves. Beyond being objects of observation or data extraction, citizens are increasingly involved in shaping administrative understanding through feedback, participation, and co-design processes.

As one respondent explains, statistical data alone are insufficient to capture non-mainstream users, requiring direct engagement:"*you can only reach them in a qualitative way by actually talking to them*" (I14). This illustrates the limits of purely quantitative or dashboard-based approaches to citizen understanding.

More broadly, citizen-centric initiatives involve iterative interactions in which citizens contribute to the refinement of administrative processes. For instance, services are redesigned based on lived experiences of users, including vulnerable groups, as illustrated by participatory design sessions mentioned by I7 and I3. In this sense, knowledge about citizens is no longer exclusively produced within administrative systems but emerges through distributed and interactive processes.

6 Discussion

6.1 Theoretical Contribution: Reconfiguration of Regimes of Knowledge in eGovernment Research

The empirical findings of this study indicate that citizen-centricity represents more than a technical or managerial reform; it reflects a deeper reconfiguration of knowledge regimes in digital government regarding the citizens. This observation aligns with an emerging strand in the literature that calls for moving beyond traditional eGov paradigms centered on efficiency and transactional service delivery toward more dynamic, adaptive, and user-focused frameworks [1].

In classic eGov research, much attention has been paid to service maturity models, channel integration, and digital inclusion [31]. Yet, these models often still implicitly presuppose a stable, universal subject; an assumption rooted in the modern episteme of the autonomous citizen as a unitary object of governance. By contrast, the present study's results suggest that citizen-centricity, through its emphasis on a set of relevant variables co-produced by and about citizens, produces knowledge about citizens that is dynamic, situated, and contingent [16]. In this sense, citizen-centricity resonates with critical eGov contributions that highlight the complexity of digital public value creation [32, 33], but it goes further by challenging the very epistemic foundation on which citizen representations in eGov are constructed. One can observe a shift from a verticality that connected the quasi-transcendental citizen as a universal unit to the government, toward a horizontality in which the government operates within co-produced, immanent multiplicities of users as multiplicities.

Importantly, the hybridization of expertise observed in the interviews, where technical, statistical, and experiential knowledge converge, echoes calls in the literature for interdisciplinary approaches in digital governance [16]. However, while existing research often frames hybridization in terms of organizational integration or stakeholder collaboration, the present findings stress that hybridization transforms knowledge production itself: citizens' profiles and the categories used to represent them are not pre-given but continuously reconfigured multiplicities.

6.2 Epistemological Contribution: Emerging Epistemic Patterns in Citizen-Centric eGov

Building on this reconfiguration, the study identifies several emerging epistemic patterns that extend current eGov theorization. First, the fragmentation and modularity of the citizen figure, appearing variously as user, co-producer, or data source, aligns with a shift in eGov studies from citizen as passive recipient to citizen as an active participant [34, 35]. Yet, most contributions frame participation largely in terms of engagement mechanisms or design practices [12]. In contrast, the present research suggests that this modularity is not merely procedural, but indicative of a fundamental transformation in how subjects are conceptualized in digital governance.

Second, the study's emphasis on knowledge regimes as dynamic and iterative complements investigations on real-time feedback incorporation and user analytics in service improvement [36, 37]. However, while traditional analytics-oriented work often frames these developments as technological enablers of better services, this study positions them as epistemic modalities: ways of seeing and saying that evolved from stable categories to multiplicities and interactions. In this sense, citizen-centricity is closer to what some scholars refer to as transformative digital government [38] but grounded in an explicit reconceptualization of the subject and knowledge relevant to citizen-centricity.

Together, these emerging patterns indicate an epistemic shift: from the modern episteme of a stable, universal human subject to a regime where the citizen is multiple, continuously reconfigured, and co-produced. In this regard, citizen-centricity should be seen not only as an organizational principle but as a marker of epistemic transformation in eGovernment scholarship. This epistemic shift compels eGov researchers to reconsider foundational concepts (such as citizen), and to explore new approaches that can accommodate fluid, emergent, and contextualized forms of knowledge. In doing so, citizen-centricity emerges not merely as a design ideal or policy directive, but as a heuristic indicator of a deeper transformation in how digital governance, the administration, and the citizen are conceptualized, studied, and practiced.

6.3 Practical Implications and Research Directions

Based on the three observed trends, namely, the multiplicities of users, the conceptualization of users as variable multiplicities, and the co-production of users, three implications for practitioners are identified. On this basis, it is possible to delineate three key avenues for future research.

The first practical implication concerns the definition and delimitation of multiplicities. The multiplicities of users to be considered vary depending on the service context, as demonstrated in studies on public e-services [39]. However, this insight does not address the practical challenge of distinguishing one multiplicity from another. Questions remain regarding the optimal granularity for defining multiplicities and the potential risk that certain users may be excluded when multiplicities are delineated. Understanding how to operationalize multiplicities without inadvertently marginalizing subsets of users is crucial for both research and practice. *How are user multiplicities constituted and bounded in citizen-centric digital governance, and what forms of inclusion and exclusion do these boundaries produce?*

The second practical implication concerns the users themselves as variable multiplicities. When users are conceptualized as configurations of multiple dimensions or attributes, it becomes essential to determine which variables are relevant for a particular service. Prior work has explored aspects of this problem in the context of digital infrastructures and user behavior [40]. An additional research question relates to the temporal dynamics of these variables: how do the dimensions that constitute user multiplicities evolve over time, and what implications does this have for the design and evaluation of public digital services? *How are citizens constructed as variable multiplicities in digital public services, and how do these configurations evolve over time?* Note that, with regard to the first two research directions, Deleuze and Guattari would argue that once multiplicities are fixed, they cease to function as dynamic assemblages and become rigid structures, thereby reproducing transcendentalism [39].

The third practical implication concerns the co-production of the concept of "citizens" by public administrations. While previous studies have examined the mechanisms enabling co-production [12], there remains a pressing need to investigate the broader institutional and organizational implications. In particular, the increasing role of administrations as mediators between public policies (intended for universal stable users) and citizens (as co-produced multiplicities) raises questions about how this intermediary function reshapes civil servant practices and the production of user knowledge. Understanding these dynamics is essential for assessing both the effectiveness and the ethical implications of citizen-oriented service design. *How do civil servants manage the tension between universal policy assumptions about citizens and the multiplicities revealed through citizen-centric governance?*

Addressing these three practical implications through the research questions we will formulate will make it possible to develop recommendations to help public administrations take into account the conceptual reconfiguration of citizens. A subsequent phase may also consider the creation or adaptation of methods for incorporating the citizen on this basis, for instance, through Design Science Research.

7 Limitations and Conclusions

It is important to acknowledge the limitations of the present study regarding the characterization of the contemporary episteme. Based on our empirical results, it is not possible to provide a definitive description of the current conditions of seeing and saying, that is, the full epistemic framework in which we currently operate. Our study does not capture all the institutional, historical, and conceptual forces that would be necessary to delineate a comprehensive episteme in Foucault's sense [8]. Nonetheless, the findings provide sufficient evidence to conclude that, in the case considered, current practices are no longer aligned with the modern episteme, which centers knowledge on the human as a stable and universal unit, and even more so with the classical episteme, structured around constitutive infinity and transcendent unity. (i) The multiplicities of users, (ii) the conceptualization of users as variable multiplicities, and (iii) the co-production of users, altogether suggest that the contemporary configuration of knowledge diverges fundamentally from these earlier regimes. This constitutes a threefold contribution: a theoretical contribution insofar as it shifts the understanding of citizen-centricity from a managerial mode to an indicator of epistemological change; an epistemological contribution in that it shows that core concepts such as the "citizen" are subject to ongoing transformation; and a practical contribution insofar as the identified characteristics raise challenges for adapting practices to account for this new conceptualization of the citizen.

Ultimately, this article does not posit that the citizen is changing; rather, it argues that the very conditions of possibility for what counts as a "citizen" as an object of administrative knowledge are changing.

References

1. Sundberg, L., Holmström, J.: Citizen-centricity in digital government research: a literature review and integrative framework. Inf. Polity., 1–18 (2023). https://doi.org/10.3233/IP-220047

2. Helbig, N., Ramón Gil-García, J., Ferro, E.: Understanding the complexity of electronic government: implications from the digital divide literature. Gov. Inf. Q. **26**, 89–97 (2009). https://doi.org/10.1016/j.giq.2008.05.004
3. Alford, J., Hughes, O.: Public value pragmatism as the next phase of public management. Am. Rev. Publ. Adm. **38**, 130–148 (2008). https://doi.org/10.1177/0275074008314203
4. Veyne, P., Murray, O.: Bread and Circuses : Historical Sociology and Political Pluralism. Penguin (1990)
5. Peters, G., Pierre, J.: Politicians, Bureaucrats and Administrative Reform. Routledge (2006)
6. Foucault, M.: Sur l'archéologie des sciences. Réponse au Cercle d'épistémologie. Cahiers pour l'analyse **9**, 9–40 (1968)
7. Foucault, M.: Archéologie du savoir (1969)
8. Foucault, M.: Les mots et les choses (1968)
9. Deleuze, G.: Foucault. Editions de Minuit (1986)
10. Dechamps, S., Simonofski, A., Burnay, C.: Citizen-centricity in digital government: a theoretical and empirical typology. Gov. Inf. Q. **42**, 102005 (2025). https://doi.org/10.1016/j.giq.2024.102005
11. Berntzen, L., Johannessen, M.R., Ødegård, A.: A citizen-centric public sector why citizen centricity matters and how to obtain it. In: CENTRIC, International Conference on Advances in Human-oriented and Personalized Mechanisms, Technologies, and Services, Rome (2016)
12. Simonofski, A., Van Den Storme, S., Meers, H.: Towards a holistic evaluation of citizen participation in smart cities. In: ACM International Conference Proceeding Series, pp. 82–9 (2020). https://doi.org/10.1145/3396956.3396985
13. Anthopoulos, L., Janssen, M., Weerakkody, V.: Introduction to the special issue on citizen centricity in smart cities. Digit. Gov.: Res. Pract. **4**, 1–4 (2023). https://doi.org/10.1145/3624556
14. Bertot, J.C., Jaeger, P.T., McClure, C.R.: Citizen-centered e-government services: benefits, costs, and research needs. In: Proceedings of the 9th Annual International Conference on Digital Government Research, Partnerships for Public Innovation, DGO 2008 (2008)
15. Alvesson, M., Karreman, D.: Varieties of discourse: on the study of organizations through discourse analysis. Hum. Relat. **53**, 1125–1149 (2000). https://doi.org/10.1177/0018726700539002
16. Willcocks, L.P.: Foucault, power/knowledge and information systems: reconstructing the present. In: Mingers, J., Willcocks, L.P. (eds.) Social Theory and Philosophy for Information Systems, pp. 238–96 (2004)
17. Economy and Society, by Max Weber. The SAGE Encyclopedia of the Sociology of Religion. 2455 Teller Road, Thousand Oaks, California 91320: SAGE Publications, Inc. (2020).https://doi.org/10.4135/9781529714401.n137
18. Rothstein, B.: The Quality of Government: Corruption, Social Trust, and Inequality in International Perspective. University of Chicago Press (2011)
19. Pollitt, C., Bouckaert, G.: Public Management Reform: A Comparative Analysis : New Public Management, Governance, and The Neo-Weberian State. Oxford University Press (2011)
20. Guglielmi, G.J., Koubi, G., Dumont, G.: Droit du service public. Montchrestien (2007)
21. Hauriou, M.: Précis de droit administratif et de droit public général. 4e ed., Larose (1900)
22. Foucault, M.: Surveiller et punir (1975)
23. Duguit, L.: Traité de droit constitutionnel. 2e éd. (1923)
24. Buttgenbach, A.: Les modes de gestion des services publics en Belgique. [Bruxelles] (1942)
25. Rittberger, V., Nettesheim, M., Huckel, C. (eds.): Authority in the Global Political Economy. Palgrave Macmillan UK, London (2008). https://doi.org/10.1057/9780230584297
26. Durrance, J.C.: The White House conference and democracy: a missed opportunity. Gov. Inf. Q. **9**, 341–346 (1992). https://doi.org/10.1016/0740-624X(92)90008-A

27. Gioia, D.A., Corley, K.G., Hamilton, A.L.: Seeking qualitative rigor in inductive research. Organ. Res. Methods **16**, 15–31 (2013). https://doi.org/10.1177/1094428112452151
28. Glaser, B., Strauss, A.: The Discovery of Grounded Theory: Strategies for Qualitative Research. Sociology Press, Mill Valley (1967)
29. Corbin, J., Strauss, A.: Basics of Qualitative Research: Techniques and Procedures for Developing Grounded Theory, 4th edn. Sage Publications, Thousand Oaks (2015)
30. Saunders, B., Sim, J., Kingstone, T., Baker, S., Waterfield, J., Bartlam, B., et al.: Saturation in qualitative research: exploring its conceptualization and operationalization. Qual. Quant. **52**, 1893–1907 (2018). https://doi.org/10.1007/s11135-017-0574-8
31. Kamaruddin, K.A., Johari, N.J.: Is it citizen-centric? A tool for evaluating e-government websites' citizen-centricity. In: Lecture Notes in Networks and Systems (2023). https://doi.org/10.1007/978-981-19-1610-6_32
32. Avgerou, C., Bonina, C.: Ideologies implicated in IT innovation in government: a critical discourse analysis of Mexico's international trade administration. Inf. Syst. J. **30**, 70–95 (2020). https://doi.org/10.1111/isj.12245
33. Martin, C.J., Evans, J., Karvonen, A.: Smart and sustainable? Five tensions in the visions and practices of the smart-sustainable city in Europe and North America. Technol. Forecast. Soc. Change **133**, 269–278 (2018). https://doi.org/10.1016/j.techfore.2018.01.005
34. Certomà, C., Rizzi, F.: Crowdsourcing processes for citizen-driven governance. In: Springer Tracts in Civil Engineering (2017). https://doi.org/10.1007/978-3-319-47904-0_4
35. Lim, S., Abdul Malek, J., Hussain, M.Y., Tahir, Z.: Citizen participation in building citizen-centric smart cities. Malays. J. Soc. Space **14**, 42–53 (2018). https://doi.org/10.17576/geo-2018-1404-04
36. Mergel, I., Kattel, R., Lember, V., McBride, K.: Citizen-oriented digital transformation in the public sector. In: Proceedings of the 19th Annual International Conference on Digital Government Research: Governance in the Data Age, pp. 1–3. ACM, New York (2018). https://doi.org/10.1145/3209281.3209294
37. Nam, T., Pardo, T.A.: Conceptualizing smart city with dimensions of technology, people, and institutions. In: Proceedings of the 12th Annual International Digital Government Research Conference: Digital Government Innovation in Challenging Times, pp. 282–91. ACM, New York (2011). https://doi.org/10.1145/2037556.2037602
38. Bannister, F., Connolly, R.: ICT, public values and transformative government: a framework and programme for research. Gov. Inf. Q. **31**, 119–128 (2014). https://doi.org/10.1016/j.giq.2013.06.002
39. Esposito, G., Clement, J., Mora, L., Crutzen, N.: One size does not fit all: Framing smart city policy narratives within regional socio-economic contexts in Brussels and Wallonia. Cities **118**, 103329 (2021). https://doi.org/10.1016/j.cities.2021.103329
40. Schelings, C., Defays, A., Elsen, C.: Profiling Citizens in the smart city: a quantitative study in Wallonia. Smart Cities (2023)

In Search of Digital Government in Home-School Relations
– A Mapping Study of Research on Digital Communication Between Home and School

Sofia Hamberg(✉) and Elin Wihlborg

Linköping University, Linköping, Sweden
{sofia.hamberg,elin.wihlborg}@liu.se

Abstract. There are, in most countries, many digital systems used for communication between home and school, concerning administration, planning, implementation, and evaluation of education and students' performance. This complexity might both be an asset and a burden to administrate for both teachers, legal guardians, and students. Education is in most countries compulsory and public funded thus digitalization in schools, and not at least home-school relations, is a form of digital government. This paper contextualises the communication between home and school as digital government through a mapping literature review with focus on their methods, analytical frameworks and cases to identify mediating components in the communication and identify research gaps. Pedagogical or didactical research was excluded to focus on education as a public service, and the digital communication between home and school as a form of digital government. The scope is explorative and aims to serve as a foundation on which to base further empirical studies in the area. The findings show that the included studies (n = 19) are predominantly qualitative and within the field of education and teaching, both considering applied theoretical models and outlets. By considering the home-school communication as a form of digital impact mediator, drawing on Helsper's corresponding fields model [8], we structure the results and suggest both theoretically and empirically generated openings for further research. By studying digital communication between home and school through a digital government lens we draw the conclusion that further research in the area is needed to explore the factors that connect digital and offline exclusion in a school context.

Keywords: Digital communication · education · digital impact mediators · digital government

1 Introduction

The ongoing digitalization of society has changed our view of communication and access to public services, and this includes both the educational system and other governmental institutions.

Published by Springer Nature Switzerland AG 2027
D. Duenas-Cid et al. (Eds.): ePart 2026, LNCS 16923, pp. 34–49, 2027.
https://doi.org/10.1007/978-3-032-36676-4_3

Educational systems are important public institutions where digitalization also in many ways changes and develops new digital government practices. Formal education and school are not optional and there is compulsory school attendance, regulated by law in most countries. As a public service, education is universal and not based on need. Digital communication in schools is interesting as it is a three-part communication, where school staff, legal guardians, and students (who are often underaged), all need to be active participants. Schools are also often a regular and daily public service, framing and forming everyday life and it might therefore affect the attitude towards public institutions in general. Because of the complexity of the setting, it is interesting to study the effects of digitalization of public services in this context.

Due to this complexity, digitalization projects in the school context seem to be a challenging matter. Digitalization of school is manyfold and includes, for example, digital tools for teaching, digital tools for administration and organization, and communication [23]. There are many varieties of how this works in practice, sometimes several functions are integrated in the same system, sometimes separate systems are used for different purposes. For example, it is common that schools have one pedagogical learning platform for pupils and teachers and a separate one for communicating pedagogical issues with parents, such as lesson planning, assessments, and evaluations. In addition, there are often platforms used for absence reporting, application for leave, grading, schedule, and different kinds of digital educational material.

Besides these official channels and systems used for communication there are also more informal ways of making contact between home and school, such as e-mail, chat groups, and social media apps. Thus, digitalization of the school sector goes beyond the pedagogical use of digital tools in the classroom, and this study will broaden the view and investigate how communication between home and school is affected from a digital government perspective.

The importance of parental engagement for pupils' school success is well established and is used as a motive in most of the studies included in this mapping review, for example in [2, 18, 19]. Even though the pupil is the primary stakeholder when it comes to school services and education, this study focuses on communication between school and the legal guardians, since the focus is on how the relationship between governmental institutions (school in this case) and the citizen (legal guardians in this case) is affected by digitalization. Parents and legal guardians have different qualifications and possibilities to use digital tools for communication offered by the school. Iacobaeus et al. [12] have shown that digital exclusion often is related to socio-economic divides, which in turn affects the possibilities for the children to succeed in school. Insufficient knowledge about digital public services has been found to be more common in socio-economic disadvantaged areas, where additionally, trust in these services is low [29].

Surveys [13] show that one in five reports a need for help to be able to use digital services (like installing apps, managing bank-ID and keywords, and understanding technical language). There are differences between groups and if living in a socio-economic disadvantaged area it is more likely that problems using the digital tools for communication might occur [16]. Together this points to the need for more research on digital communication between home and school.

Thus, digital and offline inclusion/exclusion seem to be interconnected in the home-school communication, as a form of digital government. To grasp this complexity, Helsper's [8] theory of corresponding fields in communication, where digital impact mediators can support, or hamper communication is helpful. This model builds on the classical digital divide literature [27] and argues that communication is mediated by access, skills, and attitudinal or motivational aspects, as developed in the extensively used step model [28]. However, the general model of inclusion is framing communication and shows how specific areas of digital and offline exclusion influence each other. In Helsper's [8] corresponding fields model it is argued that the fields are mediated differently through mediators, depending on the context of communication and framing of what and who is communicating. The digital impact mediators are factors that connect the digital field to the offline field and affect the possibility for the users of digital services to be included also offline. The digital mediators are the users' experiences of using digital services. The use of digital tools should be relevant, have quality, be owned by them, and be sustainable [8]. It is relevant to study digital communication between home and school in terms of mediators since functioning communication is a prerequisite for mediating education, which is the core public service in this case. The digital impact mediators will be further elaborated on in Sect. 2.2 in this paper.

1.1 Aim and Purpose of This Mapping Study

The aim of this mapping study is twofold. Firstly, it aims to investigate the research field considering digital communication between home and school, with focus on describing used methods, theoretical frameworks and to identify knowledge gaps within the area and thereby lay the foundations for further studies. Secondly, the aim is to analyse how digital communication between school (governmental service) and home (citizen) is described in previous research and discuss contributions and potential research openings for the digital government field by using Helsper's [8] digital impact mediators as a frame. The following research questions will guide this study:

RQ1: Which methods, theoretical frameworks, and knowledge gaps are described in previous studies on digital communication between home and school?

RQ2: How can the findings of previous studies on digital communication between home and school be discussed and further developed in terms of digital impact mediators [8] within the digital government field?

This study is limited to a mapping literature review. Saturation concerning research within the field has not been achieved since back- and forward citation searches have been conducted only to follow our research focus. Despite the limited number of identified studies, the usefulness of the results lies in the prospect of finding interesting ideas, findings, and aspects which can help to develop research questions and plan for further studies in the area. By relating to Helsper's [8] digital mediators, the analysis is structured to open for a discussion specifically on how home-school communication as a digital government service can be enhanced by elaborating on the digital impact mediators and to identity further research needs.

This paper proceeds by presenting our method for conducting and structuring literature searches, followed by presentation of the findings, with focus on used theoretical frameworks, methods and knowledge gaps. Finally, we discuss the findings and how they can be seen through the lens of digital government using Helsper's [8] digital impact mediators as a framing.

2 Method

In this section, we describe how the literature review was performed, followed by a paragraph where the analytical approach of this study is discussed.

2.1 Literature Review

Literature searches for this mapping study were performed using Academic search complete, Scopus, Social Science Citation Index (Web of Science), the Digital Government Reference Library (DGRL), and SwePub. The searches included articles in Swedish, but none were found. These databases were selected due to being rather broad but also with content specifically interesting from a social science point of view. The search strings used were developed with the overall topic of this study in mind and the aim was to find a selection of interesting articles focusing on digital communication (keywords such as "digital", "online", "electronic", "interaction") between home (key words such as "parents" and "legal guardian") and school (keywords such as "teachers" and other "school staff") and how it affects the users and their trust (keywords such as "legitimate", "attitude", "professional") in school. The search strings were developed in collaboration with the librarians at the university library.

The initial searches (after applying the eligibility criteria (see Table 1) below, resulted in N = 138 articles. All these were screened by titles and abstracts, doublets, and those out of scope, were removed. N = 42 articles were deemed to be especially pertinent to include in this study. The 42 articles were first skimmed through, and the 19 most relevant ones, being specifically focused on the digital communicational aspect between home and school and not primarily on learning or teaching, were selected to be included in this mapping study.

Table 1. Characteristics and eligibility criteria

Characteristics	Eligibility criteria
Topic	Digital communication between school and home (preschool/kindergarten excluded)
Research field	Social sciences, education
Year of publication	2015–2025
Language	English
Publications	Peer-reviewed articles and conference proceedings

2.2 Analytical Approach

The analysis was made in several steps within the aim of this study. Firstly, the articles were read and content relevant for this study was extracted and summarized in search tables to get an overview. Besides method, theoretical framework, and suggested further research, also theoretical implications on digital impact mediators, inspiring further research were noted.

For the theoretical analysis, we used the model of digital impact mediators (relevance, quality, ownership, and sustainability) to further deepen and discuss the selected research. The discussion builds on Helsper's [8] model and adopts it to our context of research to further develop our understanding of digital communication between home and school. The four components of mediators connecting the on-line to the off-line fields in Helsper's [8] model are supporting and enhancing communicative practices. Firstly, Helsper identifies social impact mediators such as access, skills, competencies, and attitudes that influence the risk of digital exclusion, and these are corresponded to the digital impact mediators: relevance, quality, ownership, and sustainability, that can enhance off-line inclusion through digital inclusion [8, 9]. The relevance aspect is basic for communication and motivates the user to take part in communication and use digital platforms or tools. Relevance is closely connected to experienced quality but could also include aspects relating to the format, language use and accessibility to information used in communicative practice. The third component of the model is ownership, meaning that the user gains agency and empowerment, may be complicated in this context since the communication between home and school is both general for all pupils and individual and specific regarding single pupils or a group of pupils. There is also commonly sensitive personal information exchanged here, on health, family and related social issues. Thus, the ownership and sharing of information might appear as a critical component in the found studies. The fourth mediating component, sustainability, is especially interesting in this context since the child, and his/her parents are locked-in to the school for a longer time; there are in most countries constraints for changing school and not possible to drop-out. Thus, digital communication between school and home is supposed to be sustainable and trustworthy over time. The digital impact mediators and their features are summarized in Table 2 below.

Table 2. Overview of Helsper's [8] proposed digital impact mediators.

Digital impact mediator	Features and examples
Relevance	Usefulness in everyday life
Quality	Experienced quality, easy to use and positive
Ownership	Experience of agency and empowerment, the digital tool and its content is owned by them
Sustainability	Digital experience should be socially and financially sustainable use (over time)

Helsper [8] points out that there is little research done on how the use of digital tools and digital engagement affects people's offline life and their offline inclusion and the digital impact mediators are still to be tested, both theoretically and empirically. This study is an effort to apply and discuss these mediators in a school context and thereby contribute to theoretical development.

3 Findings

The findings are structured as follows: first there will be a summary of some general aspects interesting for this study, followed by a summary of methods, theoretical frameworks, and research gaps identified in the included studies. Finally, the findings will be framed, using Helsper's proposed digital impact mediators.

3.1 The Overall Picture of the Results

The main findings accounted for below are selected with digital communication in mind. The referred to studies might have had other objectives besides digital communication between school and home, but these are not considered in this overview. Nine of the studies are showing both the perspective of teachers and parents; four studies focus mainly on parents' perspective and five studies mainly on teachers' perspective. The aims of the studies are all about exploring how digital communication between home and school functions (since this was an inclusion criterion for this study), but some studies (three of them) also have the ambition to develop frameworks for studying digital communication and how it affects the relationship between teachers and parents.

The eligibility criteria for this study only included articles from 2015 and later but also within that span we can see that most of them, 14 out of 19 were published 2020 or after, are rather recent (see Fig. 1 below). Furthermore, the included studies are diverse when it comes to country of origin, where Finland has the highest representation with four studies (see Fig. 2 below).

The lion's share of the included articles in this mapping study was published in journals within the educational research field. Of the 19 included studies 13 were published in journals (or other outlets) with "education" or "teaching" in the titles and only one article was published within the field of digital government (see Fig. 3 below).

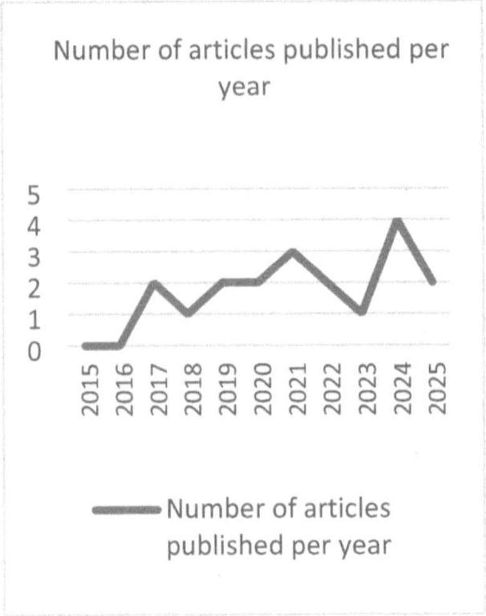

Fig. 1. Articles per year

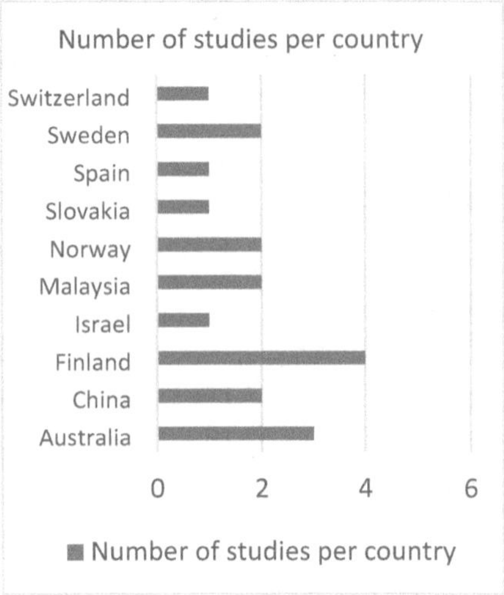

Fig. 2. Articles per country

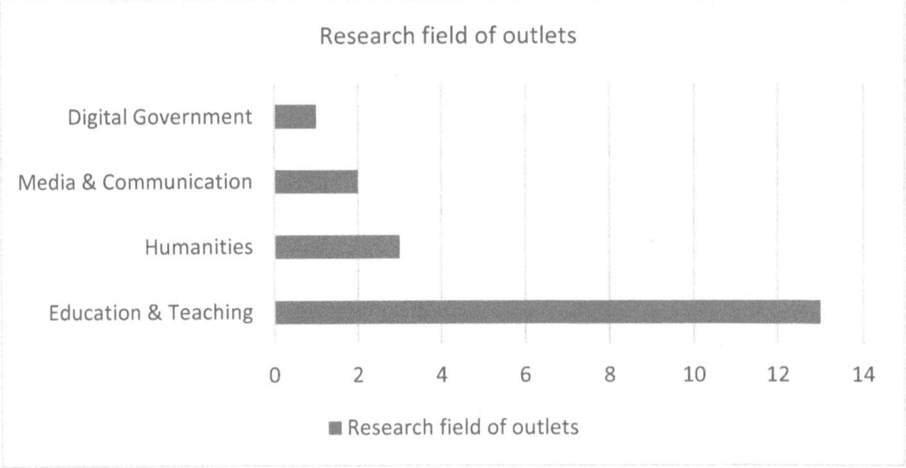

Fig. 3. Number of articles published in outlets with different focus and research fields.

Based on the 19 studies included here, some interesting aspects could be pointed to and discussed. Firstly, one aspect is that the character of the communication leans towards being one-way communication, information sent out from teachers, rather than two-way communication, interaction between two actors or parents and teachers. For example, it is concluded by Beeli-Zimmerman [1] that parents often take the role of receivers of information while schools are the distributors, sending our mass e-mails to reach as many parents as possible as effectively as possible. This is supported by Mendelova's [21] study where teachers are "the initiators of communication" which parents respond to. The challenges connected with establishing interaction (two-way communication) is also discussed by Gu [7] who found that parents not having full access to the digital communication systems, because of different barriers which will be discussed below, are hindered in their interaction with school. In line with this, Huang [10] found that the type of communication is changing over time and moves from being one-way to two-way communication. Furthermore, Gorfinkel et al. [6] discuss the differences between three types of communication: linear (one-way), interactional (two-way), and transactional (dynamic and contextual).

Another interesting finding that the included studies in several instances bring up, is the matter of barriers for communication where different types of barriers, hampering a functional digital communication between home and school are discussed. For example, Bordalaba et al. [2] discusses how "beliefs about the context", such as technical tools and technical competences, both the user's own competencies and the competence among others, and beliefs about the context, time and necessity, could constitute barriers for using digital communication tools, such as e-mails and school platforms. However, Bordalaba et al. [2] also found that the beliefs of users might strengthen the willingness to use digital communication, for example "beliefs about the medium" in terms of convenience and immediacy. Another perspective of digital communication between home and school in terms of barriers are given by Gu [7] who found that dissatisfaction with external factors, such as technical features, organisational factors, social circumstances, and negative attitudes towards the digital communication platform had negative impact on both teachers' and parents' usage. Furthermore, both Lutovac et al. [20] and Traeger-Soudry et al. [26] discuss how unclear boundaries and uncertainty considering relationship between parents and teachers could be an effect of the use of digital communication tools.

Problems with parents not using the digital tools for communication to an equal extent was also found in Huang's [10] study where only 30% of the parents use the social networking for communication with school, which Huang [10] discusses in terms of "digital reproduction of inequality" (p. 480). The differences when it comes to use of digital tools for communication affects the relationship between teachers and parents and Mendelova's [21] study shows how socioeconomic characteristics might determine the possibilities for teachers and parents to cooperate since parents with lower socioeconomic status are "difficult to reach" (p. 1405). In the Gorfinkel et al. [6] study the importance of developing a more "democratic use of digital and social media" (p. 97) is discussed. This idea, the terms used is "listening out", will be further discussed in relation to our theoretical framework in section four.

The findings of the included articles show that digital communication between home and school does not always work smoothly and without problems. One of the studies, Pangrazio et al. [23], started from another angle and investigated the use of digital platforms in schools and a mapping of these data infrastructures have been made, and it is shown that it is a "patchwork" of different platforms and that these "data infrastructures in schools [...] shape and establish social relationships, cultures, values, and beliefs" (p. 66). The notion of that there is a "patchwork" of platforms is also confirmed by this mapping study since the variety of digital platforms studied in the included articles is great and include WhatsApp [17], WeChat, Instagram, blogs, different educational ICT tools and platforms, etcetera. This is something we see a need to build further on and include in future studies on digital communication between home and school.

3.2 Methodological approaches in Previous Research

The methods used in the studies included in this literature overview are diverse, but with a clear leaning towards qualitative methodology. Eleven of the studies are based on interviews, either with teachers, parents, or both. The most common type of interview is semi-structured, with open-ended questions, and made with the ambition to get in-depth answers. The number of respondents in the studies range from 7–75, and are made with either teachers, parents, or a mix of both parents and school staff (teachers, principals for example). The interviews took place on site (face-to-face) or digitally (Zoom or Teams). In one of the studies the interviews were combined with observations [10]. The interview studies were often done within the format of case studies and in several of the included studies, methods were combined, the categories in Fig. 4 below are therefore overlapping.

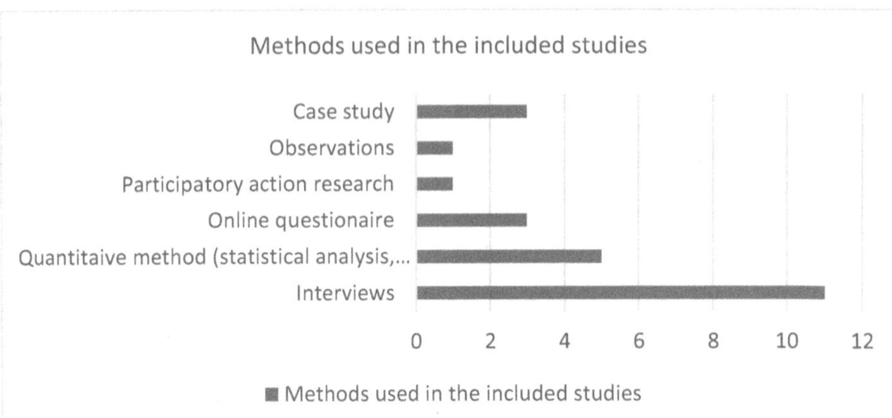

Fig. 4. Distribution of used methods in the included studies. In some studies, a combination is used and the categories are therefore overlapping.

Other methods used and discussed in the included studies were media mapping of digital platforms [6], three online questionnaires to teachers [7, 14, 21], a participatory action research/case study [11], and three statistical analyses of digital communication [18, 22, 24].

3.3 Theoretical Frameworks Used for analysis in Previous Research

The theoretical models used in the 19 included studies forming the basis of this mapping study are diverse, but some theories are reoccurring. In this section we will shortly comment on the theories referred to in the previous studies.

The most common theoretical model is the Technology Acceptance Model (TAM) which is specifically mentioned in three of the studies: Beeli-Zimmermann et al. [1], Bordalba [2], and Gu [7]. In the studies the TAM framework is adapted and/or used in combination with other theoretical models, for example "The Decomposed Theory of Planned Behaviour [2], the Media Richness Theory [1], and the Theory of Reasoned Action (TRA) [7]. The use of TAM-models in research on e-government and e-service diffusion is discussed by Sundberg [24] who points out that a TAM-model is limited to assessing the "willingness" of users and not their actual usage.

Besides the use of the TAM-model there seems to be two authors who are prevalent in the area, namely Goodall, referred to in five of the 19 articles, and Epstein, referred to in eight of the 19 articles. Both Epstein and Goodall have studied the relationship between home and school, Epstein [4] with focus on partnership and Goodall [5] with focus on parental engagement vs. parental involvement, and technology used for home-school communication. In Levinthal et al. [19] and in Traeger-Soudry et al. [26] Goodall's theoretical framework of parent-teacher partnership and parental engagement is used as a starting point for the analysis, while in other articles the theories are used only partly, or to add other dimensions to the discussion.

Other theoretical frameworks discussed and used in the included studies are, organisational communication theory in Gorfinkel et al. [6],"The politics of listening and voice" [6], "Concept of professional landscape" and "stories to live by: a narrative conception of teacher identity" in Lutovac et al. [20], and "third space theory" from post-colonial studies in Jia & Zhu [14].

3.4 Research Gaps Identified When Analysing the Included Studies

The findings of the studies included in this mapping study point to some research gaps within the area and further research is suggested, both specifically about digital communication between home and school and more generally, concerning the relationship between parents and teachers, and their roles, respectively. Primarily we see five different categories of suggested further development of studies within the field, pointed to in the included articles. Firstly, the matter of culturally and linguistically diverse (CALD) groups which are shown not to have the same access to digital school services and thereby are not fully included as in Gorfinkel et al. [6], Levinthal [19], and Gu [7] needs to be studied according to the included articles. Partly connected to this, are the findings about gendered communication, discussed both by Hutchinson [11] and Traeger-Soudry [26], who pointed out that the respondents, both teachers and parents were primarily women. Furthermore, Sundberg [24] found that predominantly women are the users of e-services connected to the educational sector. Secondly, an interesting suggestion for further studies is the notion of lacking content on digital communication within the teachers' training programs. Lutovac et al. [20] point out this gap, stating: "it appears that teacher education programs globally fail to prepare pre-service teachers for relationships with parents" (p. 6). Thirdly, the relationship and the power balance between parents and teachers in terms of for example professionalism and partnership is suggested as an area for further studies [10, 22, 25]. Fourthly, the included studies suggest that further studies aim to further develop theoretical frameworks [1, 2]. Finally, there are suggestions in the included studies to broaden further research geographically to be able to make international comparisons [18, 26]. Furthermore, several studies [3, 15] mention the importance of evaluating the effectiveness and effects of digitalisation of school related public services, even though it is not specifically expressed as a suggestion for further research.

3.5 Framing the Results into Digital Impact Mediators

When abductively analysing the included studies, we found patterns indicating that Helsper's corresponding fields model could be useful when discussing the findings presented in the previous section. The four digital impact mediators (relevance, experienced quality, ownership, and sustainability) [8] are used for analysis, and they can be seen as both barriers and enablers for communication and access to public services, in this case education. In this section, the results in the selected studies are structured and discussed in relation to the four digital impact mediators [8].

Relevance is described by Helsper [8] as that something needs to be considered useful in everyday life to be worth the effort to engage in the digital field. Basically, the content of the digital communication has to be seen as relevant (useful) by the parents and teachers. In the studies included in this mapping review we can see that parents find information from school relevant and in Levinthal [19] it is expressed by one of the interviewed parents that the teacher's communication is "the only way to get information" since her son doesn't "speak so much". Similar thoughts are expressed by parents in one study where the social media platform Instagram is seen as resource since she otherwise has to "pull all this information out" from her children and other parents expressed that they did not see the use of digital communication as relevant and expressed it like "just another thing I have to learn" or that teachers should focus on teaching and not on posting photos on social media [11]. Mendelova et al. [21] found that "interest" from both parents and teachers is crucial when it comes to establishing functioning cooperation between home and school, meaning that both parties need to see the digital communication as relevant. Taken together the studies indicate that relevance is a key for functioning digital communication between home and school. Both parties of the communication seem to actively choose relevant media for different types of communication.

Experienced Quality is, according to Helsper [8], related to user experience of a digital tool that is easy to access, use and provides the demanded information or service. Several of the included studies bring up benefits, such as speed, cost effective, and availability with digital communication between home and school. However, there are challenges and one aspect brought up by Beeli-Zimmermann et al. [1] is the diversity of offered digital communication channels which might make digital communication complicated to manage for both parents and teachers. Another aspect connected to quality, as defined by Helsper [8] are the linguistic barriers for some groups of users. As shown in the study by, Beeli-Zimmermann [1] who found that the studied schools' websites were almost exclusively monolingual, Gorfinkel et al. [6] suggest a more actively including "stakeholders across cultural and linguistic divides", and Gu [7] discusses how "disadvantaged" parents are excluded from the system due to "middle-class norms", and linguistic and technical barriers. Furthermore, Thyness [25] found that "monolingual ideologies" are more prevalent when the communication is digitally mediated. Quality here seems to mainly referee to language, and we could conclude that none of the included studies primarily focused on the quality of the digital platforms, nor on the design aspects.

Ownership is described by Helsper [8] as a factor empowering the user and thereby providing them with agency. The studies above indicate that information in general can be seen as empowering but to gain agency two-way communication (on equal premises) would be better than one-way communication. Beeli-Zimmermann [1] shows that digital communication among school and home is used for emitting information (mass e-mails for example) rather than inviting to dialogue. They conclude that parents take the role as receivers of information, rather than owners of the communication. This is supported by the findings of Mendelova et al. [21] who showed that teachers most often initiate communication and parents respond. This shows a lock of ownership in the communicative process that is not empowering the parents in this relation. This is interesting since the schools "owns" the educational platforms, but not other open communication platforms like Instagram and WhatsApp where parents could claim ownership of groups or pages in use. Thus, we can see that there is a scattered field of communication tools with diverse ownership.

Sustainability, in this context [8], is when digital communication is seen as something that will be worth the effort of learning how to use because it will be a long-term investment, both socially, cognitively, and economically. There are few findings in the included articles relating to sustainable communication over time and the reason for this might either be that the methodological approaches of the included studies are not longitudinal or that the concept of sustainability, in terms of being a digital impact mediator, is not clearly defined or adapted to an educational context. This indicates that there is a need for further studies within the area, which is further discussed below. Almost half of the articles included (n = 9) refer in these discussions to Epstein and her ideas of overlapping spheres and the importance of involvement and partnership. Epstein suggests a framework for how to "develop and sustain" [4, p. 9] these partnerships between home, school and community. This could work as a framework for further studies on digital communication in an educational context, connecting to how digital impact mediators frame and form digital public services and digital communication between home and school.

4 Conclusions and Openings for Further Research

This mapping of research on digital communication between home and school shows that there is a potential to further study this issue from a digital government perspective. Besides the overall conclusions of the study this section also discusses implications for theoretical approaches and further research.

4.1 Overall Conclusions from the Mapping Study

The main overall impact of the analysis of previous studies included in this mapping study reveals that the complexity of digital communication between home and school, in terms of both diversity of systems and platforms, and of users and their capabilities is extensive. The included studies show a large complexity in the digital public services for communication between home and school, and this points to a need for further analysis of both the short- and long-term effects for all involved actors.

The included studies mainly are found in journals and research fields within educational studies and pedagogical disciplines. Therefore, the digital component of communication was in the background in most studies. One of the included studies [24] pointed to a lack of research on actual usage of digital government platforms and that focus often is on experiences connected to usage. This is something that is confirmed by this mapping study and the conclusion is that there is a need for methodological development for studies within the field of digital government in a school context.

Instead, the educational and pedagogical context, professional roles of teachers and other professionals in schools were in focus. In this research field schools were not explicitly seen as a public service, treated as such considering funding and regulations. The discussion on trust in schools, and thereby in other public services was not articulated in the included studies. The diverse theoretical frameworks used in the included studies show the complexity of the issue and the need for further focus on schools as a public service, contributing literacy and inclusion in a digital society.

Furthermore, we can conclude that most of the included studies are based on interviews, small samples and qualitative analyses. This might explain the focus on single relations and few indications on more generalizable results. These methodological approaches also make the studies very contextual, but still they pay very limited attention to how and why the outcomes of the studies can be explained by contextual factors, such as country, legal frameworks, educational levels, etcetera. There is lack of studies considering the diversity of involved actors in these qualitative studies.

To conclude, even though, this mapping is based on a rather small ($n = 19$) selection of studies it shows some key characteristics, the educational focus and lack of problematisation of the public and contextual setting of communication between home and school, a qualitative research approach, and quite an unproblematized approach to the digital tools used in the included studies. We can also conclude that the main lesson learned is that there are diversity and a potential for development within the digital government field, which will be further discussed below.

4.2 Ideas for Further Research Based on the Mapping Study

The second aim of this study was to function as a foundation for further studies within the field of digital government in an educational context. Below we combine the theoretical frame of digital impact mediators and the concluded impact of the included studies with a focus on what we see as missing components from a digital government research perspective.

The main overall need for further research is to contextualize these topics in digital government, to highlight that the communication and public services are forming interactions among home and schools. There is a need to show that public digital services manifest in a school context, enhance rights and duties to build trust in education in particular and in public digital services in general.

There is also a need to make the digital technologies for communication explicit by opening for analysis of the design and usage of digital school platforms to identify challenges and benefits and thereby also be able to improve and adapt in relation to needs and analyse implications on digital communication in relations among users and service providers (school in this case). Also, how and why different groups face challenges and may become excluded is a critical area for further research since mandatory education is a unique compulsory public service. As it is relevant to study, through in-depth empirical case studies, critical groups of users, as for example students with special needs, parents with migrant background, and groups with less civic and language skills, highlight the effects and possibilities for functioning digital communication with school. In a longer perspective this could be developed to discuss and show implications of functioning digital communication between home and school and how it could have practical and economic effects for groups and society.

In relation to this, studies focusing on complementary digital communication and channels affecting offline possibilities in the communication between home and school and framing the choices of channels into Helsper's [8] corresponding fields model and digital impact mediators. The digital impact mediators have a potential to guide further research. The digital systems used for communication are in this perspective seen as mediators of digital public services and are thereby essential for a functioning and inclusive digital government. Thus, in the following we would suggest further deductively generated research questions based on the four digital impact mediators:

- Is digital communication between home and school perceived as useful for everyday life by the users and is it thereby **relevant**?
- How are digital tools for communication between parents and teachers designed from an inclusion perspective (languages, access, etc.) and how does it affect the **experienced quality** among users?
- Who initiates digital communication (parents (which parents?) or teachers)? And is the communication one-way (information) or two-way (dialogue)? How does this affect the feeling of **ownership**?
- Is the use of digital tools for communication between home and school seen as something worth an investment from a long-term perspective and does it promote trust in the institution, in other words, is it **sustainable**?

Finally, this study focused on digital communication between home and school with legal guardians in focus which means that we cannot say anything about how digitalization of schools affects the pedagogical aspects and pupils more than indirectly. This is something that could be further elaborated on and studies about how teaching about digital government and the use of digital public services is realized in the classroom would be interesting as a research project.

Acknowledgements. This study is part of Hamberg's PhD-project, mainly funded by the Swedish Research Council (grant number 2023-00793).

Disclosure of Interests. There are no competing interests regarding the presented study.

References

1. Beeli-Zimmermann, S., Burri, M., Ewald, A.S., Wannack, E.: The complexity of managing diverse communication channels in family-school relations in Switzerland as seen by parents and schools. Aust. Int. J. Rural Educ. **34**(2), 1–2 (2024). https://doi.org/10.47381/aijre.v34i2.714
2. Bordalba, M., Bochaca, J.: Digital media for family-school communication? Parents' and teachers' beliefs. Comput. Educ. **132**, 44–62 (2019). https://doi.org/10.1016/j.compedu.2019.01.006
3. Bönnhoff, H.: Fostering digital citizens in Norway: experiences of migrant mothers. Families, Relat. Soc. (2020)
4. Epstein, J.L., Sanders, M.G., Sheldon, S.B., et al.: School, Family, and Community Partnerships: Your Handbook for Action, 4th edn. Corwin Press (2018)
5. Goodall, J.: Parental engagement to support children's learning: a six point model. School Leadersh. Manag. **33**(2), 133–150 (2013). https://doi.org/10.1080/13632434.2012.724668
6. Gorfinkel, L., Muscat, T., Ollerhead, S., Chik, A.: The role of government's 'Owned Media' in fostering cultural inclusion: a case study of the NSW department of education's online and social media during COVID-19. Media Int. Aust. **178**(1), 87–100 (2021). Article 1329878x20968291. https://doi.org/10.1177/1329878x20968291
7. Gu, L.: Home-school communication through a web-based learning management system: experiences and lessons. In: The International Conference on Information Communication Technologies in Education (ICICTE 2018) Proceedings Chania, Greece (2018)
8. Helsper, E.J.: A corresponding fields model for the links between social and digital exclusion. Commun. Theory **22**(4), 403–426 (2012)
9. Helsper, E.J.: The Digital Disconnect. The Social Causes and Consequences of Digital Inequalities. SAGE Publications, London (2021)
10. Huang, H.: The promise and dilemma of the mediated parent-teacher relationship in the age of social networking sites. Front. Educ. China **12**(4), 468–489 (2017). https://doi.org/10.1007/s11516-017-0034-1
11. Hutchison, K., Paatsch, L., Cloonan, A.: Reshaping home–school connections in the digital age: challenges for teachers and parents. E-Learning Digit. Media **17**(2), 167–182 (2020). https://doi.org/10.1177/2042753019899527
12. Iacobaeus, H., Francisco, M., Nordqvist, C., Sefyrin, J., Skill, K., Wihlborg, E.: Digitalt utanförskap: en forskningsöversikt. DINO-rapport 2019:3. Linköpings universitet, Linköping (2019). https://www.diva-portal.org/smash/get/diva2:1374480/FULLTEXT01.pdf
13. Internetstiftelsen. Svenskarna och internet 2024. En årlig studie av svenska folkets internetvanor (2024). https://svenskarnaochinternet.se/rapporter/svenskarna-och-internet-2024/. Accessed 30 Sep 2024
14. Jia, J., Zhu, L.: Transition in teacher and parent roles, responsibilities and relationships within the "Third Space" of online Chinese EFL learning. Sage Open **14**(2) (2024). https://doi.org/10.1177/21582440241251652
15. Johari, S., Noordin, W.N.W., Mahamad, T.E.T.: WhatsApp conversations and relationships: a focus on digital communication between parent-teacher engagement in a secondary school in Putrajaya. Jurnal Komunikasi: Malays. J. Commun. **38**(2), 280–296 (2022). https://doi.org/10.17576/JKMJC-2022-3802-17
16. Järvaveckan Research & Internetstiftelsen. Digital inkludering och utanförskap i Sverige. Rapport som bygger på opinionsundersökning genomförd av Indikator opinion (feb-mars 2024) (2024)
17. Khoo, N.A.K.B., Osman, S., Daud, M.F., Kumar, J.A., Kurniati, D.: The implications of digital communication between parents and teachers from a teachers' point of view. Humanit., Arts Soc. Sci. Stud. **24**(1), 184–195 (2024)

18. Kuusimäki, A.M., Uusitalo-Malmivaara, L., Tirri, K.: Parents' and teachers' views on digital communication in Finland. Educ. Res. Int. **2019**, 8236786 (2019). https://doi.org/10.1155/2019/8236786
19. Levinthal, C., Kuusisto, E., Tirri, K.: Finnish and Portuguese parents' perspectives on the role of teachers in parent-teacher partnerships and parental engagement. Educ. Sci. **11**(6), 306 (2021). https://doi.org/10.3390/educsci11060306
20. Lutovac, S., Keränen, V., Körkkö, M., Uitto, M., Clandinin, D.J.: Teachers' experiences of being unsettled within their relationships with parents: Shifting professional knowledge landscapes. Teach. Teach. Educ. **152**, N.PAG–N.PAG (2024). https://doi.org/10.1016/j.tate.2024.104790
21. Mendelová, E., Guzikova, L.: Parents as school partners - socioeconomic characteristics of the family as a determinant of family-school relations. TEM J. **14**(2), 1398–1406 (2025). https://doi.org/10.18421/TEM142-40
22. Oinas, S., Vainikainen, M., Hotulainen, R.: Technology-enhanced feedback for pupils and parents in Finnish basic education. Comput. Educ. **108**, 59–70 (2017). https://doi.org/10.1016/j.compedu.2017.01.012
23. Pangraziano, L., Selwyn, N., Cumbo, B.: A patchwork of platforms: mapping data infrastructures in schools. Learn., Media Technol. **48**(1), 65–80 (2022)
24. Sundberg, L.: Diffusion of e-services: data from seven Swedish municipalities. In: Shefali Virkar, Marijn Janssen, Ida Lindgren, Ulf Melin, Francesco Mureddu, Peter Parycek, Efthimios Tambouris, Gerhard Schwabe, Hans Jochen Scholl (ed.), EGOV-CeDEM-ePart Conference, Proceedings of Ongoing Research, Practitioners, Workshops, Posters, and Projects of the International Conference EGOV-CeDEM-ePart 2020, pp. 185–193 (2020)
25. Thyness, H.: Teachers' construction of their roles and responsibilities for inclusive language practices with multilingual parents-a case from Norway. Lang. Educ. **38**(5), 800–817 (2024). https://doi.org/10.1080/09500782.2023.2221251
26. Traeger-Soudry, S., Rosenberg, H., Porat, C., Lowenstein-Barkai, H.: Navigating parental engagement in WhatsApp groups: a study of parent-teacher experiences. Teach. Teach. Educ. **162**, 9 (2025)
27. van Dijk, J.: The Deepening Divide: Inequality in the Information Society. Sage, Thousand Oaks (2005)
28. van Dijk, J.: The Digital Divide. John Wiley & Sons (2020)
29. Wihlborg, E., Kulanovic, A., Kaharevic, A., Elg, M.: Digital inclusion in a disadvantaged Swedish suburb–trust and participation to form quality of government. In: HICSS Conference Proceedings (2021)

Identifying Administrative Burdens in Digital Government: A Practice-Based Conceptual Framework

Ida Heggertveit(✉)

University of Agder, Kristiansand, Norway
ida.heggertveit@uia.no

Abstract. Digital self-services have become a central part of digital government, enabling citizens to apply for benefits, submit documentation, and interact with public authorities online. While these systems aim to improve efficiency and accessibility, they also shift administrative work, such as interpreting rules, navigating procedures, and demonstrating eligibility, from public officials to citizens. As a result, interacting with digital public services can generate administrative burdens for citizens in the form of learning, compliance, and psychological costs.

Although administrative burden theory offers a useful lens for understanding such experiences, research and public organizations still lack practical approaches for identifying where these burdens arise within digital service delivery. Existing research often focuses on policy design, system usability, or citizen outcomes, providing limited guidance on how burdens become visible in concrete digital service encounters and how they can be traced to the socio-technical arrangements that shape them.

This paper develops an empirically grounded, practice-based conceptual framework for identifying administrative burdens in digital self-services. The framework links citizens' experienced burdens to policy design, digital infrastructures, and organizational practices and operationalizes this perspective through a diagnostic tool for analyzing digital services. Drawing on iterative engagement with qualitative material from research of digital welfare services in Norway, the paper further demonstrates the analytical application of the framework through examples from the Norwegian Labour and Welfare Administration (NAV). The framework contributes to digital government research by providing a structured approach for diagnosing administrative burdens and identifying potential intervention points in digital public services.

Keywords: Digital self-service · Administrative Burden · Conceptual Framework

1 Introduction

Advances in information and communication technologies (ICTs) have led governments to prioritize digital service delivery to improve efficiency, accessibility, and policy implementation (Lindgren & van Veenstra, 2018), thereby reshaping citizen–state interactions (Halling & Baekgaard, 2024). In welfare administration, public organizations face

Published by Springer Nature Switzerland AG 2027
D. Duenas-Cid et al. (Eds.): ePart 2026, LNCS 16923, pp. 50–65, 2027.
https://doi.org/10.1007/978-3-032-36676-4_4

sustained pressure to deliver cost-effective services and scale service delivery through digital-first approaches, where digital channels are prioritized as the primary mode of interaction between citizens and public authorities (Lindgren & Madsen, 2025). Digital self-services are promoted as a way to provide citizens with easier and more flexible access to public benefits through online portals and digital interactions (Lindgren et al., 2022). As a result, citizens are increasingly expected to access information, complete applications, upload documentation, and communicate with authorities through digital portals (Hofmann et al., 2020).

While many citizens benefit from these developments, others struggle to self-serve (Hofmann et al., 2024). Citizens with limited digital skills, language proficiency, bureaucratic understanding, or access to social support may face difficulties completing applications, interpreting requirements, or submitting documentation (Giest & Samuels, 2023; Heggertveit & Rydén, 2024; Masood & Azfar Nisar, 2021). In welfare contexts, such challenges may be further intensified by financial stress, uncertainty, or difficult life situations that affect citizens' cognitive capacity and ability to navigate complex administrative processes (Christensen et al., 2020). As governments are obligated to provide services to *all* eligible citizens (Chan et al., 2021), these challenges raise concerns about accessibility and social inclusion.

Administrative burden theory conceptualizes these challenges as learning, compliance, and psychological costs experienced in citizen–state interactions (Burden et al., 2012; Halling & Baekgaard, 2024; Herd & Moynihan, 2019). In digital government, these costs emerge not only from policy but also from digital infrastructures (Lindgren et al., 2021) and organizational practices (Peeters, 2020) that structure how services are accessed and delivered.

Despite growing awareness of these challenges, research and public organizations still lack practical tools for identifying where administrative burdens arise in digital public services and how they might be reduced.

When digital public services impose substantial learning, compliance, or psychological costs (Heggertveit et al., 2022; Linos et al., 2022), citizens may instead seek assistance, switch channels, abandon applications, or risk exclusion (Giest & Samuels, 2023; Heggertveit & Rydén, 2024; Madsen et al., 2022; Rydén et al., 2024). This highlights the need for a practical framework for identifying administrative burdens and their sources.

The need is also academic. While administrative burden research has provided a powerful lens for understanding onerous citizen–state interactions (Burden et al., 2012; Herd & Moynihan, 2019), scholars have highlighted challenges in operationalizing the concept empirically and translating it into actionable strategies for reducing burdens (Baekgaard & Tankink, 2022; Daigneault, 2024; Daigneault, 2025; Halling & Baekgaard, 2024). Existing research often focuses on policy design or citizen outcomes but offers limited guidance on how burdens can be empirically identified in digital service encounters and traced to the socio-technical arrangements that shape them.

Rather than presenting a standalone empirical case study, this paper develops an empirically grounded conceptual framework through iterative engagement with qualitative material from research on digital welfare services in Norway. The framework emerged through an abductive and practice-oriented analytical process, where recurring citizen difficulties in digital service encounters were iteratively connected to learning, compliance, and psychological costs, as well as policy design, digital infrastructures, and organizational practices. The paper further demonstrates how the framework can be applied analytically through examples from NAV.

The rest of the paper is structured as follows. Section 2 presents the conceptual background on digital government and administrative burden. Section 3 describes the development of the practice-based conceptual framework. Section 4 presents an illustrative example of the framework's application. Section 5 discusses implications for research and practice. Section 6 concludes, and Sect. 7 outlines limitations.

2 Conceptual Background

2.1 Digital Government and Digital Self-service

Digital government refers to the use of digital technologies to transform how public administrations design, coordinate, and deliver public services (Mergel et al., 2019). It includes both digital infrastructures, such as platforms, data infrastructures, and design standards, and the organizational and policy design that structure service delivery (Lindgren et al., 2021). As citizen–state interactions increasingly move online, digital government redistributes tasks and responsibilities across citizens, technologies, and public officials (Madsen et al., 2022).

A central tool in digital government is digital self-services, which function as citizen-facing entry points through which individuals perform administrative tasks such as applying for benefits, uploading documentation, and monitoring case progress (Lindgren et al., 2019). Governments increasingly promote such services to streamline administrative processes, improve accessibility, and reduce costs (Lindgren & Madsen, 2025).

However, these services assume that citizens possess certain resources, including digital devices, language proficiency, and the ability to interpret complex information and administrative requirements. These assumptions often do not reflect citizens' life circumstances (Christensen et al., 2020; Hofmann et al., 2024), particularly for those facing unemployment, illness, migration-related barriers, or limited digital skills (Johnson & Kroll, 2023; Linos et al., 2022). As a result, digital self-services may widen inequalities when designed around assumptions of digital independence that do not align with citizens' lived realities (Alshallaqi & Al-Mamary, 2024).

2.2 Administrative Burden and Citizens' Experiences

Administrative burden refers to an individual's experience of policy design as onerous (Burden et al., 2012). This theory identifies three types of costs that citizens may encounter when interacting with public administration: learning, compliance, and psychological costs (Herd & Moynihan, 2019). Learning costs arise from the effort required to understand eligibility criteria and administrative procedures. Compliance costs refer to the work involved in demonstrating eligibility, such as providing documentation or completing applications. Psychological costs stem from experiences such as stress, stigma, uncertainty, or frustration when interacting with administrative systems.

Digitalization changes how these costs emerge. While digital self-services may reduce certain burdens, such as travel or waiting times, they may simultaneously increase learning and compliance costs as citizens must independently interpret rules, navigate complex digital environments, and translate their circumstances into standardized administrative categories (Madsen et al., 2022). Psychological costs may also arise from uncertainty, repeated application failures, or difficulties accessing assistance (Heggertveit & Rydén, 2024).

While administrative burden theory provides a useful lens for understanding citizens' experiences, focusing solely on these experiences offers limited insight into where burdens originate. Understanding their sources requires examining the socio-technical arrangements through which digital self-services are designed and delivered.

2.3 Sources of Administrative Burden in Digital Government

Administrative burdens in digital self-services emerge through the interaction of policy design, digital infrastructures, and organizational practices:

Policy design defines the rules governing access to public services, including eligibility criteria, documentation requirements, verification procedures, and reporting obligations (Moynihan et al., 2016). Digital-by-default and digital-first strategies increasingly position online channels as the primary mode of interaction between citizens and public authorities (Adelmant & Raso, 2025; Alshallaqi & Al-Mamary, 2024; Giest & Samuels, 2023). While these strategies aim to increase efficiency, they may also increase citizens responsibility for administrative task by shift responsibility for interpreting rules and demonstrating eligibility from public officials to citizens (Madsen et al., 2022).

Digital infrastructures operationalize these policy requirements through service portals, data infrastructures, and interconnected information systems (Lindgren et al., 2021). These infrastructures structure how citizens access information, authenticate their identity, submit documentation, and interact with public authorities. Design decisions embedded in digital infrastructures, such as standardized digital forms, authentication procedures, automation logics, and data architectures, translate policy requirements into concrete administrative tasks for citizens (Brown et al., 2021; Peeters & Widlak, 2018). At the same time, automation and interconnected systems may obscure decision processes and make it difficult for both citizens and public officials to understand how outcomes are produced (Peeters & Widlak, 2023; Young et al., 2024).

Administrative burdens are also shaped through organizational practices within public administration. Digitalization transforms frontline work and introduces new forms of coordination, discretion, and coping strategies (Breit et al., 2019; Breit et al., 2021). Frontline workers frequently assist citizens in navigating digital systems, interpreting administrative requirements, and resolving problems when systems fail to accommodate diverse user situations (Rydén & Hofmann, 2024). Research also suggests that administrative burdens are shaped by the administrative capacities of public organizations, including their ability to communicate requirements clearly, provide guidance, and offer personalized support to citizens (Perales-Fernandez, 2025). Variations in organizational resources and support capacities may therefore influence how burdens are experienced and managed in practice. Caseworkers must also coordinate across multiple interconnected digital systems, producing additional administrative work that often remains invisible within formal procedures (Crivellari et al., 2024).

Administrative burdens in digital government can thus be understood as socio-technical outcomes emerging from the interaction of policy design, digital infrastructures, and organizational practices.

3 Practice-Based Conceptual Framework Development

The framework was developed through an abductive and iterative research process (Andrade & Davison, 2023), drawing on qualitative material from research on digital self-services for financial assistance in the Norwegian welfare context. The empirical material included field observations of citizens using digital self-services, interviews with citizens and frontline workers, focus groups and discussions with public officials such as caseworkers, managers, designers, and legal experts.

Administrative burden theory served as an initial informing lens rather than as a fixed coding scheme. The analysis first focused on moments where citizens experienced friction, uncertainty, or breakdowns when interacting with digital services. These empirical observations were then interpreted through the concepts of learning, compliance, and psychological costs (Herd & Moynihan, 2019). Through repeated engagement with the material, it became clear that citizens' experienced burdens could not be explained as merely user-experience failures or policy rules alone but also had to be traced to digital infrastructures and organizational practices.

The framework therefore emerged through an iterative movement between empirical observations and theoretical concepts. The analytical steps were refined by repeatedly applying them to observed service encounters and discussing whether the identified burdens could be traced to policy design, digital infrastructure, organizational practice, or combinations of these. The purpose of this process is not to produce a full empirical case analysis, but to develop a conceptual framework grounded in practice and useful for identifying burdens in digital government.

3.1 Analytical Logic of the Framework

Building on this perspective, this study develops a practice-based conceptual framework for identifying administrative burdens in digital government. The framework connects four analytical dimensions that are often examined separately in existing research: citizens' experienced burdens, policy design, digital infrastructures, and organizational practices.

Administrative burden research has primarily focused on identifying the types of costs citizens encounter (Herd & Moynihan, 2019), while digital government research has often emphasized digital infrastructure and citizen-centric service design (Lindgren et al., 2019; Lindgren et al., 2021). However, these perspectives rarely provide a systematic way to connect citizens lived experiences with the socio-technical arrangements that shape them.

The framework addresses this gap by tracing citizens' experienced burdens to the socio-technical conditions that shape digital service delivery. It adopts a bottom-up diagnostic logic that starts from citizens' interactions with digital self-services.

The framework consists of four analytical steps. Step 1 identifies experienced burdens during interaction with digital services. Step 2 examines how policy design structures administrative demands. Step 3 analyzes how these demands are operationalized through digital infrastructures. Step 4 investigates how organizational practices manage, mitigate, or redistribute burdens in everyday service delivery. Figure 1 illustrates the framework and its analytical logic.

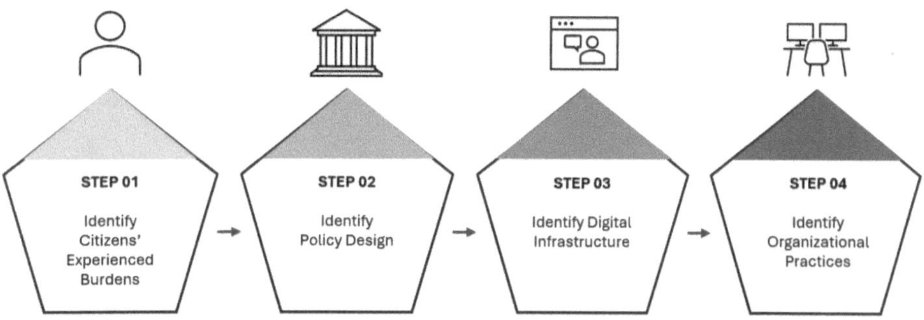

Fig. 1. Practice-Based Conceptual Framework for Identifying Administrative Burdens in Digital Government

By linking citizens' experiences to policy design, digital infrastructures, and organizational practices, the framework provides a systematic approach for identifying where administrative burdens originate in digital government systems.

3.2 Step 1: Identify Citizens' Experienced Burdens

The framework begins with citizens' experiences because administrative burdens are understood as the subjectively experienced costs individuals experience when interacting with public administration (Burden et al., 2012; Halling & Baekgaard, 2024; Herd & Moynihan, 2019). What constitutes a burden varies across individuals; the same administrative requirement may therefore be experienced differently depending on citizens' capabilities and circumstances. A digital form that appears straightforward to one citizen may therefore become confusing, stressful, or difficult to complete for another.

Not all moments of friction constitute administrative burdens. Frictions are interpreted as burdens when they are linked to interactions with government and generate costs that affect citizens' ability to access services (Christensen et al., 2020; Herd & Moynihan, 2019).

The first step therefore asks: What burdens are experienced, and where do they become visible during interaction with digital self-services?

To identify these experiences, the framework draws on the established typology of learning, compliance, and psychological costs (Herd et al., 2023; Herd & Moynihan, 2019, p. 23).

Empirically, attention is directed toward moments of friction or breakdown in citizens' interactions with digital systems. Breakdowns occur when the flow of action is interrupted and signal misalignments between digital system expectations and citizens' lived realities (Suchman, 2007). Examples include hesitation, repeated submission attempts, incomplete applications, or requests for help.

Identifying these experiences typically relies on qualitative approaches such as interviews, observations and insights from frontline workers (Alshallaqi & Al-Mamary, 2024; Daigneault, 2025; Giest & Samuels, 2023; Harvey & Myers, 1995; Heggertveit et al., 2022). Because administrative burdens are experienced subjectively and often emerge through situated interactions with digital systems, qualitative methods are particularly useful for capturing moments of confusion, uncertainty, breakdown, help-seeking, and other forms of friction that may remain invisible in formal procedures or system data.

The outcome of this step is an empirically grounded overview of where and how administrative burdens become visible during digital self-service interactions.

3.3 Step 2: Identify Policy Design Conditions

Once experienced burdens have been identified, the second step examines how these burdens are shaped by policy design.

Policy design defines the formal rules governing access to public services (Herd & Moynihan, 2019), including eligibility criteria, documentation requirements, verification procedures, and reporting obligations. These rules determine what citizens must know, document, and demonstrate to access benefits. Policy choices may influence how complex, time-consuming, or difficult administrative processes become for citizens.

This step examines how specific policy requirements align or misalign with the friction points identified in Step 1. For example, if citizens struggle to understand documentation requirements, the analysis examines how these requirements are defined in legislation or administrative guidelines.

This step typically involves document analysis of legislation, administrative guidelines, and policy documents, complemented by interviews with public officials.

The output of Step 2 is a structured overview of the policy design conditions that shape citizens' administrative tasks and contribute to experienced burdens.

3.4 Step 3: Identify Digital Infrastructure Conditions

The third step examines how policy requirements are operationalized through digital infrastructures.

Digital infrastructures include service portals, data infrastructures, and integration mechanisms that enable digital public services (Lindgren et al., 2021). These infrastructures translate policy rules into digital processes through which citizens interact with public administration. Importantly, not all administrative burdens originate in policy requirements. Some emerge from the design and functioning of digital systems themselves, such as usability issues, system errors, or limited interoperability (Schwoerer, 2025; Widlak & Peeters, 2020). These burdens may arise independently of policy intent but nonetheless affect citizens' ability to complete administrative tasks.

Design decisions embedded in digital infrastructures, such as standardized forms, authentication procedures, automated workflows, and data-sharing architectures, structure the administrative tasks citizens must perform.

This step analyzes how digital infrastructures shape, amplify, or generate burdens in practice. For example, repeated submission attempts may indicate that form logic, upload requirements, authentication systems, or missing integration create barriers to completing an application.

This step typically involves analysis of digital forms, service portals, user journeys, and system architecture, complemented by interviews with system designers or administrators.

The output of this step is a structured understanding of how digital infrastructure contributes to experienced administrative burdens.

3.5 Step 4: Identify Organizational Practices

While Steps 2 and 3 map the policy and infrastructural conditions that shape administrative burdens, these structures alone do not explain how burdens are handled in everyday service delivery. The fourth step therefore examines the organizational practices through which burdens are managed, mitigated, or redistributed in practice.

Even when policy rules and digital infrastructures are formally defined, service delivery depends on how public officials interpret rules, manage digital systems, and assist citizens in navigating administrative processes. Frontline workers often bridge gaps between system requirements and citizens' lived realities (Rydén & Hofmann, 2024). The ability to provide such support is also shaped by organizational capacities (Perales-Fernandez, 2025), including available time, staffing, coordination mechanisms, communication resources, and access to support channels. Variations in organizational capacity may therefore influence whether burdens are actively mitigated, redistributed, or left unresolved in practice.

Typical practices include repair work (e.g., rephrasing system questions or guiding citizens step by step), workarounds (e.g., using templates or reinterpreting documentation requirements), relational buffering (e.g., providing reassurance and mitigating frustration), channel switching (e.g., moving from digital to in-person support), and discretionary adjustments to accommodate citizens' situations (Heggertveit & Rydén, 2024; Hoglund Ryden & De Andrade, 2023; Rydén & Hofmann, 2024).

Not all burdens are actively managed or mitigated by others. In some cases, citizens resolve issues on their own, abandon the process, or experience unresolved burdens. The absence of organizational intervention is therefore also an important empirical finding.

These practices are typically identified through qualitative methods such as observations of service encounters, interviews with frontline workers and managers, and analysis of support interactions. Particular attention should be given to how workers respond when citizens struggle, including the strategies they use to resolve breakdowns, reinterpret rules, or compensate for system limitations.

The output of this step is a systematic overview of the organizational practices through which administrative burdens are managed, mitigated, or redistributed.

3.6 Operationalizing the Framework

To support use in research and practice, the conceptual framework is translated into an operational diagnostic tool linking empirical observations of citizen interaction with their sources in policy design, digital infrastructures, and organizational practices. The tool enables researchers and practitioners to systematically trace administrative burdens from citizen experiences to the socio-technical arrangements that shape them (Table 1).

Table 1. Operational Tool for Identifying Administrative Burdens in Digital Self-Services

Empirical Examples (what citizens say/do)	Burden Type	Policy Design	Digital Infrastructure	Organizational Practice	Potential Burden Reducers
Hesitation, confusion, repeated attempts, incomplete submissions, help-seeking, abandonment, visible frustration	Learning/Compliance/Psychological	Rules, mandates, documentation requirements, verification procedures	Forms, upload functions, authentication, automation, data-sharing, integration logics	Repair work, workarounds, buffering, channel switching, discretion	Redesign, clarification, resourcing, policy revision, integration improvements

The first column captures observable citizen behavior, such as hesitation, repeated attempts, incomplete submissions, or requests for assistance. The second column classifies these observations using the typology of learning, compliance, and psychological costs (Herd & Moynihan, 2019). The third and fourth columns identify structural sources of burdens, distinguishing between policy design and digital infrastructure. The fifth column documents how burdens are managed through organizational practices, while the final column highlights potential intervention points, such as policy revision, digital redesign, or organizational resourcing.

Together, the tool makes the analytical movement explicit: from observed citizen friction to structural conditions and potential intervention. By linking citizen experiences to their sources, the framework provides a practical method for making administrative burdens visible and actionable in digital government systems.

4 Illustrative Application of the Framework

4.1 Empirical Context

Norway provides a relevant context for examining administrative burdens in digital government due to its advanced digital infrastructure and high levels of digital adoption among citizens (Lindgren & Madsen, 2025). NAV, responsible for employment services and welfare benefits, is a flagship example of large-scale public sector digitalization (Finne et al., 2023).

Digital government reforms are often justified by the expectation that digital systems simplify administrative processes and enable citizens to complete interactions independently (Lindgren & van Veenstra, 2018). If administrative burdens emerge even in such a context, similar challenges are likely to occur in less digitally mature settings.

Despite this high level of digital maturity, research shows that some citizens still struggle to navigate digital welfare services, particularly those with limited digital skills, language barriers, or complex life situations (Heggertveit et al., 2022; Hofmann et al., 2024; Rydén & Hofmann, 2024).

The following example illustrates how the framework can be used to trace experienced burdens in a digital application for financial assistance to their underlying conditions. The example demonstrate how burdens can be linked to policy design, digital infrastructures, and organizational practices.

4.2 Illustrative Example

The following example was observed during fieldwork at a NAV office, where a citizen was completing a digital application for financial assistance with support from a frontline worker. During the process, the citizen paused at a form field asking: *"Describe your current financial situation and explain why you are unable to cover necessary living expenses."* The citizen then asked: *"What should I write here?"*.

Step 1: Identifying the experienced burden. The citizen experienced confusion and uncertainty when attempting to translate their lived situation into the system's predefined categories. Such moments are not merely minor inconveniences but reflect barriers linked to administrative requirements that shape how individuals experience government and influence access to services. The hesitation therefore indicates a learning cost related to understanding what information is required and how it should be articulated.

Step 2: Identifying the policy design. Analysis shows that the question is formulated using standardized bureaucratic language derived from legal eligibility criteria. While administratively precise, the wording leaves considerable room for interpretation regarding relevance and level of detail.

Step 3: Identifying digital infrastructure conditions. The digital form assumes that citizens can independently interpret the requirement and determine the appropriate level of detail. The burden is thus linked both to how regulatory language is translated into digital form fields and to the absence of supporting features such as examples or contextual guidance.

Step 4: Identifying organizational practices. In response, a frontline worker rephrases the question in simpler language and suggests a formulation acceptable to caseworkers. Through this repair work, the learning burden is reduced while interpretive work shifts to the frontline worker.

The recurrence of such interactions suggests that the issue is not incidental. Potential interventions could include clarifying wording, adding examples, or improving contextual guidance within the form.

While this example illustrates a learning burden, similar patterns can be observed for compliance and psychological costs, as summarized in Table 2. These include situations where citizens question repeated documentation requirements and where uncertainty leads to visible distress.

Table 2. Illustrative Application of the Framework (NAV Context)

Observations	Burden	Policy Source	Digital Infrastructure	Organizational Practice	Potential interventions
"What should I write here?"	Learning cost	Eligibility rules require description of financial situation	Open-ended form without examples or guidance	Frontline worker rephrases and suggests wording	Clarify wording; add examples and guidance
"NAV already has this information. Why do I need to send it again?"	Compliance cost (+ frustration)	Verification rules require documentation	No data reuse or pre-filled fields	Resubmission or caseworker reconstruction	Enable data reuse; clarify documentation requirements
Citizen visibly distressed, afraid of "doing something wrong"	Psychological cost	Complex requirements; unclear consequences	Limited guidance on errors and outcomes	Frontline worker reassurance; later clarification by caseworker	Clarify consequences; add in-context guidance

5 Implications for Digital Government

This study contributes to digital government research by providing a practice-based framework for identifying administrative burdens in digital self-services. First, it operationalizes administrative burden as a diagnostic tool for analyzing digital service interactions. While prior research has identified learning, compliance, and psychological costs (Herd & Moynihan, 2019), it offers limited guidance on how these can be systematically identified and traced to the socio-technical conditions that shape them. The framework addresses this gap by linking observed citizen experiences, such as confusion, repeated submission attempts, help-seeking, or visible frustration, to policy design, digital infrastructures, and organizational practices. In doing so, it provides a structured approach for identifying burdens and potential intervention points in digital self-services.

Second, the framework highlights organizational practices as a critical yet often overlooked component of digital service delivery. Administrative burden research often focused on how policy design generates learning, compliance and psychological costs (Halling & Baekgaard, 2024), while digital government research has emphasize digital strategies, infrastructures, usability, efficiency, and service delivery (Lindgren & Madsen, 2025; Lindgren et al., 2019; Lindgren et al., 2021). The framework shows how digital self-services in practice often depend on frontline worker's support. These informal practices, such as repair work, workarounds, and relational buffering, constitute a form of hidden human infrastructure that helps sustain digital service delivery but often remains invisible in formal system design and digitalization strategies. By identifying where citizens experience burdens and examining how these are managed through organizational and support practices, the framework can also help trace how responsibilities for understanding rules, completing applications, providing guidance, and managing frustration are shifted from citizens to other actors (Alshallaqi & Al-Mamary, 2024; Giest & Samuels, 2023; Rydén & Hofmann, 2024).

Third, the framework enables organizations to distinguish between different sources of administrative burdens in digital service delivery. Not all citizen friction originates in the same place. Some burdens stem from policy design, such as documentation requirements or eligibility criteria. Others emerge from digital infrastructure design, including confusing forms, or lack of system integration. Additional burdens may also arise through organizational routines and practices, such as limited support availability, inconsistent guidance, or reliance on informal workarounds to manage citizen difficulties. By distinguishing between these different sources, the framework can provide more targeted interventions.

6 Conclusion

This study developed a practice-based conceptual framework for identifying administrative burdens in digital self-services. The framework demonstrates that administrative burdens cannot be understood solely as problems of policy design, system usability, organizational support, or individual citizen capabilities in isolation. Instead, burdens emerge through the interaction between them.

By starting from citizens' experienced burdens and tracing them to underlying structural conditions, the framework provides a systematic approach for analyzing how administrative burdens become embedded in digital public services. In doing so, the study extends administrative burden research beyond the identification of citizen costs alone and contributes a practical analytical approach for connecting citizens' lived experiences with the socio-technical arrangements through which digital services are designed, organized, and delivered.

The framework further highlights that digital self-services often depend on organizational and interpersonal support practices that remain largely invisible in formal system design.

Practically, the framework provides researchers and public organizations with a structured method for diagnosing administrative burdens and identifying potential intervention points in digital government systems. By distinguishing between burdens linked to policy design, digital infrastructures, and organizational practices, the framework supports more targeted interventions, including regulatory reform, digital redesign, improved guidance, and organizational support initiatives. In doing so, it may contribute to the development of digital public services that are more efficient, accessible, inclusive, and socially responsible.

7 Limitations

Several limitations should be acknowledged. Administrative burden research highlights challenges in empirically identifying certain dimensions of burden, particularly psychological costs, which may be difficult to observe directly (Baekgaard & Tankink, 2022). As Daigneault (Daigneault, 2024) argues, rigid categorization risks overlooking the lived experience of onerousness. What ultimately matters are not whether a cost neatly fits a predefined category but whether citizens experience state interaction as burdensome.

Administrative burden literature also distinguishes between intentional and unintentional sources of burden (Peeters, 2020). The framework presented here focuses on diagnosing experienced burdens and linking them to policy design, digital infrastructures, and organizational practices. Determining whether burdens are intentionally shaped requires additional analysis of political and administrative decision-making processes beyond the scope of this study.

A further limitation is that the framework primarily focuses on citizen interactions with digital public services. While the study acknowledges that administrative burdens may extend beyond the individual citizen and responsible agency, the role of other actors is not examined in depth. Further research should therefore explore how administrative burdens are shifted and managed across broader social and institutional support networks involved in digital government interactions.

Acknowledgement. This study was financed by the Norwegian Labour and Welfare Administration (NAV) and the Research Council of Norway (project number 316246).

Disclosure of Interests. The authors have no competing interests to declare that are relevant to the content of this article.

References

Adelmant, V., Raso, J.: Data entry and decision chains: distributed responsibility and bureaucratic disempowerment in the UK's universal credit programme. Oxf. J. Leg. Stud. **45**(2), 415–446 (2025). https://doi.org/10.1093/ojls/gqaf006

Alshallaqi, M., Al-Mamary, Y.H.: Paradoxical digital inclusion: the mixed blessing of street-level intermediaries in reducing administrative burden. Gov. Inf. Q. **41**(1), 101913 (2024). https://doi.org/10.1016/j.giq.2024.101913

Andrade, A.D., Davison, R.M.: Dancing between theory and data: abductive reasoning, pp. 274–287. Edward Elgar Publishing Limited, United Kingdom (2023). https://doi.org/10.4337/9781802205398.00023

Baekgaard, M., Tankink, T.: Administrative burden: untangling a bowl of conceptual spaghetti. Perspect. Publ. Manage. Gov. **5**(1), 16–21 (2022). https://doi.org/10.1093/ppmgov/gvab027

Breit, E., Egeland, C., Løberg, I.B.: Cyborg bureaucracy: frontline work in digitalized labor and welfare services. In: Pedersen, J.S., Wilkinson, A. (eds.) Big Data, pp. 149–169. Edward Elgar Publishing (2019). https://doi.org/10.4337/9781788112352.00012

Breit, E., Egeland, C., Løberg, I.B., Røhnebæk, M.T.: Digital coping: how frontline workers cope with digital service encounters. Soc. Policy Adm. **55**(5), 833–847 (2021). https://doi.org/10.1111/spol.12664

Brown, J.T., Carey, G., Malbon, E.: What is in a form? Examining the complexity of application forms and administrative burden. Aust. J. Public Adm. **80**(4), 933–964 (2021). https://doi.org/10.1111/1467-8500.12531

Burden, B.C., Canon, D.T., Mayer, K.R., Moynihan, D.P.: The effect of administrative burden on bureaucratic perception of policies: evidence from election administration. Public Adm. Rev. **72**(5), 741–751 (2012). https://doi.org/10.1111/j.1540-6210.2012.02600.x

Chan, F.K.Y., Thong, J.Y.L., Brown, S.A., Venkatesh, V.: Service design and citizen satisfaction with e-government services: a multidimensional perspective. Public Adm. Rev. **81**(5), 874–894 (2021). https://doi.org/10.1111/puar.13308

Christensen, J., Aarøe, L., Baekgaard, M., Herd, P., Moynihan, D.P.: Human capital and administrative burden: the role of cognitive resources in citizen-state interactions. Public Adm. Rev. **80**(1), 127–136 (2020). https://doi.org/10.1111/puar.13134

Crivellari, I., Oskarsen, J.S., Grøder, C.H.: Unveiling the invisible: Caseworkers' data work in a public welfare organization. In: Proceedings of the 23nd IFIP WG 8.5 International Conference on Electronic Governance (EGOV 2024) (2024). https://doi.org/10.1007/978-3-031-70274-7_11

Daigneault, P.-M.: Reconceptualizing administrative burden around onerous experiences. Perspect. Public Manag. Gov. **7**(4), 124–136 (2024). https://doi.org/10.1093/ppmgov/gvae010

Daigneault, P.M.: Administrative burden revisited: advancing research and practice. Can. Public Adm. **68**(1), 159–172 (2025). https://doi.org/10.1111/capa.70000

Finne, J., Sadeghi, T., Løberg, I. B., Bakkeli, V., Sehic, B., Thørrisen, M.M.: Predictors of satisfaction with digital follow-up in Norwegian Labor and welfare administration: a sequential mixed-methods study. Soc. Policy Adm. (2023). https://doi.org/10.1111/spol.12936

Giest, S., Samuels, A.: Administrative burden in digital public service delivery: the social infrastructure of library programs for e-inclusion. Rev. Policy Res. **40**(5), 626–645 (2023). https://doi.org/10.1111/ropr.12516

Halling, A., Baekgaard, M.: Administrative burden in citizen-state interactions: a systematic literature review. J. Publ. Adm. Res. Theory **34**(2), 180–195 (2024). https://doi.org/10.1093/jopart/muad023

Harvey, L.J., Myers, M.D.: Scholarship and practice: the contribution of ethnographic research methods to bridging the gap. Inf. Technol. People **8**(3), 13–27 (1995). https://doi.org/10.1108/09593849510098244

Heggertveit, I., Lindgren, I., Madsen, C.Ø., Hofmann, S.: Administrative burden in digital self-service: an empirical study about citizens in need of financial assistance. In: International Conference on Electronic Participation (2022). https://doi.org/10.1007/978-3-031-23213-8_11

Heggertveit, I., Rydén, H.H.: Narratives of reality: administrative burdens and workarounds in digital self-services. In: 30th Americas Conference on Information Systems (AMCIS 2024), Salt Lake City, Utah, USA (2024)

Herd, P., Moynihan, D., Widman, A.: Identifying and reducing burdens in administrative processes. In: Report for the Administrative Conference of the United States. Published December (2023)

Herd, P., Moynihan, D.P.: Administrative Burden: Policymaking By Other Means. Russell Sage Foundation (2019). https://doi.org/10.7758/9781610448789

Hofmann, S., Madsen, C.Ø., Distel, B.: Developing an analytical framework for analyzing and comparing national e-government strategies. In: Electronic Government: 19th IFIP WG 8.5 International Conference, EGOV 2020, Linköping, Sweden, August 31–September 2, 2020, Proceedings 19 (2020). https://doi.org/10.1007/978-3-030-57599-1

Hofmann, S., Sæbø, Ø., Rydén, H.H.: Implications of digitalised welfare services from a vulnerable Citizens' perspective. Nordisk Välfärdsforskning (2), 127–141 (2024). https://doi.org/10.18261/nwr.9.2.2

Hoglund Ryden, H., De Andrade, L.: The hidden costs of digital self-service: administrative burden, vulnerability and the role of interpersonal aid in Norwegian and Brazilian welfare services. In: Proceedings of the 16th International Conference on Theory and Practice of Electronic Governance (2023). https://doi.org/10.1145/3614321.3614403

Johnson, D., Kroll, A.: The effect of electronic program applications amidst the politics of administrative burden. Gov. Inf. Q. **40**(2), 101808 (2023). https://doi.org/10.1016/j.giq.2023.101808

Lindgren, I., Madsen, C.Ø.: Digital first? Understanding Citizens' communication needs in digital public encounters. Soc. Policy Adm. (2025). https://doi.org/10.1111/spol.70017

Lindgren, I., Madsen, C.Ø., Hofmann, S., Melin, U.: Close encounters of the digital kind: a research agenda for the digitalization of public services. Gov. Inf. Q. **36**(3), 427–436 (2019). https://doi.org/10.1016/j.giq.2019.03.002

Lindgren, I., Madsen, C.Ø., Rydén, H.H., Heggertveit, I.: Exploring Citizens' channel behavior in benefit application: empirical examples from Norwegian welfare services. In: Proceedings of the 15th International Conference on Theory and Practice of Electronic Governance (2022). https://doi.org/10.1145/3560107.3560312

Lindgren, I., Melin, U., Sæbø, Ø.: What is E-Government? Introducing a work system framework for understanding E-Government. Commun. Assoc. Inf. Syst. **48**(1), 43 (2021). https://doi.org/10.17705/1CAIS.04842

Lindgren, I., van Veenstra, A.F.: Digital government transformation: a case illustrating public e-service development as part of public sector transformation. In: 19th Annual International Conference on Digital Government Research: Governance in the Data Age, DG.O 2018, Delft, The Netherlands (2018). https://doi.org/10.1145/3209281.3209302

Linos, K., et al.: How do disadvantaged groups seek information about public services? A randomized controlled trial of communication technologies. Public Adm. Rev. **82**(4), 708–720 (2022). https://doi.org/10.1111/puar.13437

Madsen, C.Ø., Lindgren, I., Melin, U.: The accidental caseworker – how digital self-service influences citizens' administrative burden. Gov. Inf. Q. **39**(1), [1–11] 101653 (2022). https://doi.org/10.1016/j.giq.2021.101653

Masood, A., Azfar Nisar, M.: Administrative capital and citizens' responses to administrative burden. J. Publ. Adm. Res. Theory **31**(1), 56–72 (2021). https://doi.org/10.1093/jopart/muaa031

Mergel, I., Edelmann, N., Haug, N.: Defining digital transformation: results from expert interviews. Gov. Inf. Q. **36**(4), 101385 [101381–101316] (2019). https://doi.org/10.1016/j.giq.2019.06.002. Article 101385

Moynihan, D.P., Herd, P., Ribgy, E.: Policymaking by other means: do states use administrative barriers to limit access to Medicaid? Adm. Soc. **48**(4), 497–524 (2016). https://doi.org/10.1177/0095399713503540

Peeters, R.: The political economy of administrative burdens: a theoretical framework for analyzing the organizational origins of administrative burdens. Adm. Soc. **52**(4), 566–592 (2020). https://doi.org/10.1177/0095399719854367

Peeters, R., Widlak, A.: The digital cage: administrative exclusion through information architecture – the case of the Dutch civil registry's master data management system. Gov. Inf. Q. **35**(2), 175–183 (2018). https://doi.org/10.1016/j.giq.2018.02.003

Peeters, R., Widlak, A.C.: Administrative exclusion in the infrastructure-level bureaucracy: the case of the Dutch daycare benefit scandal. Public Adm. Rev. **83**(4), 863–877 (2023). https://doi.org/10.1111/puar.13615

Perales-Fernandez, F.: Exploring the influence of administrative capacities on administrative burdens. J. Publ. Adm. Res. Theory, muaf011 (2025). https://doi.org/10.1093/jopart/muaf011

Rydén, H.H., Hofmann, S.: Frontline workers' role in digital self-service co-production: Channel promoters, digital helpers, or intermediators. JeDEM - eJournal eDemocracy Open Gov. **16**(2), 1–27 (2024). https://doi.org/10.29379/jedem.v16i2.919

Rydén, H.H., Hofmann, S., de Andrade, L.H.A., Heggertveit, I.: Mitigating administrative burdens: Understanding the role of intermediaries in co-producing digital self-services. In: Proceedings of the 16th IFIP WG 8.5 International Conference on Electronic Participation (ePart 2024) (2024). https://doi.org/10.1007/978-3-031-70804-6_3

Schwoerer, K.: The UX of bureaucracy: mapping user engagement to citizen experiences of administrative burdens. Int. Rev. Publ. Adm. **30**(3), 272–295 (2025). https://doi.org/10.1080/12294659.2025.2535772

Suchman, L.A.: Human-Machine Reconfigurations: Plans and Situated Actions. Cambridge University Press (2007)

Widlak, A., Peeters, R.: Administrative errors and the burden of correction and consequence: how information technology exacerbates the consequences of bureaucratic mistakes for citizens. Int. J. Electron. Gov. **12**(1), 40–56 (2020). https://doi.org/10.1504/IJEG.2020.106998

Young, M.M., Compton, M., Bullock, J.B., Greer, R.: Complexity, errors, and administrative burdens. Public Manag. Rev. **26**(10), 2847–2867 (2024). https://doi.org/10.1080/14719037.2023.2288247

Who Participates in AI Programs at Public Libraries? Exploring Demographic and Attitudinal Factors

Zong-Xian Huang[1(✉)], J. Ramon Gil-Garcia[1,2], and Mila Gascó-Hernandez[1]

[1] University at Albany, State University of New York, Albany, NY, USA
{zhuang7,jgil-garcia,mgasco}@albany.edu
[2] Universidad de Las Americas Puebla, Cholula, Puebla, Mexico

Abstract. As artificial intelligence (AI) systems rapidly expand in the public sector, scholars and practitioners advocate for inclusive civic engagement as a way to align technological advancements with public values. In line with this trend, public libraries increasingly host AI-related programs to not only foster patrons' technological competencies but also raise their awareness regarding AI and its capabilities. However, empirical research on who actually participates in these programs remains scarce. To address this gap, this study examines the demographic and attitudinal factors influencing participation in AI-related programs at public libraries. Drawing on data from an online survey conducted in the United States, this research employs a quantitative approach to analyze how demographic attributes and distinct profiles of perceived benefits and risks of AI engagement affect patrons' likelihood of participation. The findings reveal that participation is stratified along traditional socioeconomic lines, with older adults, women, individuals with lower educational attainment, and those from lower-income households being significantly less likely to engage in AI programs. Crucially, the analysis uncovers a distinct attitude-behavior gap: patrons perceiving high macro-level community benefits as opposed to individual benefits are less likely to participate. Conversely, constructively skeptical citizens harboring concrete concerns about AI demonstrate a higher propensity for active engagement. Overall, this study documents a potential divide in participation in these AI-related activities and offers actionable insights for public managers and librarians to design inclusive outreach strategies and tailored communication.

Keywords: Artificial Intelligence · Civic Engagement · Community Involvement · Public Libraries · Attitudes towards Technology

1 Introduction

Systems and applications powered by artificial intelligence (AI) are rapidly expanding in the public sector [25]. Some of these applications may pose important risks to society and greatly affect vulnerable communities. So, scholars and practitioners are advocating for more inclusive approaches to engage diverse stakeholders, ensuring that these

Published by Springer Nature Switzerland AG 2027
D. Duenas-Cid et al. (Eds.): ePart 2026, LNCS 16923, pp. 66–81, 2027.
https://doi.org/10.1007/978-3-032-36676-4_5

technological advancements align with the values and demands of the general public [2, 18]. However, given the complexity of AI technologies, citizens are expected to possess fundamental knowledge, literacy, and competencies regarding AI in order to exert meaningful participation [7, 31]. One promising strategy to achieve this is to actively involve the public in AI-related activities, such as educational lectures or interactive workshops, thereby enhancing their critical understanding, technological capabilities, and essential awareness toward the implications of these AI-powered systems [10, 14].

To respond to this trend, public libraries are seen as key actors and are increasingly expected to raise awareness and build essential competencies regarding artificial intelligence through the organization of various AI-related activities. While existing literature has discussed the design and implementation of such programs, there remains a limited empirical understanding regarding the demographic attributes of participating patrons [10, 11, 19]. Given that participation in these programs might not be uniform, it is imperative to examine which specific population segments are currently being reached and which are potentially underserved. Furthermore, the underlying perceptions that drive this engagement, specifically the perceived benefits and risks of the public involvement in AI, may also be associated with different degrees of participation in these AI-related activities held in public libraries. Therefore, in order to design AI programs that resonate with diverse demands, understanding the participation patterns of individuals with different perceptions of AI benefits and risks is crucial.

Therefore, this study examines the relationship between patrons' demographic attributes—such as gender, age, educational attainment, race, and income—and their frequency of participation in AI-related library programs. Moreover, this research also explores how individuals evaluate the impacts of engaging the community in AI technologies and how distinct profiles of perceived benefits and risks are associated with participation in AI-related activities. Two research questions guide this inquiry: (1) How does participation in AI-related programs at public libraries vary across different demographic groups? (2) How are distinct profiles of perceived benefits and risks of AI engagement associated with citizens' likelihood of participating in these programs?

2 Literature Review

As artificial intelligence becomes increasingly prevalent in both public and private organizations, recent studies have emphasized the necessity of engaging diverse stakeholders in AI initiatives through inclusive approaches. By adopting participatory mechanisms, stakeholders holding distinct perspectives and values can co-create strategies to generate public value that meets the multifaceted needs of the broader community [20, 22]. For example, Matias and Price highlight how public involvement can improve AI initiatives in five key areas [18]: negotiating disagreements and achieving equipoise, improving the quality and precision of AI training data, providing vital intuitions and explanations regarding human-algorithm behavior, spotting potential dangers, and offering unique insights into preliminary findings. Moreover, public involvement in AI initiatives can break the power imbalance in traditional technology development, empowering historically marginalized communities and affected stakeholders to meaningfully participate in and shape the design of AI systems [2].

Taking into account the complex and opaque nature of AI technologies, individuals are expected to possess a fundamental understanding and foundational knowledge of AI to facilitate meaningful engagement in AI initiatives [1, 3, 5, 7]. Past studies have emphasized that people can gain deeper knowledge and develop critical digital competencies by participating in AI-related activities. For instance, Korte et al. found that university students can enhance their AI literacy and conceptual understanding of machine learning by engaging in interactive and participatory workshops regarding AI [14]. Furthermore, hands-on involvement in AI applications extends beyond acquiring technical skills to include enhanced self-efficacy and reduced anxiety toward emerging technologies, which are strongly associated with the affective dimension of AI competence underscored by prior studies [15, 29]. As a result, whether through cognitive skill-building or psychological empowerment, participation in AI-related programs or activities can be viewed as a critical pathway for equipping individuals with the essential capabilities required to actively engage in broader AI initiatives.

In line with this increasing demand for AI literacy and public engagement, public libraries have stepped forward to provide these critical educational opportunities. In the past few years, public libraries have devoted resources to hosting AI-related activities designed to educate patrons on how to navigate and utilize these emerging tools effectively. Studies have noted that these AI-related programs encompass not only activities that raise basic awareness but also those specifically aimed at building practical digital competencies [10, 11]. Furthermore, public libraries serve as crucial venues for cultivating critical technological agency and promoting ethical frameworks in the era of AI and big data, as well as empowering individuals to critically assess algorithmic outputs while safeguarding their intellectual and creative autonomy [19].

More importantly, as inclusive and safe spaces, public libraries offer AI-related and other technology-based programs for diverse populations to achieve the goals of digital equity [4, 23, 26]. Studies have highlighted the potential extension and exacerbation of the digital divide in the era of artificial intelligence, especially for demographic groups historically characterized as digital late adopters [6, 12, 13]. Moreover, given that marginalized populations disproportionately bear the adverse risks and consequences of automated systems, developing targeted strategies to empower these communities is essential to the democratization of AI [9, 21]. Therefore, public libraries can serve as critical social and physical spaces for realizing this objective, particularly by providing inclusive AI-related activities.

Nevertheless, the question of who actually participates in these AI activities within public libraries remains under-examined, lacking empirical investigation. Prior research on digital inequality also suggests that participation in AI-related programming is unlikely to be evenly distributed across the population, as older adults and lower-income and less-educated populations often face barriers to adopting new technologies [8, 17, 29]. This echoes the concerns of Sieber et al., who highlight that despite the growing importance of public engagement in AI, researchers and practitioners spend remarkably little time reflecting on who these "publics" actually are [24]. Furthermore, from practical aspects, without a comprehensive understanding of the demographic attributes and attitudinal profiles of those attending AI-related activities, public managers and librarians face significant challenges in designing tailored outreach strategies and inclusive programming [10, 11, 27]. That is, although public libraries are viewed as promising institutions for educating the public in an inclusive manner, there is a critical need to gain a deeper understanding of who exactly utilizes these services and the underlying factors that are associated with their participation behaviors.

3 Research Design

3.1 Survey Design

The research team initially developed the survey measures based on existing literature and the research questions. The insights collected from several focus group sessions conducted with public library participants were then incorporated to refine the questionnaire. Moreover, as part of a pre-test, the research team invited external experts to provide feedback on the structure and wording of the survey instrument. Before conducting the formal survey, the questionnaire was submitted for Institutional Review Board review and received approval prior to data collection. All procedures complied with relevant ethical standards, and participants' involvement was entirely voluntary, with the option to withdraw at any time.

After finalizing the instrument, the research team constructed the online questionnaire and collected the data through the Qualtrics platform. The survey was administered in two stages: a pilot survey was conducted on January 17–18, 2026 to ensure the clarity of the questions and the correctness of the survey logic, and the formal survey was launched on January 20, 2026 and reached the targeted sample and quotas on January 28, 2026. We set up several quota criteria with the assistance of the Qualtrics team to align the achieved sample with selected demographic characteristics of the U.S. adult population, including quotas for gender, age, race, Hispanic ethnicity, and household income. However, given that the survey was conducted on an online platform, these quotas are better interpreted as an approach to improving demographic balance instead of ensuring full demographic representativeness through probability sampling. A total of 2,010 complete responses were collected.

3.2 Measurements

The primary dependent variable in this study is respondents' participation in AI-related programs or activities held by public libraries. It was originally measured using a seven-point ordinal scale ranging from "Never" to "Very frequently." Because our primary research interest lies in understanding the initial drivers of civic engagement rather than the frequency of repeated behavior, and given that roughly half of the sample reported no prior participation, we dichotomized the variable. Specifically, responses indicating "Never" were coded as 0 to represent non-participants, while all other responses indicating any frequency of participation were coded as 1 to represent participants. This binary operationalization allows us not only to better compare the distinct profiles of engaged and disengaged individuals but also to avoid imposing interval-scale assumptions on an ordinal measure.

Aligned with the research questions, we constructed two sets of independent variables: sociodemographic variables and variables capturing perceived benefits and risks. In the sociodemographic variables, we included age, gender, race, educational attainment, and household income as critical explanatory variables for citizens' participation behaviors in AI-related activities held in public libraries. Age was recoded into three distinct age groups: 18 to 34, 35 to 54, and 55 and older. Gender was captured as a binary variable including male and female respondents. Race was collapsed into a dichotomous indicator of White and Non-White backgrounds, with the latter including Black, Asian, Hispanic, Native American, and other respondents. Educational attainment was recoded into two categories differentiating those with at least a four-year college degree from those without. Finally, annual household income was grouped into three tiers: under 50,000 USD, between 50,000 and 100,000 USD, and over 100,000 USD. Taking into account the categorical nature of these variables, these groupings were used to provide parsimonious and interpretable comparisons across major sociodemographic categories while maintaining sufficient sample sizes within each subgroup for statistical analysis.

In terms of variables reflecting respondents' perceptions regarding AI, perceived benefits were directly captured by asking participants: "What do you consider to be the benefits for public libraries and the community at large of engaging the community in AI initiatives?" Designed as a check-all-that-apply question, this measure provided eight distinct options: enhancing the quality of AI initiatives, strengthening transparency and accountability, improving digital literacy among community members, increasing a sense of community and empowerment, driving innovation in library services and programming, promoting inclusive and equitable access to technology, generating greater community support, and attracting new patrons to the library. A parallel check-all-that-apply question was used to measure perceived risks, covering five specific concerns: unequal participation among individuals, a lack of expertise among participants, inaccurate public input, infeasible public input, and data privacy and security.

Importantly, rather than relying on simple additive scores that assume equal weighting for each selected option, we utilized Latent Class Analysis (LCA) to identify underlying typologies of respondents based on their unique selection patterns across these benefit and risk items. This analytical approach for handling check-all-that-apply formats is commonly used to uncover unobserved heterogeneity within the sample, allowing respondents to be categorized into specific classes based on their distinct configurations of selected items rather than a simple aggregate score [30]. We conducted separate LCA models for the perceived benefits and risks, incorporating these newly derived categories into our main models as key explanatory variables.

3.3 Estimation Strategy

We adopt a probit regression model to estimate individuals' probability of participating in AI-related activities held in public libraries. The probit regression is estimated using maximum likelihood estimation, which constrains the predicted probabilities to be between zero and one for binary outcomes.[1] To offer a clearer interpretation of the results, the Average Marginal Effects (AME) are reported as the primary quantities of interest. These values reflect the average change in the probability of the outcome, expressed in percentage points, resulting from a one-unit increase in a continuous explanatory variable or a discrete shift from a reference category for categorical variables.

As mentioned above, five sociodemographic variables are incorporated to examine the demographic determinants of participation. The latent classes representing respondents' perceived benefits and risks serve as other categorical predictors to investigate how distinct typologies of AI perceptions influence the likelihood of engaging in AI-related activities in public libraries. Robust standard errors are applied to account for potential heteroskedasticity in the data. The final analytic sample retained 1,824 observations, which differs from the full survey total because respondents were excluded if they did not provide usable selections on the perceived benefit and perceived risk items needed for latent class assignment.

4 Results

This section highlights three aspects of the results. First, latent class analysis is reported to identify distinct typologies of respondents based on their perceived benefits and risks regarding community engagement in AI initiatives. Second, descriptive statistics are discussed. Third, probit regression models examine how sociodemographic factors and cognitive profile are associated with individuals' actual participation in AI-related programs in public libraries.

4.1 Latent Class Analysis

This study first conducted a latent class analysis to identify distinct typologies of respondents based on their perceived benefits and risks of engaging the community in AI initiatives. Multiple fit indices were used to guide the model selection, including the Akaike Information Criterion (AIC), the Bayesian Information Criterion (BIC), relative entropy (a measure of how distinct the classes are), and the estimated population share of the smallest class. Widely accepted guidelines prefer lower AIC and BIC values, a relative

---

[1] While respondents' participation in AI-related programs or activities was initially measured using a seven-point Likert scale, we converted it into a binary variable to provide a more direct interpretation of initial participation rather than changes in the frequency of repeated behavior. However, as a robustness check, in addition to the binary probit regression reported in the following analyses, we re-estimated the models using three other model specifications: (1) binary logistic regression, (2) ordered logistic regression, and (3) ordered probit regression. The substantive direction of the major findings remains unchanged across these alternative link functions.

entropy of at least 0.80, and ensuring that no class contains less than 5% of the sample [30].

Table 1 presents the fit statistics for the specified models. For perceived benefits of engaging in AI for the community and public libraries, the four-class model can be viewed as the optimal solution, given that it yielded the lowest BIC value (17,838.2) and maintained an acceptable relative entropy value (> 0.80). For perceived risks, while the five-class model showed superior AIC and BIC values, it presents a potential drawback of overfitting considering the limited number of items we have in the questionnaire. Specifically, extracting five classes from a five-item binary indicator set approaches a saturated model configuration; therefore, the four-class model is viewed as a more robust and parsimonious classification for the categories of perceived risks. Based on these fit statistics, we determined four classes for both perceived benefits and risks to serve as the categorical predictors in our following analysis.

Table 1. Latent class analysis model diagnostics

Category	Classes	Min. class share	AIC	BIC	Entropy
Benefits	2	0.136	17854.53	17948.18	0.878
	3	0.054	17748.65	17891.88	0.886
	4	0.049	17645.41	17838.22	0.806
	5	0.046	17618.84	17861.23	0.724
Risks	2	0.061	11905.75	11966.34	0.944
	3	0.055	11786.87	11880.52	0.832
	4	0.067	11744.51	11871.21	0.860
	5	0.068	11669.93	11829.68	0.902

Note: AIC = Akaike Information Criterion; BIC = Bayesian Information Criterion. Min. class share refers to the estimated population share of the smallest latent class

Table 2 displays the estimated item-response probabilities for both perceived benefits and risks regarding community engagement in AI initiatives. In terms of the four classes of perceived benefits, Class 1 can be labeled as "Benefit Optimists", characterized by a high probability of perceiving all listed benefits. Conversely, Class 2 can be labeled as "Benefit Pessimists", which exhibits consistently low probabilities across all benefit items. Members of Class 3 can be identified as "Community and Empowerment Advocates", who anticipate that public involvement in AI initiatives will improve digital literacy and community support. Lastly, Class 4 is a highly distinct group that exclusively focuses on the benefit of attracting new patrons through these engagement activities in AI initiatives. We therefore labeled it as "Patron Growth Advocates" to reflect their unique perceptions.

The LCA results of perceived risks associated with public participation also revealed four subgroups. Class 1 can be categorized as "Overall Skeptics," as these individuals exhibit high concern across all risk indicators related to public input and participation.

Class 2 displays a singular concern regarding inaccurate public input during the engagement process and is thus labeled as "Input Accuracy Skeptics." Individuals classified in Class 3 show a primary concern regarding data privacy and security in the engagement of AI initiatives and can be identified as "Privacy Skeptics." Finally, Class 4 represents individuals who are primarily worried about the public's lack of expertise and the operational infeasibility of their feedback, so we labeled them as "Expertise and Feasibility Skeptics" to reflect their pragmatic concerns about the public's readiness.

Table 2. Item-response probabilities

Category	Items	Class 1	Class 2	Class 3	Class 4
Benefits	Enhance quality of AI initiatives	1.000	0.263	0.341	0.010
	Strengthen transparency and accountability	1.000	0.352	0.461	0.047
	Improve digital literacy	1.000	0.397	0.771	0.054
	Increase empowerment	1.000	0.328	0.584	0.064
	Innovation in library services	0.997	0.281	0.678	0.101
	Promotion of inclusive and equitable access	0.949	0.219	0.500	0.028
	Enhance community support for the library	1.000	0.236	0.772	0.182
	Attract new patrons	0.950	0.104	0.678	1.000
Estimated population share of class		**4.9%**	**70.4%**	**17.2%**	**7.6%**
Risks	Unequal participation	0.932	0.227	0.132	0.327
	Lack of expertise among participants	0.988	0.423	0.277	0.580
	Inaccurate public input	1.000	1.000	0.000	0.000
	Not feasible public input	0.947	0.354	0.044	0.423
	Data privacy and security	0.968	0.467	1.000	0.287
Estimated population share of class		**6.7%**	**38.1%**	**20.2%**	**35.1%**

Source: Author(s)

4.2 Descriptive Analysis

Table 3 details the descriptive statistics for the sample used this study, segmented into non-participants (n = 887) and participants (n = 937) of AI-related programs in public libraries. Overall, slightly more than half of the respondents have participated in AI-related activities in public libraries; however, these participants exhibit distinct demographic profiles compared to those who have never attended such events. Age exhibits a clear divide. Participants are notably younger, with 79.5% under the age of 55 (40.0% aged 18–34 and 39.5% aged 35–54), whereas the majority of non-participants (53.2%) fall into the 55 and older category. The gender distribution is also different across the groups, with non-participants including more females (61.7%) and participants including more males (58.3%). Regarding race, while both groups consist of a majority of

White respondents (79.9% of non-participants and 70.2% of participants), the proportion of non-White respondents is higher among the participant group (29.8%) compared to non-participants (20.1%).

Moreover, when measured by educational attainment and annual household income, the results regarding socioeconomic status show that participants generally possess higher socioeconomic standing. The participant group reports higher levels of formal education, with 52.8% holding a four-year college degree or higher, compared to 40.8% of non-participants. Income distributions reflect a similar trend toward higher socioeconomic status among participants in AI-related programs in public libraries. Non-participants are most concentrated in the lowest income bracket (38.8% earning under $50,000), and participants are more heavily represented in the upper brackets, with their largest segment (36.2%) falling into the highest income tier (>$100,000).

The distributions across the latent classes of perceived benefits highlight a counterintuitive pattern regarding civic attitudes toward engagement in AI initiatives. The vast majority of actual participants (79.5%) are classified as "Benefit Pessimists"—a group that perceives minimal macro-level community benefits from these programs—which is notably higher than the 64.7% observed among non-participants. In contrast, "Community and Empowerment Advocates" and "Patron Growth Advocates" make up 17.7% and 13.2% of non-participants, respectively, but only comprise 11.1% and 3.1% of participants.

Furthermore, the perceived risk profiles illustrate how specific concerns align with engagement behaviors. Among non-participants, there is a stronger presence of overarching fears and privacy concerns, indicated by higher proportions of "Privacy Skeptics" (26.5%) and "Overall Skeptics" (8.2%). Participants, however, are concentrated in classes defined by targeted and operational concerns. A combined 76.1% of the participant group falls into the "Input Accuracy Skeptics" (41.0%) and "Expertise and Feasibility Skeptics" (35.1%) categories. In the subsequent section, we leverage probit regression models to examine the extent to which these distinct profiles of perceived benefits and risks, as well as demographic factors, are associated with the likelihood of participating in AI-related programs held in public libraries.

Table 3. Descriptive table

	Non-participants (n = 887)	Participants (n = 937)
Age		
18–34	20.2%	40.0%
35–54	26.6%	39.5%
55+	53.2%	20.5%
Gender		
Male	38.3%	58.3%
Female	61.7%	41.7%
Race		

(continued)

Table 3. (*continued*)

	Non-participants (n = 887)	Participants (n = 937)
White	79.9%	70.2%
Non-White	20.1%	29.8%
Annual Household Income		
Under $50,000	38.8%	30.2%
$50,000–$99,999	34.9%	33.6%
>$100,000	26.3%	36.2%
Educational Attainment		
Less than a 4-year college degree	59.2%	47.2%
A four-year college degree or Higher	40.8%	52.8%
Perceived Benefits Classes		
Benefit optimists	4.4%	6.3%
Benefit pessimists	64.7%	79.5%
Community and empowerment advocates	17.7%	11.1%
Patron growth advocates	13.2%	3.1%
Perceived Risks Classes		
Overall Skeptics	8.2%	4.4%
Input accuracy skeptics	35.9%	41.0%
Privacy skeptics	26.5%	19.5%
Expertise and feasibility skeptics	29.4%	35.1%

Source: Author(s)

4.3 Inferential Analysis

Table 4 presents the average marginal effects (AME) derived from four probit regression models estimating the likelihood of participation in public library AI programs. Specifically, these AME values reflect the difference in the probability of the outcome; for instance, a value of 0.05 in Table 4 indicates that a one-unit increase in the explanatory variable is associated with a 5-percentage-point higher probability of participating in AI-related activities in public libraries, ceteris paribus. Model 1 estimates the baseline coefficient of demographic characteristics, which remain highly stable across the subsequent specifications. Models 2 and 3 individually incorporate the distinct latent profiles of perceived benefits and risks, respectively. Model 4 integrates all covariates to present the complete model. The following analysis will primarily interpret the results based on the estimates from this complete model (Model 4).

Regarding the demographic factors, the results show that there might be a fundamental stratification between those who participate and those who do not. For example, compared to the younger people (aged 18–34), respondents aged 35–54 are associated with a 7.9-percentage-point lower probability of participation, and those aged 55 and older demonstrate a large 32.1-percentage-point decrease in their likelihood to engage. Gender also shows a clear difference, with females being 9.8% points less likely to participate in these activities than males. Interestingly, regarding race, identifying as non-White is associated with a significantly higher propensity to participate, reflecting a 10.4-percentage-point increase compared to White respondents.

Socioeconomic status also plays an important role. Holding a four-year college degree or higher is associated with a 7.1-percentage-point greater probability of participation compared to those with less than a college degree, and individuals from high-income households (earning over $100,000 annually) are 9.4% points more likely to attend than those in the lowest income bracket. These results suggest that participation in public library AI programs remains stratified, favoring younger, highly educated, and higher-income individuals, while concurrently revealing a distinct and active engagement pattern among non-White people.

The distinct profiles of perceived benefits of engaging the community in AI initiatives for public libraries and the community at large reveal a counterintuitive relationship with actual participation. With "Benefit Pessimists" as the reference group, the results show that high expectations for macro-level community benefits correlate with a lower likelihood of individual participation. Specifically, being classified as a "Community and empowerment advocate" or a "Patron growth advocate" is significantly associated with a lower probability of participation by 11.7 and 27.7% points, respectively. While "Benefit Optimists" do show a positive propensity to engage, the overarching negative trend among the other two groups suggests that the perceptions of broad and abstract community benefits often fail to translate into individual participation.

Conversely, the analysis of perceived risks illustrates how specific concerns show a positive association with participation in AI-related activities in public libraries. Compared to the reference group of "Overall skeptics," who selected most of the risk items in the survey, respondents holding targeted and operational concerns are significantly more likely to participate. For example, individuals classified as "Expertise and Feasibility Skeptics" exhibit the highest propensity to engage, being associated with an 18.8-percentage-point higher probability of participation. Skepticism regarding inaccurate public input and privacy issues shows a similar trend, with "Input Accuracy Skeptics" and "Privacy Skeptics" demonstrating a 16.4-percentage-point and an 11.7-percentage-point increase in their likelihood to attend, respectively. These findings suggest that pragmatic concerns regarding inaccurate public inputs and privacy may compel individuals to actively engage in AI-related activities at public libraries.

Table 4. Average marginal effects on participation of AI-related activities

DV = Participation in AI-related activities in public libraries	Model 1	Model 2	Model 3	Model 4
Age (Ref = 18–34)				

(*continued*)

Table 4. (*continued*)

DV = Participation in AI-related activities in public libraries	Model 1	Model 2	Model 3	Model 4
35–54	−0.087**	−0.078**	−0.087**	−0.079**
	(0.029)	(0.028)	(0.029)	(0.028)
55+	−0.353***	−0.330***	−0.342***	−0.321***
	(0.028)	(0.028)	(0.028)	(0.028)
Female (Ref: Male)	−0.104***	−0.099***	−0.101***	−0.098***
	(0.023)	(0.023)	(0.023)	(0.023)
Non-White (Ref: White)	0.105***	0.098***	0.111***	0.104***
	(0.026)	(0.025)	(0.026)	(0.025)
College or Higher (Ref: Less than college)	0.078**	0.071**	0.079**	0.071**
	(0.025)	(0.024)	(0.025)	(0.024)
Income (Ref: <$50,000)				
$50,000–$99,999	0.036	0.031	0.037	0.031
	(0.028)	(0.028)	(0.028)	(0.028)
>$100,000	0.104***	0.095**	0.105***	0.094**
	(0.031)	(0.030)	(0.031)	(0.030)
Perceived Benefits (Ref: Benefit pessimists)				
Benefit optimists		0.061		0.117*
		(0.048)		(0.048)
Community and empowerment advocates		−0.129***		−0.117***
		(0.032)		(0.032)
Patron growth advocates		−0.284***		−0.277***
		(0.039)		(0.039)
Perceived Risks (Ref: Overall skeptics)				
Input accuracy skeptics			0.128**	0.164***
			(0.047)	(0.048)
Privacy skeptics			0.076	0.117*
			(0.049)	(0.050)
Expertise and feasibility skeptics			0.158***	0.188***
			(0.047)	(0.049)
Observations	1824	1824	1824	1824
Log-Likelihood	−1114.89	−1083.87	−1107.16	−1074.95

(*continued*)

Table 4. *(continued)*

DV = Participation in AI-related activities in public libraries	Model 1	Model 2	Model 3	Model 4
p-value (LR Test)	<0.001	<0.001	<0.001	<0.001
Goodness-of-Fit	0.156	0.185	0.163	0.193

Notes: Numbers in parentheses indicate delta-method standard errors. Goodness-of-fit is reported using Tjur's coefficient of discrimination [28]. * $p < .05$, ** $p < .01$, *** $p < .001$

5 Discussion

The findings from the two research questions shed light on both the theoretical and practical implications of civic engagement in AI initiatives, as well as the critical role of public libraries. Regarding the first research question, consistent with prior literature on technology adoption and the digital divide [6, 16, 29], the results indicate that participation remains stratified along traditional socioeconomic lines. Older adults, women, individuals with lower educational attainment, and those from lower-income households are significantly less likely to engage in these AI-related programs at public libraries. This justifies the necessity of targeted programming for specific demographic groups who are marginalized in this current wave of AI training, as failing to proactively engage these populations risks further exacerbating systemic digital inequalities [9, 21]. Notably, the positive association between non-White respondents and program participation may highlight a strong grassroots demand for digital upskilling within these specific communities, a phenomenon that warrants further empirical exploration.

Furthermore, regarding the second research question concerning the association between the perceived benefits and risks of engagement in AI initiatives and patrons' participation behaviors in AI-related activities at public libraries, the findings reveal a complex and nuanced dynamic. In terms of perceived benefits, the data presents a counterintuitive picture. Patrons who hold high expectations for macro-level community benefits, such as community empowerment and patron growth, exhibit a lower likelihood of participation compared to the baseline group who perceive no benefits at all (benefit pessimists). This finding suggests that while individuals might view engagement in AI initiatives as beneficial to community and public libraries at a macro level, they may feel no immediate personal relevance to participate themselves.

In contrast, the findings regarding perceived risks reveal that individuals with well-defined concerns show a higher propensity to engage in AI-related activities at public libraries, compared to those possessing broad and generalized fears about the risks of engaging the community in AI initiatives. This pattern suggests that targeted and concrete reservations compel informed citizens to step forward, particularly those with concerns regarding uninformed and inaccurate public input. As a result, it is crucial for public managers and librarians to recognize that concerns regarding community engagement in AI are frequently voiced by constructively skeptical citizens who actively participate in these programs. The task of maintaining the sustained enthusiasm of these individuals, while simultaneously helping those with only generalized fears develop a more nuanced and concrete perspective, is therefore essential.

Two practical implications arise from these findings. First, policymakers and public librarians could consider cultivating strategic partnerships and associations with specific community stakeholders to render their outreach strategies more targeted and effective. In doing so, public libraries can not only diversify their patron base participating in AI-related programs, but also facilitate the development of tailored curricula for specific demographic groups, such as older adults. Moreover, communication regarding the benefits of civic engagement in AI needs to explicitly connect abstract community benefits to tangible personal utility, while actively inviting critical feedback from skeptical citizens. Ultimately, by addressing these demographic and attitudinal dimensions, public libraries can better fulfill their mandate as hubs for public value co-creation in the era of artificial intelligence by facilitating civic engagement in AI initiatives in a truly inclusive manner. From a broader perspective, the results carry ethical implications. If participation in AI-related library programs is socially stratified, then opportunities to build AI literacy, shape public narratives about AI, and develop confidence for later civic engagement in AI initiatives may become stratified as well. In this sense, the inclusiveness of AI programs in public libraries is not only urgent but also central to civic engagement in AI.

6 Conclusion

Employing a quantitative approach, this study draws on data from an online survey conducted in the United States to examine the demographic and attitudinal factors influencing participation in AI-related programs held in public libraries. The findings indicate that participation remains stratified along traditional socioeconomic lines. Older adults, women, individuals with lower educational attainment, and those from lower-income households are significantly less likely to engage in such activities. Furthermore, regarding the attitudinal profiles of participants, patrons who perceive high macro-level community benefits from AI are surprisingly less likely to participate personally, whereas constructively skeptical citizens with targeted and concrete concerns demonstrate a higher propensity to actively engage in these AI-related programs.

This study makes a threefold contribution to theory and practice. First, although prior literature has heavily emphasized the importance of AI-related programs, there remains a dearth of empirical research investigating the actual demographic composition of the participants. Second, this research enriches the theoretical understanding of civic engagement in AI initiatives by uncovering a distinct attitude-behavior gap. Third, this study re-emphasizes the importance of inclusive outreach strategies for public libraries and the necessity of tailored communication regarding the impacts of AI, offering actionable insights for public managers and librarians.

This research is not without limitations. First, given the use of cross-sectional data, establishing causal relationships is difficult. Future research could consider employing a panel data design to address this shortcoming. Second, we measured patrons' participation activities through self-reported survey questions. While self-reporting is a widely accepted proxy for assessing behaviors, future studies could consider utilizing administrative data, such as library attendance records, to provide a more objective assessment of actual participation behaviors. Third, while we measured respondents' perceptions of engaging communities in AI initiatives, indicators regarding more generalized attitudes toward AI technologies are not included in the models. It would therefore be beneficial for future studies to include direct measures of AI attitudes to assess how these dimensions shape citizens' willingness to participate in AI initiatives and their evaluations of public libraries' roles. Fourth, although this study relies on survey data representative of the United States, future research can examine these dynamics in other national and cultural contexts to test the generalizability of the findings.

Acknowledgments. This study was supported by the Institute of Museum and Library Services (IMLS) under grant No. LG-254889-OLS-23. The views and opinions expressed in this paper are those of the authors and do not necessarily reflect the views of IMLS.

Disclosure of Interests. There is no conflict of interest.

References

1. Bell, A., et al.: Think about the stakeholders first! Toward an algorithmic transparency playbook for regulatory compliance. Data Policy **5**, e12 (2023). https://doi.org/10.1017/dap.2023.8
2. Birhane, A., et al.: Power to the people? Opportunities and challenges for participatory AI. In: Proceedings of the 2nd ACM Conference on Equity and Access in Algorithms, Mechanisms, and Optimization, pp. 1–8 Association for Computing Machinery, New York (2022). https://doi.org/10.1145/3551624.3555290
3. Buhmann, A., Fieseler, C.: Towards a deliberative framework for responsible innovation in artificial intelligence. Technol. Soc. **64**, 101475 (2021). https://doi.org/10.1016/j.techsoc.2020.101475
4. Buyannemekh, B. et al.: The expanding role of public libraries in tribal settings: from traditional services to (advanced) digital skills, citizen engagement, and innovation. Publ. Libr. Q. **44**(6), 684–710 (2025). https://doi.org/10.1080/01616846.2024.2430060
5. Chee, H., et al.: A competency framework for AI literacy: variations by different learner groups and an implied learning pathway. Br. J. Edu. Technol. **56**(5), 2146–2182 (2025). https://doi.org/10.1111/bjet.13556
6. Cotter, K., Reisdorf, B.C.: Algorithmic knowledge gaps: a new horizon of (digital) inequality. Int. J. Commun. **14**, 745–765 (2020)
7. Gilman, M.E.: Beyond window dressing: public participation for marginalized communities in the datafied society. Fordham Law Rev. **91**, 503–555 (2022)
8. Gran, A.-B., et al.: To be or not to be algorithm aware: a question of a new digital divide? Inf. Commun. Soc. **24**(12), 1779–1796 (2021). https://doi.org/10.1080/1369118X.2020.1736124
9. Huang, Z.-X., et al.: Enabling inclusive civic engagement in AI initiatives: the case of queens public library. In: Proceedings EGOV-CeDEM-ePart Conference (2025)
10. Huang, Z.-X., et al.: How could public libraries empower communities to participate in government AI initiatives? A practical view of benefits and challenges. In: Conference on Digital Government Research (2025). https://doi.org/10.59490/dgo.2025.1034
11. Huang, Z.-X., et al.: Public libraries and their role in raising awareness about AI and fostering inclusive civic engagement: current practices and future development. In: Proceedings of the 25th Annual International Conference on Digital Government Research, pp. 703–710. Association for Computing Machinery, New York (2024). https://doi.org/10.1145/3657054.3657137

12. Huang, Z.-X., et al.: The role of public libraries in addressing the digital divide in the era of artificial intelligence: the case of Frisco public library. In: Proceedings of the 59th Hawaii International Conference on System Sciences (2026). https://doi.org/10.24251/HICSS.2026.293
13. Kitsara, I.: Artificial Intelligence and the digital divide: from an innovation perspective. In: Bounfour, A. (ed.) Platforms and Artificial Intelligence, pp. 245–265. Springer International Publishing, Cham (2022). https://doi.org/10.1007/978-3-030-90192-9_12
14. Korte, S.-M., et al.: Enhancing artificial intelligence literacy through cross-cultural online workshops. Comput. Educ. Open **6**, 100164 (2024). https://doi.org/10.1016/j.caeo.2024.100164
15. Long, D., Magerko, B.: What is AI literacy? Competencies and design considerations. In: Proceedings of the 2020 CHI Conference on Human Factors in Computing Systems, pp. 1–16 ACM, Honolulu (2020). https://doi.org/10.1145/3313831.3376727
16. Lutz, C.: Digital inequalities in the age of artificial intelligence and big data. Hum. Behav. Emerg. Technol. **1**(2), 141–148 (2019). https://doi.org/10.1002/hbe2.140
17. Lythreatis, S., et al.: The digital divide: a review and future research agenda. Technol. Forecast. Soc. Chang. **175**, 121359 (2022). https://doi.org/10.1016/j.techfore.2021.121359
18. Matias, J.N., Price, M.: How public involvement can improve the science of AI. Proc. National Acad. Sci. **122**(48), e2421111122 (2025). https://doi.org/10.1073/pnas.2421111122
19. Michalak, R., Ellixson, D.: Reimagining libraries in the AI era: ethical AI literacy and human-centered strategies. AI Soc. **41**(5), 5057–5077 (2026). https://doi.org/10.1007/s00146-025-02827-4
20. Moon, M.J.: Searching for inclusive artificial intelligence for social good: participatory governance and policy recommendations for making AI more inclusive and benign for society. Public Adm. Rev. **83**(6), 1496–1505 (2023). https://doi.org/10.1111/puar.13648
21. Queerinai, O.O., et al.: Queer in AI: a case study in community-led participatory AI. In: Proceedings of the 2023 ACM Conference on Fairness, Accountability, and Transparency, pp. 1882–1895. Association for Computing Machinery, New York (2023). https://doi.org/10.1145/3593013.3594134
22. Shams, R.A., et al.: AI and the quest for diversity and inclusion: a systematic literature review. AI Ethics **5**(1), 411–438 (2025). https://doi.org/10.1007/s43681-023-00362-w
23. Shen, L.: Out of information poverty: library services for urban marginalized immigrants. Urban Libr. J. **19**(1), 4 (2013)
24. Sieber, R., et al.: Who are the publics engaging in AI? Public Underst. Sci. **33**(5), 634–653 (2024). https://doi.org/10.1177/09636625231219853
25. Sousa, W.G.D., et al.: How and where is artificial intelligence in the public sector going? A literature review and research agenda. Gov. Inf. Q. **36**(4), 101392 (2019). https://doi.org/10.1016/j.giq.2019.07.004
26. Strover, S.: Public libraries and 21st century digital equity goals. Commun. Res. Pract. **5**(2), 188–205 (2019). https://doi.org/10.1080/22041451.2019.1601487
27. Taylor, K., Corrall, S.: Personalized service? Changing the role of the government librarian. J. Inf. Sci. **33**(3), 298–314 (2007). https://doi.org/10.1177/0165551506070728
28. Tjur, T.: Coefficients of determination in logistic regression models—a new proposal: the coefficient of discrimination. Am. Stat. **63**(4), 366–372 (2009). https://doi.org/10.1198/tast.2009.08210
29. Wang, C., et al.: The artificial intelligence divide: who is the most vulnerable? New Media Soc. **27**(7), 3867–3889 (2025). https://doi.org/10.1177/14614448241232345
30. Weller, B.E., et al.: Latent class analysis: a guide to best practice. J. Black Psychol. **46**(4), 287–311 (2020). https://doi.org/10.1177/0095798420930932
31. Young, M., et al.: Participation versus scale: Tensions in the practical demands on participatory AI. First Monday (2024). https://doi.org/10.5210/fm.v29i4.13642

What Does Access to Digital Public Services Require from Citizens?

Christian Østergaard Madsen[1,3](✉), Ida Lindgren[2], and Øystein Sæbø[3]

[1] IT University of Copenhagen, 2300 Copenhagen, Denmark
christian.o.madsen@uia.no

[2] Linköping University, 581 83 Linköping, Sweden
ida.lindgren@liu.se

[3] University of Agder, Kristiansand, Norway
oystein.sabo@uia.no

Abstract. For over a century, the public sector has used information technology to enable and streamline administrative tasks. Today, welfare benefit administration in Scandinavia predominantly occurs through a combination of automation and digital self-service, both of which rely heavily on national data registries. This has led to increased government efficiency in many areas, and improved service quality for most citizens. However, automation and digital self-service also impose several demands on citizens regarding IT access, skills and competencies, as well as their personal characteristics and life history. In this paper, we explore and illustrate these demands through a series of vignettes based on in-depth qualitative studies of citizens and frontline employees in Scandinavia. We conclude by classifying the various types of demands and discussing them in relation to existing research and current practices for public service provision.

Keywords: Citizens · automation · digital self-service · skills and competencies · administrative visibility · classification

1 Introduction

In 1890, Herman Hollerith invented the tabulating machine to store and process data from census taking in the US. This is an early example of information technology (IT) being developed to improve the efficiency of public sector administration [1]. To facilitate this use of IT, citizens were classified into predefined categories, covering aspects such as their address, sex, age, occupation, marital status, number of children, and length of time living in the US. This development would continue in the following decades. Personal identification numbers were initiated as early as the 1940s in Sweden, and in the 1960's in Denmark and Norway. In Denmark, personal identification numbers were electronic from the start, elsewhere they were digitized later. The 1960's also saw the establishment of national data registries to improve the efficiency of tax administration through automation [2]. The combination of data registries and personal identification

D. Duenas-Cid et al. (Eds.): ePart 2026, LNCS 16923, pp. 82–97, 2027.
https://doi.org/10.1007/978-3-032-36676-4_6

numbers are a cornerstone of digital government and key in public service provision in Scandinavia. However, the public sector's reliance on data registries has severe consequences for those who do not fit into the predefined categories or whose data are missing in the registries [3–7].

A little over a century after Hollerith's tabulating machine e-government emerged. Again, IT was presented as means for the public sector to improve its efficiency and effectiveness. However, this time the goal was not only to achieve internal improvements through automation, but also external improvements through the provision of digital self-service. Hereby, the public sector outsources tasks to citizens, by utilizing the internet and web-based technologies as the infrastructure [8, 9].

Today, access to most public services in the Nordic countries is primarily offered online. Considerable parts of public administration have been automated or outsourced to citizens. While this has brought many benefits, it also places numerous demands on the citizens who wish to access these services. Researchers have addressed some of these demands, such as the need for internet access and digital and administrative competencies [10–15]. Others have pointed out how benefits are unevenly distributed when accessed through digital services, how digitalization may lead to unforeseen consequences, and how automation does not cover all citizens equally [4, 6]. However, to the best of our knowledge, there is no combined overview of the different types of requirements and demands that digital government places on citizens.

Our research objective, therefore, is to identify and classify the demands and requirements digital government places on citizens. We draw on a recent large-scale Scandinavian research project for empirical data, combining in-situ interviews and observations from contact centers with interviews and workshops with citizens and frontline workers [16, 17]. From this material, we develop a series of vignettes, which we use analytically and illustratively to identify and exemplify the different types of requirements and demands citizens face. The research question guiding our study is: *What is required from citizens to gain access to digital public services?*

This work is important in several ways. We contribute empirically to digital government research through examples of actual use situations from the frontline, and theoretically by exploring and synthesizing the different requirements that citizens must fulfill to successfully interact with government digitally.

Having presented our research goal, the remaining sections are organized as follows. In Sect. 2, we present previous studies from the literature on digital government that address the skills and competencies citizens require, as well as research on the digital divide and automation and administrative blindness. In Sect. 3, we describe our empirical studies and analytical strategy. In Sect. 4, we present our findings, which we discuss in relation to previous research in Sect. 5. Finally, we conclude the paper and discuss limitations and further research in Sect. 6.

2 Previous Research

For decades, researchers have studied the transition towards digital government and what this requires from citizens. In the following, we will present a brief outline of this research to position our study and inform our analysis. Table 1 presents an overview of the requirements that are known from previous studies.

Table 1. Citizen requirements for access to digital public services

Requirement	Description	Previous studies
IT access	Access to physical information technologies (hardware), i.e. computers, the internet, printer, scanner etc. and software i.e. MS Office, e-mail	[14, 18]
Digital skills and competencies	Having the ability to use information technology, browse the internet, search and assess information, upload documentation etc	[10]
Administrative skills and competencies	Having knowledge of how the public sector works, understanding bureaucratic language, rules and regulation within and across domains	[11–13, 15]
Social network and skills	Access to warm experts in one's personal network, or the ability to get help with the application from public sector intermediaries such as librarians or healthcare workers	[19, 20]
Administrative visibility	Being visible to the government by appearing in central data registries	[4, 6, 21]
Administrative compatibility	Being the type of citizen and in the type of life situation that fits with the underlying logic and design of the IT systems	[22, 23]
Trust	Having trust in government, the internet, and the public IT systems	[24, 25]

Initially, the discussion centered around how digitalization might create a digital divide between citizens who had the access and skills to use digital technologies and those who had not. Scholars found clear relationship between people's socio-economic background, level of education, and their ability to use the internet [10, 14, 26]. Early research also found that citizens' trust, both in government and the internet were important prerequisites for their willingness to adopt of digital public services [24, 25].

While digital access and skills are necessary to access digital public services, they are not the only prerequisites, however. Citizens must also have a general understanding of how the public sector works and knowledge of the specific services in question to succeed in their digital interactions with government [11–13, 27]. Grönlund, Hatakka and Ask [12, p. 217] use the term *administrative competencies*, which they define as: *"the ability to navigate bureaucracy, which includes having a good idea of how society's institutions work, the terminology involved and hence being better able to know where to go to find the forms, procedures, contact information etc. necessary, and indeed understand the information once found and being able to act upon it."*. Skaarup [27] argues that the silo structure of the public sector increases the requirements from citizens, as their experiences from one public service interaction may not be transferable to other areas. Therefore, citizens must therefore attain new knowledge each time they interact with a new service area. Döring [11] has synthesized the different types of literacies required using the umbrella term *administrative literacy*. He argues that in addition to digital skills and knowledge of the public sector, citizens must also possess the ability to critically evaluate information. In a time where mis- and disinformation is easily shared online, or generated by large language models, this becomes more important than ever.

Several recent examples have shown that automated public services do not affect everyone equally and bring their own set of requirements on citizens [4, 6, 21]. Most importantly, citizens must be *administratively visible* to the state [3]. For the purposes of automation, this entails not only appearing in central databases but being fully represented through a complete dataset that covers all the government's needs. Larsson [4] studied a highly automated service, child benefits in Norway using data from the national registry. His findings show that certain groups of citizens are excluded from the automated services even though they are eligible. They must apply manually in some cases multiple times placing a higher burden on them than other citizens. Larsson further raises the concern that *"the groups not covered disproportionately comprised lower income families"* [4, p. 7]. Similar findings have been reported from the Netherlands, where certain groups of citizens where not only more likely to be excluded but also to be accused of fraud [6, 21].

Exclusion from public service provision is also related to the next concept, namely *administrative compatibility*. Digital services are developed with certain ideal types of citizens in mind. Consequently, the extent to which citizens "fit" the underlying logic of digital systems (be it self-service applications or automation) differ considerably [17, 22, 23]. While administrative visibility relates to the extent to which a citizen's data is present in central registries, administrative compatibility concerns the extent to which the citizen and the life situation they are in, align with the intentions of the designers of the IT systems. Lindgren et al. [17] show how administrative compatibility can differ across public benefit areas. For universal benefits, such as pensions and parental leave, between 70 and 90 percent of the population are typically covered by the IT systems. However, for means tested benefits, this share is often considerably lower, because the target group may not have lived for long in the country, and the administrative decisions involve high levels of discretion. Consequently, the recipients of means-tested benefits such as financial assistance are likely to experience higher levels of administrative burdens related to their applications and may not be able to complete the application at all [22].

Finally, many citizens get assistance in their digital interaction with government. Sometimes this assistance is provided by close friends or relatives, so-called *warm experts* [19, 20]. This assistance can also be provided by NGOs, interest groups (e.g., patience groups, associations for seniors), or employees from other public organizations, such as librarians, or social workers, that act as intermediaries between citizens and government. These instances show that a citizen's close personal network, or social skills, may be very useful to solve problems that occur related to digital public services.

We will apply the concepts presented here in our analysis and return to them when we discuss our findings in Sect. 5. First, however, we present our empirical studies.

3 Method

The purpose of our study is to explore and describe what access to digital government, especially through automation and digital self-service, requires from citizens. This entails generating new knowledge and discovering new concepts; therefore, we chose a qualitative approach [28]. From 2021 to 2024, we conducted empirical studies for a research project that assessed which public services are suitable for digitalization from both citizen and organizational perspectives [16, 17]. The project was a collaboration with agencies and municipalities who administer welfare services in Scandinavia.

3.1 Empirical Studies

We followed Blaikie's [28] and Creswell's [29, pp. 185–186] guidelines for qualitative research. *Multiple sources of data* were collected and generated through 1) observations and short in-situ interviews with citizens and frontline workers, 2) semi-structured interviews with employees from the different organizations (frontline workers, caseworkers, IT specialists and managers); and 3) workshops and discussions of findings with citizens and employees. When possible, empirical studies were conducted in *natural settings*, e.g. contact centers where citizens can receive assistance with their applications. We acknowledged our personal backgrounds and experiences to be *reflexive of our own roles* as researchers. Two of the researchers have backgrounds as caseworkers with in-depth knowledge of welfare provision in Scandinavia. Furthermore, we present a *holistic account* of the research problem from the perspectives of citizens through *rich descriptions* using composite vignettes. Findings were *validated* through presentations and discussions with the participants [30]. The overall, comprehensive dataset and analysis for the larger research project is presented in detail in [31, 32]. Table 2 presents the empirical studies for the study presented in this paper. [30, 31].

Table 2. Empirical studies

Organization	Research technique	Participants
The Norwegian Labor and Welfare Administration (Nav)	In-situ interviews, observations	Front office employees n = 14, Citizens, n = 84
A Swedish municipalility	Observations and structured in-situ survey, interviews	Observations of citizens, n = 60. Interviews with front- and back-office workers, n = 17

The interviews generated insight into citizens' applications for public benefits and services. We explored user journeys from the perspectives of citizens (as applicants) and the frontline workers who assisted them throughout the process. Applications for public benefits are increasingly conducted online and partially automated, using pre-filled forms and data from large registries. Nevertheless, many citizens still encounter problems and consequently contact frontline workers, either online via e-mail or chat, on the telephone, or in person at contact centers. The in-situ interviews and observations with citizens provide deep insight into individual cases from their perspective.

3.2 Data Analysis

We employ an abductive research strategy, whereby we analyzed interview statements and observational data from the perspectives of the participants. Next, we sought to identify common themes and patterns in these statements and thereby raise the level of abstraction from the specific and empirical to the abstract and theoretical level. We applied the findings from previous studies on different types of requirements, to inform our analysis, while remaining open to new insights from the data. Hereby, our analysis iteratively moved between the empirical data and theoretical concepts presented earlier.

We employed vignettes to analyze our data and present our findings. According to Miles and Huberman: *"A vignette is a focused description of a series of events taken to be representative, typical, or emblematic in the case you are doing. It has a narrative, storylike structure that preserves chronological flow and that normally is limited to a brief time span, to one or two key actors, to a bounded space, or to all three"* [33, p. 81]. Willis [34, pp. 476–477] presents several arguments for presenting qualitative data in narrative form. First, they allow the researchers to present complicated, situated accounts and rich stories from the data, rather than merely summarizing the results in abstract categories. Second, the narratives are a valuable way of presenting the results of a research project to non-academic partners. We found this to be especially valuable, since we conduct engaged scholarship in collaboration with public agencies to generate new knowledge and improve public service delivery.

In the next section, we present our vignettes of citizens' interaction with public agencies and the problems experienced. From these vignettes and previous studies, we synthesize the different types of requirements citizens must possess to get access to digital public services.

4 Findings

In the following, we present five vignettes from two different research settings. The vignettes illustrate how citizens who lack certain prerequisites are unable to access digital public services. We apply the theoretical concepts presented in Sect. 2, to identify and discuss the different types of requirements that we have found in the empirical data.

The first two vignettes are taken from a Swedish municipality and concern application of financial assistance. Financial assistance is granted one month at a time, and many applicants hand in new applications every month for an extended period. There is a digital service for submitting applications, but only half of the applications to the municipality are made digitally, the rest are made using paper forms. We observed and interviewed people handing in paper applications, to ask why they did not use the digital form. Most interviewees were immigrants that did not have Swedish as their mother tongue and had severe problems filling out the digital form online on their own. In agreement with their case workers, they used the paper form instead. In addition to these problems, other interesting and more unique causes for using the paper form were also observed.

The last three vignettes are based on empirical studies conducted at a local contact center for the Norwegian Labor and Welfare Administration (Nav). Here we interviewed and observed citizens visiting the contact centers to understand their motivation for coming to the centers. Most visitors have tried to access services online, while some prefer to meet someone to talk with for various reasons. The vignettes are based on observations made by us researchers, followed by interviews with the citizens observed.

4.1 Vignette 1

We meet a man in his thirties. He has a closed envelop in his hand and is waiting in line to put the envelop, containing his application for financial assistance, in the mailbox at the Contact Center in a Swedish municipality. We ask him why he is there, and if he knows about the digital application form available online. Yes, he is aware of the digital form, but he cannot use it. He does not have any digital devices, including the BankID application needed to be identified online. After some conversation, it turns out that he has just been released from prison and is without income. He cannot afford digital devices of his own and, because of his criminal record, he is not entrusted with Bank ID. For these reasons, he must manage all government contacts manually, using physical identification.

This is an example of a person has the necessary language and administrative skills, but who lacks the most fundamental resources to use digital services, i.e., the digital devices needed for IT access and the necessary digital identification required for digital access. The first issue can quite easily be resolved by e.g., using one of the openly accessible computers at the Contact Center, or lending a device from a friend. Digital access, however, cannot be compensated for by neither the municipality, nor others. BankID is a digital identification tool, issued by the banks in Sweden, necessary for gaining access to most public e-services. This is a personal digital artefact, and without it, the applicant was forced to apply manually, regardless of IT access and other necessary skills. Lacking a digital identity thus makes him incompatible with the administrative logic behind the digital public service design, calling for a manual application process.

4.2 Vignette 2

A young man in his twenties arrive at the Contact Center. He is there to hand in his application for financial assistance and to use a computer to print documents that he needs to append to his application. He is new in the city and finds it difficult to navigate the municipality services, since they are organized differently from what he is used to from the municipality in which he lived before. He finds the personnel at the Contact Center helpful in explaining how the municipality is organized and what he needs to do to gain access to necessary services. We ask why he is handing in the application on paper, and he tells us that he is homeless. He has a smartphone and BankID and thus digital access, but for some unknown reason, his case worker insists on sending him the application for financial assistance on paper. For this purpose, he had to give the case worker his friend's address to receive the application form, even though he could have made the application through the digital public service. He does not understand why the public official does not allow him to apply digitally.

This is an example of a person who does have access to IT, and digital access. He has the necessary language skills and social network to help him, but he is asked by his case worker to apply manually. The reason why the public official insists on a manual process, using a paper form, is not clear to him. In interviews with back-office employees, we were told that the case workers can allow for paper applications when they deem that a particular applicant is unlikely to succeed with the digital service. The paper form is more flexible than the digital form, in terms of not including forcing functions, and allowing the applicant to write more freely and make notes in the margins. Possibly, the paper form was deemed more flexible in this case, allowing the applicant to omit a permanent address. But, submitting the form on paper was made difficult because the applicant was homeless. This can be an example of lack of administrative visibility, as well as his case being incompatible with the internal logic of the administration. This placed additional burdens on the applicant in relation to gaining access to the form, having to print all documentation on paper, and handing in the application in person.

4.3 Vignette 3

At a contact center in Norway, a lady in her thirties approaches us and encourages us to listen to her story. She is frustrated and would like the world to know her story. Before discussing her specific case, she begins to talk about the embarrassment of not being able to use digital services to accomplish her tasks, even though she has a master's degree and decent digital skills. She feels incompetent and "small", facing the "big troll" (as she puts it) of online public services. She tells us that she has been through a difficult divorce with a violent husband who is resourceful and has a far better income than her.

She is here to apply for social support to take care of her children. Digital services do not align with the complex situation and level of conflict she is facing. The digital categories do not match her real-life situation, nor does she have all the information and documentation needed to apply for services online. Moreover, the digital services presume the existence of neutral and correct information concerning the matter. When applying, the user must confirm that the information is correct. "But", she tells us, "how can I confirm that the information is correct when I know that my ex-partner provides contradictory information and is lying to the authorities? He is the one with more resources; they will for sure trust him, not me. Hence, I am the one who might be prosecuted for providing information I know is correct while he, the one who is lying, is the one they will trust. There is not trust between us and no correct information for Nav to trust either. The digital services do not work properly if we can't agree on what is correct or not.

I think of myself as a resourceful member of Norwegian society, and I feel embarrassed to realize that I need help here in the office to apply for basic services instead of being able to do so without asking others for help."

Here we meet a person who, despite possessing digital competence and access to technology, finds herself excluded from using digital self-services. The complexity of her life situation cannot be addressed by the categories provided through online services. This is an example of where digital services are designed to address standard users without the flexibility needed to accommodate more exceptional cases. Moreover, the lack of a common agreed-upon truth and shared agreement on the correct information makes her case incompatible with existing digital solutions. For her to be able to apply online, she needs to negotiate with her ex-partner to achieve a common understanding and reporting of the required information. The burden of negotiating is on her and her ex-husband, not on the public authorities. Finally, the feeling of not being trusted represents a key reason for not using digital self-services, resulting in the embarrassment of needing further support. This may exemplify how citizens feel obliged to have the capacity to use digital services, leading to the stigma involved when not being able to succeed.

4.4 Vignette 4

We observe an elderly man with an immigrant background visiting a social services office in Norway (NAV) to apply for financial social assistance. When he enters, the service personnel direct him to a self-service computer without asking about the purpose of his visit or his digital skills. He approaches the computer hesitantly. We (the observers) approach him and ask for permission to observe. "Could you please help me out?" he asks us. "I would very much like to apply using a paper form. Can you please help me find out how?" We ask him to try his best first for us to observe. He enters the system and is soon asked to provide information from his bank. After several attempts, he logs into his bank and finds the information, but then he gets lost and approaches us for help: "Where do I apply?" We tell him he needs to go back to Nav.no to apply there. He ignores the "Important information before you start" page and goes directly to the form. There, he is stuck on the first page when asked to provide some keywords: "What is a keyword?" he asks. Observing his increasing frustration, we decide to help and find a link to a page entitled "If you prefer the paper-based version." The link leads to a new webpage with a lot of information but no option to print a paper form.

We withdraw to reduce stress in the situation. After 20 min, the man leaves the office looking sad. On the way out, he tells the counter officer that he failed to accomplish his task and is told that he is welcome to come back next week to try again. The officers later tell us that the man opened five different versions of the form, most likely causing a system error.

The person lacks appropriate digital skills required to complete his intended task at the self-service computer. Even though he requested a paper form, he was asked to try online. Furthermore, he demonstrates insufficient language skills and becomes frustrated by the large amount of information he must read and understand to apply online. Notably, he fails to notice information that he clearly needs to complete his task. No one at the office attempts to clarify whether he has the necessary knowledge of the service he is applying for, the language skills, or the digital skills required to complete the application. He is not helped during his visit and is asked to return next week to try again, without explanation of why he failed or how he might succeed next time.

4.5 Vignette 5

A woman and her daughter arrive at the NAV office. We ask the woman for an interview but realize that she does not understand Norwegian. Her daughter, who is 14 years old and was born in Norway, speaks Norwegian fluently and is acting as a translator. The mother is in her fifties and came from Somalia around 20 years ago. They come to the office to hand in documentation regarding her income, household and living expenses for her five children. The woman says she cannot read or write and lacks digital skills, while her sons and daughters are doing well in school and have access to, and competence with, digital services. She is aware that she could ask them to submit the documents online, but she has reasons for coming to the office once a month to hand them in herself, despite the travel and the need to bring a family member to translate.

First, she does not trust digital services when she cannot understand how the information is handled. Seeing someone in person is important for her so she can be sure she has done what is expected. Second, based on her experiences in Somalia she does not trust government in general. After years of war, corruption, and unstable political regimes, she has no reason to trust public authorities. She is therefore afraid of making a mistake that could come back to harm her later. When using digital services, she must confirm that the information is correct. She fears she might be prosecuted for providing incorrect information: "I can't risk going to jail for making mistakes online. Who will take care of the children if I'm not around?". For her, it is better to come to the office and see someone who can confirm that the information is correct and received.

The woman has limited language skills and low digital literacy and depends on friends and family for assistance. Her minor daughter is directly involved in handling her mother's dealings with public authorities. The daughter's responsibilities, submitting monthly documentation about the family's economic situation, could readily be completed online, which she recognizes. However, the mother prefers in-person visits because she does not trust digital services she cannot understand and lacks trust in government institutions due to her personal experiences.

5 Discussion

When we combine our empirical insights with the extant literature, we get an overview of what is required by citizens to gain access to digital public services. We summarize these requirements in Fig. 1 below.

Overall, the different types of requirements can be placed into three groups: 1) *Access to information technology* covers both hardware and software, as well the digital services needed to identify oneself and interact with government, 2) the various types of *skills and competencies* needed, and 3) *personal characteristics*, which include a person's ability to be seen and administered by the government (administrative visibility and compatibility), but also a person's trust in government and digital technologies.

1) **Access to information technology (hardware and software).** As stated in the introduction, citizens must have *access to IT hardware* to be able to access digital public services [14, 18, 26]. However, we also found that citizens must also have access to specific types of government information technologies. Without the necessary *digital identification* interaction with most public organizations is almost impossible, because so many services are predominantly offered online. Further, in Denmark especially, and increasingly also in Norway and Sweden, written communication takes place digitally, through secure e-mail, referred to as *digital post.* Thus, citizens must have access to this as well, to receive and send digital written information from government [35]. Finally, Scandinavian countries are fast becoming cashless societies. Welfare benefits are not paid out physically, but instead transferred to people's designated bank accounts, e.g. NemKonto in Denmark. Just as citizens need digital ID to access the digital public services, they must have a *bank account* and the ability to receive the output from their applications.

Fig. 1. Citizen requirement for access to digital public services.

2) **Skills and competencies.** As illustrated in the vignettes, we found that citizens need a variety of skills and competencies to successfully interact with government digitally. While several of these areas are well-known from previous studies, we were surprised by their total extent when we combine all of these. First, the citizen is required to have *digital and administrative skills and competencies* needed to find the relevant digital public service online, and to fill out the digital form with the right information and documentation [11–13]. Second, and somewhat surprisingly, we found a prominent problem throughout our empirical data, regardless of country or context, namely the importance of language skills. In Scandinavia, the interaction between applicants and government is expected to be in the native language (Norwegian, Swedish, Danish). User interfaces are seldom translated into any other language. When they are translated, it is mostly into English. Therefore, citizens must have a high *mastery of the local language* to be able to read and understand information in letters, websites and self-service applications. The technological development and outsourcing of tasks to citizens via self-service application has generally increased the level of digital and administrative skills required. However, we may forget that citizens also need basic *reading and writing skills*. It is estimated that between 5 to 8 percent of the population in Scandinavia suffer from dyslexia [36, 37]. The increased globalization has also led to immigration from countries with higher levels of illiteracy. During

the empirical studies, we were surprised by how many of those that showed up in person at the contact center had reading and writing difficulties. When a person lacks the necessary skills (language, digital, administrative), much can be compensated through the help of social networks, as previously addressed by [22, 23]. Having a *social support network*, and *the social skills* to use this network, is therefore also an important requirement for accessing digital public services. The vignettes illustrate different situations in which the citizen is required to align with the digital service logic and compensate for lacking the necessary resources to do so – either by applying in alternative parallel manual processes or asking for help from front-line personnel.
3) **Personal characteristics.** The final set of requirements are related to the individual who applies for the public services, the situation they are in, and their personal characteristics and life history. These aspects have not received a lot of attention in the past but are becoming increasingly addressed in more recent digital government and public administration studies [4, 5, 21]. This is partly due to several recent scandals, where certain population groups have either been excluded from public services or accused of committing fraud.

These results are in line with our empirical studies, where we have identified several requirements that follow on extensive use of forcing functions in digital forms and automation technologies in the internal case handling process. In order to handle applications automatically and thus increase government efficiency, the public organization can either rely on registry data already available for the organization, or on structured data supplied by the applicant. We refer to being covered by the registry data as *administrative visibility* [4, 6, 9]. In the second vignette, we illustrated an applicant who lacked necessary information (home address) and therefore was not allowed by this case worker to use the digital service. Similarly, the third vignette illustrated a case where an applicant was in conflict with an ex-partner, and where they are providing conflicting information to the authorities. When information is not standardized and structured, but must be submitted by the applicant, and there are multiple interpretations on what is correct information, digital service interactions are made difficult. The applicant's life situation must also be understood and acknowledged in the digital service design – here called *administrative compatibility* [20, 21]. For example, Larsson [4] has illustrated how applicants not aligned with Nav's view of the typical applicant was excluded from automated case handling.

Finally, we have identified another type of personal characteristic that is important for people's willingness to use digital public services, namely their *trust* in the government and the digital services. The importance of trust for citizens' adoption of digital public services is well-known [24, 25] However, as Scandinavian researchers, we tend to overlook this, because of the general high level of social and institutional trust in Scandinavia [38]. Once again, we see how globalization influences citizens' characteristics, as exemplified by the fifth vignette, with a woman who shows up in person because she does not trust the government or the digital technologies. Our results have important implications for research and practice. We have identified numerous requirements for citizens' access to digital public services. Many of these were known from previous studies [3, 4, 6, 7, 10, 12–15, 24–26]. However, we also identify new requirements, and some that have become more important due to global events such as rising inflation and increased immigration. Notably, we also find that many of the requirements do not operate in isolation, but also appear together, thereby having significant consequences for the citizens who are not able to fulfil these requirements.

We recommend that practitioners take our results into consideration when designed and testing public service provision. As we demonstrate, that are numerous exclusionary mechanisms besides the mere access to information technology. Our results show that it is especially important that public authorities provide alternate means of accessing their services, as well as help, for those citizens who do not meet the requirements.

6 Conclusion

The study presented here was guided by the research question *What is required from citizens to gain access to digital public services?* We answered this question by analyzing data from in-situ interviews and observations with citizens and frontline employees from public welfare organizations in Scandinavia. We have identified a total of nine different requirements in the empirical data and from the extant digital government and public administration literature and summarized these in Fig. 1.

We find that these are general requirements that cover all citizens who seek access to digital public services in Scandinavia. However, we also claim that many of these requirements are invisible in the sense that many citizens do not experience them as barriers during their public encounters. Scandinavian citizens may take some requirements for granted, (i.e. being able to read and write, speak the language, and having a computer and smartphone with internet access) or be altogether unaware of them (appearing in data registries and having trust in government). However, when we observe and analyze situations where citizens cannot get access to the digital public services, the underlying requirements and their consequences become visible to us. Our findings also indicate that the requirements appear to affect people in vulnerable situations more than the general population.

Limitations, Transferability of Findings, and Future Studies

While several of our results are in line with previous digital government research, this study is nevertheless undertaken in a specific context, which must be considered. This research was conducted in Scandinavian countries, which are often regarded as frontrunners in public sector digitalization. Moreover, the Scandinavian welfare systems differ from other parts of the world, both concerning the type of services and benefits available [39], but especially with regards to the high reliance on data registries and automation [2, 40]. By conducting our empirical studies at local service centers, we get insight into problems and exclusionary requirements that cause people to show up in person. We recognize the limitations of our approach and choice of empirical setting. Notably, we do not get insight into the many digital public service encounters that are fully digital, occur without problems, or where citizens get help by other means than showing up in person, i.e. by calling, mailing, chatting or through their personal network. We therefore ask the readers to consider these contextual factors when assessing the transferability of our findings to other settings [41].

As for all explorative and qualitative studies, we cannot make statistical generalizations concerning the extent of our findings [26]. Nor can we make a prioritized list of which requirements are more common or pose the biggest barriers. As such, we invite scholars to critically assess our results, repeat the study in different settings, measure the extent of the requirements, and test their effect in more controlled settings.

Acknowledgments. Our study applies data from two research projects funded by the Norwegian Labor and Welfare administration (Nav), and the Norwegian Research Council. We are grateful to Nav, ATP, and the anonymous Swedish municipality for providing access to contact centers and employees, and for discussing and validating our interpretations and findings with us at multiple meetings and workshops. Further, we wish to thank our colleagues Hanne Höglund Rýden, Ida Heggertveit, Maria Booth, and Sara Hofmann who participated in the data collection and many fruitful discussions in the projects. We are also grateful to the three reviewers for their constructive feedback.

References

1. Chadwick, A., May, C.: Interaction between states and citizens in the age of the internet: 'e-Government' in the United States, Britain, and the European Union. Gov. Int. J. Policy Administrat. Inst. **16**(2), 271–300 (2003). https://doi.org/10.1111/1468-0491.00216
2. Jæger, B., Löfgren, K.: The history of the future: changes in Danish e-government strategies 1994–2010. Inf. Polity: Inte. J. Gov. Democracy Inf. Age **15**(4), 253–269 (2010). https://doi.org/10.3233/ip20100217
3. Scott, J.: Seeing like a State: How Certain Schemes to Improve the Human Condition Have Failed. Yale University Press (1999)
4. Larsson, K.K.: Digitization or equality: When government automation covers some, but not all citizens. Gov. Inf. Q. **38**(1), 1–10, 101547 (2021). https://doi.org/10.1016/j.giq.2020.101547
5. Peeters, R., Widlak, A.: The digital cage: administrative exclusion through information architecture – the case of the Dutch civil registry's master data management system. Gov. Inf. Q. **35**(2), 175–183 (2018). https://doi.org/10.1016/j.giq.2018.02.003
6. Ranchordás, S.: The Invisible Citizen in the Digital State: Administrative Law Meets Digital Constitutionalism, pp. 15–40 (2024). https://doi.org/10.1007/978-94-6265-647-5_2
7. Bowker, G.C., Star, S.L.: Sorting Things Out. Classification and Its Consequences. The MIT Press, Cambridge (2000)
8. Lindgren, I., Madsen, C.Ø., Hofmann, S., Melin, U.: Close encounters of the digital kind: a research agenda for the digitalization of public services. Gov. Inf. Q. **36**(3), 427–436 (2019). https://doi.org/10.1016/j.giq.2019.03.002
9. Madsen, C.Ø., Lindgren, I., Melin, U.: The accidental caseworker – How digital self-service influences citizens' administrative burden. Gov. Inf. Q. **39**(1), 101653 (2022). https://doi.org/10.1016/j.giq.2021.101653
10. van Deursen, A., van Dijk, J.: Improving digital skills for the use of online public information and services. Gov. Inf. Q. **26**(2), 333–340 (2009). https://doi.org/10.1016/j.giq.2008.11.002
11. Döring, M.: How-to bureaucracy: a concept of citizens' administrative literacy. Adm. Soc. **53**(8), 1155–1177 (2021). https://doi.org/10.1177/0095399721995460
12. Grönlund, A., Hatakka, M., Ask, A.: Inclusion in the e-service society – investigating administrative literacy requirements for using e-services. Springer Berlin/Heidelberg (2007). https://doi.org/10.1007/978-3-540-74444-3
13. Skaarup, S.: The role of domain-skills in bureaucratic service encounters. In: International Conference on Electronic Government, pp. 179–196. Springer, Linköping (2020). https://doi.org/10.1007/978-3-030-57599-1_14

14. Helsper, E.J., Reisdorf, B.C.: The emergence of a 'digital underclass' in Great Britain and Sweden: changing reasons for digital exclusion. New Media Soc. **19**(8), 1253–1270 (2017). https://doi.org/10.1177/1461444816634676
15. Skaarup, S.: Beyond substantive goals–a framework for understanding citizens need and goals in bureaucratic encounters. In: International Conference on Electronic Government, pp. 86–102. Springer, Heidelberg (2021). https://doi.org/10.1007/978-3-030-84789-0_7
16. Hofmann, S., Madsen, C.Ø., Lindgren, I., Verne, G.: A citizen-centered analysis of what public services are suitable for digital communication channels. In: Proceedings of Ongoing Research, Practitioners, Posters, Workshops, and Projects at EGOV-CeDEM-ePart 2021 (2021)
17. Lindgren, I., Madsen, C.Ø., Rydén, H.H., Heggertveit, I., Hofmann, S., Sæbø, Ø.: What public services are suitable for digitalization? - A classification of public service characteristics. In: Lecture Notes in Computer Science (including subseries Lecture Notes in Artificial Intelligence and Lecture Notes in Bioinformatics) , pp. 201–217. Springer Science and Business Media, Deutschland (2024). https://doi.org/10.1007/978-3-031-70274-7_13
18. van Dijk, J.A.G.M.: One Europe, digitally divided. In: Chadwick, A., Howard, P.N. (eds.) Routledge Handbook of Internet Politics, vol. 21, pp. 288–304. Routledge, London , New York (2009)
19. Bakardjieva, M.: Internet Society. The Internet in Everyday Life. SAGE Publications, London (2005)
20. Madsen, C.Ø., Kræmmergaard, P.: Warm experts in the age of mandatory e-government : interaction among danish single parents regarding online application for public benefits. Electron. J. e-Gov. **14**(1), 87–98 (2016). http://www.ejeg.com/issue/download.html?idArticle=447
21. Peeters, R., Miller, S.M., Schuilenburg, M.: Digital government inclusion: Exploring strategies for inclusive government automation. Elsevier Ltd. (2025). https://doi.org/10.1016/j.giq.2025.102028
22. Heggertveit, I., Lindgren, I., Madsen, C.Ø., Hofmann, S.: Administrative Burden in Digital Self-service: An Empirical Study About Citizens in Need of Financial Assistance, pp. 173–187 (2022). https://doi.org/10.1007/978-3-031-23213-8_11
23. Lindgren, I., Madsen, C.Ø., Rydén, H.H., Heggertveit, I.: Exploring Citizens? Channel behavior in benefit application: empirical examples from norwegian welfare services. In: ACM International Conference Proceeding Series, pp. 416–423. Association for Computing Machinery (2022). https://doi.org/10.1145/3560107.3560312
24. Bélanger, F., Carter, L.: Trust and risk in e-government adoption. J. Strat. Inf. Syst. **17**(2), 165–176 (2008). https://doi.org/10.1016/j.jsis.2007.12.002
25. Carter, L., Bélanger, F.: The utilization of e-government services: citizen trust, innovation and acceptance factors. Inf. Syst. J. **15**(1), 5–25 (2005). https://doi.org/10.1111/j.1365-2575.2005.00183.x
26. Helsper, E.J.: The digital disconnect: the social causes and consequences of digital inequalities (2021)
27. Skaarup, S.: Beyond substantive goals: a framework for understanding citizens' needs and goals in bureaucratic encounters. eJournal eDemocracy Open Gov. **14**(2), 1–31 (2022). https://doi.org/10.29379/jedem.v14i2.731
28. Blaikie, N.: Designing Social Research. Polity (2012)
29. Creswell, J.W.: Research Design. Qualitative, Quantitative, and Mixed Methods Approaches, 4th edn., SAGE, London (2014)
30. Creswell, J.W., Miller, D. L.-: Determining validity in qualitative inquiry. Theory Pract. **39**(3), 124–130 (2000). https://doi.org/10.1207/s15430421tip3903_2

31. Hofmann, S., Sæbø, Ø., Heggertveit, I., Rýden, H.H., Madsen, C.Ø., Lindgren, I.: Hvilke tjenester bør digitaliseres og hvordan? (2024). Accessed 05 Mar 2026. https://www.uia.no/forskning/prosjekter/sv/hvilke-tjenester-bor-digitaliseres-og-hvordan/
32. Lindgren, I., Madsen, C.Ø.: Digital first? Understanding citizens' communication needs in digital public encounters. Soc. Policy Adm. (2025). https://doi.org/10.1111/spol.70017
33. Miles, M., Huberman, M.: Qualitative data analysis. An expanded sourcebook (1994)
34. Willis, R.: The use of composite narratives to present interview findings. Qual. Res. **19**(4), 471–480 (2019). https://doi.org/10.1177/1468794118787711
35. Jansen, A., Berger, J.B., Goldkuhl, G.: First choice, free choice or no choice - differences in secure digital post in the scandinavian countries. IOS Press, Guimarães (2016). https://doi.org/10.3233/978-1-61499-670-5-135
36. Ebro, C., Gellert, A.S.: Dyslexia. Lex. Danmarks Nationalleksikon (2025)
37. Svenska Dyslexiföreningen. Accessed 19 Mar 2026. https://www.dyslexiforeningen.se/
38. Sønderskov, K.M., Dinesen, P.T.: Danish exceptionalism: explaining the unique increase in social trust over the past 30 years. Eur. Sociol. Rev. **30**(6), 782–795 (2014). https://doi.org/10.1093/esr/jcu073
39. Esping-Andersen, G.: The Three Worlds of Welfare Capitalism. John Wiley and Sons, Inc. (2013)
40. Madsen, C.Ø., Hjelholt, M.: Digitalisering og administrative byrder. Politica 56(4), 296–317 (2024)
41. Drisko, J.W.: Transferability and generalization in qualitative research. Res. Soc. Work. Pract. **35**(1), 102–110 (2025). https://doi.org/10.1177/10497315241256560

Transferring and Integrating Citizens' Practice-Based Knowledge for Digital Service Development in the Public Sector

Michaela Schmidt(✉), Tangni C. Dahl-Jørgensen, Babak A. Farshchian, and Morteza Moalagh

Norwegian University of Science and Technology, Trondheim, Norway
{michaela.schmidt,tangni.c.dahl-jorgensen,babak.a.farshchian, morteza.moalagh}@ntnu.no

Abstract. Digitalization of public services promises improved effectiveness and transparency of service delivery, as well as more inclusive governments. However, challenges in accessing and using digital public services are widely recognized. To create more inclusive systems and prioritize citizens' needs, developers need to understand citizens' practices of accessing and using digital public services. This knowledge about citizens and their practices resides outside the organization and is not easily captured through interviews or surveys. In this paper, we use a knowledge transfer and integration lens to understand how citizens' practice-based knowledge is transferred and integrated into the development of digital public services. Based on insights from a case study of a large public sector organization, we answer the following research question: How is citizens' practice-based knowledge transferred and integrated into the development process of digital public services? Our findings show how citizens' practice-based knowledge is acquired and transferred through different channels before reaching the development teams. These channels include direct interactions between citizens and frontline employees, both in person and by phone, as well as development teams actively reaching out to citizens to gather feedback. Further, knowledge gets transferred formally and informally. The integration of knowledge depends on development teams' responsiveness and prioritization of feedback.

Keywords: Digital Public Service Development · Knowledge Transfer · practice-based knowledge

1 Introduction

There is global interest in digitalizing public services to enhance the effectiveness and transparency of public service delivery [1, 2], as well as to improve accessibility and create more inclusive governments [3, 4]. Yet, the majority of digitalization initiatives in the public sector focus primarily on internal efficiency and contribute only indirectly to other societal values, such as citizen relations [5], while also outsourcing many of

D. Duenas-Cid et al. (Eds.): ePart 2026, LNCS 16923, pp. 98–113, 2027.
https://doi.org/10.1007/978-3-032-36676-4_7

the tasks related to service provision. This can lead to the exclusion of citizens and exacerbate the digital divide, as has been documented, e.g., in citizen self-service [6, 7]. Developing more inclusive digital systems and prioritizing citizens' needs requires understanding citizens' practices of accessing and using digital public services, including their background, experiences, and skills [8, 9]. Such knowledge about citizens and their practices resides outside the organization and is not easily captured through interviews or surveys [10, 11]. Other interaction channels, e.g., face-to-face public encounters between frontline employees (FLE) and citizens, can provide more insights into how citizens use the system in practice [12, 13]. However, the prevalence of digital channels reduces these interactions and can lead to negligence of citizens' practice-based knowledge when developing digital solutions.

We therefore seek to understand how this knowledge can be made available to system developers. We use the concept of knowledge transfer, which includes the processes of identifying, acquiring, and integrating knowledge to develop new or enhance existing solutions [14]. Through this lens, we aim to investigate how citizens' practice-based knowledge contributes to system development, i.e., in what forms it exists, and how it is transferred and integrated into the development of digital public services. We therefore pose the following research question:

How is citizens' practice-based knowledge transferred and integrated when developing digital public services?

Our findings are based on a case study of a large public organization, the Norwegian Labor and Welfare Administration (NAV), which develops most of its digital solutions in-house. This in-house development offers a unique opportunity to observe the knowledge transfer from citizens to development teams, eliminating the need to navigate inter-organizational barriers. At the same time, it has a great societal impact as NAV's digital solutions are used by more than 2 million users every year [15].

Our findings show how citizens' practice-based knowledge is acquired and transferred through direct interactions between citizens and FLEs, both in person and by phone, as well as development teams actively reaching out to citizens to gather feedback. Further, knowledge gets transferred formally and informally. The integration of knowledge depends on development teams' responsiveness and prioritization of feedback.

We contribute to the literature on knowledge transfer in the public sector in several ways. First, our findings shed light on the day-to-day activities of FLEs and product developers that support or hinder knowledge transfer in the development and maintenance of digital public services. We further highlight the importance of practice-based knowledge for system development, as is recognized in the literature [8, 9], but often neglected in practice. Additionally, while earlier research has mainly focused on activities such as citizen co-creation workshops and engagement in idea generation [16, 17], we point to the importance of citizens' practice-based knowledge for the ongoing revision and maintenance of digital public services.

The paper is structured as follows: Sect. 2 provides an overview of the related literature on Digital Public Service Development and Knowledge Transfer; Sect. 3 describes the Case and our methodological approach; Sect. 4 presents our findings; and Sect. 5 provides the discussion.

2 Background

2.1 Digital Public Service Development

Public service delivery is increasingly digitalized to enhance effectiveness, accessibility, and create more inclusive governments [1–4].

As digital public services become accessible at any time and from any place, concerns about accessibility rise, impacting the quality of these services. While key challenges presented often relate to structural barriers, such as insufficient infrastructure or a lack of skills [18], putting the onus on societal issues and individuals, we focus on the organizational development of digital systems.

It is widely recognized that including users in the development of systems has many benefits, such as a more accurate definition of requirements or improved user interfaces [19]. Including citizens, however, presents challenges such as citizens' willingness and ability to participate [20], designing for diverse citizen needs, and problems organizations face in facilitating participatory activities [21]. Yet, not including citizens poses the risk of flawed results and project failures [22]. Additionally, citizen participation is often limited to the idea generation phase and is absent when it comes to providing feedback during later implementation efforts [16, 17].

Considering that the implementation of digital systems in public services provision deeply impacts both administrative and citizen practices [23, 24], effective digital systems ought to be developed with a deep understanding of the practices in which they are embedded [8, 9]. Practice, in a broad sense, relates to "situated doing" that is often collaborative and is shaped by an actor's knowledge, skills, and experiences [24].

Other researchers have examined citizen practices in accessing digital services [7, 24], 25] as the drive towards digital self-service means that acquiring knowledge about public services is increasingly entrusted to citizens [25].

Although citizen practices have been addressed by researchers, few have examined knowledge transfer in the development of public digital systems. Our study highlights the importance of citizens' practice-based knowledge and empirically demonstrates how organizations can transfer and integrate it into the development of digital solutions.

2.2 Knowledge Transfer

Defining and understanding the nature of knowledge has occupied philosophers and researchers since early history, when knowledge was considered a state of mind that could range from mere guessing to knowing [26]. Knowledge is commonly divided into tacit and explicit knowledge, where tacit refers to knowledge gained through experience, and explicit refers to knowledge that can be easily articulated in formal language [14]. Tacit knowledge is practice-based, embedded in situated activity, and shaped by an individual's background, experience, and ability to sense and feel. Accessing this practice-based knowledge requires focusing on the actual activities of work, fostering skills of knowing and supporting engaged participation in the practice [10]. For it to be useful to an organization, it must be transferred between actors and integrated into existing processes and practices. Transferring knowledge describes the process of identifying, acquiring, and applying this knowledge to develop new or enhance existing ideas, processes, and solutions [14]. In practice, this means that an organization needs to know where the required knowledge is located, be able and willing to acquire it, and have the skills and capacity to make the acquired knowledge useful by relating it to the organization's needs.

Knowledge transfer further includes moving knowledge between individuals and often across boundaries of specialized knowledge domains. While transferring knowledge between individuals is easy when they share the same practice or expertise, it becomes "sticky", i.e., harder to move, across the boundaries of different practices [27]. People who work across different practices become knowledgeable about the other practice without being professionals in it, and can therefore act as knowledge-brokers [28]. Knowledge can further be transferred via different modes, such as formal or informal, personal or impersonal. Informal modes include unscheduled meetings or friendly discussions, while formal modes include established routines or the transfer of staff to other units. Ultimately, knowledge transfer is only successful if the acquired knowledge is applied and integrated, and thereby can lead to improved performance and value creation [14]. In the case of system development, knowledge integration relates to the process of coding knowledge into the system's new functionalities [29]. Several factors can influence the knowledge transfer process. These include the nature of the knowledge (e.g., tacit or explicit), the motivation and reliability of the knowledge source, the motivation and ability to apply and retain knowledge of the knowledge recipient, and the organizational context in which the knowledge transfer takes place, e.g., the use of communication tools between source and recipient [30].

In our case of digital public services, the practice-based knowledge resides primarily with the citizen, and therefore outside the organization that develops those services. Their knowledge must be transferred to the development teams, where it can be integrated into the development of digital systems. Acquiring knowledge from citizens can be challenging [20, 22], and frontline employees in particular have been recognized as valuable resources in service development, as they understand the organization's goals and limitations as well as the service users' perspective [12, 13].

3 Case Description and Method

3.1 Case Description

This paper builds on a case study of the Norwegian Labor and Welfare Administration (NAV). NAV is a nationwide public agency that provides both municipal and state services for social and financial security, as well as facilitating job-market transitions. Around 30% of the state's budget is administered by NAV to provide the different services [31]. Between 2012 and 2016, the administration underwent internal restructuring to better align with the Norwegian government's strategy to digitalize public services and improve citizens' access to them [32]. Initially, NAV relied on external vendors for digitalization but shifted its strategy to in-house development in 2016 [33]. By 2019, NAV's IT department had expanded to approximately 800 employees across eight distinct product areas, enabling more frequent system updates and releases. Each product area has a leader and consists of development teams that support specific NAV services. The first digital service for citizens was launched in 2013 [34], followed by the personal portal "Ditt NAV," which enabled citizens to access various service applications and case documents, communicate with their caseworker, and more. While this shift to digital service provision improved efficiency and access, it also led to reduced opening hours and closures of some local offices [35, 36].

NAV's structure and the in-house development of digital public services present a unique opportunity to explore knowledge transfer in system development.

3.2 Method

We conducted a case study grounded in an interpretivist epistemological view with the aim of identifying, exploring, and explaining the relationship between different factors of a given social setting [37]. Our findings are based on data from ten guidance sessions between citizens and FLEs, as well as 18 in-depth interviews with frontline employees, members of development teams, and other employees involved in the development process of digital systems for service provision.

The guidance sessions (GS) between citizens and FLEs (here, caseworkers) lasted between 20 min and 2 h and were aimed at discussing the usability and design of a digital application form, as well as capturing citizens' experiences regarding the comprehensibility of the bureaucratic language, rules, and eligibility criteria. In addition to the observations of the GS, we conducted semi-structured interviews with FLEs and citizens before and after each guidance session (pre- and post-interviews), lasting between 15 and 60 min.

We further conducted semi-structured in-depth interviews with employees in NAV, holding different positions such as caseworker, contact center specialist, lawyer, tech lead, agile coach, product area lead, and systems designer. The interviews focused on collaboration between the employees and development teams, as well as challenges and barriers affecting the collaboration. These interviews lasted between 1–2 h each.

All guidance sessions and interviews were conducted in Norwegian or English, recorded, and transcribed. The quotes presented in this paper are the author's translations. Table 1 gives an overview of the data collection.

The data was analyzed using abductive thematic analysis [38], which integrates inductive and deductive reasoning. Abductive thematic analysis allows the identification of patterns in empirical data while at the same time linking them to existing theories.

The interview transcripts and observation notes were imported into NVivo for coding. The first authors conducted several rounds of open coding, developing a codebook that captures meaning from the data. In collaboration with the second author, the codebook was revised, and codes were further summarized in categories, guided by our theoretical perspective of knowledge transfer. A summary of the established categories and subcategories is presented in Fig. 1.

Table 1. Summary data collection

Participant / Role	**Abb**	**Data Source**
Guidance Session 1–10	GS 1–10	Observation
FLE / Caseworker	FLE1	GS 1 and 2, Pre- and post-interview

(continued)

Table 1. (*continued*)

Participant / Role	Abb	Data Source
FLE / Caseworker	FLE2	GS 3, 4, 5, 6, Pre- and post-interview
FLE / Caseworker	FLE3	GS 7, 8, 9, 10, Pre- and post-interview
Citizen	C1	GS 1, Pre- and post-interview
Citizen	C2	GS 2, Pre- and post-interview
Citizen	C3	GS 3, Pre- and post-interview
Citizen	C4	GS 4, Pre- and post-interview
Citizen	C5	GS 5, Pre- and post-interview
Citizen	C6	GS 6, Pre- and post-interview
Citizen	C7	GS 7, Pre- and post-interview
Citizen	C8	GS 8, Pre- and post-interview
Citizen	C9	GS 9, Pre- and post-interview
Citizen	C10	GS 10, Pre- and post-interview
FLE / Caseworkers and Change Agent	FLE 4–6	Semi-structured interview
FLE / Contact Centre Specialist	FLE 7–12	Semi-structured interview
Tech Lead	TL1, TL2	Semi-structured interview
Jurist	Jurist	Semi-structured interview
Product Area Lead	PL1, PL2	Semi-structured interview
Subject Matter Expert	SME	Semi-structured interview

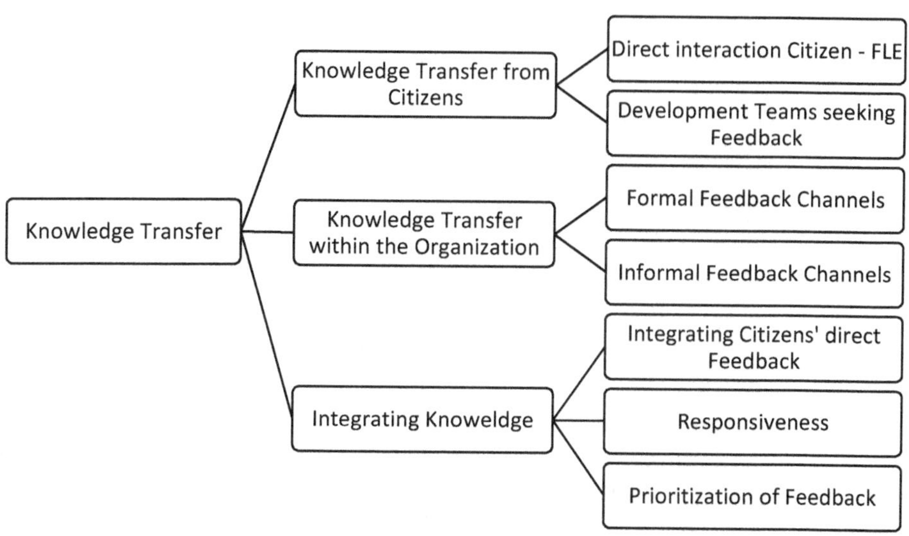

Fig. 1. Coding results

4 Findings

4.1 Knowledge Transfer from Citizens

As NAV is seeking to improve the digital systems for service provision, they can acquire citizens' practice-based knowledge through different channels. From our data, we identified two such channels: a) direct, often physical, interaction between FLE and citizens, and b) the development teams actively reaching out to citizens.

Direct Interaction with FLE. Direct interaction includes in-person meetings and phone conversations between citizens and FLEs, such as caseworkers, counselors, or the contact center.

Findings in this section stem from interviews with FLEs and from observations of guidance sessions in which FLEs help citizens use NAV's digital systems. Through the guidance session, they gain first-hand insights into citizens' practices of accessing and using digital solutions, as they would in regular consultation meetings.

One of the main challenges for citizens is navigating the NAV website, which serves as the starting point for most interactions with NAV and also contains information related to citizens' cases. *"It's difficult, when you log in to My Page in NAV, to see how many applications you have and where each application is located. And to find them, checking if you forgot an answer or if you need to provide information that is missing [...] it's not easy." (C6).* To help navigate the website, one citizen suggests adding a shortcut or link on My Page that takes you to your applications or the application form if you have applied for this service in the past (GS8). FLEs who interact with citizens regularly recognize this problem "*It's very hard to get an overview, especially on NAV's website [...] It becomes a bit of a jungle to get through*" (FLE5).

In addition to the general challenges of navigating the website, citizens often struggle to file digital application forms. Difficulties include a general understanding of the application form and retrieving and uploading relevant documentation (Observation GS). Comprehension is often affected by how questions are phrased; for example, formulations such as "if this is NOT correct, please provide ..." are often confusing for citizens and are misunderstood (observations GS5 and GS6). These examples show the detail of feedback that can be gathered from guidance sessions.

While documents should be attached as PDFs, "*Many send screenshots directly from their phones, and that's not always the best solution. Sometimes it works, sometimes it doesn't. [...] As it is now, you never really know whether what's been submitted is readable or not.*" (FLE6). Occasionally, citizens need to submit additional documentation at a later stage, yet "*[How to upload] additional documentation is very difficult to find*" (FLE2). When citizens can't find the right function, they often submit new applications, creating additional work for caseworkers.

Development Teams Seeking Feedback. A second channel for accessing citizens' practice-based knowledge is when development teams reach out to them for feedback on specific systems and functions. Within NAV, "*[Development] teams are responsible for contact with the citizen and for gathering insight, and they are the ones who deliver services to the citizens*" (PL1). As development teams work on separate IT-solutions or services within NAV requiring different levels of citizen feedback, there is no standardized way of collecting feedback from the citizens. Thus, the extent of a development team's interaction with citizens differs across product areas.

The technical lead of one product team admits they do not have much communication with citizens and that it is also *"kind of an expensive way for [them] to get feedback"* (TL1). Their main strategy is to collaborate with the citizens' contact councils of organizations representing different user groups (e.g., visually impaired). There, the development teams can present new ideas or system developments and receive direct feedback. Other methods to get qualified input from citizens include Hotjar (customer-focused surveys), reaching out to friends of friends, and asking for feedback on early versions of what is being developed to see how they actually use the tools through analytics. A project focusing on changing the application process for unemployment benefits did a lot of preliminary work, such as *"interviews with users and user journeys. There have been surveys on nav.no with, like "answer this and answer that"*. (FLE10).

However, the technical lead points out that this feedback does not always match how the systems will be used in practice, "*In the long run, it's more important to look at what people do rather than only what they say*" (TL2), therefore other approaches like user tests may be more successful "*So there's a test lab where we showed customers (i.e., citizens) the application dialogue, [...], we could invite users in and have them filmed while they were doing it. And then we interviewed them [...] while they were sitting and going through the application, and then a bunch of us sat in a back room there and watched camera footage and listened to audio recordings [...]. And noted down all the feedback they had about whether there was a word they got stuck on that they didn't understand [...] and yes, everything really that could be relevant.*" (FLE10).

4.2 Knowledge Transfer Within the Organization

Once the citizens' practice-based knowledge has entered the organization, it must be transferred from the frontline to the development teams who are responsible for the digital systems. Knowledge can be transferred through formal and informal channels.

Formal Feedback Channels Tickets. One formal channel to transfer feedback across the organization is through so-called tickets. Tickets are digital text messages that can be submitted by employees in NAV directly to the development teams. They often consist of questions regarding specific functions in the system, change requests, or suggestions for improvement.

FLEs can open a ticket to the development team to report when the system does not work as expected.

Organizational Structure. In addition to tickets, NAV has formally established feedback structures through which development teams manage feedback on the systems they develop. This includes two semi-formally established roles: the implementation coordinator and change agents. The implementation coordinators manage a group of change agents who are members of local NAV offices and serve as a link between the development teams and the change agents representing local offices. While the change agents are usually FLEs who assist citizens, this feedback structure is primarily for internal NAV systems, e.g., case management systems, but can also be used for citizen-facing systems, e.g., the NAV website.

These designated roles (change agents and implementation coordinators) structure the feedback that goes to the development teams: "*It is mainly me, as a change agent and the implementation coordinator in the management unit, who has the direct dialogue with the product teams. (..) We are the mouthpiece for the product team, to keep the coordination as simple as possible, so that not everyone has to give their input directly to them. Then we collect it and send in a combined product from us (i.e., NAV contact center).*" (FLE9). Communication with the development team takes place via phone, scheduled meetings, e-mail/Teams messages, among others.

Contact Center. For general questions, citizens usually contact the contact center by phone or chat. They handle most incoming requests and often serve as a mediator between citizens and the development team by identifying issues arising from changes to Nav's website or service applications.

"*So, we are a kind of temperature gauge for the solutions that are put into practice, as we are the first to be contacted by a user if something is not working. [...]. So we are very concerned with being close to the product areas and product teams [...] that deliver the solutions, so we can provide our input and provide user insight. Because it is quite unique. We talk to many 1,000 users every single day, and I also experience that the product teams appreciate that we are involved and can provide input.*" (FLE9).

The number of questions about the use of digital systems has decreased in recent years, but inquiries usually increase after changes to the system or the launch of a new version.

Sometimes, development teams rely on contact center employees' access to citizens' experiences and practices when developing new systems or changing existing ones. "*When they (the development teams) have piloted a system, we are the ones who are contacted with a kind of user experience of that system. And then our job is to report it back to the developers [...] we collect the insights on behalf of others and can report it back to those who are creating the solution.*" (FLE9).

Informal Feedback Channels: Networks, Meetings, Personal Contact

In addition to formalized channels for knowledge transfer, more informal ones exist: "*We have networks at the office where we talk together, and there are systems and local resources, like people who work a bit more with IT, who make help sheets and guidance. [...] Locally, we don't really have anyone who is formally dedicated to IT [...] NAV hasn't defined or funded a specific responsible IT role at each office; instead, we rely on central user support.*" (FLE6).

Sometimes, developers who had previously been in contact with FLEs would call directly if they wanted specific feedback on something. "*Yes, it can happen that they (development team) call directly and want to have a chat. It is also quite natural that if there are very specific things that they want feedback on, then it is perhaps not so natural to bring it up in such a joint meeting where many others are present, but rather contact me or the implementation coordinator directly. [...] So we also have one-on-one contact with relevant people where there is a need for it.* (FLE9).

One domain expert working directly with the development teams emphasized the benefit of close proximity to their colleagues at the contact center: *"In addition to that, I try to listen to what they say, because I sit in the office [...] I sit at a contact center with those (working) on the phone, so I try to hear what they say in the hallways, which is what might be a little different than what they say to us through the formal feedback points. [...] because it may be that we haven't got any errors on something, or we haven't gotten any feedback in teams, but I hear something by the coffee machine, and then I think I've heard it for a while, and then I may be like, maybe we should have a user testing on this, or we should have a workshop on this.*" (SME).

These findings highlight the benefits of continuous contact and non-official channels between development teams and FLEs, who possess knowledge of citizens' feedback and needs.

4.3 Integrating Knowledge

Integrating Citizens' Direct Feedback. For some larger initiatives, citizens were invited to provide information directly, which led to significant changes. "*I was in a different team where we worked on the content on Nav.no. There, we based a lot of what was done on the preparatory work, on the questions and answers that were sent to users and user interviews. [...] we summarized that it is difficult to navigate, and when they do find it, they don't understand what is written. So there was a lot of work to do, but we chose first to looked at the structure, that is, get it in a logical structure so that what is relevant to your situation is found in a place that is logical to find it. So, we sorted a lot, deleted a lot of content, and rewrote some content.*" (FLE12).

Sometimes, gathering feedback can lead to abandoning development ideas, as the solution would not provide much additional value to citizens. "*We discussed creating a [...] page where people could follow the status of their cases and see their [application] history. For that, we recruited people we could collaborate with through user organizations [...] by having them come in for interviews and sometimes test the solution, where we observed them. Designers would sit with them and ask questions along the way, like why they did something, what they were thinking when they saw a page, and what they would expect to do* next *[...] but it turned out that the hypothesis didn't hold. Developing that solution wouldn't have given them much additional value*" (PL1).

Responsiveness. When FLEs provided feedback on errors, they might experience different levels of responsiveness from the development teams. In many cases, they receive confirmation and updates on their feedback. "*I find that most errors that we report are handled relatively quickly, but of course, it depends a bit on how precarious the errors are, and I fully understand that those who receive these errors also have to make their priorities. (...) I find that we get a status (update) in a way. (...) So, I find that they are good at keeping us in the loop once they have received the error and have to figure it out. And then we always get feedback that it is now solved.*" (FLE10).

However, on more complex inquiries, FLEs sometimes experience frustration when forwarding them to the development teams.

"I had a user who had been told to pay back much more money than she had actually received in her account, so I reopened the case. [...] We were supposed to look into it.

After three weeks, I got a response saying that they didn't know why the amount was the way it was, and the inquiry was closed again. I reopened the case several times. [...] In the end, I actually had to bring in network managers to take over the correspondence on this issue. Only then did they move forward and looked into how the system was built and try to figure it out." (FLE4).

Prioritization of Feedback. FLEs at the contact center noted that development teams would prioritize changes that might lead to neglecting citizens, as many systems were primarily for NAV's employees' internal use. "S*o, I think it's a bit easy to forget that at the other end of the case management system is an end user who will receive a decision and a payment. Then they are very concerned with creating a solution that works well for those who will process the case, and that is important. I completely agree with that. But we cannot forget that in that process, there are also external users who will receive the result of that case processing. So, it is just as important that they get a good experience of that case processing and what information they will have available based on that delivery.*" (FLE11).

Additionally, the development has been moving so fast that some citizens fall behind, and features that could help some citizens, like content in different languages, are not being prioritized: *"And maybe language is something that is not always prioritized as number one in product development. At the same time as working on product development and new information on nav.no, the development happens so smoothly and continuously that it may not be prioritized to create content in languages like English or Polish, or otherwise adapt it to the users."* (FLE7).

A product lead in a development team confirms that some services are prioritized over others, often due to outside pressure. "*What is difficult is that we all have different priorities, and especially in the [big] projects. They have received many millions from the ministry to create something very specific (...) they have a lot of eyes on them to deliver, to create value and create benefits, so there's that. Getting them to prioritize differently is extremely demanding, and I fully understand that it is like that, just to be clear, but that makes our job very difficult.*" (PL2).

As presented, several internal and external factors can influence the prioritization of feedback and therefore its integration.

5 Discussion

This paper addresses the problem of transferring and integrating citizens' practice-based knowledge into the development of digital public services. We apply a knowledge transfer perspective to show how developing organizations can acquire knowledge from citizens and integrate it into their development processes. With findings from a case study of the Norwegian Labor and Welfare Administration (NAV), we answer the following research question:

How is citizens' practice-based knowledge transferred and integrated into the development process of digital public services?

Our findings show how citizens' practice-based knowledge is acquired and transferred through various channels before reaching development teams, as well as the challenges associated with knowledge transfer. Knowledge transfer channels include direct interaction between citizens and FLE, both in person and by phone, as well as development teams actively reaching out to citizens to gather feedback. Further, knowledge is transferred formally and informally. The integration of knowledge can depend on development teams' responsiveness and prioritization of feedback. We will now discuss these findings by outlining how knowledge transfer occurs in NAV, highlighting the importance of practice-based knowledge and the role of citizens' knowledge in not only developing but also maintaining digital infrastructures for public services.

First, our findings give a comprehensive overview of how an organization can access citizens' knowledge through different channels. While literature on citizen involvement for system development often focuses on facilitated, participatory activities [20, 21], our findings show that knowledge is already transferred to the organization as citizens interact with FLEs. Direct interaction, as observed in the guidance sessions, gives FLEs deep insights into citizens' tacit knowledge of understanding and using digital solutions. They often know what citizens struggle most with when navigating the website or applying for services digitally, turning FLEs into knowledge brokers within the organization [28]. At the same time, FLEs see that easier and better solutions for citizens can reduce their own workload. These findings support earlier research that highlights FLEs as valuable resources for service development, due to their deep understanding of the organization as well as the service users' practice [12, 13].

The FLEs' knowledge and valuable insights have been further recognized by the development teams, as they actively reach out to caseworkers and contact center employees. The knowledge transfer from the frontline to the development teams can happen through formal and informal channels, as is also described in the literature [14]. Formal channels include a ticketing system and organizational routines for delivering feedback, while informal channels include networks and personal contact. While formal channels are important for collecting feedback, they also become more filtered and mediated as feedback is structured and combined along the way. Informal channels, on the other hand, are often used for very specific inquiries but can also lead to feedback by "chance" as conversations are overheard at the coffee machine (SME). This highlights the importance of encouraging informal channels and close contact between employees of different units.

Once the knowledge has reached the development teams, it can be codified and integrated into the digital system. Factors such as the organizational context and characteristics of the knowledge source and recipient can influence the knowledge transfer process [30]. Our findings show that citizens and FLEs feedback can lead directly to significant changes in the solution, as well as to the abandonment of development ideas. Sometimes, however, the feedback seems to get stuck. A development team's responsiveness can vary and may depend on the seriousness of the error, as well as its complexity and the effort required to fix it. Additionally, development teams set their own prioritization of handling the received feedback. As the majority of developed systems at NAV are for internal use, the feedback addressing external use, i.e., citizen-facing, may not be prioritized when it does not affect internal operations. External factors, such as public funding for some development projects, may further divert attention from other projects and thereby down-prioritize their maintenance and development.

Second, we want to highlight the importance of practice-based knowledge. The literature on system development recognized the importance of understanding the practices in which systems are embedded [9] but also the difficulties in accessing such practice-based knowledge, as it cannot simply be acquired through interviews or surveys [10, 11]. Our findings demonstrate that development teams in NAV have had experiences that align with this literature. The development teams have different strategies to collect insights directly from citizens, including customer-focused surveys, interviews, user journeys, or user tests. However, they realize that the feedback they receive does not always align with how the system will be used in practice, and that it is more valuable to look at what people do rather than what they say. Sometimes, user tests can simulate practice and yield relevant insights. A more accurate insight into citizens' use of the digital solution, however, is gained by FLEs who are contacted by citizens when they get stuck while using it. Our observations of guidance sessions and the interviews with FLEs revealed that, through their interactions with citizens, they gain insight into the citizens' practice-based knowledge.

Third, we want to point to the role of citizens in the development and maintenance of digital public infrastructures. Citizen involvement is often limited to idea generation and testing of prototypes, lacking evaluation at later stages [16, 17]. Digital systems, however, are constantly in need of revision and maintenance, especially in the public sector, where poorly developed systems can contribute to digital exclusion. Our findings show that citizens have valuable knowledge on the use of digital solutions, which can be used to improve the existing system. These improvements may not only lead to improved usability and access for citizens but also reduced workloads for FLE.

To conclude, our findings show that organizations can acquire citizens' practice-based knowledge through various channels and day-to-day activities, provided that formal and informal feedback structures are in place to support knowledge transfer from the frontline to the development team. Additionally, we highlight the value of practice-based knowledge over explicit knowledge, and the importance of citizens not only for the development but also for the maintenance of public digital infrastructures.

6 Limitations and Future Research

This study is not without limitations. While our study includes interviews and observations with citizens, FLE, and development teams, other relevant actors involved in the development of digital public services may be missing from our sample. We further assume that there may be additional strategies and channels for acquiring and transferring knowledge that are not represented in our data. Although the focus of this study is on citizens' knowledge, we acknowledge that the citizens' perspective on how their knowledge is accessed or used is not addressed.

Therefore, we recommend extending this study, including more actors and organizations, to further explore the topic of citizens' practice-based knowledge for the development of digital public services. Including a citizen perspective on the topic can give additional insights into the knowledge transfer process.

Acknowledgements. This research was funded by the Sustainable Digital Transformation (SustainDiT) project at NTNU as well as through the Norwegian Labor and Welfare Administration

(NAV). We further thank the anonymous reviewers for their constructive feedback, and all study participants who generously contributed through interviews, enhancing the quality and rigor of this study.

Disclosure of Interest. The authors have no competing interests to declare that are relevant to the content of this article.

References

1. Anthopoulos, L.G., Siozos, P., Tsoukalas, I.A.: Applying participatory design and collaboration in digital public services for discovering and re-designing e-Government services. Gov. Inf. Q. **24**(2), 353–376 (2007). https://doi.org/10.1016/j.giq.2006.07.018
2. Coles-Kemp, L., Ashenden, D., Morris, A., Yuille, J.: Digital welfare: designing for more nuanced forms of access. Policy Des. Pract. **3**(2), 177–188 (2020). https://doi.org/10.1080/25741292.2020.1760414
3. Castro, C., Lopes, C.: Digital government and sustainable development. J. Knowl. Econ. **13**(2), 880–903 (2022). https://doi.org/10.1007/s13132-021-00749-2
4. Torfing, J., Sørensen, E., Røiseland, A.: Transforming the public sector into an arena for co-creation: barriers, drivers, benefits, and ways forward. Adm. Soc. **51**(5), 795–825 (2019). https://doi.org/10.1177/0095399716680057
5. Twizeyimana, J.D., Andersson, A.: The public value of E-Government – a literature review. Gov. Inf. Q. **36**(2), 167–178 (2019). https://doi.org/10.1016/j.giq.2019.01.001
6. Madsen, C.Ø., Lindgren, I., Melin, U.: The accidental caseworker – How digital self-service influences citizens' administrative burden. Gov. Inf. Q. **39**(1), 101653 (2022). https://doi.org/10.1016/j.giq.2021.101653
7. Heggertveit, I., Lindgren, I., Madsen, C.O., Hofmann, S.: Administrative burden in digital self-service: An empirical study about citizens in need of financial assistance. In: Electronic Participation, EPART 2022. LNCS, vol. 13392, pp. 173–187. Springer, Heidelberg (2022). https://doi.org/10.1007/978-3-031-23213-8_11
8. Bratteteig, T., Wagner, I.: Unpacking the notion of participation in participatory design. Comput. Support. Coop. Work CSCW **25**(6), 425–475 (2016). https://doi.org/10.1007/s10606-016-9259-4
9. Suchman, L.: Supporting articulation work. In: Computerization and Controversy, pp. 407–423. Elsevier (1996). https://doi.org/10.1016/B978-0-12-415040-9.50118-4
10. Dougherty, D.: Organizing Practices in Services: Capturing Practice-Based Knowledge for Innovation (2004). https://doi.org/10.1177/1476127004040914
11. Orlikowski, W.J.: Knowing in practice: enacting a collective capability in distributed organizing. Organ. Sci. **13**(3), 249–273 (2002). https://doi.org/10.1287/orsc.13.3.249.2776
12. Engen, M., Magnusson, P.: Casting for service innovation: the roles of frontline employees. Creat. Innov. Manag. **27**(3), 255–269 (2018). https://doi.org/10.1111/caim.12263
13. Karlsson, J., Skålén, P.: Exploring front-line employee contributions to service innovation. Eur. J. Mark. **49**(9/10), 1346–1365 (2015). https://doi.org/10.1108/EJM-10-2012-0568
14. Liyanage, C., Elhag, T., Ballal, T., Li, Q.: Knowledge communication and translation – a knowledge transfer model. J. Knowl. Manag. **13**(3), 118–131 (2009). https://doi.org/10.1108/13673270910962914
15. NAV. Utbetalingene fra Nav økte med 51 milliarder kroner i 2024. nav.no. Accessed: 16 Mar 2026. https://www.nav.no/no/nav-og-samfunn/statistikk/flere-statistikkomrader/nyheter/utbetalingene-fra-nav-okte-med-51-milliarder-kroner-i-2024

16. Rodriguez Müller, A.P., Casiano Flores, C., Albrecht, V., Steen, T., Crompvoets, J.: A scoping review of empirical evidence on (digital) public services co-creation. Adm. Sci. **11**(4), 130 (2021). https://doi.org/10.3390/admsci11040130
17. Voorberg, W.H., Bekkers, V.J.J.M., Tummers, L.G.: A systematic review of co-creation and co-production: embarking on the social innovation journey. Public Manag. Rev. **17**(9), 1333–1357 (2015). https://doi.org/10.1080/14719037.2014.930505
18. European Commission. Access to essential services: key challenges for the most vulnerable – report - Employment, Social Affairs and Inclusion. Accessed 07 Oct 2025. https://employment-social-affairs.ec.europa.eu/news/access-essential-services-key-challenges-most-vulnerable-report-2023-06-28_en
19. Karlsson, F., Holgersson, J., Söderström, E., Hedström, K.: Exploring user participation approaches in public e-service development. Gov. Inf. Q. **29**(2), 158–168 (2012)
20. Holgersson, J., Karlsson, F.: Public e-service development: understanding citizens' conditions for participation. Gov. Inf. Q. **31**(3), 396–410 (2014). https://doi.org/10.1016/j.giq.2014.02.006
21. Axelsson, K., Melin, U., Lindgren, I.: Exploring the importance of citizen participation and involvement in e-government projects: practice, incentives, and organization. Transform. Gov. People Process Policy **4**(4), 299–321 (2010). https://doi.org/10.1108/17506161011081309
22. OECD. Focus on Citizens: Public Engagement for Better Policy and Services. OECD Stud. Public Engagem. (2009). https://doi.org/10.1787/9789264048874-en
23. Boulus-Rødje, N.: In search for the perfect pathway: supporting knowledge work of welfare workers. Comput. Support. Coop. Work CSCW **27**(3), 841–874 (2018). https://doi.org/10.1007/s10606-018-9318-0
24. Verne, G., Bratteteig, T.: Do-it-yourself services and work-like chores: on civic duties and digital public services. Pers. Ubiquitous Comput. **20**(4), 517–532 (2016). https://doi.org/10.1007/s00779-016-0936-6
25. Breit, E., Salomon, R.: Making the technological transition - citizens' encounters with digital pension services. Social Policy Administrat. **49**(3), 299–315 (2015). https://doi.org/10.1111/spol.12093
26. Jain, V.: Analyzing the nature of knowledge in the IS field. In: AMCIS 2003 Proceedings (2003). https://aisel.aisnet.org/amcis2003/374
27. Brown, J.S., Duguid, P.: Knowledge and organization: a social-practice perspective. Organ. Sci. **12**(2), 198–213 (2001). https://doi.org/10.1287/orsc.12.2.198.10116
28. Wenger-Trayner, E., Fenton-O'Creevy, M., Hutchinson, S., Kubiak, C., Wenger-Trayner, B. (eds.): Learning in Landscapes of Practice: Boundaries, Identity, and Knowledgeability in Practice-Based Learning. Routledge, London (2014). https://doi.org/10.4324/9781315777122
29. Patnayakuni, R., Rai, A., Tiwana, A.: Systems development process improvement: a knowledge integration perspective. IEEE Trans. Eng. Manag. **54**(2), 286–300 (2007). https://doi.org/10.1109/TEM.2007.893997
30. Szulanski, G.: Exploring internal stickiness: Impediments to the transfer of best practice within the firm. Strateg. Manag. J. **17**(S2), 27–43 (1996). https://doi.org/10.1002/smj.4250171105
31. regjeringen.no. Statsbudsjettet 2025: Statens inntekter og utgifter. Regjeringen.no. Accessed 16 Mar 2026. https://www.regjeringen.no/no/statsbudsjett/2025/statsbudsjettet-2025-statens-inntekter-og-utgifter/id3055676/
32. Statskonsulent. Strategier og planer for innovasjon og fornyelse i offentlig sektor (2006). Accessed 08 Oct 2025. https://kudos.dfo.no/documents/9640/files/9536.pdf
33. Mikalef, P., Parmiggiani, E. (eds.): Digital Transformation in Norwegian Enterprises. Springer, Cham (2022). https://doi.org/10.1007/978-3-031-05276-7

34. NAV. NAV Årsrapport 2014 (2015). Accessed 21 Mar 2023. https://www.nav.no/_/attachment/inline/bb396e02-a59f-4ed4-b6aa-6f4ee1ee8a22:d6304522cfd8b257c5a2becdb36884e3b728f46a/arsrapport-2014.pdf
35. Ringnes, T.I.F.: Store omstillinger ved halvparten av NAV-kontorene. Accessed 08 Oct 2025. https://memu.no/artikler/store-omstillinger-ved-halvparten-av-nav-kontorene/
36. Saeed, M.: Reduserte åpningstider – til brukernes beste? (2019). Accessed 08 Oct 2025. https://oda.oslomet.no/oda-xmlui/bitstream/handle/10642/7732/Saeed_Moniba.pdf?sequence=2
37. Walsham, G.: Interpretive case studies in IS research: nature and method. Eur. J. Inf. Syst. **4**(2), 74–81 (1995). https://doi.org/10.1057/ejis.1995.9
38. Thompson, J.: A guide to abductive thematic analysis. Qual. Rep. **27**(5), 1410–1421 (2022). https://doi.org/10.46743/2160-3715/2022.5340

AI-Mediated Presentation of Persistent Government Health Data: Effects on Usability, Cognitive Load, and Task Accuracy

Vasiliki Siougioudi(✉) and Teta Stamati

Harokopio University of Athens, Athens, Greece
vsiougioudi@hua.gr

Abstract. As governments across Europe expand citizen access to electronic health records, access alone does not guarantee comprehension. This study asks whether AI-mediated presentation of the same persistent government health data produces better outcomes than a conventional EHR interface, and whether this matters differently for different user groups. A within-subject experiment with 36 participants across three groups (clinicians, health AI/IT professionals, and non-expert citizens), was conducted and each evaluated a mock EHR and an AI-mediated presentation of the same synthetic patient records. AI-mediated presentation produced statistically significant advantages in usability, cognitive load, and task accuracy. Medium-to-large effect sizes were also reported, along with strong Bayesian support for the two primary outcomes. Citizens reported the largest gains. Qualitative findings also revealed that trust in AI emerged as a distinct dimension. Presentation format appears to be a variable that may affect who can effectively use citizen-facing government health data.

Keywords: Electronic Health Records · AI-Mediated Presentation · Usability · Cognitive Load · eGovernment

1 Introduction

One of the most consequential forms of citizen-facing government data is electronic health records (EHRs). They accumulate over a lifetime span, are maintained by public institutions, and are increasingly mandated by law to be accessible to citizens. The European Commission's Digital Decade Program sets a target of 100% citizen access to EHRs by 2030. This target is largely being met at infrastructure level, as 83% of EU member states provided some form of access by the end of 2023 [1]. The Greek National Electronic Health Record (NEHR), was launched in March 2025 by the Ministry of Health and HDIKA. In December 2025, an AI-powered citizen health assistant (myHealth Agent) was introduced, constituting one of the first national EHR systems in Europe to embed an AI presentation layer directly into citizen-facing health data access [2]. This study was conceptually motivated by this development. The access indicators do not measure or study whether citizens make sense of what they access. Research on

Published by Springer Nature Switzerland AG 2027
D. Duenas-Cid et al. (Eds.): ePart 2026, LNCS 16923, pp. 114–129, 2027.
https://doi.org/10.1007/978-3-032-36676-4_8

patient portals consistently indicates that health literacy and digital literacy are stronger engagement predictors than access itself. Also, comprehension is a binding constraint for non-expert users [3, 4].

The question of how government data is presented has received little systematic attention. EHR usability literature has established that presentation format affects both cognitive load and task performance in clinical populations [5], but has not examined non-expert users. Emerging AI literature on AI-mediated clinical summarization has focused on accuracy and clinical efficiency [6], leaving open the question of whether AI-mediated presentation changes the citizens' outcome. This study addresses that gap.

Using a within-subject design with 36 participants across three groups (clinicians, health AI and IT professionals, and non-expert citizens), we compare a mock EHR interface with an AI-mediated presentation of the same synthetic records on usability, cognitive load, and task accuracy. The study is positioned as Phase 1 of a two-stage program, as findings are treated as hypothesis-generating for the between-group comparison.

2 Literature Review

2.1 Persistent Government Data and the Limits of Access as a Policy Goal

Electronic health records (EHRs) are a form of data managed by governments that accumulates over time, is maintained by public health institutions, and is accessed by multiple actors for multiple different purposes (clinicians, patients, system administrators, and researchers). In this study, EHRs are treated as state-managed information infrastructures as they are owned by public institutions, carry obligations of access and accountability to citizens, subject to the same design pressures as any citizen-facing government data service. This type of data is especially challenging because of the need for the same record being utilized by all actors, despite significant differences in their background prior knowledge and familiarity with clinical terminology. This study treats EHRs as a representative case of what we term persistent government data: information growing longitudinally, serving multiple actor categories simultaneously, while remaining institutionally managed.

Open government data literature has focused primarily on data availability, treating access as a central policy goal. Janssen et al. (2012) [7] challenged this assumption early on, as the study showed that availability alone does not produce benefits. Adoption and understanding depend on factors that the access-oriented approach has ignored in the past. What these indicators do not measure, is whether citizens are able to make sense of what they access. Lindgren et al. (2019) [8] identified this gap between service delivery and user experience as one of the main issues in public service digitalization. The present study is concerned with this gap, in the specific context of health data.

2.2 EHR Usability: What We Know

A substantial body of research has examined how usable EHRs are for clinical professionals. The findings seem to be consistent: EHR systems as currently designed, impose cognitive burden on their users. Poor interface design contributes to information overload, clinical errors, and burnout among healthcare professionals [9]. Mazur et al., (2019) [5], compared two versions of an EHR in a controlled experiment - one with standard usability and one enhanced with targeted improvements – and measured cognitive workload utilizing the NASA Task Load Index. The enhanced interface produced a 30%-point increase in performance and significantly lower cognitive load. The key finding is that same clinical data, if presented differently, produce different outcomes.

The limitation of this literature is its prime focus on clinical professionals. This is a reasonable starting point, however, as more countries implement citizen access to health records as a policy target, there is another question becoming equally important: how non-expert users interact with the same data?

2.3 When Citizens Access Their Health Records

Research on patient portals (the systems through which citizens access their health records), demonstrate a different story. The study of Irizarry et al. (2015) [3], reviewed 120 studies and showed that health literacy and digital literacy were the strongest predictors of whether patients engaged with portals and benefited from them. Despite availability, only 15–30% use portals routinely [4] while the most common barrier is comprehension rather than access [5]. Patients could not understand the medical content they had been provided with. These studies suggested that access has been insufficient when records remain expressed in clinical language or organized through formats designed for professional use.

2.4 Cognitive Load Theory

Theoretically, the explanation on the reason the presentation format affects comprehension, comes from Cognitive Load Theory [12]. The core idea describes how human working memory has limited capacity, and any information environment makes demands on this capacity. Demands rise both from content complexity and from how content is organized. Cognitive Load Theory distinguishes between extraneous load, imposed by poor information design, and germane load, supporting schema formation and deep understanding. Sweller et al. (2019) [13] reviewed thirty years of evidence confirming this framework. In the present context, the relevance of Cognitive Load Theory lies in the possibility that AI-mediated summarization may reduce load by reorganizing information into a more interpretable form. Usability is measured using the System Usability Scale [14] and cognitive load using the unweighted Raw NASA-TLX [16].

2.5 AI-Mediated Presentation of Clinical Data

A growing body of research has examined whether Large Language Models (LLMs), are able to present clinical information more effectively than structured EHR formats. LLM-generated summaries of patient records are comparable in quality to summaries written by physicians, but in this case with a substantial time advantage [17, 18]. Bednarczyk et al. (2025) [6] reviewed the field systematically, and noted that existing studies focus almost exclusively on accuracy and efficiency as assessed by clinicians. However, evidence on how non-expert users experience AI-mediated clinical information is nearly absent.

The most relevant study for this present work is that of Stadler et al. (2024) [19]. The research compared LLMs with traditional search engines in a controlled experiment with 91 university students who completed a health information task. Students that used LLMs reported significantly lower cognitive load. However, they also produced lower-quality reasoning. A theoretical tension is created, as the study suggests that we would expect lower load and better performance to be correlated positively. These findings however suggest that this relationship is not automatic, but rather depends on whether the cognitive load reduction results from clearer information organization or from oversimplification that removes meaningful complexity. In this present study, we measure usability, cognitive load, and task accuracy simultaneously, allowing direct examination whether the two outcomes move together or diverge. Rebitschek et al. (2025) [20] studying 300 lay participants using LLMs for health information, found that output quality depended strongly on how well users could formulate their questions (and that non-expert users were consistently poor at this). The present study removes this variable, as four fixed expert-authored prompts are being employed, in order to compare presentation format itself rather than the quality of user prompting.

2.6 Trust and Multi-stakeholder Access in eGov

In eGovernment contexts, AI introduction in citizen-facing services raises questions beyond usability. When public data access is AI-mediated, the questions concern who controls the representation, on what basis, and with what accountability. Perceived trustworthiness is a primary predictor of citizen engagement with AI-driven government interfaces, alongside accessibility [21]. Digital administrative burden falls disproportionately on less digitally experienced [22]. Alon-Barkat & Busuioc (2023) [23] found that in public sector decision-making contexts users seem to over-rely on AI output when it is aligned with their expectations and disengaged when it is not. In health data contexts, this suggests direct governance implications, as an AI layer that renders records more accessible may simultaneously affect how citizens perceive their own health situation in manners they cannot verify or contest. The presentation layer here is not neutral and acts as an interpretive intervention by the state on citizen-facing public data, while carrying the same accountability obligation as any other form of government communication available. Accountability obligation extends to algorithmic transparency, as citizens interacting with AI-mediated public data have a legitimate interest in understanding how these outputs are produced, and also have the right to contest them. Wachter et al. (2017) [29] identify this right to explanation as a core requirement for algorithmic accountability. Diakopoulos (2015) [30] frames transparency as a prerequisite for democratic legitimacy in decision systems containing automations.

2.7 What Remains Unknown

To the best of the authors' knowledge, no controlled experiment has compared a conventional EHR interface with an AI-mediated presentation of the same underlying health data on usability, cognitive load, and task accuracy combined, across a sample of non-expert citizens alongside clinical professionals. EHR usability literature [5] and LLM cognitive load literature [19] address adjacent questions in separate contexts, with separate populations, without resolving tensions between their findings.

This study addresses the question of whether an AI presentation layer applied to government-managed health data produces effects in usability, cognitive load, and task accuracy compared to a conventional EHR interface, and whether these differences are consistent across user groups with different levels of clinical and AI expertise. Governments across Europe are investing in citizen access to persistent public data (health records, tax histories, social benefit records), assuming that engagement is a by-product of access and value is a by-product of engagement. This study examines whether the representation layer determines who is actually able to utilize this access.

3 Methodology

3.1 Research Design

This study employed a mixed-methods, within subject comparative design to examine how alternative representation layers applied to the same persistent government data affect perceived usability, cognitive workload, and user sensemaking.

Two presentation formats were compared, while the underlying patient information was kept constant. In Condition A, the information was presented through a mock electronic medical record interface, organized in a conventional record-based structure. Condition B presented the same information through AI-mediated responses derived from the same patient record, simulating an AI-agent tool employed in a hypothetical electronic health record interface. The patient data were held constant while the representational interface was the only variable, allowing the examination of the effects of presentation format on user interaction and comprehension. The use of static outputs was deliberate to isolate presentation format from confounds (e.g., prompt variability).

The study used a within-subject structure so each of the 36 participants evaluated both conditions. The questionnaire was distributed in two versions in order to mitigate potential order effects. In the first version, the sequence was A- > B, while in the second it was B- > A. The distribution among stakeholder groups was approximately balanced: clinicians and health workers, as well as health AI/ health informatics were allocated evenly (6/6) while citizens were distributed 7/5.

The evaluation combined four complementary components: task-based comprehension questions, perceived usability measurement using the System Usability Scale (SUS), cognitive workload measurement using the Raw NASA-TLX Task Load Index (NASA-TLX), and open-ended questions derived from sensemaking theory and relevant literature to examine how participants interpreted the presented information.

3.2 Study Materials

The study was situated in a digital public health context and was conceptually informed by the structure and interaction logic of an already operational electronic health record environment, the Greek national health record setting, serving as initial practice study reference point. The direct use of a live institutional system was not feasible for controlled experimentation, resulting in the development of purpose-built mock interfaces.

Two patient records were derived from Synthea, an open-source synthetic health record database, to ensure realistic and internally consistent patient data. Each record included demographic information, chronic conditions, encounter details, medication and prescription related information, and alerts or warnings (where applicable).

Two representations of the same underlying data were created; in Condition A, the patient information was displayed through a mock electronic record user interface organized into structured sub-sections that included patient details, encounters, conditions, medication and alerts. In Condition B, the same data were represented through outputs of four fixed prompts generated by a general use LLM (GPT-4.1), simulating an AI-agent embedded in the EHR system. The model was prompted with the following questions:

- Show the patient's chronic conditions.
- What is the patient's current hospital encounter or admission?
- List the medications recorded in the patient's record.
- Are there any alerts or important warnings in the patient's record?

The aim was to isolate the information while avoiding additional interaction variables such as system latency or prompt formulation.

Prior to distribution, the questionnaire underwent expert review by digital health and health informatics specialists. The review assessed patient record clarity, the relevance of the task-based questions, and the overall survey instrument structure. Feedback derived from this process informed minor refinements to wording and instructions so that the tasks accurately reflected the elements presented.

3.3 Participants and Recruitment

Participants were recruited during February 2026 using purposive sampling targeting predefined stakeholder groups that are normally involved with the digital health systems. Recruitment was done through professional networks, collaborators, and associated contacts employed in the healthcare, digital health, and AI development industries.

The survey was conducted online using Google Forms and remained open for ten days. The final sample consisted of 36 participants, distributed among three stakeholder groups: clinicians and health professionals ($n = 12$), citizens/ non-expert users ($n = 12$), health AI and health informatics professionals involved in similar systems' design and development ($n = 12$). Participation in the survey was voluntary and anonymous.

3.4 Procedure

The questionnaire was created around two synthetic patient records. For every case, participants were shown both representational formats according to the order assigned in each counterbalanced questionnaire version.

Participants first viewed the presentation format and then completed four task-based comprehension questions designed to assess whether key factual case elements could be identified. The tasks were derived from the underlying patient record, independently from the prompts and they focused on information identification, rather than clinical judgement.

Following the task questions, each participant completed the System Usability Scale (SUS) and the unweighted Raw NASA-TLX Load Index (NASA-TLX) referring to each condition they had just interacted with. Then, the same sequence followed for the alternative condition. This procedure was repeated for both patient records.

Completing the questionnaire required 12–15 min based on pilot testing and it was conducted in the Greek language, corresponding to the participants' native language.

3.5 Measures

Perceived Usability was measured using the SUS Scale [14]. The participants were asked to complete the ten-item instrument after each condition with a five-point Likert scale. Scores were calculated according to standard procedure while internal consistency was assessed using Cronbach's alpha.

Cognitive Workload was measured with the unweighted Raw NASA-TLX [16] scale. Participants rated six dimensions on a 0–10 scale.

Comprehension tasks were employed to examine how participants identify key factual elements from both presentations. Their responses were coded as correct or incorrect to form a total accuracy score per condition. Correct answers were determined against a predefined answer key derived from the underlying patient records. Each task had one correct answer corresponding to factual information present in both formats.

Open-ended questions were included to identify how participants described their experience corresponding with the information presented, and which features of the data supported higher or lower understanding.

3.6 Data Analysis

Quantitative analysis focused on comparisons between the two formats presented. SUS, NASA-TLX and task accuracy scores were summarized with the use of descriptive statistics.

Normality was assessed prior to analysis. Paired t-tests were used where normality held; On the other hand, where the assumptions were not met, the non-parametric Wilcoxon signed-rank test was applied. Effect sizes were also calculated as to assess the magnitude of the differences observed. Bayesian paired-samples t-tests were additionally conducted using a Cauchy prior ($r = 0.707$). As an indicative analysis, Spearman correlations were computed between prior familiarity scores and condition difference scores.

Regarding the stakeholder groups, this research examines them descriptively to provide contextual insight on how different actor populations approach persistent records. However, given the limited subgroup sizes, the analysis does not attempt to draw statistically generalizable conclusions between the groups.

The participants' responses to the open-ended questions were analyzed with inductive thematic analysis with iterative coding. Repeated reading of the answers resulted in initial codes, which were then grouped into broader themes describing recurring patterns.

4 Results

4.1 Participants and Data Quality

All 36 participants completed both conditions and there were no dropouts. The sample included clinicians ($n = 12$), health AI/IT professionals ($n = 12$), and citizens ($n = 12$). Counterbalancing was applied (AB: $n = 19$; BA: $n = 17$).

4.2 Instrument Reliability

SUS internal consistency was assessed separately for each patient case and condition, before aggregations. Cronbach's alpha ranged from .878 to .945. Reverse-scored items (2, 4, 6, 8, 10) were recoded before estimation. All four values exceed the threshold for good reliability [28].

4.3 Descriptive Statistics

Scores across the two patient cases were aggregated. Descriptive statistics are demonstrated in Table 1.

The AI condition shows advantages on all three outcomes. SUS distributions are narrower (AI SD = 14.85 vs EHR SD = 19.90), and task accuracy clusters near ceiling ($Mdn = 1.00$), under the AI condition (Fig. 1).

Table 1. Descriptive Statistics ($N = 36$)

Variable	N	Mean	SD	Median	IQR	Range
A_SUS (EHR)	36	66.08	19.90	72.50	28.44	25.00–100.00
B_SUS (AI)	36	77.74	14.85	81.25	19.38	43.75–98.75
A_TLX (EHR)	36	3.896	0.931	3.875	1.125	2.333–6.500
B_TLX (AI)	36	3.468	0.983	3.375	0.958	1.667–7.250
A_tasks (EHR)	36	0.885	0.132	0.875	0.167	0.583–1.000
B_tasks (AI)	36	0.938	0.097	1.000	0.135	0.708–1.000

Note. SUS = System Usability Scale (0–100); TLX = Raw NASA Task Load Index (1–10); Tasks = mean proportion correct (0–1)

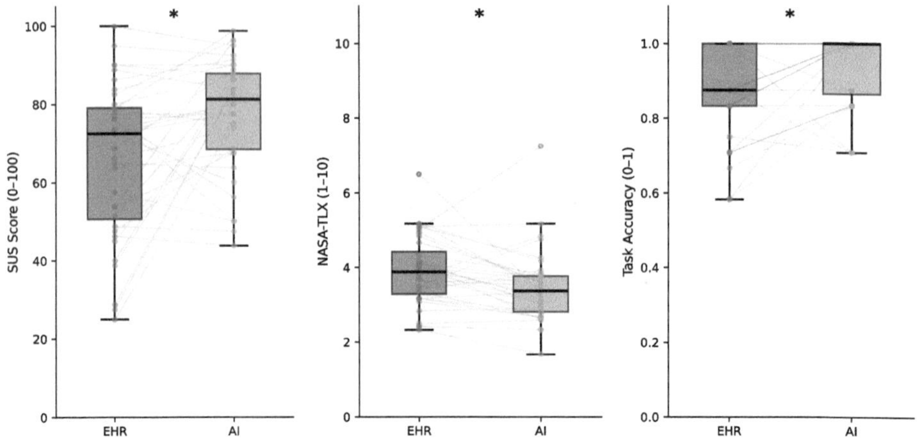

Fig. 1. EHR vs. AI-mediated Presentation across usability (SUS), cognitive load (NASA-TLX), and task accuracy. Individual data points shown with paired observations connected. * $p < .05$.

4.4 Normality Assessment

Shapiro- Wilk tests on difference scores (B – A) demonstrated deviations from normality for SUS ($W = 0.905$, $p = .005$) and task accuracy ($W = 0.886$, $p = .001$). Regarding NASA-TLX, the difference score was normally distributed ($W = 0.954$, $p = .139$). Wilcoxon signed-rank tests were used for SUS and task accuracy; a paired samples t-test was used for cognitive load.

4.5 Main Comparisons

Results are shown in Table 2. The three outcomes showed statistically significant differences in favor of the AI-mediated condition, with medium-to-large effect sizes across usability, cognitive load, and task accuracy (order effects were non-significant, all $p >$ .080).

Table 2. Normality Tests and Paired Comparisons

Comparison	S- W W	p (S – W)	Test	Statistic	p-value	ES	95% CI (ES)
A_SUS vs B_SUS	0.905	.005	Wilcoxon	W = 132.0	.005	r = .56	[.24, .77]
A_TLX vs B_TLX	0.954	.139	Paired t-test	t (35) = 2.76	.009	d = .45	[.10, .79]
A_tasks vs B_tasks	0.886	.001	Wilcoxon	W = 35.5	.029	r = .59	[.14, .83]

Note. SUS = System Usability Scale (0–100); TLX = Raw NASA Task Load Index (1–10); Tasks = mean proportion correct (0–1)

Bayesian paired-samples t-test using a Cauchy prior ($r = 0.707$) provided strong evidence for the usability difference ($BF_{10} = 14.34$) and moderate evidence for the cognitive load difference ($BF_{10} = 4.57$). Evidence for task accuracy difference was anecdotal ($BF_{10} = 2.37$), a result consistent with the variability observed for this outcome. This suggests that a larger sample is needed in order to establish this effect with confidence.

Medians of both SUS conditions exceed the conventional "acceptable" threshold of 70 [15], and task accuracy under the AI condition reached perfect median score.

4.6 Role of Prior Familiarity

As an indicative analysis of the relationship between prior familiarity and condition difference scores, Spearman correlations were computed. Given the sample size ($N = 36$), these findings should be interpreted with caution and treated as hypothesis-generating and not confirmatory. Difference scores also reveal a strong negative correlation between SUS and NASA-TLX ($\rho = -.770$, $p < .001$), suggesting usability and cognitive load improvements co-vary substantially – a relationship Phase 2 should examine more formally (Table 3).

Table 3. Spearman Correlations: Familiarity and Condition Difference Scores

Variable	fam_EHR	fam_AI	diff_SUS	diff_TLX	diff_tasks
fam_EHR	–				
fam_AI	.255	–			
diff_SUS	−.099	.472**	–		
diff_TLX	.154	−.346*	−.770***	–	
diff_tasks	−.018	−.057	.050	.233	–

*Note. fam_EHR = EHR familiarity (1–5); fam_AI = AI/LLM familiarity (1–5); diff = B – A. * p < .05, ** p < .01, *** p < .001*

AI familiarity was associated positively with the SUS difference score ($\rho = .472$, $p = .004$) and associated negatively with the TLX difference score ($\rho = -.346$, $p = .039$). There was no significant association found with task accuracy score ($\rho = -.057$, $p = .742$). Finally, EHR familiarity was not associated significantly with any outcome (all $p > .36$). This pattern indicates that prior experience with AI tools does not affect task performance, but rather amplifies the benefits of the AI-mediated condition on usability and workload.

4.7 Exploratory Descriptive Analysis by Profession Group

Given the three equally stratified groups ($n = 12$), a descriptive breakdown was conducted. Due to insufficient per-group power, between-group inferential tests were not performed.

The largest SUS gain was shown in citizens ($\Delta = +12.8$ points), followed by clinicians/health workers (+11.5) and health AI/IT professionals (+10.6). In cognitive load measurements, health AI/IT professionals showed the greatest reduction (ΔTLX = -0.53), compared to citizens (-0.48) and clinicians/health professionals (-0.27). The gains of task accuracy were the largest for citizens (+7.3 percentage points; clinicians/health workers and Health AI/IT: 4.2 pp each). These descriptive patterns motivate targeted-between group comparisons in Phase 2.

4.8 Qualitative Findings: Thematic Analysis

Five open-ended questions were included in the questionnaire to capture navigation strategies and/or format preferences. The questions were optional, which resulted in varying response rates (Q1–2: $n = 32$; Q3: $n = 25$; Q4: $n = 31$; Q5: $n = 35$). The questions were informed from sensemaking research (cue extraction, bracketing, frame building) [25, 26], but they were not designed as a formal sensemaking instrument. After the completion of the questionnaire, the responses were analyzed with the thematic analysis methodology of Braun and Clarke (2006) [27]. The analysis resulted in five themes. All quotes below are presented in the original Greek format with English translations in brackets.

Theme 1: Medications as primary focus
The most consistently mentioned items across Q1 and Q2 by all professional groups were medications and chronic conditions. Clinicians linked medications to diagnosis:

> *"τη φαρμακευτικη αγωγης για να εντοπισω χρονια νοσηματα"* [the medication list in order to identify chronic conditions]

Theme 2: Holistic vs. sequential reading strategies
Most participants had combined approaches when forming an overall opinion (Q3). One participant also noted that the AI-mediated format guided them step-by-step, while at the same time the EHR drew attention to another elements:

> *"Στην περίπτωση AI πήγε με τη σειρά ενώ στην 2η περισσότερα στα Alerts λόγω χρώματος"* [In the AI case I went through the data in order, while in the second I focused more on Alerts because of the color]

Relevantly, sequential reading may reduce the risk of overlooking information, while attention-driven scanning depends on interface design choices.

Theme 3: Low perceived difficulty
19 out of 31 participants reported no difficulty in completing the tasks (Q4). However, every time the difficulty was mentioned, the reason for visit was the common item.

> *"Αν προκειται για επείγουσα ή τακτική"* [Whether it was an emergency or routine visit]

Theme 4: AI preferred; EMR for at-a-glance overview

Almost half of the participants (18/36, 50%) preferred the AI-mediated format. They also cited sequential structure and easier use.

> *"η δευτερη, πιο ξεκαθαρα παρουσιασμενη στοιχεια για γρηγορη αντιμετωπιση των προβληματων"* [the second view, more clearly presented information for faster problem management]

Seven participants preferred the EHR version for its consolidated view.

> *"Στην πρώτη οι πληροφορίες ήταν μαζεμένες και βοηθούσαν στη δημιουργία συνολική εικόνας"* [In the first, the information was consolidated and helped in building a fuller picture]

Three participants proposed a combined workflow: AI for navigation and EHR for verification.

> *"Τα chatbots – μπορώ να ρωτάω ακριβώς αυτό που θέλω. Πρέπει να λειτουργούν συνδυαστικά με το φάκελο"*[Chatbots- I can ask exactly what I want. They need to work in combination with the record]

Theme 5: Trust and algorithmic transparency
Two participants raised concerns about trusting AI output, without knowing the algorithm deployed.

> *"φοβαμαι ότι επειδη ο αλγοριθμος δεν είναι γνωστος φοβαμαι την παρουσιαση των ντατα αν θα είναι ορθα"* [I am afraid that because the algorithm is not known, the data presentation may not be accurate]

A third framed the EHR as a check on AI-informed deployment:

> *"η σύγχρονη καταγραφή μπορεί να σε κατευθύνει στο να λάβεις απόφαση χωρίς να είναι σωστή – θα πρέπει να είναι περισσότερο εργαλείο όταν έχεις βγάλει συμπέρασμα"* [Modern recording methods may lead you into making a wrong decision – it should be more of a tool once you have already come to a conclusion]

These concerns appear independently of format preference, while trust appears to be a distinct dimension from perceived usability.

5 Discussion

This study's results are consistent across all three outcome measures: AI-mediated presentation of the same persistent health data produces higher usability, lower cognitive load, and higher task accuracy, in regard to the conventional EHR interface. The effect sizes are medium-to-large ($r = .56$ for SUS, $d = .45$ for NASA-TLX, $r = .59$ for task accuracy), and the Bayesian analysis provided strong supporting evidence for the usability and cognitive load differences. This pattern is consistent with the core prediction of the Cognitive Load Theory: the same data, structured in a different manner, produces different cognitive outcomes. This finding also extends the work of Mazur et al. (2019) [5], by demonstrating that a qualitatively different presentation format produces comparable effects across a mixed, non-clinical sample. Critically, the concerns of the study of Stadler et al. (2024) [19] that AI-mediated presentation reduces load at the cost of reasoning quality is not borne out. Accuracy improved alongside load reductions, suggesting that the AI format in this context reduced extraneous rather than germane cognitive load.

The descriptive analysis by profession group adds a finding that, while not inferentially testable with this sample size, aims to be policy-relevant. Descriptively, citizens showed the greatest gains across all three outcomes (most notably in task accuracy), though this pattern should be interpreted with caution, as part of the advantage may reflect language accessibility differences between conditions rather than presentation format alone. This is consistent with the patient literature's well documented comprehension gap between clinical professionals and non-expert users [4, 5] while suggesting that the AI presentation layer may function as a partial equalizer across actor groups. If this pattern holds at scale in Phase 2, it may imply that the representation layer may be a key variable in determining whether citizen access to government health data translates into meaningful use. This is a governance question, as decisions on how public data is presented carry distributional consequences on who is able to participate in their own health management.

The qualitative findings introduce a dimension the quantitative instruments do not capture. Participants who preferred the AI-mediated format and performed better under it, still expressed concerns about its output reliability and algorithmic opacity. Trust in AI-generated government data appears as a distinct construct, as users seem to find a system easier to use but at the same time doubt whether what it shows them is accurate. This finding is consistent with Alon-Barkat & Busuioc (2023) [23] on automation bias in public sector AI, and aligns with algorithmic transparency literature suggesting that trustworthiness warrants formal measurement as an independent construct in Phase 2 [29, 30]. Moreover, it has direct implications for deployment: usability gains alone are not enough for adoption in a context where misunderstanding stakes are high. Necessary conditions for responsible deployment include transparency on how AI-summaries are produced, and mechanisms for citizens to verify, or even contest them.

5.1 Theoretical and Practical Contributions

This study makes two main contributions. Theoretically, it provides controlled empirical evidence that AI-mediated presentation of persistent government health data improves usability, reduces cognitive load, and increases task accuracy across a multi-stakeholder sample that includes non-expert citizens. The findings are consistent with Cognitive Load Theory into a citizen-facing government data infrastructure and support treating the representation layer as a variable separable from data content and access. Practically, the findings suggest that governments invest more in citizen access to EHRs and treat presentation format as a design and possible policy decision with distributional consequences.

5.2 Limitations

Several limitations apply. Firstly, the sample ($N = 36$) supports paired comparisons but still is insufficient for inferential between-group analysis. As a result, the profession group findings are descriptive. Thirdly, the participants were recruited purposively, which may overrepresent digitally literate citizens, comparing to the general population. Moreover, the study used a single AI model (GPT – 4.1) generating fixed prompts and static screenshots. This limits ecological validity and generalizability to other models. Finally, the study was conducted in Greek with Synthea-generated records containing English medical terminology. In Condition B, the LLM rendered clinical terms into accessible Greek, while Condition A maintained the original English terminology. This introduces a language accessibility confound that may have inflated the AI-condition advantage for non-expert citizens. Phase 2 should control this by using source records with consistent language use.

5.3 Future Work

A second study with a larger, stratified sample is being prepared to support inferential comparisons and to examine whether the results observed here are robust across diverse populations. Future work should also examine whether the usability advantage of AI-mediated presentation of public data is able to be deployed responsibly.

6 Conclusion

This study provides empirical evidence that AI-mediated presentation of persistent health data improves usability, reduces cognitive load, and increases task accuracy relative to conventional EHR interface, across a diverse sample of expert and non-expert actors. The findings are exploratory at this scale, while suggesting that how government data is presented matters, and in a different manner for diverse user groups.

Disclosure of Interests. The authors have no competing interests to declare that are relevant to the content of this article.

References

1. European Commission: Digital Decade 2024: eHealth Indicator Study. https://digital-strategy.ec.europa.eu/en/library/digital-decade-2024-ehealth-indicator-study. Accessed 14 Mar 2026
2. HDIKA: National Electronic Health Record—MyHealth Agent. https://www.ehealthrecord.gov.gr/. Accessed 14 Mar 2026
3. Irizarry, T., Dabbs, A.D., Curran, C.R.: Patient portals and patient engagement: a state of the science review. J. Med. Internet Res. **17**(6), e4255 (2015)
4. Lyles, C.R., Nelson, E.C., Frampton, S., Dykes, P.C., Cemballi, A.G., Sarkar, U.: Using electronic health record portals to improve patient engagement: research priorities and best practices. Ann. Int. Med. **172**(11 Suppl), S123–S129 (2020)

5. Mazur, L.M., Mosaly, P.R., Moore, C., Marks, L.: Association of the usability of electronic health records with cognitive workload and performance levels among physicians. JAMA Netw. Open **2**(4), e191709 (2019)
6. Bednarczyk, L., et al.: Scientific evidence for clinical text summarization using large language models: scoping review. J. Med. Internet Res. **27**(1), e68998 (2025)
7. Janssen, M., Charalabidis, Y., Zuiderwijk, A.: Benefits, adoption barriers and myths of open data and open government. Inf. Syst. Manag. **29**(4), 258–268 (2012)
8. Lindgren, I., Madsen, C.Ø., Hofmann, S., Melin, U.: Close encounters of the digital kind: a research agenda for the digitalization of public services. Gov. Inf. Q. **36**(3), 427–436 (2019)
9. Middleton, B., Sittig, D.F., Wright, A.: Clinical decision support: a 25 year retrospective and a 25 year vision. Yearb. Med. Inf. Suppl **1**, S103-116 (2016)
10. Asgari, E., et al.: Impact of electronic health record use on cognitive load and burnout among clinicians: narrative review. JMIR Med. Inf. **12**(1), e55499 (2024)
11. Ancker, J.S., et al.: Use of an electronic patient portal among disadvantaged populations. J. Gen. Intern. Med. **26**(10), 1117–1123 (2011)
12. Sweller, J.: Cognitive load during problem solving: effects on learning. Cogn. Sci. **12**(2), 257–285 (1988)
13. Sweller, J., van Merriënboer, J.J.G., Paas, F.: Cognitive architecture and instructional design: 20 years later. Educ. Psychol. Rev. **31**(2), 261–292 (2019)
14. Brooke, J.: SUS: a quick and dirty usability scale. In: Jordan, P.W., Thomas, B., Weerdmeester, B.A., McClelland, A.L. (eds.) Usability Evaluation in Industry, pp. 189–194. Taylor & Francis, London (1996)
15. Bangor, A., Kortum, P., Miller, J.: Determining what individual SUS scores mean: adding an adjective rating scale. J. Usability Stud. **4**(3), 114–123 (2009)
16. Hart, S.G.: NASA-task load index (NASA-TLX); 20 years later. In: Proceedings of the Human Factors and Ergonomics Society Annual Meeting, vol. 50, no. 9, pp. 904–908 (2006)
17. Van Veen, D., Van Uden, C., Blankemeier, et al.: Adapted large language models can outperform medical experts in clinical text summarization. Nat. Med. **30**(4), 1134–1142 (2024)
18. Schoonbeek, R.C., et al.: Quality and efficiency of integrating customised large language model-generated summaries versus physician-written summaries: a validation study. BMJ Open **15**(9), e099301 (2025)
19. Stadler, M., Bannert, M., Sailer, M.: Cognitive ease at a cost: LLMs reduce mental effort but compromise depth in student scientific inquiry. Comput. Hum. Behav. **160**, 108386 (2024)
20. Rebitschek, F.G., Carella, A., Kohlrausch-Pazin, S., Zitzmann, M., Steckelberg, A., Wilhelm, C.: Evaluating evidence-based health information from generative AI using a cross-sectional study with laypeople seeking screening information. npj Digit. Med. **8**(1), 343 (2025)
21. Androutsopoulou, A., Karacapilidis, N., Loukis, E., Charalabidis, Y.: Transforming the communication between citizens and government through AI-guided chatbots. Gov. Inf. Q. **36**(2), 358–367 (2019)
22. Peeters, R.: Digital administrative burdens: an agenda for analyzing the citizen experience of digital bureaucratic encounters. Perspect. Public Manag. Gov. **6**(1), 7–13 (2023)
23. Alon-Barkat, S., Busuioc, M.: Human–AI interactions in public sector decision making: automation bias and selective adherence to algorithmic advice. J. Public Adm. Res. Theory **33**(1), 153–169 (2023)
24. Walonoski, J., et al.: Synthea: an approach, method, and software mechanism for generating synthetic patients and the synthetic electronic health care record. J. Am. Med. Inf. Assoc. **25**(3), 230–238 (2018)
25. Klein, G., Moon, B., Hoffman, R.F.: Making sense of sensemaking 1: alternative perspectives. IEEE Intell. Syst. **21**(4), 70–73 (2006)

26. Weick, K.E.: Sensemaking in Organizations. Sage, Thousand Oaks (1995)
27. Braun, V., Clarke, V.: Using thematic analysis in psychology. Qual. Res. Psychol. **3**(2), 77–101 (2006)
28. Tavakol, M., Dennick, R.: Making sense of Cronbach's alpha. Int. J. Med. Educ. **2**, 53–55 (2011)
29. Wachter, S., Mittelstadt, B., Russell, C.: Counterfactual explanations without opening the black box: automated decisions and the GDPR. Harv. J. Law Technol. **31**(2), 841–887 (2017)
30. Diakopoulos, N.: Algorithmic accountability: journalistic investigation of computational power structures. Digit. Journal. **3**(3), 398–415 (2015)

A Guided Framework for Designing an Interactive Dashboard of Local E-Participation

Anchal Dua(✉) and Maria A. Wimmer

University of Koblenz, Universitätsstraße 1, 56070 Koblenz, Germany
{adua,wimmer}@uni-koblenz.de

Abstract. Dashboards are increasingly deployed in the public sector to support transparency and data-driven decision-making. However, existing practices of dashboard design often lack a holistic, end-to-end approach that integrates public-value considerations throughout the dashboard development lifecycle. This study develops a research-grounded framework using the Design Science Research Methodology (DSRM). We synthesize from a literature review and consolidate methods, techniques, design principles, development guidelines, existing frameworks, and implementation cases. These insights are integrated into a structured dashboard design lifecycle comprising multiple phases and sub-phases that guide the transformation of raw data into interactive dashboards. The framework is operationalized through the ErLE dashboard, which visualizes e-participation offers in German municipalities.

Keywords: Dashboard design framework · Interactive dashboard · E-Participation in local government

1 Introduction

Dashboards have evolved into powerful tools that integrate data visualization, interactivity, and analytics to transform raw data into actionable knowledge. Initially emerging in the private sector as instruments of business intelligence and performance management, dashboards were developed to support managerial control, monitor key performance indicators (KPIs), and enhance operational efficiency [1, 2]. Over the past two decades, their use has expanded significantly into public administration, where the goals differ fundamentally. In the public sector, dashboards are not merely decision-support tools but mechanisms of good governance promoting transparency, accountability, and citizen participation by making complex information accessible and interpretable for both policymakers and the public [2]. This shift reflects broader developments in digital government and open-data governance, where information systems increasingly support transparency, responsiveness, and inclusivity in public institutions. The OECD's Data-Driven Public Sector Framework highlights the need for governments to use data not only for efficiency but also for public value creation and citizen trust [3]. At the same

Published by Springer Nature Switzerland AG 2027
D. Duenas-Cid et al. (Eds.): ePart 2026, LNCS 16923, pp. 130–148, 2027.
https://doi.org/10.1007/978-3-032-36676-4_9

time, the expansion of open government data initiatives [4, 5] and smart city programs [6] has reinforced expectations for data-driven decision-making. The widespread use of public dashboards during the COVID-19 pandemic further demonstrated their role in communicating real-time information and supporting public engagement [7–10]. These developments position dashboards as key infrastructures for transparent and accountable governance.

Despite these advances, the design and development of public sector dashboards remain methodologically fragmented. Existing frameworks and studies make useful contributions but are often domain-specific, data pipeline focused or limited to stages of the design process. For instance, technical frameworks concentrate on data integration and pipeline automation [10, 11], while others emphasize visualization design patterns [12] or are task specific [13. In domains such as healthcare [7–10] or urban planning [4], dashboards have been developed as context-specific applications, yet these approaches rarely extend into generalizable methodological guidance. Even where development phases such as requirement gathering, prototyping, or evaluation are identified, practical guidance on how these stages connect across the full dashboard lifecycle remains limited [14]. This fragmentation emphasizes the pressing need for a general, end-to-end methodological guideline that systematically guides public sector dashboard projects from conception to dashboard deployment and maintenance, ensuring coherence across phases.

To address this gap, we investigate two research questions: (RQ1) *What methods, techniques, and design principles identified in literature support the development of interactive dashboards?* (RQ2) *How can a structured framework guide the design of interactive dashboards aligned with public sector values?* Reviewing existing research, we propose a structured, phase-based framework that outlines the major phases of the design process and specifies the key activities, outputs, supporting techniques, relevant design principles and guidelines associated with each phase. The framework supports researchers and practitioners in designing dashboards in government and related fields.

The remainder of the paper is structured as follows: Sect. 2 outlines the research design and the methods used in the study. Section 3 examines existing dashboard design methods, principles, guidelines, frameworks, and implementation strategies reported in the literature to establish the theoretical background. Section 4 presents the proposed framework, its validation and practical demonstration. We conclude in Sect. 5 by discussing the key contributions and implications for theory and practice and by outlining the study's limitations and directions for future research.

2 Research Design

This study adopts the Design Science Research (DSR) paradigm by applying the rigor, relevance and design cycles, thereby following the DSRM (phases: problem identification, definition of solution objectives, artefact design and development, demonstration, evaluation and finally communication) proposed in [15]. Based on insights from a systematic literature review (rigor cycle), we develop, instantiate and evaluate (relevance and design cycles of DSR) a phase-based dashboard design framework.

To address RQ1, a Systematic Literature Review (SLR) is conducted following the protocol in [16], ensuring transparent and reproducible evidence synthesis. The review scope and eligibility criteria are defined using the PICOC framework [18]. Based on this definition, a Boolean search string combining dashboard-related terms with design methods, visualization techniques, usability guidelines, design principles, and frameworks is applied across seven databases: ACM Digital Library, Scopus, ScienceDirect, Taylor & Francis Online, Wiley Online Library, IEEE Xplore, and SpringerLink. These databases are selected due to their strong coverage of research in information systems, visualization, and HCI. Studies are included if they are peer-reviewed, written in English, available in full text, and address interactive dashboard design or development practices. Studies focusing exclusively on backend technologies, static dashboards, BI tool implementations (e.g., Power BI or Tableau), or lacking methodological clarity are excluded. The initial search retrieves 2,453 records. After duplicate removal and title–keyword screening, 439 studies remain. Abstract screening reduces this set to 142 studies, which are assessed through full-text review, resulting in 53 studies eligible for quality assessment. To improve completeness, backward and forward snowballing following Wohlin [19] are applied and identify 7 additional studies, of which 3 meet the inclusion criteria. After quality assessment, 46 primary studies are retained. The screening and data management activities are supported by the Parsifal platform, which facilitates duplicate removal, study screening, quality assessment, and structured data extraction. The full review protocol, screening decisions, and extracted data are publicly available via the Parsifal project dashboard. The selected studies span 2012–2025, with a noticeable increase after 2018, reflecting the growing relevance of dashboards in domains like smart cities, healthcare monitoring, public governance, and decision support.

Following the literature review, the extracted findings are synthesized to identify recurring dashboard development activities, lifecycle phases, design principles, guidelines, and implementation practices across the reviewed studies. Through iterative comparison and thematic consolidation, related concepts are grouped into broader phases and sub-phases, while the identified principles and guidelines are mapped to the corresponding development stages. The resulting framework structure is iteratively refined within the DSR design cycle and instantiated through the ErLE dashboard.

3 Literature Review

By applying the protocol of [16] in the following, we summarize the insights of the literature analysis along the overall dashboard concept, the methodological landscape, dashboard design principles, design guidelines, and existing dashboard design and development frameworks. We finally review implementation strategies from case studies. These insights build the foundation for the subsequent dashboard design and development framework presented in Sect. 4.

Dashboards and their Benefits in Public Sector. The concept of a dashboard originates from the automotive domain, where instrument panels allow drivers to monitor key indicators while maintaining focus on navigation. In information systems, dashboards have evolved into visual interfaces that consolidate and present critical information to support situational awareness and data-driven decision-making. Few defines a dashboard as *"a visual display of the most important information required to achieve specific objectives, organized on a single screen so that it can be monitored at a glance"*

[1]. More broadly, dashboards are interactive user interfaces that integrate data from multiple sources and present aggregated information through visual elements such as charts, graphs, and indicators to support monitoring and analysis [20]. Originally developed within executive information systems to monitor organizational performance [21], dashboards have since expanded across domains including education [22], healthcare [5, 23, 24], and governmental domains [2, 9, 25].

Across domains, dashboards provide several recurring benefits. Operationally, they enable real-time monitoring, rapid detection of anomalies, and improved organizational oversight [26, 27]. Strategically, dashboards link organizational objectives with measurable indicators, supporting benchmarking [2, 26], policy evaluation [20], and evidence-based decision-making [2, 5, 25]. In public governance, dashboards additionally generate social and governance benefits, including improved transparency, communication of government performance, and opportunities for citizen engagement with public data [2]. Internally, they assist public organizations in monitoring performance and allocating resources, while externally they function as civic technologies that enhance accountability, transparency, and public trust by making government performance information accessible to citizens [2, 27]. From the synthesis of literature, we therefore define public sector dashboards as *interactive, role-aware visual interfaces that consolidate curated administrative and (open) government data on a single screen, providing both officials and citizens with at-a-glance, yet drill-down-capable, insights into key indicators. Such dashboards support activities including monitoring, communication, transparency, and data-informed decision-making.*

Methodological Landscape of Dashboard Development Approaches. The increasing reliance of public and non-profit sectors on data-intensive decision environments has intensified the demand for dashboards that translate complex datasets into cognitively accessible and actionable insights [2]. Over the past decades, scholars have proposed diverse dashboard development methods, ranging from exploratory urban informatics prototypes to highly structured, user-centered engineering cycles in healthcare and policy monitoring. However, the methodological landscape remains fragmented: terminology varies, phase boundaries are inconsistently defined, and cross-domain knowledge transfer is limited. To systematize this landscape, we synthesize the reviewed research along two dimensions: dominant methodological characteristics shaping dashboard development (Table 1), and recurring lifecycle phases across these approaches, revealing shared process structures and points of divergence.

Across the reviewed methodologies, dashboard development follows a structured, time-ordered lifecycle, although phases may overlap or run in parallel:

1. Needs assessment and requirement gathering [10, 17, 20, 25, 27, 28, 30–35], where stakeholder goals, KPIs, and use cases are identified.
2. Some approaches include environmental scanning/benchmarking [35] and tool selection or development decisions [10, 25, 28–30, 32, 34].
3. Data acquisition [10, 27, 28, 34]. Technical and analytical structuring activities unfold in parallel: data pre-processing and integration [10, 28, 29, 32, 34], alongside KPI and metadata definition [4, 20, 25, 29–32, 34].

4. Design-oriented phases, including user profiling/persona development [25, 33] and low-fidelity prototyping [4, 17, 20, 27, 30–33, 56], which inform early conceptual validation.
5. High-fidelity prototyping [17, 25, 28, 31–33], accompanied by visualization design [25, 29, 30, 34, 35] and iterative design and co-creation [4, 17, 20, 27, 30, 32, 33, 56], reflecting the strongly iterative character of dashboard development.
6. Evaluation mechanisms such as usability testing and validation [4, 10, 17, 20, 25, 27, 31–35], which often overlap with implementation.
7. Technical development [25, 28–32, 34] and deployment [4, 10, 29–32, 35].
8. Maintenance and refresh cycles [4, 10, 17, 25, 30, 32–35], ensuring continuous adaptation and feedback integration.

Table 1. Methodological approaches across literature

Methodological Characteristics	Key Focus Areas where these are Applied
Design-informed / exploratory	Environmental scanning, benchmarking, rapid mock-ups for exploratory contexts [27]
User-centered / participatory	Stakeholder engagement, personas, co-design, iterative usability testing [20, 28]
Data-centric / pipeline-driven	Scalable data pipelines, pre-processing, modular integration [10, 29]
BI lifecycle approaches	Structured stages (Define–Prototype–Build–Deploy) and KPI governance [4, 30, 31]
Hybrid / design science	Combines participatory design, data engineering, and formal usability evaluation [25, 32]

The temporal structure demonstrates that dashboard development is neither strictly linear nor ad hoc, but a phased yet flexible process in which analytical, technical, and user-centered activities interact dynamically.

Although the reviewed methodologies differ in emphasis, several common patterns emerge: dashboard development generally follows recurring lifecycle phases. Trade-offs are evident across approaches. User-centered methods enhance usability and stakeholder alignment but require greater coordination, whereas data-centric and BI-oriented approaches prioritize technical robustness and scalability, sometimes limiting continuous stakeholder involvement. Hybrid approaches combine participatory design with engineering practices, increasing methodological complexity. Overall, no single method is universally suitable; effective dashboard development requires adapting methodological elements to the institutional context, data environment, and user needs. These insights inform the flexible dashboard design framework proposed in Sect. 4.

Dashboard Design Principles. Design principles provide foundational guidance for how dashboards are conceptualized, structured, and refined. Beyond aesthetic considerations, they give cognitive, technical, and functional directions that improve usability, reduce interpretation errors, and ensure that dashboards remain relevant for diverse user groups. Across the reviewed literature, recurring design orientations emerge, including user-centered design, cognitive clarity, modular interaction, and transparency in data communication. Building on the literature synthesis, we identify and synthesize 25 design principles for public sector dashboards (Table 2). For clarity and consistent reference, each principle is labelled sequentially (P1–P25), which facilitates discussion and traceability throughout the proposed framework in Sect. 4.

Table 2. Core dashboard design principles in literature

Code	Principle	Short Description
P1	User-Centered Design	Involve end-users throughout the design process to ensure relevance and usability [2, 4, 24]
P2	Accessibility and Inclusiveness	Design dashboards usable by diverse users, accounting for literacy, ability, and cultural diversity [4, 9, 24]
P3	Feedback & Iterative Improvement	Incorporate continuous user input for iterative refinement and enhancement [2, 52]
P4	Data Literacy Consideration	Tailor data presentation for users with varying statistical understanding [9, 24]
P5	Stakeholder Diversity Representation	Reflect the informational needs of varied societal and institutional stakeholders [4, 24]
P6	Visual Simplicity and Clarity	Use minimal and context-appropriate visualizations, selecting visual elements that best support the represented data and analytical task to enhance understanding [2, 9, 46, 49]
P7	Clear Information Hierarchy	Structure content logically to aid comprehension and prioritize key data [8]
P8	Color Effectiveness	Apply colors purposefully to differentiate and draw attention without confusion [9, 49]
P9	Contextual Interpretation Support	Provide metadata, labels, and cues to prevent misinterpretation [4, 24]
P10	Visual Coherence and Consistency	Maintain consistent styling, layout, and iconography across dashboard [4, 24, 46]
P11	Minimalism	Eliminate non-essential elements to highlight important content [24, 46]
P12	Responsiveness Across Devices	Ensure usability across desktop, mobile, and other digital platforms [4]
P13	Interactive Functionality	Enable filters, drill-downs, and responsive elements for deeper exploration [2, 10]

(continued)

Table 2. (*continued*)

Code	Principle	Short Description
P14	Performance Metric Linkage	Align dashboard indicators with policy or strategic goals [8]
P15	Transparency and Data Provenance	Disclose data sources, updates, and methodologies clearly [53]
P16	Security and Privacy	Ensure protection of sensitive or personal data [4]
P17	Error Prevention	Design safeguards to avoid misinterpretation or misuse [24]
P18	Standardization	Use uniform indicators, layouts, and formats across dashboards [2, 24]
P19	Modularity	Allow reusable, rearrangeable dashboard components for flexibility [25]
P20	Data Granularity	Support both overview and detailed data layers [2, 4]
P21	Contextual Awareness	Adapt content to geographic, institutional, or cultural settings [8]
P22	Local Relevance & Personalization	Reflect locally meaningful information or enabling user personalization [2, 8]
P23	Real-Time and Updated Data	Deliver timely and current data for decision relevance [8]
P24	Creating Public Value	Align dashboards with democratic accountability, equity, and transparency [2, 8, 46]
P25	Customizing Metrics	Allow custom definition of indicators based on local or institutional needs [2, 8]

Design Guidelines. While principles refer to high-level normative ideals such as accessibility, clarity, or user empowerment that provide a conceptual foundation for dashboard development [2]. Guidelines, by contrast, are practical recommendations that translate these principles into specific design decisions suited to different contexts and user groups [36]. While principles articulate the underlying rationale of dashboard design, guidelines translate these principles into concrete design recommendations. To ensure dashboards are both functional and meaningful, the literature review reveals several recurring categories of dashboard design guidelines. Table 3 summarizes the main guideline categories and their primary focus. Each guideline is associated with a reference code (e.g., U1, C1, M1), which is used in Sect. 4 to link the guidelines to the corresponding sub-phases of the proposed framework.

Table 3. Categorized guidelines from the review literature

Guideline category	Focus
User-centered design (U1–U10)	Ground dashboard design in empirical user research to understand stakeholder roles, tasks, and data literacy levels, informing personas, stakeholder segmentation, and task-oriented functionalities [2, 6, 11, 31, 32, 43, 47, 51, 54]
Cognitive-aware design (C1–C8)	Reduce cognitive load and improve perceptual clarity through prioritizing key metrics, limiting visual clutter, applying Gestalt principles, and aligning visualizations with analytical tasks [23, 47, 48, 51, 55]
Modularity & task-based interaction (M1–M8)	Structure dashboards into modular components aligned with analytical tasks such as monitoring, comparison, or exploration, supported by drill-downs and coordinated views [17, 23, 28, 34, 45, 50]
Feedback & interactivity (F1–F8)	Enable responsive exploration through interactive filtering, drill-downs, details-on-demand, dynamic data updates [28, 34, 45, 47, 55]
Transparency & data ethics (T1–T5)	Communicate data provenance, intent, and limitations through metadata, responsible visual signaling, and human-in-the-loop decision support [26, 34, 43]
Data literacy support (DL1–DL5)	Support diverse analytical capabilities through combined visual and textual explanations and interfaces that accommodate both novice and expert users [6, 45, 50, 55]
Visual communication (V1–V5)	Promote clear interpretation through consistent visual semantics, structured layouts, and clear labelling [26, 34, 44, 45, 55]
Cross-platform accessibility (A1–A3)	Ensure inclusive access across devices and compliance with accessibility standards such as WCAG [44, 45, 55]

Bach et al. describe design patterns as structured, reusable solutions to recurring dashboard design challenges [36]. Positioned alongside guidelines and annotated examples, these patterns represent an important form of design knowledge that supports dashboard development. By offering consistent solutions to common layout, interaction, and content trade-offs, design patterns help reduce fragmentation and provide greater coherence in dashboard design practice.

Existing Dashboard Design and Development Frameworks. Among the reviewed literature, four frameworks provide relevant methodological perspectives on dashboard development [10–13]. Table 4 summarizes their application domains, core phases, and methodological focus. While these approaches offer valuable guidance, they are largely domain-specific and emphasize different aspects of the development process. They highlight recurring phases like requirement analysis, data preparation, visualization design, and evaluation, but treat these elements with varying levels of emphasis and integration. These frameworks therefore serve as methodological baseline and source of inspiration for synthesizing the development phases and design considerations incorporated into the framework proposed in Sect. 4.

Table 4. Comparative summary of dashboard development frameworks.

Framework	Applic. Dom	Core phases	Key Focus
Cepero et al. [12]	Smart city	Requirements analysis → Data preparation → Visualization design → Evaluation	User-centered design and participatory usability validation
Conrow et al. [11]	Mobility	Data selection → Transformation → Comparison → Visualization → Assembly	Data-processing pipeline separating backend processing from interface design
Olszak & Kisołek [10]	Crisis monitoring (Dash-Cov)	Data acquisition → Pre-processing → Analysis → Visualization → Dissemination	Real-time integration of heterogeneous data sources
Lempinen [13]	Organizational performance	Decision needs identification → KPI modelling → System design → Dashboard construction → Evaluation	Links dashboards to organizational decision-making processes

Implementation Strategies: Lessons From the Cases. Prior studies present practical dashboard implementations across public sector contexts, illustrating how design principles are applied in practice. A COVID-19 dashboard in Sudan emphasizes simple visualizations and streamlined data pipelines in resource-constrained settings [7]. BarcelonaNow highlights modular architectures and participatory interaction for exploring heterogeneous municipal data [38]. A UAE traffic dashboard demonstrates the value of interactive filtering and structured data preprocessing for policy analysis [39], while India's Ministry of Tribal Affairs dashboard shows how integrating multiple datasets enables cross-departmental monitoring [40]. Across these cases, common priorities emerge: user-centered visualization, integration of heterogeneous data, and robust data preparation. These insights inform the dashboard design framework in the next section.

4 Dashboard Design and Development Framework

The proposed framework for dashboard design and development provides a structured yet flexible lifecycle that captures the full process from initial scoping to long-term maintenance. Building on the insights discussed in Sect. 3, the framework adopts a hybrid dashboard design approach that integrates participatory design principles with systematic data engineering and evaluation practices, thereby balancing methodological trade-offs and providing comprehensive guidance across the dashboard development lifecycle. The framework is organized into 5 umbrella phases: scoping & requirement gathering, design & planning, prototyping & validation, development & deployment, and maintenance & continuous improvement which are further broken down into 15 sub-phases. Each sub-phase outlines the key activities, expected outputs, supporting tools, as well as the design guidelines and principles that should be applied to ensure quality and alignment with public sector values. The supporting tools and techniques listed in this framework phases are intended as examples rather than a prescriptive catalogue. They are provided to offer practitioners a starting point and inspiration for selecting appropriate methods in their own contexts. Depending on project scope, organizational environment, and available resources, alternative or additional tools may be equally valid. Comprehensive documentation is treated as a cross-cutting concern embedded within all phases: maintaining clear records of decisions, requirements, and changes not only supports public values such as transparency and accountability but also ensures knowledge transfer, reproducibility, and easier future enhancements. While the phases are presented sequentially for clarity, the framework is inherently iterative: a feedback loop allows any phase to be revisited as new requirements emerge, user insights are gathered, or contextual conditions change.

Phase 1: Scoping and Requirement Gathering. Dashboard development begins with a scoping phase that defines the project's purpose, stakeholders, requirements, and governance boundaries. As socio-technical systems embedded in organizational and regulatory contexts, dashboards require early clarification of objectives and constraints to avoid misaligned solutions. This phase involves four key tasks: defining objectives, identifying stakeholders, eliciting preliminary requirements, and assessing legal, ethical, and organizational constraints. Together, these establish the foundation for subsequent design decisions, including indicator selection, data preparation, and visualization strategies. Table 5 summarizes the sub-phases of this stage, their focus, key outputs, and the guidelines and principles that inform each activity.

Scoping activities rely primarily on qualitative and standards-based methods. Objective definition may use structured approaches such as SMART criteria, OKRs, or Goal-Question-Metric (GQM). Stakeholder analysis can be supported through interviews, workshops, power–interest mapping, and RACI matrices. Requirements documentation may follow ISO/IEC/IEEE 29148, while ISO 9241-210 supports early human-centered context analysis. Legal and risk assessments can draw on GDPR DPIA procedures or PIA or HIPAA (Health Insurance Portability and Accountability Act) guidelines (if dealing with health data in USA) and ISO 31000 risk management practices.

Phase 2: Design and Planning. Following scoping, the design and planning phase translates the identified objectives and requirements into operational design specifications as shown in Table 6.

Table 5. Operationalization of phase 1: Sub-phases, key outcomes, principles, and guidelines

Sub-phases	Focus	Key Outcome	Key PrinciplesGuidelines
Objective definition	Clarify the purpose, scope, and success criteria of the dashboard, aligning it with organizational or policy goals and public value objectives such as transparency or accountability	Defined dashboard purpose and scope boundaries	P15, P24–P25; U8
Stakeholder identification	Identify relevant stakeholder groups (e.g., policymakers, administrators, researchers, citizens) and map their roles, information needs, and decision contexts	Stakeholder map and engagement plan	P1–P5; U1, U7
Requirements elicitation	Translate objectives and stakeholder needs into high-level functional and analytical requirements, incl. Expected analytic tasks (monitoring, comparison, exploration,…)	Initial requirements specification	P1–P5; U1–U6, U9, U9, U10, DL1, DL3, DL5
Legal, ethical, and risk assessment	Identify regulatory, ethical, and organizational constraints, including data protection obligations, risks of misinterpretation, and resource limitations	Risk register and governance constraints	P15–P23; M6

Typical techniques include technology evaluation matrices and pilot deployments for platform selection; data cataloguing and Extract, Transform and Load (ETL) workflows for data preparation; and metadata standards such as DAMA-DMBOK for KPI and metadata modelling. User profiling may employ persona modelling and journey mapping, while accessibility checks may follow WCAG 2.2 guidelines. Information

Table 6. Operationalization of phase 2: Sub-phases, key outcomes, principles, and guidelines

Sub-phase	Core Focus	Key Output	Key Guidelines/Principles
Technology & platform selection	Evaluate BI tools or custom solutions based on interoperability, scalability, security, and cost	Approved technology stack and procurement plan	P15–P23; M6
Data landscape assessment	Map available data sources, ownership, formats, update cycles, and quality dimensions	Data catalogue and access agreements	P15–P23; T3
Data preparation & integration	Clean, transform, and harmonize heterogeneous data sources using profiling, schema matching, and transformation workflows	Integrated dataset with documented transformation steps	P15–P23; T3
KPI & metadata modelling	Define indicators, formulas, aggregation rules, and metadata elements linking KPIs to objectives	KPI dictionary and metadata schema	P15, P24–P25; T1–T3
User profiling & accessibility	Develop personas and user journeys reflecting diverse stakeholder needs and literacy levels; address accessibility standards	Personas, user journeys, accessibility checklist	P1–P5; U2, U3, U6, U9, DL1, DL3, DL4, DL5, A1, A2
Information architecture & narrative design	Structure dashboard modules and navigation flows guiding users from overview to detailed insights	Information architecture diagrams and storyboards	P6–P13; C1, C3, C4, VC1–VC3
Visualization & interaction design	Select visual encodings and interaction mechanisms aligned with analytical tasks and user capabilities	Visual design system and interaction specifications	P6–P13; C1, C2, C4, C5, C6-C8, M1, M5, M7, M8, F1, F3, F4, F7, VC1, VC4, VC5

architecture can be developed through card sorting and diagramming tools, and visualization concepts are commonly explored through low-fidelity prototypes and design systems.

Phase 3: Prototyping and Validation. Operationalizes the validated prototype as a functioning system in the organizational environment, translating design into production-ready software and practice (Table 7). Prototyping moves from low-fidelity sketches and wireframes to high-fidelity interactive models for realistic testing. Validation runs in parallel, assessing usability, interpretability, and alignment with objectives.

Table 7. Operationalization of phase 3: Sub-phases, key outcomes, principles, and guidelines

Phase	Focus	Key Outcome	Key Principles/ Guidelines
Prototyping & Usability Validation	Develop low-fidelity wireframes and high-fidelity interactive prototypes. Test prototypes with representative users using SUS, SEQ, or think-aloud protocols. Gather feedback, refine design, and document iterations	Prototypes with usability test results and iteration log	P1–P5; U1, U5, U9, M5, F1, F5, F6, F8

Prototyping typically employs wireframing and design tools (e.g., Figma, Balsamiq) and interactive dashboard environments (e.g., Tableau and Power BI). Validation methods include moderated think-aloud protocols, task-based usability testing, SUS and SEQ questionnaires, and participatory workshops. Supporting tools such as screen-recording software, issue trackers, and accessibility testing utilities help document findings and maintain traceability between evaluation results and design revisions.

Phase 4: Development and Deployment. Operationalizes the validated prototype and releases it as a functioning system within the organizational environment. While the previous focused on design and evaluation, this stage translates the prototype into production-ready software and organizational practice, combining technical implementation with strategies that support adoption and sustainable use. Table 8 summarizes the sub-phases, their key outcomes and applied principles and guidelines.

Development may employ custom frameworks (e.g., Streamlit and Dash) or integrated BI platforms (e.g., Power BI, Tableau, Qlik). Open source alternatives such as Metabase, Apache Superset, or Grafana are frequently used, where transparency or cost considerations are important. Deployment typically relies on containerization and orchestration tools (e.g., Docker), CI/CD pipelines (GitHub Actions, GitLab CI), and version control (GitHub, GitLab). Hosting is usually through cloud infrastructures.

Table 8. Operationalization of phase 4: Sub-phases, key outcomes, principles, and guidelines

Sub-Phases	Focus	Key Outcome	Key Principles/ Guidelines
Technical development & system integration	Implement backend data pipelines, APIs, and frontend interfaces based on validated design. Configure authentication, role-based access, security controls; Conduct unit, integration, and performance testing	Pre-production dashboard with operational pipelines and APIs	P15–P23; M6-M8, F2, F6, A1, A2
Deployment & adoption	Release the dashboard into production infrastructure and embed it in organizational practice through onboarding, training, documentation, and communication activities	Live dashboard, training materials, and support structures	P14, P24-P25; A1–A3, T1–T3, T5,

Phase 5: Maintenance and Continuous Improvement. Ensures that the deployed dashboard remains reliable, relevant, and responsive to evolving needs. This stage focuses on sustaining performance and incorporating improvements over time (Table 9). Continuous monitoring safeguards the stability of data pipelines and system performance, while structured feedback mechanisms support iterative refinement of indicators, visualizations, and functionality.

The presented framework provides a structured and practical guide for the end-to-end design and development of public sector dashboards. By combining insights from multidisciplinary literature with clearly defined phases, activities, tools, design principles and guidelines that can be applied in each phase. The framework translates theoretical recommendations into an actionable process that addresses technical, organizational, and user-centered considerations. Its emphasis on documentation, iteration, and long-term maintenance further reflects the dynamic nature of dashboards as evolving socio-technical systems. These elements position the framework as both a conceptual contribution and a practical reference for future dashboard initiatives.

Framework Validation. To assess the clarity, relevance, and completeness of the proposed dashboard design and development framework, a content validity assessment is conducted using peer review with 10 review panel participants (advanced master students and early-career researchers with prior exposure to dashboard design, data visualization,

Table 9. Operationalization of phase 5: Sub-phases, key outcomes, principles, and guidelines

Sub-phases	Focus	Key Outcome	Key Principles/ Guidelines
Monitoring & support	Monitor data pipelines, refresh cycles, and system performance; manage incidents and user-reported issues; maintain documentation and support channels	Operational monitoring logs, issue tracking records, updated documentation	P15–P23; M5, M6, M8, F2, F6
Iterative enhancement	Incorporate user feedback, new data sources, and evolving policy needs through controlled update cycles	Updated dashboard versions, extended KPIs, improvement records	P1–P5; A1–A3, T1–T5,

or digital government projects). Content validity evaluates whether the elements of an artefact adequately represent the construct they are intended to capture, focusing on conceptual coverage rather than performance outcomes. We use peer review [12] and draw on established procedures for content validity assessment, such as those described in [37]. The reviewers receive an eight-page summary of the framework and evaluate individual framework components including phases, sub-phases, and associated design principles and guidelines using a five-point scale across two dimensions: relevance and clarity. The feedback indicates strong overall content validity according to established thresholds. Qualitative feedback suggests minor refinements related to terminology and diagram clarity, which are incorporated into the framework.

Framework Demonstration Along the ErLE Dashboard. To demonstrate the applicability of the proposed framework, it is instantiated in the Erfolgsfaktoren Lokaler E-Partizipation (ERLE) project[1] through the design and development of the dashboard. The ERLE project investigates e-participation success factors across municipalities through a nationwide survey (see the data report in [41]) and makes the resulting insights accessible to researchers, policymakers, and municipal practitioners through an interactive data dashboard. In this context, the development of the dashboard followed the phases of the proposed framework, guiding the process from requirement scoping and data preparation to visualization design and interactive implementation.

The real-world application illustrates how the framework can support the systematic design and development of dashboards within a public sector context. In line with the DSRM proposed by [15], the ERLE dashboard therefore serves as demonstrative instantiation of the framework, providing a practical basis for its subsequent evaluation. Further

[1] https://lokale-partizipation.de, last accessed: 12.05.2026.

details on the operationalization of the framework and the technical implementation of the dashboard are documented in a technical report [42].

5 Discussion and Conclusion

This study addressed a methodological gap in public sector dashboard development: Although dashboards are increasingly used in public administration, existing knowledge on how to design and develop them remains fragmented across domain-specific methods, visualization approaches, and technical implementation strategies. Furthermore, relevant design principles and guidelines are dispersed throughout the literature, making them difficult to identify and apply systematically in practice.

To address this gap, this study developed a context-neutral dashboard design and development framework that consolidates methodological phases, design principles, guidelines, and implementation considerations into a structured yet iterative lifecycle. Rather than providing isolated checklists or domain-specific recipes, the framework outlines what activities should be performed and when, and what outputs should be produced throughout the dashboard development process, from project scoping to long-term maintenance. The demonstration of the framework through the ErLE dashboard and its content validity assessment suggest that the framework provides a conceptually coherent and practically understandable structure for guiding dashboard development activities. The findings indicate that the framework can serve as a methodological guide that helps developers and stakeholders navigate the complexity of dashboard projects while maintaining a clear connection between design decisions, implementation activities, and public sector values.

Nevertheless, limitations should be acknowledged. The framework was demonstrated within a single case, limiting conclusions regarding its applicability across different organizational, institutional, and policy contexts. Moreover, the evaluation focused primarily on content validity and conceptual coherence rather than long-term organizational adoption, operational performance, or project outcomes. Consequently, the framework should be regarded as a flexible methodological guide rather than a prescriptive process model, as its implementation is likely to vary depending on organizational capacity, stakeholder involvement, project scope, and data environments. Future research should further evaluate the framework through its application in additional dashboard projects across different policy domains and organizational settings. Such studies could examine how practitioners use the framework during real-world development processes and assess its adaptability to varying institutional contexts. Furthermore, validation involving practitioners, domain experts, and other relevant stakeholders would provide additional insights into the framework's practical relevance, completeness, usability, and clarity.

Acknowledgments. This study was conducted along the ErLE research project, which was funded by Stiftung Mercator GmbH. We thank the project partners for their valuable feedback along the dashboard development.

Disclosure of Interests. The authors have no competing interests to declare that are relevant to the content of this article.

References

1. Few, S.: Information Dashboard Design: The Effective Visual Communication of Data. O'Reilly Media, Inc. (2006)
2. Matheus, R., Janssen, M., Maheshwari, D.: Data science empowering the public: data-driven dashboards for transparent and accountable decision-making in smart cities. Gov. Inf. Q. **37**(3), 101284 (2020)
3. OECD. 2019. The Path to Becoming a Data-Driven Public Sector. OECD Digital Government Studies, OECD Publishing, Parishttps://doi.org/10.1787/059814a7-en
4. Chokki, A.P. et al.: Engaging citizens with open government data: the value of dashboards compared to individual visualizations. Dig Gov: Res. Pract. **3**(3) (2022)
5. Concannon, D., Herbst, K., Manley, E.: Developing a data dashboard framework for population health surveillance: widening access to clinical trial findings. JMIR Formative Res. **3**(2), e11342 (2019). https://doi.org/10.2196/11342
6. Young, G.W., Kitchin, R., Naji, J.: Building city dashboards for different types of users. J. Urban Technol. **28**(1–2), 289–309 (2021)
7. Abdelsamad, A.M.O., Karrar, A.Z.: An interactive dashboard for monitoring the spread of COVID-19 in Sudan. In: Proceedings of 2020 ICCCEEE (Feb. 2021), pp. 1–6 (2021)
8. Bos, V.L.L.C., et al.: Development and actionability of the Dutch COVID-19 dashboard: descriptive assessment and expert appraisal study. JMIR Public Health Surveill. **7**(10), e31161 (2021), https://doi.org/10.2196/31161
9. Monkman, H., et al.: Opportunities to improve COVID-19 dashboard designs for the public. Stud. Health Technol. Inf. 16–20 (2021)
10. Olszak, C., Kisołek, A.: Dashboard framework. A tool for threat monitoring on the example of Covid-19. In: Proceedings of 55th HICSS, pp. 276–285 (2022)
11. Conrow, L., et al.: A conceptual framework for developing dashboards for big mobility data. Cartogr. Geogr. Inf. Sci. **50**(5), 495–514 (2023)
12. Cepero, T., Montané-Jiménez, L.G., Maestre-Góngora, G.P.: A framework for designing user-centered data visualizations in smart city technologies. Technol. Forecast. Soc. Change, 210 (2025). https://doi.org/10.1016/j.techfore.2024.123855
13. Lempinen, H.: Constructing a design framework for performance dashboards. In: Keller, C., Wiberg, M., Ågerfalk, P.J., Eriksson Lundström, J.S.Z. (eds): Nordic Contributions in IS Research. SCIS 2012. LNBIP, vol. 124, pp. 109–130. Springer, Heidelberg (2012)
14. Delgado, M., Pando, D., Maldonado-Mahauad, J.: Dicrev-dash: proposal for the design, creation and evaluation of a dashboard for data visualization. In: Proceedings of LALA conference, CEUR-WS.org, paper 10 (2021)
15. Peffers, K., Tuunanen, T., Rothenberger, M.A., Chatterjee, S.: A design science research methodology for information systems research. J. Manag. Inf. Syst. **24**(3), 45–77 (2007)
16. Carrera-Rivera, A., Larrinaga, F., Lasa, G.: Context-awareness for the design of Smart-product service systems: literature review. Comput. Ind. **142**, 103730 (2022)
17. Catalino Piquet, B.U., et al.: Dashboard design for key performance indicators visualization of STEAM government initiatives: a case study. Proceedings of the XI Latin American Conference on Human Computer Interaction (New York, NY, USA), pp. 1–9 (2023)
18. Petersen, K., Vakkalanka, S., Kuzniarz, L.: Guidelines for conducting systematic mapping studies in software engineering: an update. Inf. Softw. Technol. **64**, 1–18 (2015)
19. Wohlin, C.: Guidelines for snowballing in systematic literature studies and a replication in software engineering. In Proceedings of 18th International Conference on Evaluation and Assessment in Software Engineering, New York, NY, USA, pp. 1–10 (2014)
20. Aprillia, D.A., Noranita, B., Nugraheni, D.M.K.: Dashboard design as monitoring system distribution of government assistance at balai besar perikanan budidaya air payau (BBPBAP)

Jepara, central java. In: Proc 5th International Conference on Informatics and Computational Sciences, pp. 187–192 (2021)
21. Eckerson, W.: Performance Dashboards. Wiley (2012)
22. Verbert, K., et al.: Learning dashboards: an overview and future research opportunities. Pers. Ubiquit. Comput. **18**(6), 1499–1514 (2014)
23. Lechner, B., Fruhling, A.: Towards public health dashboard design guidelines. In: Proceedings of International Conference on HCI in Business, pp. 49–59. Springer, Cham (2014)
24. Rabiei, R., et al.: Developing public health surveillance dashboards: a scoping review on the design principles. BMC Public Health **24**(1) (2024) 2024
25. Contreras, V., Montané, L., Cepero, T., Benitez, E., Mezura, C.: Building adaptable dashboards for smart cities: design and evaluation. Prog. Comput. Softw. **48**(8), 534–551 (2022)
26. Young, G.W., Kitchin, R.: Creating design guidelines for building city dashboards from a user's perspectives. Int. J. Hum. Comput. Stud. **140** (2020)
27. Kitchin, R., Saalsen, S., McArdle, G.: The praxis and politics of building urban dashboards. Geoforum **77**, 93–101 (2016). https://doi.org/10.1016/j.geoforum.2016.10.006
28. Tsouros, I., et al.: From raw data to informed decisions: the development of an online data repository and visualization dashboard for transportation data. Procedia Comput. Sci. **231**, 161–167 (2024). https://doi.org/10.1016/j.procs.2023.12.188
29. Joshi, A., Amadi, C., Katz, B., Kulkarni, S., Nash, D.: A human-centered platform for HIV infection reduction in New York: Development and usage analysis of the ending the epidemic (ETE) dashboard. JMIR Public Health Surveill. **3**(4) (2017)
30. Komarudin, M., Despa, D., Septama, H.D., Yulianti, T.: Design of key performance indicator dashboard for Indonesian higher education based on one data. In Proceedings International Conference on Converging Technology in Electrical and Information Engineering (ICCTEIE), pp. 6–11 (2023)
31. Orlando, T.M., Sunindyo, W.D.: Designing dashboard visualization for heterogeneous stakeholders (case study: ITB central library). In: Proceedings of ICoDSE, pp. 1–6 (2017)
32. Felizzola, H., Gomez, C., Arrieta, N., Jerez, V., Erazo, Y., Camacho, G.: Enhancing transparency in public procurement: A data-driven analytics approach. Inf. Syst. **125**, 102430 (2024). https://doi.org/10.1016/j.is.2024.102430
33. Dolatabadi, S.H., Gatial, E., Budinská, I., Balogh, Z.: Integrating Human-Computer Interaction Principles in User-Centered Dashboard Design: Insights from Maintenance Management. In Proceedings of 28th International Conference on Intelligent Engineering Systems (INES), pp. 000219–000224. IEEE (2024)
34. Simonofski, A., Fink, J., Burnay, C.: Supporting policy-making with social media and e-participation platforms data: a policy analytics framework. Gov. Inf. Q. **38**(3), 101590 (2021)
35. Smith, V.S.: Data dashboard as evaluation and research communication tool. New. Dir. Eval. **140**, 21–45 (2013). https://doi.org/10.1002/ev.20072
36. Bach, B., Freeman, E., Abdul-Rahman, A., Turkay, C., Khan, S., Fan, Y.: Dashboard design patterns. IEEE Trans. Vis. Comp Graph. **29**(1), 342–352 (2023)
37. Almanasreh, E., Moles, R., Chen, T.: Evaluation of methods used for estimating content validity. Res. Social Adm. Pharm. **15**(2), 214–221 (2019)
38. Marras, M., Manca, M., Boratto, L., Fenu, G., Laniado, D.: BarcelonaNow: empowering citizens with interactive dashboards for urban data exploration. In: Companion of The Web Conference 2018, pp. 219–222 (2018)https://doi.org/10.1145/3184558.3186983
39. Awad, M., Al Redhaei, A., Fraihat, S.: Using business intelligence to analyze road traffic accidents. In: Proceedings of CEEeGov, pp. 83–92 (2022)
40. Kapoor, N.J., Maurya, A.P., Raman, R., Govind, K., Nedungadi, P.: Digital India eGovernance initiative for tribal empowerment: performance dashboard of the ministry of tribal affairs. In: Proceedings of ICTCS, pp. 1017–1027. Springer, Singapore (2021)

41. Kneuer, M., et al.: Data Report: Survey "Digital Participation on the Local Level". ErLE consortium (2025). https://doi.org/10.17605/OSF.IO/FQNV7
42. Wimmer, M.A., et al.: Das Dashboard zum Verbundprojekt "Erfolgsfaktoren lokaler E-Partizipation: Mapping und Stärkung digitaler bürgerschaftlicher Beteiligung in deutschen Kommunen", ErLE consortium (2026). https://doi.org/10.34657/29305
43. Hadiprawoto, T.R.; Ridley, A.L.: Transparent dashboards: open data practices for promoting competition-as-motivation in business dashboards. In: Proceedings of IEEE 16th Pacific Visualization Symposium (PacificVis), pp. 142–146 (2023)
44. Huang, Y.W., Yang, Y.J., Jeng, W.: User perception and eye movement on a pandemic data visualization dashboard. Proc. Assoc. Inf. Sci. Technol. **59**(1), 121–131 (2022)
45. Ma, Q., Millet, B.: Design guidelines for immersive dashboards. In: Proceedings of the Human Factors and Ergonomics Society Annual Meeting, vol. 65, no. 1, pp. 1524–1528 (2021)
46. Martins, N., Martins, N., Martins, S., Brandãoet, D.: Design principles in the development of dashboards for business management. In: Perspectives on Design II: Research, Education and Practice, pp. 353–365. Springer, Cham (2022)
47. Nasir, A.H.M., Surin, E.S.M., Ahmad, M.N.: Towards unified information dashboard design - common principles, practices, and challenges. Procedia Comput. Sci. **234**, 1586–1592 (2024)
48. Netek, R.: Map-Oriented Dashboards Online—From Minor Method to GIScience Trend. Advances in Geoinformatics Technologies. Earth and Environmental Sci Lib, 33–57 (2024)
49. Pastushenko, O., Hynek, J., Hruška, T.: Evaluation of user interface design metrics by generating realistic-looking dashboard samples. Expert Syst. **38**(5) (2021)
50. Vuckovic, M., Wolosiuk, D., Schmidt, J.: Designing interactive analytics dashboards for diverse target groups: the process and decision-making. In: Proceedings of 9th Int Conf on Smart and Sustainable Technologies (SpliTech), pp. 1–4 (2024)
51. Yigitbasioglu, O.M., Velcu, O.: A review of dashboards in performance management: implications for design and research. Int. J. Account. Inf. Syst. **13**(1), 41–59 (2012)
52. Steinmetz-Weiss, C., et al.: Design thinking for city dashboard development: recommendations from a study of smart asset management in Sydney Australia. Can. Geogr./Géographies canadiennes **68**(1), 129–141 (2024). https://doi.org/10.1111/cag.12868
53. Vila, R.A., Estevez, E., Fillottrani, P.R.: The design and use of dashboards for driving decision-making in the public sector. In: Proceedings of 11th Int Conference on Theory and Practice of Electronic Governance (New York, NY, USA), pp. 382–388 (2018)
54. Borukhson, D., Glaum, L., Weinhardt, C., Fegert, J.: Visualizing social sentiment: designing an interactive dashboard for informed decision-making. In: Proceedings of Hawaii International Conference on Systems Science (2025).https://doi.org/10.24251/HICSS.2025.180
55. Jin, W., Shan, X., Ma, K.: Research on information visualization design for public health security emergencies. In: Proceedings of International Conference on Human-Computer Interaction, pp. 325–336. Springer, Cham (2021)
56. Vannieuwenhuyze, J.: Dashboards for Input-Evaluation of Policy Programs: Lessons Learned from an Antwerp Dashboard for Garden Streets. ISPRS Annals of the Photogrammetry, Remote Sensing and Spatial Information Sciences VI-4/W2–2020, pp. 173–179 (2020). https://doi.org/10.5194/isprs-annals-VI-4-W2-2020-173-2020

Institutionalising AI into Democratic Deliberations: Tensions in Practice

Sem Nouws(✉) and Marijn Janssen

Delft University of Technology, Delft, The Netherlands
{s.j.j.nouws,m.f.w.h.a.janssen}@tudelft.nl

Abstract. AI-enabled mass deliberation is a new attempt to revive deliberative democracy with digital technology. AI applications such as summarisation and automated moderation provide opportunities to scale up deliberative processes that currently are constrained to discussions in small groups. Notwithstanding, the AI push underestimates the difficulties in sustainably integrating digital technologies (including generative AI) into sedimented organisational and normative practices of decision-making. This paper aims to understand the gap between institutionalization of AI-enabled deliberation as a conceptual notion and its effectuation in practice. For this, we collected and synthesised well-established conditions and requirements for institutionalisation through co-creation sessions with experts. Thereafter, we asked practitioners to assess these conditions and requirements on importance, contribution to institutionalisation, feasibility, and easiness of implementation. Our research identifies five tensions in practice that contribute to the gap between the conceptual understanding of institutionalisation and practitioners' perceptions of whether it is achievable in practice. These tensions need to be overcome for institutionalisation to be accomplished.

Keywords: Artificial Intelligence · AI · deliberative democracy · mass deliberation · institutionalisation

1 Introduction

The emergence of Generative AI (GenAI) applications, especially large language models (LLMs), has sparked a new cycle of tackling presupposed barriers to effective democratic deliberation and participation [3, 29, 33]. Democratic deliberation revolves around the exchange of arguments to reach consensual decisions and is often put into practice in mini-publics, which are resource-intensive. This hinders scaling up deliberation to expand public engagement. Conceptually, GenAI applications offer functionalities that may overcome known barriers (such as resource constraints and capacity limitations) to scale up mini-publics and realise mass deliberation [19, 33]. GenAI can potentially automate moderation and, thereby, lessen the need for human moderators. In this way, it increases efficiency in public administration, improves response speed, and allows for creating overview of discussions in a short time frame. Functionalities like summarisation and text analysis facilitate synthesising the substantial volumes of information generated in mass deliberations. As such, GenAI prevents information overload and enables

Published by Springer Nature Switzerland AG 2027
D. Duenas-Cid et al. (Eds.): ePart 2026, LNCS 16923, pp. 149–164, 2027.
https://doi.org/10.1007/978-3-032-36676-4_10

asynchronous discussions. Recent experiments with AI-enabled deliberation platforms show that deliberators are positive about the input provided by AI applications [cf. 29].

Nevertheless, the implementation of GenAI into deliberation is likely to repeat the failures of earlier attempts to digitalise participative and deliberative processes. The anticipated potential of preceding digital innovations in participation and deliberation is still unfulfilled [13]. Scholars argue that initiatives of digital participation and deliberation are often preoccupied with adoption. As a result, practitioners and researchers tend to disregard the slow and more difficult process of institutionalisation [24, 27]. This paper defines institutionalisation as the process in which AI-enabled mass deliberation becomes a sustainably integrated activity in policy- and decision-making in such a way that it is meaningful to participating deliberators [cf. 24]. Approaching deliberation as a technological project and the disregard for institutionalisation are probably going to be repeated for AI-enabled mass deliberations. The literature on AI-enabled deliberation tends to focus narrowly on technological artefacts [4, 26], neglecting behavioural and institutional factors such as assigned responsibilities and safeguards that curb the risks of using GenAI. This technology push, already observed in earlier waves of digitalising deliberative processes, narrows the view on transforming deliberative practices [5]. Starting from the AI features that might advance democratic deliberation obscures the non-technical efforts needed to realise institutionalisation. The conceptual conditions and requirements for institutionalisation have been identified before. However, what impedes the actual institutionalisation of AI-enabled mass deliberations in organisational practices is poorly understood.

This paper aims to understand the gap between conceptual notions of institutionalisation and its realisation in practice, as this gap may impede the sustainable integration of AI in meaningful deliberative practices. We compared the conceptual understanding of such institutionalization in theory with practitioners' perceptions of that conceptual understanding. The focus on this gap is motivated by the vast literature on the success factors of institutionalising online deliberation but the limited accomplishments in practice. The insights on the manifestation of institutionalisation can be used by organisations when transforming existing deliberative practices by integrating GenAI applications.

This paper is structured as follows. Section 2 discusses our perspective on both AI-enabled mass deliberation as well as institutionalising digital technologies in public organisations. Section 3 describes the research methodology. Section 4 presents the results of the study, starting with an overview of conditions and requirements for institutionalising AI-enabled mass deliberation found in literature and formulated in co-creation sessions with experts. Thereafter, we discuss the assessment of these conditions and requirements by practitioners. Section 5 presents our conclusions.

2 Institutionalisation of AI-Enabled Mass Deliberations

Democratic deliberation concerns the argumentative practices enacted to justify decisions made in a democratic setting [12]. In deliberation, arguments are formulated and improved as participants learn from one another's ideas and perspectives [10]. In other words, argumentation transforms in order to arrive at a well-supported decision in a deliberative process. This basic idea underlying deliberation is used in all sorts of democratic innovations within public organisations. This section discusses a particular democratic innovation, AI-enabled mass deliberations, and describes the core aspects of institutionalising these democratic innovations.

2.1 AI-Enabled Mass Deliberations

Despite the consensual and educational affordances of a deliberative notion and practice of democracy, it suffers from a scaling problem. Democratic deliberations have been successfully conducted in practice in so-called mini-publics and citizen assemblies. Participants in mini-publics or assemblies are either selected through random sampling or sortition (i.e., creating a group of deliberators that represent all possible perspectives on the discussed topic). But these deliberations by small groups of citizens sideline large parts of the public. Authors have argued that mini-publics, therefore, do not live up to democratic values [2].

To counter the issues of mini-publics, authors have argued to scale up the number of citizens engaged in deliberations. Mass deliberations afford openness and accessibility. Deliberators are not selected, and engagement is maximized as the possible number of involved citizens is not capped. Organising mass deliberation in the traditional way (offline) would raise the costs for human moderation, and the needed time and space to levels that cannot be covered by public organisations.

At the same time, mass deliberation poses challenges that GenAI can address. The main challenge is information overload [3]. The amount of information (e.g., text or audio recordings) produced in a deliberative process increases when more people participate, obscuring the overview of all that has been discussed. GenAI functionalities like summarization, synthesis, and analysis of textual and audiovisual input can reduce this information overload. In addition, GenAI applications can facilitate learning among deliberators by providing background information suggesting directions for consensus, broadening perspectives in a discourse by playing devil's advocate, or augmenting expression by assisting deliberators in generating content [33].

2.2 Institutionalising AI into Deliberative Practices

As discussed in the introduction, the adoption of a new technological application should be followed by institutionalisation to actually create new deliberative practices [24]. The creation of a new practice means that new routines of activities emerge within an organisation from adapted interactions between people [21]. In the case of AI-enabled deliberation, these interactions are mediated by AI applications. Institutionalisation, then, refers to the sustainable integration of AI applications in deliberative practices and, therefore, decision-making processes in a public organization.

At the same time, institutionalisation is a process of infusing value(s) into an organisation or practice [25]. An institutionalised practice conveys or manifests a specific view on how things should be done by actors enacting that practice. When institutionalising AI-enabled mass deliberation, the normative goal is to establish meaningful deliberation – i.e., meaningful to both engaged citizens as well as public organisations receiving deliberative output [6]. Combined, institutionalisation means the sustainable integration of digital technologies into deliberative practices while upholding its main normative goals (i.e., meaningful deliberation).

Guaranteeing meaningful deliberation when AI is integrated into deliberative practices is not self-evident. A large and growing body of literature has investigated the threat of AI to democracy and democratic deliberation. Shortall et al. [26] doubt whether AI-enabled deliberation can address the democratic deficits of mass deliberation: lack of participation, reduced argument quality, empathy loss, power distortions, and negative perceptions. Ercan et al. [8] observes how communication at scale instigates a shift in the interaction between deliberators towards expressing, while moving away from listening and reflecting among participants. Other authors have argued that by integrating AI into deliberative processes they are, by definition, not democratic anymore [1]. In general, when designing AI-enabled deliberation, all trade-offs and design choices made can undermine the meaningfulness of deliberation. Therefore, this research aims to provide a better understanding of the gap between conceptual notions of institutionalisation and the perceptions of those notions practice to know whether AI-enabled mass deliberation is a viable proposition.

3 Research Methods

Studies of online and digital deliberation have produced significant insight into the success factors of institutionalising deliberation platforms [see 11, 22], is often not realised in practice [13, 27]. Moreover, the integration of new AI applications might require new forms of, or approaches towards institutionalisation. Therefore, we conducted an exploratory study of the differences between conceptual notions of institutionalisation of AI-enabled deliberation and practitioners' perceptions of those notions. Section 3.1 discusses our approach towards synthesising expert knowledge on institutionalisation applied to AI-enabled mass deliberation. Section 3.2 presents the research methods for assessing those conditions and requirements by practitioners.

3.1 Synthesis of Conceptual Notion of Institutionalisation

We identified (theoretical) conditions and requirements for the institutionalisation of AI-enabled mass deliberation through co-creation sessions with experts and practitioners in deliberative democracy, online deliberation, AI applications for deliberation, and institutionalisation of digital technologies. The experts and practitioners provided their perspectives and input regarding conceptualisations of institutionalising AI-enabled deliberation. These were synthesized into conditions and requirements for institutionalization. Conditions were defined as elements that must be guaranteed in order to achieve institutionalisation. Conditions can be considered a need-to-have. Requirements were defined as elements that can be pursued to increase the likelihood of institutionalisation. Requirements can be considered a nice-to-have. The co-creation sessions started from the understanding of institutionalisation in online deliberation literature as discussed in Sect. 2.

We organised five co-creation sessions as part of the AI4Deliberation project. We performed two asynchronous and one synchronous session within a research project on

AI-enabled mass deliberation. 8 unique participants provided input in the asynchronous sessions. In the synchronous session, 14 participants (of whom 10 had not yet participated in the asynchronous sessions) engaged in co-creation. Apart from the co-creation sessions within the research project, we organised two sessions at conferences. First, we facilitated a synchronous session at the EGOV2025 conference. 14 e-government scholars participated in this session. The other external session was organised at a practitioner-oriented conference on digital democracy in The Hague. This synchronous co-creation session included 31 participants. In total, 63 unique participants (mostly researchers) participated in our co-creation sessions.

We arranged the co-creation sessions as deliberative processes. We also used two deliberation platforms, as in the case studies presented in Sect. 3.2. As such, participants of the co-creation were informed about digital and AI-enabled deliberation and could experience the platforms themselves. For the asynchronous sessions, we used DebateGraph: an online platform for collectively creating argumentation graphs developed by ThoughtGraph. The synchronous sessions were facilitated through ECHO. ECHO is an AI-enabled deliberation platform by the company Dembrane. ECHO records live discussions between deliberators, transcribes the audio recordings, and the transcriptions using prompts. The co-creation sessions resulted in a synthesised list of 22 conditions and 16 requirements for institutionalisation of AI-enabled mass deliberation.

3.2 Assessment by Practitioners

Since institutionalisation happens through the enactment of practices by practitioners, we asked practitioners in four case studies to assess the identified conditions and requirements. Each individual participant filled out a survey. They scored each condition and each requirement on four criteria using a five-point Likert scale (1 = totally disagree; 2 = disagree; 3 = neutral; 4 = agree; 5 = totally agree). The criteria tested for two factors. First, the criteria 'importance' and 'contribution to meaningful deliberation/sustainable integration' evaluated the conceptual coverage of institutionalisation. Second, the criteria 'feasibility' and 'easiness of implementation' evaluated the relation of each condition or requirement to practice. We introduced the four criteria to participants as follows:

1. *Importance*: The extent to which the condition/requirement is important for achieving institutionalisation of AI-enabled mass deliberation.
2. Contribution to the Related Subgoal: The extent to which the condition/requirement contributes to the subgoal (meaningful deliberation or sustainable integration).
3. *Feasibility*: The extent to which the condition/requirement itself can be achieved by public organisations.
4. *Easiness of implementation*: The extent to which public organisations will be able to translate the condition/requirement to their daily work practices.

In addition to scoring the individual conditions and requirements, participants answered four sets of open questions about the full set of conditions and requirements:

1. Do the two subgoals (sustainable integration and meaningful deliberation) cover everything that is needed in your understanding of institutionalisation? Why?

2. To what extent are the two subgoals complementary OR contradictory? What contradictions do you see in the full set of conditions and requirements? What can be done about these contradictions?
3. To what extent does the full set of conditions and requirements reflect all necessary elements within the subgoals? Why?
4. Do you have any remarks, comments, additions, or improvements regarding the conditions and requirements?

Using the qualitative data collected from the open questions the scores provided by the practitioners could be interpreted. The triangulation of the explorative scores of the conditions and requirements with the answers provides insight into the perceptions of practitioners on whether the conceptual notion of institutionalising AI-enabled mass deliberation is a viable proposition in practice.

The conditions and requirements for institutionalisation were assessed by practitioners in four case studies. All four organisations in the cases were experimenting with the integration of AI into their deliberative processes. As such, they are also confronted with the institutionalisation process. In line with the explorative nature of our study, we selected cases that represent the different types of organisations involved in deliberation (governments, industry, NGOs), the different administrative levels on which mass deliberation can take place (local, national, international), and different deliberative forms (audio-based, text-based). The variety of cases provides a comprehensive understanding of institutionalising AI-enabled mass deliberation.

The four cases were selected to cover local, national, and international levels, and different countries. The local case is situated within a German municipality that aims to upgrade its current deliberative practices on the Consul platform by integrating AI applications. One national case is conducted by an Italian NGO that sets up an AI-enabled mass deliberation to formulate policy recommendations for Italy's climate adaptation plan. This case utilises the platform ECHO for audio-based AI-enabled deliberation. The other national case is rolled out in the official consultation procedure of national laws in Greece. The existing opengov.gr platform will be expanded with AI features. The international case tests the integration of AI applications into the graph-based discussion platform DebateGraph on the topic of Long Covid open to deliberators from all over the world. The numbers of survey participants in each case are presented in Table 1.

Table 1. Participants evaluating conditions and requirements for institutionalization of AI-enabled mass deliberations

Case	#participants
Local (Germany)	3
National (Italy)	6
National (Greece)	4
International	2

4 Results

4.1 Conditions and Requirements for Institutionalizing AI-Enabled Mass Deliberations

Section 2.2 conceptualised institutionalization of AI-enabled mass deliberation as the sustainable integration of routines in human actions (mediated by GenAI applications) to facilitate meaningful deliberation. This section elaborates on the meaning of both sustainable integration as well as meaningful deliberation. The refinement of these concepts is based on the conditions and requirements for institutionalising mass deliberations supported by generative AI applications identified in the co-creation sessions with experts.

Meaningful Deliberation. According to Deligiaouri [6], meaningful deliberation pertains to both the participants and organisers of deliberation. Citizens should feel engaged and their voices should be heard. For governments, deliberation should provide possibilities to aggregate opinions and interests and translate them into laws and policies. Accordingly, we consider meaningful deliberation to be achieved when deliberators (i.e., citizens participating in the deliberation) do have influence on actual decision-making within a public organisation. To achieve this, deliberators should have equal positions in the deliberative process, and the deliberative output should be of enough quality to be useful to policymakers. More generally, meaningful deliberation requires that (deliberative) democratic values are fostered [1, 4, 15, 26] and that the effectiveness of mass deliberation is guaranteed [5].

First, in the case of AI-enabled deliberation, the fostering of deliberative democratic values requires a shift away from the technology push currently observed in initiatives that integrate GenAI in deliberative practices [26]. The arrangement of digital technologies imposes specific forms of (rational) deliberation or determines the 'most appropriate ways' to deliberate [5], which can also widen traditional divides between citizens in deliberation [7] or introduce new forms of digital divides [18]. Jungherr & Rauchfleish [18] show the possibility of a so-called AI penalty; their empirical study on willingness to engage in AI-enabled deliberation indicates that people who distrust AI are less willing to participate in AI-enabled deliberative processes. Accordingly, the integration of AI applications undermines democratic values such as representativeness and engagement. In response, authors argue that both research and design of AI-enabled deliberation should be socio-technical- or value-oriented [4, 26]. Table 2 presents the conditions and requirements identified in the co-creation sessions fostering (deliberative) democratic values.

Table 2. Conditions and requirements related to fostering democratic values (meaningful deliberation)

Conditions	Requirements
Ensure accessibility of mass deliberation for all possible deliberators	Put deliberators in the lead of organising the mass deliberation

(continued)

Table 2. (*continued*)

Conditions	Requirements
Enlarge the capacities of deliberators to engage in mass deliberation	Respond to concerns regarding AI among potential deliberators
Guarantee influence on official decision-making	Guarantee meaningful human control over AI by involved and affected actors
Ensure transparency about the use of AI in mass deliberation	Guarantee the traceability of deliberative output's influence on official decision-making
Develop AI and digital literacy among deliberators	

Second, the effectiveness of mass deliberation needs to be guaranteed for deliberation to be meaningful. Mass deliberation can be considered effective when its output is adopted in official decision-making [5, 28]. Adoption is encouraged by sufficient quality of deliberative output [30]. It is essential to empower citizens so they can produce output of sufficient quality [32]. From the perspective of a public organisation, the output of deliberation should be useful, understandable, and actionable for decision-makers who, in the end, have to translate deliberative output to policies. AI applications can both increase and undermine deliberative quality [1, 26, 33]. At the same time, decision-makers should demonstrate commitment to deliberative output [32], as outlined in Table 2. The co-creation sessions resulted in three conditions and two requirements for guaranteeing the effectiveness of mass deliberation, see Table 3.

Table 3. Conditions and requirements related to guaranteeing effectiveness of mass deliberations (meaningful deliberation)

Conditions	Requirements
Establish a structured deliberative process that allows for mass deliberation	Measure deliberative quality by monitoring discussions
Balance the added value of AI applications with its costs	Allow for updating the deliberative output
Address safety problems caused by AI (such as manipulation, hallucination, and bias) that undermine deliberative quality	

Sustainable Integration. Second, we examine the elements of sustainable integration. We define sustainable integration as AI-enabled mass deliberation that has become a routine or standard activity in the daily practice of a public organisation. In that case, mass deliberation has passed the pilot phase and is no longer dependent on specific individuals organising the deliberation. To sustainably integrate AI-enabled mass deliberation, public organisations should cultivate long-term support for such deliberations by all actors involved in or affected by mass deliberation [22, 28], situate AI-enabled mass deliberation in their existing organisational practices [14], and continuously respond to changing deliberative practices and AI applications through adaptive governance [16, 17].

Ensuring long-term support is needed to bring AI-enabled mass deliberation beyond the pilot phase. Support is about engaging politicians and government in the deliberative process [28], which involves establishing political will to accept the consequences of deliberation and to allocate structural resources to sustain deliberative processes [22, 32]. Table 4 presents the conditions and requirements related to cultivating long-term support identified in the co-creation sessions.

Table 4. Conditions and requirements related to cultivating long-term support (sustainable integration)

Conditions	Requirements
Demonstrate commitment as public organisation to AI-enabled mass deliberation	Dedicate organisational capacity for AI-enabled mass deliberation
Confer legitimacy to the AI-enabled mass deliberative practice	Identify possible tensions between stakeholders
Adopt a deliberative stance/attitude within the public organisation	
Establish political will to engage in AI-enabled mass deliberation	
Guarantee the needed resources to enact AI-enabled mass deliberation	

The new AI-enabled deliberative practices will emerge in an already existing organisational context including established decision-making practices. When institutionalising AI-enabled mass deliberation, new activities need to be integrated into existing practices, and the old practices might need to be transformed [27]. Organisational practices provide stability to the behaviour of public organisations and, therefore, ensure trust, predictability and efficiency [23]. At the same time, such stable practices are hard to change [9]. Accordingly, adapting practices takes time. A smooth integration of AI-enabled mass deliberation is more likely to succeed when it aligns with existing practices. Transforming practices needs institutional (re)design alongside technological and process design [11]. Table 5 shows the conditions and requirements derived from the need to situate AI-enabled mass deliberation in existing organisational practices.

Table 5. Conditions and requirements related to situating AI-enabled mass deliberation in existing organizational practices (sustainable integration)

Conditions	Requirements
Situate mass deliberative practices in official decision-making processes	Combine AI-enabled mass deliberation with classic (mini public) deliberation

(continued)

Table 5. (*continued*)

Conditions	Requirements
Issue a formal mandate for deliberative practices	Relate AI-enabled deliberative practice to representative democratic practices (if applicable)
Respect the Rule of Law within the deliberative practice	Limit dependency on third parties
Establish data and AI governance	
Embed the AI applications in a broader technical infrastructure	

Democratic innovation goes along with governance innovation [15]. Here, we consider governance to be the management and control activities performed to ensure the effectiveness of deliberative practices. Especially the volatility of innovations in Gen AI applications asks for a governance approach aligned with that uncertainty. Janssen [16] positions adaptive governance as approach to evolve practices in response to changes in AI applications. Moreover, governance of AI-enabled practices should deal with the role of framing of AI functionalities [31] and should consider the dependencies of AI applications on a broader technological infrastructure [20]. The co-creation sessions resulted in the conditions and requirements for adaptive governance presented in Table 6.

Table 6. Conditions and requirements related to practicing adaptive governance (sustainable integration)

Conditions	Requirements
Design both the technological and governance parts of AI-enabled mass deliberation concurrently	Take time to organise AI-enabled mass deliberation
Establish spaces for (mutual) learning about AI-enabled mass deliberation	Formulate a vision on AI-enabled mass deliberation
React to changing interests and needs among involved and affected actors in AI-enabled mass deliberation	Enlarge capacity of public organisations to intervene in deliberative practice
React to continuous developments and innovations in technology, especially AI	Map the risks of using AI in mass deliberation
	Encourage monitoring changes in deliberative practices

4.2 Perceptions of Conditions and Requirements by Practitioners

Scoring the Individual Conditions and Requirements. To obtain an indication of how the practitioners participating in the assessment perceive the requirements and conditions we asked participants to score them on each of the four criteria. The conditions and requirements were presented in the tables presented in Sect. 4.1. Considering the small sample of practitioners, we provide an explorative and descriptive analysis of the results. This means that we focus on the frequencies of scores and the medians of these scores[1].

In general, the frequencies and medians of the scores show a difference between the assessment of the criteria related to conceptual coverage and the assessment of criteria related to translation to practice. The medians of the criteria 'importance' and 'contribution to the related subgoal' have medians of 4 or 5 (only one condition and one requirement have a median of 3). This suggests that the practitioners agree with the conceptual framing of institutionalisation of AI-enabled deliberation. In contrast, the scores of the practice-related criteria show more variety. The medians of the 'feasibility' of conditions and requirements vary between 2 and 5. Most conditions and requirements have a median of 3 – a neutral assessment. Regarding 'easiness of implementation', participants seem to be more pessimistic. Again, most conditions and requirements have a median of 3, but eight conditions and requirements have a median of 2. The difference between the two categories of criteria suggests that participants observe a gap between institutionalisation as a concept and its realisation in practice.

Here, we zoom in on the eight conditions and requirements with a median of 2. These are the following:

- Enlarge the capacities of deliberators to engage in mass deliberation (condition)
- Guarantee influence on official decision-making (condition)
- Put deliberators in the lead of organising the mass deliberation (requirement)
- React to continuous development and innovations in technology, especially AI (condition)
- Adopt a deliberative stance/attitude within the public organisation (condition)
- Establish political will to engage in AI-enabled mass deliberation (condition)
- Guarantee the needed resources to enact AI-enabled mass deliberation (condition)
- Dedicate organisational capacity for AI-enabled mass deliberation (requirement)

The low medians for these conditions and requirements suggest that participants have low confidence in the openness of public organisations towards meaningful deliberation in general. Moreover, the scores imply that guaranteeing influence of deliberators on official decision-making is difficult, that political will is hard to encourage, and that public organisations will not provide the needed resources.

Reflection on the Full Set of Conditions and Requirements. The scores provided by practitioners point out a gap between institutionalisation understood conceptually and in practice. This section uses the answers to the four open questions to arrive at explanations for this gap and how the gap may be a problem in the special case of AI-enabled mass deliberation. This section presents five tensions in practice perceived by practitioners that explain the lower scores on 'feasibility' and 'easiness of implementation' for all conditions and requirements.

Deliberative Attitude Versus Resistance to Deliberation. The participants argued that the general attitude within public organisations is not conducive to integrating new

[1] The dataset of frequencies and medians per criteria for each condition or requirement can be requested from the authors.

forms of democratic practices such as AI-enabled deliberation. Participants doubted whether public organisations have genuine interest in real and meaningful deliberation. They argued that public organisations tend to have reservations towards the pressure that follows from deliberative outputs. One participant even stated that a deliberative attitude is contrary to the current administrative culture within public organisations. Moreover, participants indicated that elements related to culture or informal interactions between actors are missing in the understanding of institutionalisation presented to them. For example, several participants mentioned the need for cultural change in public organisations to address the scepticism often observed among public servants.

Capacity and Competencies Versus Constrained Resources. Regarding AI-enabled mass deliberation, much attention goes to the empowerment of citizens and what that means in terms of strengthening their competencies and skills to engage in digital deliberation. Participants stated that they would like to see more emphasis on the importance of competencies, skills and capabilities among public servants, facilitators, moderators and other roles involved in organising or governing deliberation. Both competency and capacity at the organiser's side (particularly towards the use of AI) needs to be guaranteed to ensure deliberative quality. This means that institutionalisation includes training of facilitators and technical staff, incentives for public servants and politicians to engage in AI-enabled mass deliberation, and the allocation of necessary resources. However, the practitioners mentioned that these resources (mostly financial) are often not allocated or insufficient. The lack or shortage of these resources can also be interpreted as little commitment at the side of public organisations in establishing AI-enabled deliberative practices.

Meaningful Deliberation Versus Sustainable Integration. Participants confirmed the complementarity between sustainable integration and meaningful deliberation but also anticipated tensions between the organisational implications of both sides of institutionalisation. Several participants felt that sustainable integration requires formal and more rigid procedures, while meaningful deliberation asks for flexibility and adaptation. As one participant stated: 'several conditions under "sustainable integration" emphasize embedding deliberation in existing processes and formal mandates, while "meaningful deliberation" requires responsiveness to participants. If existing processes are rigid, these can conflict. The adaptive governance conditions address this, but implementers may find the balance difficult in practice.' Another participant stated: 'high standards [regarding mass deliberation] clash with limited organisational capacity, fragmented responsibilities and mismatches of competences between levels of government.' Remarkably, one participant gave an opposite interpretation to the tension. They argued that to achieve meaningful deliberation demands 'resource-intensive and complex designs', while attaining sustainable integration 'requires easiness, agility, cost-efficiency, and long-term managing.' This tension refers to the ambidextrous capacity of organisations.

Potential of AI for Scaling Versus Unknown Implications of Mass Deliberation. Mass deliberation was an unrealistic goal for a long time because of physical and mental constraints, and its resource-heavy nature. Consequently, there is little experience with scaling deliberation, and its organisational implications are unknown. Similarly, not all implications of using GenAI in deliberation are known. Participants were particularly concerned about the effects on deliberative quality. They foresaw the implementation of new control measures such as monitoring the influence of AI applications on deliberative quality, required efforts towards communicating how AI processes and summarised inputs, safeguards against 'consensus manufacturing' through AI summarisation, requirements for human facilitators alongside AI tools, and multi-language requirements. Moreover, public organisations will need to redefine their role in handling low-quality or off-topic submissions. This tension implies that changing deliberative practices will instigate many new and unforeseen activities, especially when AI-enabled deliberation is implemented for the first time.

Emergent Institutionalisation Versus Detailed Guidance. As discussed before, institutionalisation is a continuous and emergent process without a clear end goal or defined consequences. Moreover, institutionalisation is inherently context-dependent. Therefore, different adequate and successful deliberative practices can emerge. One participant argued to conceptually reframe the conditions and requirements to 'ongoing practices or vectors of continuous improvement.' This framing also aligns with the interpretation of institutionalisation as 'entering a reflexive process [rather] than achieving a state.' However, this conflicts with clarity needed at the side of both deliberators and organisers of deliberations. Practitioners commented that public organisations need more information about the consequentiality of conditions and requirements, for example, in the form of success indicators or specified consequences of not meeting the conditions and requirements. At the same time, as another participant commented, this position can result in a situation in which public organisations only focus on satisfying the conditions and requirements, without being concerned about actual institutionalisation of meaningful deliberation.

5 Conclusions

This paper identified tensions between the conceptual notions of and the practice of institutionalisation that may impede the sustainable integration of AI in meaningful deliberative practices. We compared conditions and requirements formulated by online deliberation experts (i.e., the conceptual notion of institutionalisation) with practitioners' perceptions of these conditions and requirements. Our research confirmed the gap between the conceptual notion of and the practice of institutionalisation. This gap manifests in five inherent tensions in practice: 1) deliberative attitude versus resistance to deliberation; 2) capacity and competencies versus constrained resources; 3) meaningful deliberation versus sustainable integration; 4) potential of AI for scaling versus unknown implications of mass deliberation; and 5) emergent institutionalisation versus detailed guidance.

Our findings imply that the tensions in practice need to be resolved or addressed before AI-enabled mass deliberation becomes a feasible proposition. Moreover, the findings suggest that the tensions are not necessarily related to the technological aspects of AI-enabled mass deliberation, but more so to its institutional aspects. The findings raise the question of whether the tensions are surmountable or reconcilable. If not, new deliberative practices enabled by AI applications that do foster the normative goal behind democratic deliberation will not emerge and these initiatives will not move beyond adoption. Accordingly, investments made in AI-enabled mass deliberation would be misused or wasted.

Our research indicates that more research is needed on the implications or consequences of introducing AI into deliberations. The mass scale and the use of AI in the type of deliberations discussed in this paper make deliberative processes more complex and institutionalisation even more complicated. For example, it will be harder to keep the overview of what happens in the deliberation and more efforts are needed to check the quality of deliberations. Especially, the transition from mini-public deliberations to AI-enabled mass deliberation asks for more investment of resources before the efficiency gains of AI-enabled deliberation pay off.

The research in this paper is of exploratory nature. The research has focused on conditions and requirements, which implies that there is an identifiable end state of institutionalisation. However, institutionalisation is a continuous process of human agents enacting a specific practice [21] that cannot be fully anticipated and controlled. The conditions and requirements in this study provide a general understanding of institutionalisation, but ethnographic research is needed to grasp the granularity of institutionalisation of AI-enabled mass deliberations. Another limitation of this research is that we did not provide the practitioners with detailed explanations of each condition or requirement. This kept the survey concise but reduced comprehensibility and increased the interpretability of the conditions and requirements. This asks for more explanatory research of the perceptions of practitioners on institutionalisation and whether frames of prejudices are hindering institutionalisation.

Acknowledgments. The research presented in this paper was conducted within the AI4Deliberation project. The authors would like to acknowledge the contribution of all consortium partners to formulate the project. AI4Deliberation is funded by the European Union under grant agreement No. 101178806. Views and opinions expressed are however those of the author only and do not necessarily reflect those of the European Union or European Research Executive Agency. Neither the European Union nor the granting authority can be held responsible for them.

Disclosure of Interests. The authors have no competing interests.

References

1. Alnemr, N.: Emancipation cannot be programmed: blind spots of algorithmic facilitation in online deliberation. Contemp. Polit. **26**(5), 531–552 (2020). https://doi.org/10.1080/13569775.2020.1791306
2. Böker, M.: Justification, critique and deliberative legitimacy: the limits of mini-publics. Contemp. Polit. Theory **16**(1), 19–40 (2017). https://doi.org/10.1057/cpt.2016.11
3. Bono Rossello, N., Simonofski, A., Castiaux, A.: Artificial intelligence for digital citizen participation: design principles for a collective intelligence architecture. Gov. Inf. Q. **42**(2), 102020 (2025). https://doi.org/10.1016/j.giq.2025.102020
4. Carstens, J.A., Friess, D.: AI within online discussions: rational, civil, privileged?: ethical considerations on the interference of AI in online discourse. Mind. Mach. **34**(2), 10 (2024). https://doi.org/10.1007/s11023-024-09658-0

5. Coleman, S., Moss, G.: Under construction: the field of online deliberation research. J. Inform. Tech. Polit. **9**(1), 1–15 (2012). https://doi.org/10.1080/19331681.2011.635957
6. Deligiaouri, A.: Open governance and E-rulemaking: online deliberation and policy-making in contemporary Greek politics. J. Inform. Tech. Polit. **10**(1), 104–124 (2013). https://doi.org/10.1080/19331681.2012.746109
7. Epstein, D., Newhart, M., Vernon, R.: Not by technology alone: the "analog" aspects of online public engagement in policymaking. Gov. Inf. Q. **31**(2), 337–344 (2014). https://doi.org/10.1016/j.giq.2014.01.001
8. Ercan, S.A., Hendriks, C.M., Dryzek, J.S.: Public deliberation in an era of communicative plenty. Policy Polit. **47**(1), 19–35 (2019). https://doi.org/10.1332/030557318X15200933925405
9. Farjoun, M.: Beyond dualism: stability and change as a duality. Acad. Manag. Rev. **35**(2), 202–225 (2010). https://doi.org/10.5465/amr.35.2.zok202
10. Fishkin, J.: Democracy When the People Are Thinking, vol. 1. Oxford University Press (2018). https://doi.org/10.1093/oso/9780198820291.001.0001
11. Friess, D., Eilders, C.: A systematic review of online deliberation research: a review of online deliberation research. Policy Internet **7**(3), 319–339 (2015). https://doi.org/10.1002/poi3.95
12. Gutmann, A., Thompson, D.F.: Why Deliberative Democracy? Princeton University Press (2004)
13. Hartz-Karp, J., Sullivan, B.: The unfulfilled promise of online deliberation. J. Deliberative Democracy, **10**(1) (2014). https://doi.org/10.16997/jdd.191
14. Hendriks, C.M.: Coupling citizens and elites in deliberative systems: the role of institutional design. Eur J Polit Res **55**(1), 43–60 (2016). https://doi.org/10.1111/1475-6765.12123
15. Hendriks, F.: Key values for democratic governance innovation: two traditions and a synthesis. Public Admin. **100**(4), 803–820 (2022). https://doi.org/10.1111/padm.12738
16. Janssen, M.: Responsible governance of generative AI: conceptualizing GenAI as complex adaptive systems. Policy Soc. **44**(1), 38–51 (2025). https://doi.org/10.1093/polsoc/puae040
17. Janssen, M., Van Der Voort, H.: Adaptive governance: towards a stable, accountable and responsive government. Gov. Inf. Q. **33**(1), 1–5 (2016). https://doi.org/10.1016/j.giq.2016.02.003
18. Jungherr, A., Rauchfleisch, A.: Artificial Intelligence in deliberation: the AI penalty and the emergence of a new deliberative divide. Gov. Inf. Q. **42**(4), 102079 (2025). https://doi.org/10.1016/j.giq.2025.102079
19. Landemore, H.: Can artificial intelligence bring deliberation to the masses? In: Chang, R., Srinivasan, A. (eds.), Conversations in Philosophy, Law, and Politics 1st ed, pp. 39–69. Oxford University Press, Oxford (2024). https://doi.org/10.1093/oso/9780198864523.003.0003
20. Mueller, M.L.: It's just distributed computing: rethinking AI governance. Telecommun. Policy, 102917 (2025).https://doi.org/10.1016/j.telpol.2025.102917
21. Orlikowski, W.J.: Sociomaterial Practices: Exploring Technology at Work. Organ. Stud. **28**(9), 1435–1448 (2001). https://doi.org/10.1177/0170840607081138
22. Panopoulou, E., Tambouris, E., Tarabanis, K.: Success factors in designing eParticipation initiatives. Inf. Organ. **24**(4), 195–213 (2014). https://doi.org/10.1016/j.infoandorg.2014.08.001
23. Peters, B.G.: Institutional Theory in Political Science: The New Institutionalism (4th ed). Edward Elgar publishing (2019)
24. Randma-Liiv, T.: Adoption is not enough: institutionalization of e-participation initiatives. Public Policy Admin. **38**(3), 329–351 (2023). https://doi.org/10.1177/09520767211069199
25. Selznick, P.: Leadership in Administration: A Sociological Interpretation. University of California Press (1984)

26. Shortall, R., Itten, A., Meer, M.V.D., Murukannaiah, P., Jonker, C.: Reason against the machine? Future directions for mass online deliberation. Front. Polit. Sci. **4**, 946589 (2022). https://doi.org/10.3389/fpos.2022.946589
27. Steinbach, M., Sieweke, J., Süß, S.: The diffusion of e-participation in public administrations: a systematic literature review. J. Organ. Comput. Electron. Commer. **29**(2), 61–95 (2019). https://doi.org/10.1080/10919392.2019.1552749
28. Stromer-Galley, J., Webb, N., Muhlberger, P.: Deliberative E-rulemaking project: challenges to enacting real world deliberation. J. Inform. Tech. Polit. **9**(1), 82–96 (2012). https://doi.org/10.1080/19331681.2012.635971
29. Tessler, M.H., et al.: AI can help humans find common ground in democratic deliberation. Science, **386**(6719), eadq2852 (2024). https://doi.org/10.1126/science.adq2852
30. Towne, W.B., Herbsleb, J.D.: Design considerations for online deliberation systems. J. Inform. Tech. Polit. **9**(1), 97–115 (2012). https://doi.org/10.1080/19331681.2011.637711
31. Veale, M., Matus, K., Gorwa, R.: AI and global governance: modalities, rationales, tensions. Ann. Rev. Law Soc. Sci. **19**(1), 255–275 (2023). https://doi.org/10.1146/annurev-lawsocsci-020223-040749
32. Velikanov, C., Prosser, A.: Mass online deliberation in participatory policy-making—Part I. In: Paulin, A.A., Anthopoulos, L.G., Reddick, C.G. (eds.), Beyond Bureaucracy, vol. 25, pp. 209–234. Springer, Cham (2017).https://doi.org/10.1007/978-3-319-54142-6_13
33. Zeginis, D., Patsouras, C.-F., Promikyridis, R., Tambouris, E., Tarabanis, K.: Review of AI features to support mass deliberations. In: Lindgren, I., et al. (eds.) Electronic Government, vol. 15944, pp. 336–350. Springer, Cham (2026). https://doi.org/10.1007/978-3-032-01589-1_21

Can AI Representatives Be Legitimate? An Exploratory Study in Citizen Participation

Richard J. Specker[1(✉)], Anna K. Boos[1], Birgit Schenk[1,2], Lena N. Fischer[1], Wenyuan Wu[1], Mateusz Dolata[1,3], and Gerhard Schwabe[1]

[1] University of Zürich, 8050 Zürich, Switzerland
specker@ifi.uzh.ch
[2] University of Applied Sciences Ludwigsburg, 71634 Ludwigsburg, Germany
[3] Zeppelin University, 88045 Friedrichshafen, Germany

Abstract. While citizen participation provides essential local expertise for effective policies, many initiatives struggle to move beyond consultation. Progressing toward policy implementation is often hindered by limited resources and technical complexity. This study investigates the potential of AI representatives to bridge this gap. Using exploratory interviews with an embedded vignette, we explore how 52 citizens and public servants in a German municipality perceive the legitimacy of AI agents acting as digital representatives during internal policy formulation meetings. Our findings provide empirical evidence that AI representatives are perceived as a potentially legitimate mechanism to ensure that citizens' voices persist throughout the policy cycle. We identify the specific conditions and rationales under which AI representatives are seen as a viable compromise to shift participation from simple consultation toward a more structural role in governance.

Keywords: Citizen participation · AI representatives · Legitimacy

1 Introduction

Citizen participation aims to leverage local expertise for democratic legitimacy and effective policy [2,17]. Nevertheless, citizen influence often diminishes along the policy cycle, as their role is commonly limited to providing information while authorities retain decision-making control [25,29]. Prior research identified practical barriers that prevent the inclusion of citizens throughout the policy-making process: for citizens, the process is time-intensive, feels pointless if input is ignored in final decisions, and requires specialized knowledge in later technical stages [17,22]. Simultaneously, public authorities face high costs and risks of increased hostility if processes appear insincere or outcomes lack authority. Without a meaningful link to formal policy-making, these initiatives risk becoming formalities that result in poor policies and discouraged future engagement [17].

Published by Springer Nature Switzerland AG 2027
D. Duenas-Cid et al. (Eds.): ePart 2026, LNCS 16923, pp. 165–181, 2027.
https://doi.org/10.1007/978-3-032-36676-4_11

Artificial Intelligence (AI) present an opportunity to address current challenges in citizen participation [30,39]. Recent studies show that AI systems can generate authentic content and amplify unaddressed or minority viewpoints [15,19,20]. Within the realm of citizen participation, this technology could offer novel mechanisms: data collected during initial participatory or deliberative processes could power AI agents that act as representatives (henceforth *AI representatives*) for citizens [11,39]. These agents function as intermediaries, advocating for citizens' interests within post-consultation processes where public involvement typically decreases [24]. Theoretically, such an implementation could transition citizen input from episodic consultation into a structural, albeit non-decisory, presence within the technical stages of policy formulation and decision-making. Despite this potential, the introduction of AI representatives not only brings new actors into the policy process but also has the potential to reconfigure established procedures that result in decisions binding on citizens. This raises questions regarding the democratic legitimacy of such AI-assisted governance processes:

RQ1: *How do citizens and public servants evaluate the legitimacy of such AI representatives within internal administrative policy-making processes?*

RQ2: *Why and when do these actors regard AI representatives as legitimate?*

To address these questions, we conducted an exploratory qualitative study [34,43] during an Association Summit in a German municipality. Instead of evaluating a functional prototype, our methodology relied on 52 semi-structured interviews with citizens and public servants that centered on a vignette [16]. The vignette described a narrative in which an AI agent, provisioned with data from the summit, would represent citizens during internal policy-making processes.

Our study makes two contributions: (1) We provide empirical evidence that both citizens and public servants perceive AI representatives as a legitimate mechanism to ensure that civic voices persist throughout the policy-making process. (2) We explain the rationale behind these evaluations. By documenting these perceptions, we highlight the conditions under which AI representatives are viewed as a compromise that could shift citizen involvement from consultations towards forms of participation with greater power redistribution.

2 Related Work

2.1 Citizen Participation in Democracies

Democracy is a political regime grounded in procedures of collective decision-making by the citizenry, in which all citizens hold an equal right to participate in shaping common action and to control the exercise of political authority [7]. Given the scale and complexity of politics, it is largely uncontested that democratic decision-making must rely on a division of political labor, and that democratic rule must in practice be exercised through political representatives. This raises the question of how such representation can best be understood in order to do justice to the democratic ideal of including all citizens as equal participants. In this context, Hanna Pitkin's [33] distinction between *trustees* and *delegates* marks the main point of contention: A representative acting as a trustee exercises independent judgment. This approach assumes that the division of political labor leaves officeholders are typically better placed than ordinary citizens to reach well-informed and well-reasoned decisions. A representative acting as a delegate, by contrast, defers to the expressed preferences of those represented. The latter understanding foregrounds the idea of democratic accountability: because citizens authorize representatives to act on their behalf, representatives are answerable for enacting citizens' judgments, and a failure to do so erodes popular control over governance and consequently, democratic legitimacy.

A growing body of research has turned to *democratic innovations* as reforms that seek to broaden citizen participation beyond the periodic elections and occasional referendums or citizens' initiatives characteristic of modern representative democracies. These innovations seek to strengthen citizens' role in governance by broadening their opportunities to participate, deliberate, and exert influence [9]. Examples include deliberative mini-publics such as citizens' assemblies, juries, deliberative polls, participatory budgeting as well as direct-democratic instruments such as referendums and citizens' initiatives. While these instruments intend to draw on local knowledge to ensure public values are incorporated into policies and to strengthen democratic legitimacy [29,32,40], they are not without drawbacks. Risks include (1) domination by vocal groups who may exploit the process to advance personal agendas rather than the public interest, and (2) lack of expertise among participants, which can result in technically flawed proposals that delay projects and strain local budgets [17].

Beyond these general risks, the effectiveness of these participatory instruments depends on the specific stage of the policy cycle (agenda-setting, formulation, decision-making, implementation and evaluation) in which they are used [17,18,22]. Participation is often established during agenda-setting but diminishes during formulation and decision-making, creating a "participation gap" [22,24]. As public servants have considerable discretion within these closed administrative structures, the nuances of the original deliberative intent are often filtered out or lost [24]. Without mechanisms that maintain influence in the subsequent phases of the policy-making process, participation risks devolving into what Arnstein [2] describes as tokenism (i.e., citizens are informed and/or heard but lack decision-making power). Ultimately, the challenge for modern governance is finding a balance between citizen involvement and the need to avoid project delays and high administrative costs [17].

A more recent strand of research turns to *digital democratic innovations* which specifically explores how the use of digital technologies can enhance the scale, transparency, inclusiveness, and deliberative quality of democratic procedures [10,19,26,32,44,45]. While earlier technologies primarily supported democratic procedures, advancements in Natural Language Processing (NLP) have shifted the focus toward content-based support. By utilizing sentiment analysis and text summarization, NLP tools can assist public servants in interpreting community perceptions at scale [40,41]. However, these tools serve to condense data for internal administrative use, instead of actively advocating for the citizen's perspective. In contrast, generative AI systems introduce the possibility of active representation [19,30,40]. It is precisely here that our research on AI representatives in administrative decision-making can be situated. We envision AI representatives that make citizens present at stages of the policy cycle from which they are normally distant. Whereas citizens usually elect their legislators directly, the officials who staff the executive are not directly chosen by the citizenry. At this comparatively insulated stage of internal policy formulation, public involvement typically diminishes [24]. In light of the democratic ideal sketched above, namely, an equal entitlement to participate in shaping common action and to control the exercise of political authority, AI representatives may, therefore, in principle, enhance democratic legitimacy. By embedding citizen input directly into these later stages, such technology could potentially bridge the gap between initial public consultation and final policy implementation.

While the technical feasibility of such solutions is increasing, the introduction of AI representatives challenges traditional notions of how citizen interests are advocated. Consequently, the success of this innovation is not merely a technical question, but relies also on perceived legitimacy. Perceived legitimacy is tied to the acceptance of the decision-making process and the final outcome [12]. Current studies suggest that the public remains skeptical of using such technologies in the public domain, generally preferring an advisory role assisted by humans over a decisive one [13]. This necessitates a closer examination of what constitutes legitimacy in the context of artificial representation and the conditions under which such a representative is accepted as a valid democratic voice.

2.2 The Concept of Political Legitimacy

The concept of political legitimacy refers to the normative status of political institutions and their decisions as justified and permissible [31]. A political authority is considered legitimate if political power is exercised by the rightful agents and through the rightful means [42]. In democratic contexts, political legitimacy is often equated with democratic legitimacy, as this authority derives from the equal participation of citizens in political decision-making. When assessing the political or democratic legitimacy of a political system, this is often done along three dimensions: input, throughput, and output legitimacy [37].

Input legitimacy is established through the participation and representation of relevant stakeholders (e.g., citizens) [37]. It requires meaningful opportunities to influence democratic institutions and is strengthened when these institutions reflect citizens' interests and provide accessible channels for voicing their views [14,27,35]. Input legitimacy, therefore, depends on the inclusion of diverse perspectives and demographic groups [12], as well as on citizens' trust that they have a fair opportunity to shape decision-making and outcomes [8]. Input legitimacy is undermined, however, when citizens cannot see how their contributions actually feed into the decision-making process [27]. A meaningful participatory instrument must therefore show that citizen input has tangible effects on outcomes, otherwise, access may be perceived as mere tokenism [8,27].

Throughput legitimacy focuses on the internal quality and integrity of the governance process [37]. It captures the participants' assessment of how decisions are actually formulated, processed, and implemented within institutions, emphasizing whether the process is perceived as proper, just, and fair [12,27]. For a process to be considered legitimate in its throughput dimension, it must not only exhibit a high degree of transparency and efficacy but also openness to interest consultation, which, in turn, must be perceived as adequate and constructive [27,37]. Strong throughput legitimacy tends to enhance public trust and support acceptance of the resulting outcome [8].

Output legitimacy represents the performance dimension of a political system [37]. It is rooted in the idea that a process earns its right to exist based on how effectively it fulfills the collective purposes of the community [8,35,37]. For citizens to view a policy or decision as legitimate in its output, the results must not only be efficient and effective but must also be perceived as fair, responsive, and just [8,12,14,27,36]. Output legitimacy can thus be seen as a feedback loop tied to input legitimacy, in which citizens assess whether governing bodies have met their needs and sanction those that fail to produce outcomes consistent with their wishes [37].

The three dimensions of legitimacy are not standalone assessment criteria but intertwined and mutually reinforcing. A policy-making process may be carefully designed and procedurally balanced (throughput legitimacy), yet without fair opportunities for citizen participation (input legitimacy), its democratic legitimacy remains limited. Conversely, participatory formats alone are insufficient if different views are not weighed fairly in the process (throughput legitimacy) or if poor policy outcomes result (output legitimacy) [27]. By introducing new actors and reconfiguring established procedures, AI representatives raise new normative questions as they may affect all three dimensions of legitimacy.

3 Method

To explore the legitimacy of AI representatives in policy processes, we adopted a qualitative exploratory-interpretive paradigm [34,43]. This approach prioritizes understanding citizens' and public servants' perceptions by reaching a shared interpretation of the data, rather than an objectively correct one [23]. Our research design employs semi-structured interviews with an embedded vignette [16,28].

Vignettes serve as stimuli, typically consisting of text or images, to which participants respond [16]. These systematically constructed descriptions of personas, objects, or situations are used as input for judgment-making processes [1,3]. Our study used a vignette to explore a hypothetical situation grounded in digital trends. The aim was to elicit reactions to a plausible future in which AI agents represent civic interests during internal administrative deliberations. This approach works in three ways: (1) It uses an imaginary scenario to spark discussions about events that do not happen in daily life. (2) It uncovers how people might react to these situations in principle. (3) It allows us to study beliefs and perceptions that are hard to capture through direct questioning alone [16].

We combined this approach with qualitative coding [4,34] and sensemaking sessions during interpretation workshops, which involved a multidisciplinary research team with expertise in Information Systems, Public Administration, and Political Philosophy [38].

3.1 Data Collection

Data collection relied on semi-structured interviews with an embedded vignette. This methodological combination is particularly well-suited for exploring participants' opinions concerning complex and potentially sensitive issues, such as democratic legitimacy [16,21]. The semi-structured interviews provided a standardized framework for exploring core themes while maintaining the necessary flexibility to probe deeper into unexpected participant insights [28].

Field. We conducted our research in a German municipality known for its active citizen engagement in policy-making. The context for our study is based on an "Association Summit" held in October 2025. This summit served as a platform where members of local associations (e.g., sports clubs, music clubs, churches, and civic initiatives) could interact directly with municipal departments. During the summit, these association members could express their needs and voice dissatisfaction with existing administrative procedures. For instance, participants could highlight areas where routine tasks, such as applying for event permits or reserving public facilities, were causing unnecessary difficulties for their association work. To efficiently manage and process the large volume of citizen input, the municipality used note-taking software. This digital documentation served as the foundation for subsequent internal administrative processes, enabling municipal departments to deliberate and issue formal, coordinated responses to concerns reported at the summit.

Participants. We employed a purposive and convenience sampling strategy to recruit participants at the Association Summit. Our participant pool consisted of two cohorts: (1) 34 citizens recruited directly at the Association Summit. To ensure the inquiry remained grounded in the immediate experience of participation while maintaining data quality, interviews were conducted in person at dedicated tables positioned at the periphery of the event hall. (2) 18 public servants who had been professionally involved in the summit. These participants were approached via email within one week of the event to ensure the experience remained salient. Fifteen interviews were conducted via Microsoft Teams, and three were held in person.

This sampling strategy allowed us to compare the perceptions of citizens (who generated the data) with those of public servants (who processed the data). The combined sample of 52 participants exhibited a broad age range (21–86 years) and a balanced gender distribution, as detailed in Table 1.

Table 1. Demographic Characteristics of Interviewees

Category		Association Members ($n = 34$)	Public Servants ($n = 18$)	Total ($N = 52$)
Age	Mean	53	40	49
	Min–Max	21–86	23–63	21–86
Gender	Female	14	11	25
	Male	20	7	27

Vignette Study. Our vignette was designed to address a specific practical gap in the municipality's internal administrative processes following the Association Summit. Although the note-taking software captured association members' concerns for departmental deliberation, these citizens remain excluded from the subsequent decision-making phase. The vignette addressed this gap by presenting the following scenario:

Following the initial Association Summit, municipal departments gather for a workshop to further prioritize concerns and develop solutions. In the current process, administrative staff make these decisions based on professional expertise, strategic goals, and available resources. While citizen concerns provide the foundation for this work, the citizens themselves are not directly involved in this internal phase.
In this new scenario, an AI agent participates in the aforementioned workshop. This AI agent is provisioned with the specific opinions and prioritized perspectives previously gathered from the association members, acting as their representative. The agent contributes ideas, comments on proposed solutions, and participates in voting to keep the citizens' interests at the center of the deliberation.

By grounding the vignette in the association summit, we ensured the scenario was relevant to those involved. Also, we intentionally left out technical details of the AI agent in order to keep the focus on the core concepts and avoid distracting participants with jargon. This selective context allowed participants to respond based on their own real-world experiences and expectations of local authorities.

Interview Procedure. Data collection was conducted by a five-member research team (three female, two male), consisting of one senior researcher, three research assistants, and one graduate student. All interviews followed a pretested interview guide designed to remove ambiguities and ensure the questions elicited the intended information.

The interview procedure followed a five-phase protocol designed to transition participants from their lived experiences to their perspectives on a future-oriented scenario: (1) Participants shared their personal experiences at the Association Summit. (2) They were then presented with a hypothetical scenario to trigger thoughts on how an AI agent might represent them in the administrative policy-making process (see Subsection on Vignette Study). (3) Using open-ended questions, we explored how this technology could affect trust, accountability, and the relationship between the municipality and its residents. (4) Then, we invited participants to share any final thoughts not already covered. (5) Finally, we collected basic demographic information.

On average, interviews with citizens lasted 16 min and those with civil servants 31 min. In total, these amount to 18 h of audio-recorded interview material. All participants provided written consent. Each session was transcribed verbatim before being anonymized. To ensure confidentiality and support analysis, we used a token-based identification system: "AM-XX" denotes association members (citizens), while "EM-XX" denotes public servants. All quotes used in the subsequent result section have been edited for readability (removing filler words and self-corrections) while preserving the original intent.

3.2 Data Analysis

We adopted an exploratory-interpretive paradigm [34,43], prioritizing the attainment of a shared interpretation of the data over an objectively correct truth [23]. We managed data in MaxQDA and coded it following Bingham and Witkowsky's five-phase framework [4], which incorporates Saldaña's two-cycle coding [34].

In the first phases of organization and sorting, we conducted a deductive "pre-coding" workshop where researchers from information systems, political philosophy, and public administration identified broad categories of interest. These categories served as an initial conceptual guide while the primary coder concurrently identified significant passages for granular analysis. The first subsequent coding cycle employed an abductive approach, allowing the integration of established theoretical models with new concepts emerging from the transcripts. We utilized three primary coding methods: descriptive coding to achieve an organizational grasp of the data; in vivo coding to capture participants' authentic linguistic expressions; and emotional coding to label affective responses, providing deep insight into participants' subjective perspectives on AI representation [34]. Parallel to this process, a codebook and memos were maintained to ensure consistency across the 52 transcripts.

During the interpretation phase (second coding cycle), the focus shifted from initial description to synthesis through two interpretation workshops involving the full multidisciplinary team. They moved beyond descriptive summaries to axial and pattern coding, specifically focusing on legitimacy dimensions. This iterative process yielded five pattern codes (e.g., "Ambiguous and Contested Attribution of Responsibility in AI Representation" and "AI as a Functional Tool in Contrast to Human Actors") and eleven nuanced themes. In the final stage, the analysis bridged inductive findings with deductive academic inquiry by applying theoretical codes from the literature.

4 Results

4.1 Input Legitimacy

Input legitimacy depends on the inclusiveness and openness of the processes involving the AI representatives. Participants noted that these representatives lower participation barriers, such as social status or specialized knowledge, which typically marginalize certain groups. The AI representative provides these groups with the formal representation necessary for political influence. As one public servant noted *"[With AI representatives] we have people at the table and, [...] we include those who might not dare to sit at the table themselves. [...] They might dare to [participate], because perhaps language skills are lacking"* (EM-08). Beyond social hesitation, AI representatives also address structural temporal constraints that restrict access for those with standard working hours. As one citizen observed, *"citizens are usually not available during the day from 8:00 to 16:00 because they also have to work [...] there is certainly an added value here"* (AM-26). However, some participants cautioned that if the AI representatives are solely powered by data from physical events, it may inherit existing exclusionary patterns: *"At an Association Summit like this, always only a certain clientele from the associations attends"* (AM-27). Furthermore, a few public servants expressed concern that the citizens might interpret AI integration as administrative withdrawal, fearing citizens might think *"now [the authorities] are closing themselves off completely. Now they only want to do things via AI, and now we [as citizens] have no say at all"* (EM-10).

The openness of a policy process is fundamentally shaped by the perceived neutrality and anonymity it affords. Participants suggested that AI representatives can express ideas that are not influenced by human psychological biases, serving as *"a positive influence, because I, as a human [...] maybe because of my current state of mind, just cannot express this idea, and the AI is free from something like that. It just brings it"* (AM-34). From a strategic perspective, the AI representative was viewed as a shield for voicing unpopular opinions without personal risk, as one interviewee put it: *"I can use it anonymously [...] I don't have to say it personally, but someone else says it. I can push someone [the AI] forward"* (EM-01). Finally, another participant suggested that the AI representatives might mitigate the dominance of vocal groups, hoping that *"it acts as a kind of corrective [...] where people who are loud and report frequently simply prevail, and through an AI that might be put into perspective"* (AM-29).

4.2 Throughput Legitimacy

Throughput legitimacy rests on transparency, deliberation quality, accountability and representation fidelity in policy processes. Regarding transparency, (AM-26) argues for public disclosure when AI representatives act in democratic bodies: *"I actually consider it sensible that the AI [...] is disclosed, especially if a real decision that affects real life results from it"*. Beyond mere disclosure of the AI's presence, (EM-05) highlights the need for clear boundaries between human and machine labor to maintain accountability: they would always state the *"AI played a role [...] Ultimately, though, I must check it and then make it available"*.

The AI representative introduces a trade-off for deliberation quality. On one hand, it could increase information usage and broaden the solution space. As one citizen noted: *"I do believe that solutions might emerge that we, with our perspectives and our limitations, would never have thought of"* (AM-12). In addition, it serves to rationalize and objectify internal discussions. (EM-18) assumes that the AI representative would improve the deliberation environment: *"I would have a counterpart who would represent the opinions of the volunteers more objectively [...] the volunteers would be better represented because, of course, every person enters such a process with their own perspective [...] and limitations"*. By acting as a corrective, the AI representative neutralizes polarized or emotional voices, ensuring substantive issues, rather than aggressive rhetoric, dominate the discourse. As (EM-07) explains: *"It can only work with what it's been provisioned [...], and it's not emotional. It doesn't have any feelings. We often find that a singular party, which is particularly aggressive, tends to stand out, but that doesn't apply here because everything is the same"*.

Conversely, some fear this neutrality loses human nuance. For instance, (AM-21) notes that while AI systems can summarize facts, the ability to sense if something is *"formulated a bit awkwardly [...] is just missing for me"*. In addition, some public servants raised concerns about the AI representative's capacity for normative compromise inherent to policy-making: *"the artificial intelligence must be capable of compromise [...] otherwise, it will be a very difficult discussion"* (EM-01). Expanding on this, (AM-16) argues that compromise requires a feedback loop that AI representatives lacks: *"Without feedback, it won't work [...] I need to know which arguments make my ideal solution unfeasible so I can change my position. [...] I need intermediate results and must communicate them to the AI [representative]"*.

Regarding accountability, the integration of AI representatives does not absolve public servants of their responsibilities. (AM-26) clarifies that even if municipal decisions are influenced by AI-generated information, the human staff remain the final authority: *"if city employees sit down and make a decision based on, say, information influenced by an AI, the employees are still responsible for that decision. They are also the ones who might later realize it wasn't a good decision and can reverse or change it"*. (AM-01) seconds this view, asserting that *"if AI is used to perform tasks that would otherwise be done by a person in the city, then of course the city is responsible"*.

The representation fidelity remains a point of tension, particularly regarding the potential for information dilution. (AM-01) insists on the necessity of an unadulterated contribution, fearing that AI representatives might distort the original intent of a citizen's input: *"I want my concern to go in there unadulterated [...] if it means that my contribution is changed by it, I want it without [AI representatives]"*. This concern over fidelity is echoed by (AM-17), who asks: *"What information gets lost in the process? [...] If I send a representative, that person isn't me"*. This risk of distortion includes the potential for misinterpretation or technical error; as (AM-30) notes, there is a danger that the AI representatives might *"interpret things incorrectly or be prone to errors, perhaps inflating issues that are actually small or vice versa"*. Nevertheless, many participants acknowledge AI representatives as a pragmatic compromise: it is *"better than no one [...] you cannot constantly involve all the people"* (AM-08).

4.3 Output Legitimacy

The primary conditions for perceived output legitimacy are the performance and quality of results, alongside the subsequent acceptance of those outcomes. Outcome legitimacy may increase because AI representatives are closely tied to expectations of superior political and technical results. For instance, participants specifically value the influence of AI representatives as a means to introduce diverse perspectives and reduce groupthink, thereby enhancing deliberation quality and leading to improved policy outcomes. This stimulation is seen as a prerequisite for high-quality outputs. As one association member noted: *"If there are actually different, nuanced positions represented by the associations, then it makes sense for those to be incorporated. Otherwise, there is a risk that the group reaches an agreement too quickly [...] without getting any counter-arguments that might provide food for thought"* (AM-09). Public servants further perceive AI representatives as strategic partners in problem-solving, with one public servant suggesting that by incorporating more voices, authorities can reach a *"If we, the specialist departments, were to take it seriously, then, as in any democratic process where more voices are heard, you would hopefully end up with a better solution. A solution that makes sense for most of the participants involved"* (EM-14). Ultimately, output legitimacy is derived from the AI representative's ability to synthesize a broader range of perspectives during the policy formulation phase than is possible through manual processing.

The use of AI representatives could enhance the general acceptance of the final policy. When citizens perceive that their input has been faithfully carried into the technical stages of the policy process via AI representatives, their sense of ownership over the result may increase. This relationship was highlighted by an association member who stated: *"the results could then have better acceptance, because those who 'fed' the AI would feel more involved in the process"* (AM-27).

5 Discussion

5.1 Perceptions and Conditions of Legitimacy

While AI systems are often framed as efficiency tools intended to facilitate institutional processes [40,41], the current results reveal a conceptual openness to AI systems acting as a representative of citizens' interests. Specifically, this interview study uses a vignette to examine how citizens and public servants evaluate the legitimacy of AI representatives within administrative policy processes *(RQ1)* and to explore the underlying rationales for these perceptions *(RQ2)*. The findings indicate that citizens and public servants perceive the integration of AI representatives into policy-making processes as potentially enhancing the democratic legitimacy of these processes. This perception is driven by the rationale that AI representatives can contribute to more inclusive and reflective internal administrative deliberations, potentially improving policy outcomes.

Regarding input legitimacy, our findings suggest that AI representatives are valued for their potential to increase inclusiveness by lowering traditional barriers to participation [17]. By removing the need for citizens to be physically or synchronously present in all stages of policy-making, the AI representatives enable the inclusion of individual interests to be aggregated into structured policy input, even from groups that are usually underrepresented. This aligns with the theoretical requirement that input legitimacy depends on the inclusion of diverse perspectives to ensure a wide range of voices is heard [12]. However, this potential is limited by skepticism regarding the authenticity of AI representatives. For a process to be legitimate, institutions must reflect the actual, nuanced interests of the public [35]. If the AI representative is seen as oversimplifying or distorting these needs, it risks distancing the policy outcome from the citizens it intends to serve. However, participants also identified a risk: administrators might use the AI's presence as a reason to decrease direct, face-to-face engagement with citizens. To maintain legitimacy, the AI representative should therefore be positioned as a tool that supplements and amplifies citizen voices during the internal phases of policy-making, rather than a replacement for direct human interaction.

In evaluating throughput legitimacy, the focus shifts to the internal quality of the deliberation process. Our analysis reveals that the AI representative is seen as a medium that can encourage more rational and objective discussion in introducing perspectives that might otherwise be overlooked. It acts as a constructive partner that improves the overall deliberation environment [26,45]. This reflects the theoretical claim that inclusive governance strategies increase throughput legitimacy and enhance overall performance [6]. However, a critical part of this legitimacy is maintaining clear lines of responsibility. Although AI representatives can contribute to deliberative quality, citizens and public servants in our study agreed that the municipality, acting on behalf of the citizens, must remain accountable for its decisions. This requires that the procedures be open to scrutiny, which in turn depends on transparency: it must be clear how the AI representative weighs opinions, how the input of the AI representative is handled by the public servants, and where ultimate decision-making authority resides [5,12].

Output legitimacy is tied to the perceived quality and acceptance of the final policy. Participants in our study suggested that an AI representative could improve the quality of outcomes by diversifying the perspectives available to public servants and thereby reducing the risk of groupthink during internal meetings. By ensuring that specific citizen inputs remain accessible throughout the decision-making process, the AI representative prevents the loss of detail that typically occurs when qualitative feedback is manually summarized into high-level guidelines [24]. This persistence of data aligns with the requirement that legitimate outcomes must visibly reflect the interests of those they affect [8,37].

In addition, we must evaluate these findings through the lens of Arnstein's ladder of participation [2]. While research often describes e-participation as a way to align policies with citizens' needs, our results indicate that AI representatives might be able to increase the level of citizens' influence. By including AI representatives in internal policy-making stages, citizens gain a tangible degree of influence that was previously inaccessible (i.e., from *consultation* to other levels of involvement, such as *partnership*) [2]. However, we acknowledge that the redistribution of power is still limited, as the AI representative can (and likely should) be overruled by public servants.

From a democratic perspective, this intervention addresses a systemic challenge highlighted in recent governance surveys, which suggest that many states fall short of citizen expectations for meaningful participation [30]. Public authorities often struggle to include citizens in policy-making due to specialized knowledge and time-intensive processes [17]. AI representatives provide a mechanism for citizen preferences to be voiced and accounted for across the policy cycle. This addresses the accountability deficit noted in the post-consultation phase [24,30]. By lowering the barriers of technical jargon and time commitment for citizens, AI representatives ensure that citizens' interests remain visible across the policy cycle [18,30].

5.2 Design for Legitimacy

While our findings indicate a general consensus that AI representatives can enhance democratic legitimacy of policy processes, citizens and public servants diverge on the specific conditions under which these agents should operate. Both groups agree that for AI representatives to be legitimate, they must yield more efficient and effective processes that produce objectively improved outcomes. The rationale here is that AI representatives can neutralize emotional discussions by providing a synthesis of data and introducing unaddressed perspectives. However, a significant disagreement exists regarding how this objective should be achieved *(RQ2)*. This tension is primarily manifested in the conflict between authenticity (representation fidelity) and efficiency (minimizing cognitive load).

An interesting finding of this research lies in the distinction between the representation models desired by different stakeholders. We observe a tension between what political theory describes as the *Delegate* model and the *Trustee* model of representation [33]. Citizens in our study favored the *Delegate* model, where AI representatives act as strict intermediaries, providing inputs based solely on earlier citizen consultations. This preference reflects a desire for high fidelity and a fear of information loss during the transition from consultation to formulation. Public servants, however, preferred the *Trustee* model, where AI representatives have autonomy to reason, evaluate policy alternatives, and estimate public acceptance of compromises. For public servants, the agent's legitimacy is tied to its utility. An agent that merely repeats verbatim summaries offers little help in navigating the complex trade-offs inherent to policy-making.

This divergence in expectations introduces a design trade-off for public administration: whether AI representatives should function as knowledge-retrieval systems or as systems capable of reasoning. This debate extends to the appropriate level of data aggregation. Representing every individual citizen's stance separately would maximize fidelity and fulfill the citizen's desire to avoid the dilution of their specific interests, which is a common critique of traditional participation [2,24]. However, such a granular approach risks increasing administrative complexity to a point that exceeds the utility of the intervention. Public servants, therefore, prefer aggregated stances that identify synergies and dominant narratives to streamline decision-making.

Similarly, the treatment of extreme positions presents a dilemma. Excluding these voices risks violating democratic pluralist values, while including non-constructive input can undermine the process. Whether the AI representatives are designed to act as a consensus filter or a neutral mirror of all input is a value-based choice that directly impacts their perceived utility.

6 Conclusion

This research investigated the perceived legitimacy of AI representatives in policy-making of local governance. We examined a futuristic yet plausible scenario in which AI agents serve as digital representatives for citizens during internal administrative deliberations. Through 52 semi-structured interviews in a German municipality, we assessed how citizens and public servants evaluate this intervention. Our analysis confirms that stakeholders perceive these AI representatives as a potentially legitimate mechanism for ensuring that civic voices persist throughout the later stages of the policy cycle.

The significance of these findings lies in their potential to address persistent challenges in citizen participation. Traditional methods often struggle to move beyond consultation in policy-making due to high costs and time constraints. AI representatives offer a viable compromise to lower these barriers. They provide a persistent presence for citizen perspectives during the critical stages of policy formulation and decision-making. For public servants, AI representatives may help ground internal discussions in citizens' needs. As such, this intervention could shift citizen involvement from merely being heard toward a more structural presence in the policy-making process. By ensuring inclusiveness and amplifying minority viewpoints, AI representatives may facilitate a more meaningful redistribution of power.

However, this study is subject to at least two limitations that provide potential areas for future research. (1) The study relied on a vignette-based scenario. This captures initial perceptions rather than the practical frictions of actual implementation. (2) In addition, the interviewed group may not represent the broader public because association members are already representatives for a special group of citizens. Future research must broaden the demographic scope to determine if perceived legitimacy extends to those traditionally disengaged from political processes.

Future work should shift from vignette-based exploration to the development and evaluation of functional prototypes in laboratory settings. This progression is essential to derive design principles that ensure AI representatives effectively support policy-making. Ultimately, this research highlights the potential for AI representatives to facilitate a shift in citizen participation, moving from simple consultation toward forms of participation with greater redistribution of power.

Acknowledgments. The research described in this article was funded by the Swiss National Science Foundation (SNSF) under grant number 208013.

Disclosure of Interests. The authors declare that they have no conflicts of interest and no financial or personal relationships with third parties that may have influenced the conduct of this research.

References

1. Alexander, C.S., Becker, H.J.: The use of vignettes in survey research. Public Opin. Q. **42**(1), 93–104 (1978)
2. Arnstein, S.R.: A ladder of citizen participation. JAPA **35**(4), 216–224 (1969)
3. Atzmüller, C., Steiner, P.M.: Experimental vignette studies in survey research. Methodology **6**(3), 128–138 (2010)
4. Bingham, A.J.: From data management to actionable findings: a five-phase process of qualitative data analysis. Int. J. Qual. Methods **22** (2023)
5. Bovens, M.: Analysing and assessing accountability: a conceptual framework. Eur. Law J. **13**(4), 447–468 (2007)
6. Caby, V., Frehen, L.: How to produce and measure throughput legitimacy? Lessons from a systematic literature review. Politics Gov. **9**(1), 226–236 (2021)
7. Cheneval, F.: Demokratietheorien zur Einführung. Junius (2015)
8. Curtin, D., Meijer, A.J.: Does transparency strengthen legitimacy? Inf. Polity **11**(2), 109–122 (2006)
9. Elstub, S., Escobar, O.: Defining and typologising democratic innovations. In: Elstub, S., Escobar, O. (eds.) Handbook of Democratic Innovation and Governance. Edward Elgar Publishing (2019)
10. Fischli, R., Muldoon, J.: Empowering digital democracy. Perspect. polit. **22**(3), 819–835 (2024)
11. García-Marzá, D., Calvo, P.: Algorithmic Democracy - A Critical Perspective Based on Deliberative Democracy. Springer, Cham (2024)
12. Haesevoets, T., Roets, A., Steyvers, K., Verschuere, B., Wauters, B.: Towards a multifaceted measure of perceived legitimacy of participatory governance. Governance **37**(3), 711–728 (2024)
13. Haesevoets, T., Verschuere, B., van Severen, R., Roets, A.: How do citizens perceive the use of Artificial Intelligence in public sector decisions? Gov. Inf. Q. **41**(1), 101906 (2024)
14. Haggart, B., Keller, C.I.: Democratic legitimacy in global platform governance. Telecommun. Policy **45**(6), 102152 (2021)
15. Herbold, S., Trautsch, A., Kikteva, Z., Hautli-Janisz, A.: Large Language Models can impersonate politicians and other public figures (2024)
16. Hughes, R., Huby, M.: The construction and interpretation of vignettes in social research. Soc. Work. Soc. Sci. Rev. **11**(1), 36–51 (2004)

17. Irvin, R.A., Stansbury, J.: Citizen participation in decision making: is it worth the effort? Public Adm. Rev. **64**(1), 55–65 (2004)
18. Jann, W., Wegrich, K.: Theories of the policy cycle. In: Fischer, F., Miller, G.J., Sidney, M.S. (eds.) Handbook of Public Policy Analysis: Theory, Politics, and Methods, pp. 69–88. Routledge (2017)
19. Jarrett, D., et al.: Language agents as digital representatives in collective decision-making (2025)
20. Lee, S., Kim, M., Hwang, S., Kim, D., Lee, K.: Amplifying minority voices: Ai-mediated devil's advocate system for inclusive group decision-making. In: Proceedings of IUI '25, p. 17–21 (2025)
21. Louise Barriball, K., While, A.: Collecting data using a semi-structured interview: a discussion paper. J. Adv. Nurs. **19**(2), 328–335 (1994)
22. Macintosh, A.: Characterizing e-participation in policy-making. In: Proceedings of the 37th HICSS, pp. 1–10 (2004)
23. Merriam, S.B.: Qualitative Research in Practice Examples for Discussion and Analysis, 2nd edn. Jossey-Bass, San Francisco (2019)
24. Michels, A.: Citizen participation in local policy making: design and democracy. Int. J. Public Adm. **35**(4), 285–292 (2012)
25. Michels, A., Graaf, L.D.: Examining citizen participation: local participatory policy making and democracy. Local Gov. Stud. **36**, 477–491 (2010)
26. Mikhaylovskaya, A.: Enhancing deliberation with digital democratic innovations. Philos. Technol. **37**(1), 3 (2024)
27. Mosley, J.E., Wong, J.: Decision-making in collaborative governance networks: pathways to input and throughput legitimacy. J. Public Adm. Res. Theory **31**(2), 328–345 (2021)
28. Myers, M.D., Newman, M.: The qualitative interview in IS research: examining the craft. Inf. Organ. **17**(1), 2–26 (2007)
29. Nabatchi, T., Leighninger, M.: Public Participation for 21st Century Democracy. John Wiley & Sons, Hoboken (2015)
30. OECD: Exploring new frontiers in citizen participation in the policy cycle, OECD Publishing. Technical report (2025)
31. Peter, F.: Political legitimacy. In: The Grounds of Political Legitimacy, 1st edn., pp. 3–22. Oxford University Press, Oxford (2023)
32. Pislaru, M., Vlad, C.S., Ivascu, L., Mircea, I.I.: Citizen-centric governance: enhancing citizen engagement through artificial intelligence tools. Sustainability **16**(7), 2686 (2024)
33. Pitkin, H.F.: The Concept of Representation. University of California Press (1967)
34. Saldaña, J.: The coding manual for qualitative researchers. SAGE (2021)
35. Scharpf, F.W.: Demokratische Politik in der internationalisierten Ökonomie. In: Greven, M.T. (ed.) Demokratie — eine Kultur des Westens? 20. Wissenschaftlicher Kongreß der Deutschen Vereinigung für Politische Wissenschaft, pp. 81–103. VS Verlag für Sozialwissenschaften, Wiesbaden (1998)
36. Schmidt, V.A.: Democracy and legitimacy in the European union revisited: input, output and 'throughput'. Polit. Stud. **61**(1), 2–22 (2013)
37. Schmidt, V.A.: Conceptualizing legitimacy: input, output, and throughput. In: Europe's Crisis of Legitimacy, pp. 25–55. Oxford University Press (2020)
38. P Schwartz-Shea, P., Yanow, D.: Interpretive Research Design: Concepts and processes. Routledge (2013)
39. Seeber, I., et al.: Machines as teammates: a research agenda on AI in team collaboration. Inf. Manag. **57**(2), 103174 (2020)

40. Sieber, R., Brandusescu, A., Sangiambut, S., Adu-Daako, A.: What is civic participation in artificial intelligence? Environ. Plan. B: Urban Anal. City Sci. **52**(6), 1388–1406 (2025)
41. Sprenkamp, K., Dolata, M., Schwabe, G., Zavolokina, L.: Data-driven intelligence in crisis: the case of Ukrainian refugee management. Gov. Inf. Q. **42**(1), 101978 (2025)
42. Sreenivasan, G.: Oxford studies in political philosophy. In: Presented at the Three Concepts of Legitimacy (2024)
43. Stebbins, R.A.: Exploratory Research in the Social Sciences. SAGE (2025)
44. Sæbø, Ø., Rose, J., Flak, L.S.: The shape of eparticipation: characterizing an emerging research area. Gov. Inf. Q. **25**, 400–428 (2008)
45. Tessler, M.H., et al.: AI can help humans find common ground in democratic deliberation. Science **386**(6719) (2024)

Gamifying Deliberation: Towards a Process-Sensitive Framework

Yannick von Ostheim[1(✉)], Maurice Flechtner[2,3], and Francesco Veri[2]

[1] Universität Bamberg, Bamberg, Germany
Yannick.von-ostheim@uni-bamberg.de
[2] Universität Zürich, Zürich, Switzerland
[3] ETH Zürich, Zürich, Switzerland

Abstract. Digital platforms increasingly use gamification mechanisms such as points, badges, and leaderboards to boost participation in democratic deliberation. Yet deliberation is a multi-stage process, each with distinct cognitive demands and normative priorities, including equality, inclusion, plurality, authenticity, and reflection. Applying a uniform points-badges-leaderboards (PBL) gamification layer can therefore create design-process mismatches, where incentives that are functional in one stage become counterproductive in another. Based on a stage-by-stage analysis of common deliberative modules, the paper maps how standardized gamification can distort participant acquisition, group building, learning, problem definition, open discussion, self-reflection, decision-making, and output generation and validation - for example by amplifying homophilous recruitment, privileging speed over comprehension, or threatening plurality through quantity biases and gatekeeping effects in problem framing. The paper then introduces a normative process-sensitive alignment framework and a "stage-fit test" that link each deliberative module to suitable extrinsic- or intrinsic-leaning mechanisms, associated risks and affordances, and concrete design safeguards. The framework is theory-driven and requires empirical validation in concrete deployments. Finally, the paper derives policy implications for treating gamification as a form of governance, proposing stage-bound design standards, quality-first evaluation criteria, and transparency obligations for public-sector deliberation platforms. This paper offers a normative, process-sensitive framework and stage-fit test.

Keywords: Deliberative democracy · Gamification · Democratic innovation

1 Introduction

The move of deliberative formats onto digital platforms has intensified interest in how specific design elements shape participation and deliberative quality [32]. Gamification—the use of isolated game elements in non-gaming contexts to stimulate motivation and foster engagement [21,22,28]—is increasingly proposed to

D. Duenas-Cid et al. (Eds.): ePart 2026, LNCS 16923, pp. 182–198, 2027.
https://doi.org/10.1007/978-3-032-36676-4_12

mobilize citizens with low political interest or low cognitive involvement [17,31]. Yet the same incentive logic that can increase engagement may also amplify persistent problems of online deliberation, including quality loss under scale, strategic participation, dominance dynamics, and fragile civility [34]. To realize gamification's potential without undermining deliberative quality, we must understand how incentives interact with the *modular* structure of deliberative processes.

Democratic deliberation is not a homogeneous activity but a process divisible into interconnected steps [2], each implying different cognitive demands and normative design priorities. In practice, however, gamification is often deployed as a platform-wide "engagement layer" largely independent of process stage [17,18], creating design–process mismatches: incentives that trigger activity early on may be ill-suited in later stages requiring reflection, reciprocal uptake, and collaboration. Moreover, deliberative success cannot be reduced to participation rates but depends on high-quality, inclusive participation [15,30]. Incentive regimes therefore have legitimacy implications because they shape who participates, whose contributions become visible, and whether interaction remains compatible with deliberative norms.

Systematic mappings and reviews of gamified e-participation classify studies by context (e-participation, citizen participation), method, and generic "forms of participation," and list common elements (points, badges, leaderboards, missions), but do not systematically tie particular mechanics to specific deliberative stages [23,47,56]. We address this gap by (i) specifying deliberative process stages and their normative priorities, (ii) diagnosing stage-specific design–process mismatches, (iii) presenting a consolidated framework table that integrates stage priorities, risks and affordances, and the resulting mechanism orientation, design recommendations, and suitability assessments and (iv) discussing governance implications for public-sector deliberation platforms.

2 Deliberation and Its Principles

We define deliberation as a process of reasoned discourse among free and equal stakeholders [7]. To assess whether gamification mechanisms are compatible with deliberative quality, we first introduce a normative reference framework that specifies what deliberation is meant to achieve and at which analytical level each requirement operates. We identify five core principles—equality, inclusion, plurality, authenticity, and reflection—organised across three analytical levels. The principles were selected through a structured reading of foundational deliberative theory [10,14,20,35] and specifications of deliberative quality [3,11,54], focusing on those principles that (i) recur across major theoretical accounts, (ii) are analytically distinguishable from one another, and (iii) govern aspects of the participant-facing process that gamification mechanisms can directly affect.

Three Analytical Levels. At the **structural level**, *equality* ensures that all participants possess equal standing to speak, be heard, and have their contributions taken seriously [20,35], while *inclusion* ensures that the deliberative body reflects the demographic, experiential, and perspective diversity of the affected public [59]. At the **discursive level**, *plurality* guarantees exposure to a genuinely diverse range of perspectives and viewpoints [31]. Plurality is analytically distinct from inclusion as inclusion concerns *persons* (who participates) and plurality concerns *reasons* (the cognitive diversity actually present in exchange). A demographically representative body may nonetheless exhibit discursive homogeneity if participants adapt similar frames or self-censor dissenting views. *Authenticity* requires that communication remains sincere, transparent, and non-coercive [3,20,54]. At the **cognitive-reflective level**, *reflection* captures participants' capacity to weigh competing arguments, reconsider initial positions, and integrate diverse reasons into more informed judgments [14,41,42]. This is the feature that most clearly separates deliberation from debate, bargaining, or preference aggregation. The three levels are functionally interdependent: structural equality and inclusion create the conditions for discursive plurality, authenticity enables genuine engagement with that plurality, and reflection transforms plural exchange into reasoned judgment. While all of these principles play an important role throughout a deliberative process, we will identify priority-principles per deliberative module, focusing on those principles that play the most important role in each module.

3 Process Stages and Modules

Deliberative processes can generally be divided into three phases: Pre-Deliberation, Deliberation and Post-Deliberation [9,12]. Each phase comprises several modules with distinct functions and normative priorities. The phases and modules described here are derived from deliberative theory [13,20,40] and practice [42,58] and informed by deliberative process scholarship [19,54]. To complement this conceptual structuring, we conducted an exploratory analysis of descriptions drawn from 106 documented deliberative methods scraped from Participedia, a crowdsourcing platform for democratic innovations. During the analysis we extracted process features and clustered them to identify the most prevalent deliberative process stages. Full technical details are reported in Appendix A.

Table 1 summarises the core modules, their priority principles, and their deliberative function.

4 Design Process Mismatch

These stage-specific task demands imply that incentive design is inherently conditional: what counts as "meaningful engagement" changes across the deliberative process. This section, therefore, examines how uniform gamification layers can overlook these shifts and thereby generate systemic design-process mismatches. We define a design-process mismatch as the situation in which a platform-wide incentive regime promotes behaviors that are functional in one deliberation stage but conflict with cognitive demands and normative priorities of

Table 1. Deliberative process modules, normative priorities, and core functions.

Phase	Module	Priorities	Description
Pre-Delib.	Acquisition	Equality; Inclusion	Recruit a participant pool that reflects the diversity of affected stakeholders and ensures equal chances to participate
	Group Building	Authenticity	Build trust and community among participants through collaborative activities (e.g. collective rule-setting) to support later deliberative exchange
	Learning	Reflection; Plurality	Level the knowledge playing field via balanced learning materials that present diverse perspectives and prompt initial reflection
Deliberation	Problem Definition	Plurality	Frame the issue at hand, ensuring a wide range of perspectives are considered before discussion is structured
	Open Discussion	Plurality; Reflection; Authenticity	Exchange arguments in moderated settings to build shared understanding; all stakeholder perspectives should be expressed and heard
	Self-Reflection	Reflection	Provide a safe space for individual reasoning about encountered arguments and flaws in one's own position
	Decision-Making	Plurality; Equality	Aggregate perspectives into a collective outcome while safeguarding minority viewpoints and equal voice
Post-Delib.	Generating Outputs	Plurality; Authenticity	Draft a report that faithfully represents both the plurality of perspectives and participants' reasoning
	Output Validation	Equality	Participants review the draft to detect misrepresentation; each individual's objections carry equal weight
	Communication	Plurality	Convey outcomes to the broader public without skewing or oversimplifying the range of perspectives

Note: Since drafting outputs and communicating outcomes fall predominantly to organizers, the subsequent analysis focuses on **Output Validation** as the primary participant-facing post-deliberation module.

another. To make this argument tractable, we normatively diagnose mismatches stage by stage, focusing on how common gamification mechanisms along the lines of points, badges and leaderboards (PBL) [21] reshape participation, attention, and interaction in ways that can diverge from deliberative quality criteria.

4.1 Pre-deliberation Mismatches

Acquisition. Acquisition prioritizes *inclusion* and *equality* in who enters the process. Uniform PBL designs often reward easy-to-count recruitment actions (e.g., sharing invitations, bringing in participants) [48]. The mismatch arises when invitations follow homophily: incentives amplify recruitment within existing social and political backgrounds, privileging already engaged groups over harder-to-reach participants [36]. Thus, participation may increase while inclusion and equality decline, skewing the participant pool and weakening deliberative legitimacy and epistemic gains.

Learning. Learning prioritizes *reflection* and *plurality* through balanced information uptake. PBL typically rewards completion/progression (points, streaks, level-ups), which can shift effort toward speed and task completion rather than comprehension and engagement with diverse perspectives [24, 57]. The risk is superficial uptake that weakens reflection and plurality before deliberation begins.

Group Building. Group building aims to foster *authenticity* by cultivating trust, psychological safety, and a cooperative tone. A platform-wide PBL layer makes participation visible and comparable (badges, ranks, leaderboards), and social comparison can introduce status pressure and anxiety that reframes interaction as competition rather than community [44]. This may undermine authenticity and raise barriers for less confident participants.

4.2 Deliberation Mismatches

Problem Definition. Problem definition should protect *plurality* in how issues are framed and which perspectives become salient. Under uniform PBL, rewards tend to target either the *quantity* of framings or the most *visible/endorsed* framings. This creates a twofold mismatch: quantity rewards privilege premature, low-quality framings [37], while "quality" rewards can become gatekeeping if evaluation criteria are naively restrictive, narrowing legitimate perspectives [6]. Plurality is thus threatened through either fragmentation (quantity) or exclusion (gatekeeping).

Open Discussion. Open discussion prioritizes *plurality*, *reflection*, and *authenticity* in reason exchange. Uniform gamification typically incentivizes visible activity (posting frequency, reactions, rank improvement) [37]. Reward salience can induce strategic participation: optimizing for points/visibility, prioritizing easily rewarded contributions, and avoiding minority positions that attract fewer endorsements [5, 37]. The risk is higher interaction volume alongside declining reason-giving, reciprocal uptake, and equality of voice.

Self Reflection. Self-reflection aims to enable *reflection* without public-performance pressure, yet platform-wide gamification often adds immediate feedback, progression cues, and frequent reward signals. Gamification frameworks for cognitive tasks warn that poorly task-matched game elements can add irrelevant cognitive load and distort the targeted mental process [27]. Accordingly, reward salience may crowd out introspection by pulling attention toward completion and immediate gains rather than critical self-examination and engagement with counterarguments [26]. Reviews further caution that shallow PBL aligned primarily to progression can undermine deeper reflective goals, motivating process-sensitive selection of elements [39].

Decision-Making. Decision-making must reconcile *plurality* with *equality* in aggregation and justification. A non-specified PBL approach can reward convergence (majority alignment, "consensus" achievements) and make agreement signals more salient than reasons. When rewards are tied to agreement, dissent becomes costly: participants may strategically align with majorities or suppress minority views to avoid status loss, reputational penalties, or foregone rewards [55]. This conforms to critiques of overjustification and extrinsic reward focus as drivers of conformity and strategic behavior [8,51]; under salient badges/leaderboards, social comparison steers behavior toward reward-maximizing norms rather than truthful expression, especially when status is at stake [37,44]. If assimilation occurs, dissent may be marginalized and equal say weakened, even if the process appears smoother or more efficient.

4.3 Post-deliberation Mismatches

Output Validation. Output validation prioritizes *equality* in reviewing and correcting the draft report. Under uniform PBL, validation is often rewarded by quantity or speed (e.g., "review completed" badges), which incentivizes superficial, check-the-box reviewing: participants skim to finish quickly, and unequal time resources translate into unequal influence over what is contested or corrected [46]. This aligns with mappings showing that many gamification systems reward fast/frequent actions while rigorous evaluation of quality effects remains rare, and with design-framework reviews warning that poorly aligned PBL can compromise output quality when users rush for rewards [27,39]. The resulting risk is weakened equality and error correction, lowering confidence in the final output.

4.4 Cross-Cutting Mismatch Patterns

Across stages, three mechanisms explain why platform-wide PBL layers often misalign with process-specific priorities. First, *metric capture* follows countability: rewards target what is easy to measure (posts, clicks, reactions), shifting effort toward activity rather than deliberative contribution; mappings and reviews show that most implementations operationalize success via action counts and PBL mechanics, which strongly shapes behavior [46,48,51], and experiments find robust quantity increases with small, inconsistent, or absent effects on depth and intrinsically motivated engagement [37]. Second, *capacity and cognitive overload* arises when reward-driven volume and salience reduce attention per contribution and strain moderation, consistent with cognitive-task frameworks warning that poorly task-matched game elements impose irrelevant load and degrade cognitive performance and output quality [27], and with meta-analytic evidence showing behavioral engagement gains alongside smaller or inconsistent cognitive gains [24,57]. Third, *sustainability vulnerability* reflects that short-term incentive spikes often decay: reviews link shallow PBL and overjustification to initial gains followed by plateau/decline, especially when rewards are removed [1,8,51], and longer-term mappings highlight dependence on external rewards and limited sustained effects as recurring failures [39,46,51].

5 Process Sensitive Framework

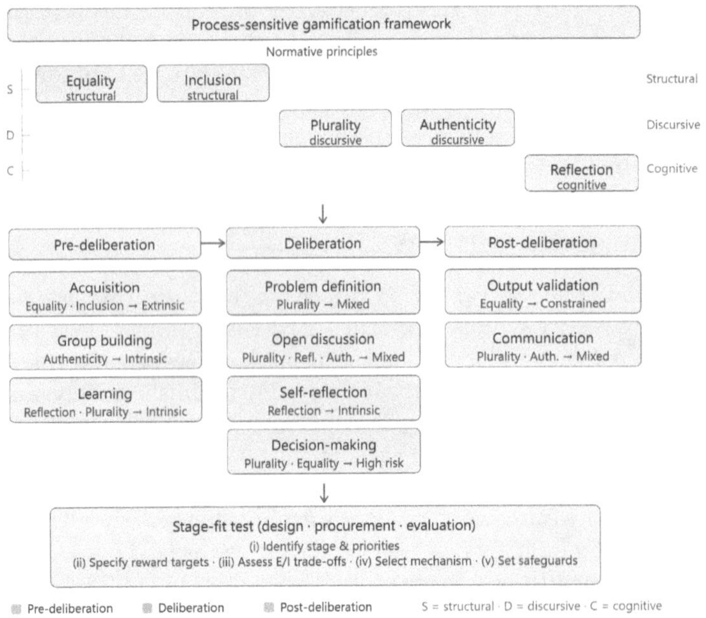

Fig. 1. Process-sensitive gamification framework

Before specifying stage-sensitive recommendations, we distinguish motivation as a psychological state from gamification as a design intervention. Here, "intrinsic" (I) vs. "extrinsic" (E) motivation denotes motivational regulation—whether participation is experienced as inherently meaningful or as instrumentally driven by separable outcomes (e.g., rewards, status, approval) [49]. Gamification mechanisms are design affordances that *tend* to elicit such regulations (e.g., points/rankings foreground external evaluation; scaffolding, mastery feedback, and cooperative roles support autonomy, competence, and relatedness), yet their effects are conditional on stage-specific tasks, priorities, and implementation. This distinction enables the framework to assess stage-fit at the level of incentive design while remaining explicit about motivational pathways to deliberative quality. An overview of the framework is provided in Fig. 1.

5.1 Stage-Fit Test (How to Use the Framework)

We operationalize process-sensitive gamification as a *stage-fit test* for design, procurement, and evaluation.
Operationally, the test proceeds as follows:

(i) **Identify stage & priorities.**
(ii) **Specify reward targets** (which behaviors are rewarded in practice).
(iii) **Assess trade-offs** (risks/affordances of extrinsic-leaning vs. intrinsic-supportive options).
(iv) **Select mechanism orientation** (extrinsic-leaning, intrinsic-supportive, or mixed).
(v) **Set safeguards & tests** (guardrails, success metrics, rollback triggers).

Table 2 instantiates the stage-fit test by mapping each deliberative step to its normative priorities, the key risks and affordances of extrinsic-leaning versus intrinsic-supportive mechanisms, and the resulting mechanism orientation, design recommendations, and suitability assessment.

5.2 Design Recommendations

In the remainder of this section, we unpack Table 2 into concrete, phase- and stage-specific design recommendations, highlighting the safeguards and suitability constraints implied by the mapped E/I trade-offs.

Pre-deliberation: Mobilize Without Skewing Inclusion. In the pre-deliberation phase, the goal is to acquire and prepare the ideal set of participants. Overall, it is most important to ensure that no single group of stakeholders is excluded or drops out of the process early. Across the entire pre-deliberation phase, the guiding principle is that gamification should widen the door to participation and deepen preparatory engagement without sorting participants into visible hierarchies before deliberation has even begun [52,61].

While participant acquisition has the clearest case for extrinsic incentives—referral bonuses, invitation rewards, or milestone badges for sign-up—these mechanisms must be constrained by diversity-sensitive design. Specifically, recruitment rewards should be calibrated to incentivize outreach beyond participants' immediate social networks, for instance by offering higher rewards for invitations accepted by individuals from underrepresented demographic or attitudinal groups, or by capping rewards from within-network referrals. Without such safeguards, gamified recruitment risks amplifying homophily and producing a participant pool that is larger but less inclusive [52,53]. Such strategies should however not translate into restrictive selection rules or participation thresholds tied to specific profiles.

In the learning module, the framework recommends a shift toward intrinsic-supportive mechanisms: rather than awarding points for module completion or speed, designers should reward demonstrated comprehension—for example through short reflection prompts, perspective-taking exercises, or quizzes that

Table 2. Process-sensitive gamification framework: stage priorities, risks/affordances (extrinsic-leaning vs. intrinsic-supportive), and design guidance.

Phase	Step	Priorities	Key risks (E/I)	Key affordances (E/I)	Orientation	Design recommendation	Suitability
Pre-delib.	Participant acquisition	Equality; Inclusion	**E:** Homophily (bubble recruitment); **I:** —	**E:** Low-threshold mobilization; **I:** —	Extrinsic	Incentivize *diverse* invitations	Yes
Pre-delib.	Group building	Authenticity	**E:** Social comparison; social anxiety; **I:** Social anxiety (if misdesigned)	**E:** —; **I:** Trust/community → collaboration	Intrinsic	Use collaborative group effort with shared goals and roles	Yes
Pre-delib.	Learning	Reflection; Plurality	**E:** Superficial completion (points over learning); **I:** —	**E:** Short-term engagement; low access barrier; **I:** Enhanced learning; information seeking	Intrinsic	Reward comprehension and engagement with perspectives, not speed	Yes
Delib.	Problem definition	Plurality	**E:** Superficial framing; herding; **I:** Overidentification with one problem	**E:** Short-term engagement; **I:** Enjoyment/playfulness → creativity	Mixed	Avoid herding; do not show prior contributions	Constrained
Delib.	Open discussion	Authenticity; Plurality; Reflection	**E:** Strategic participation; short-termism; cognitive load; repetition; **I:** Overidentification with one problem	**E:** Short-term engagement; goal setting (if aligned); **I:** Sustained deep engagement; openness to counterarguments; norm compliance	Mixed (I>E)	Quality > quantity; avoid rewarding repetition; reward respectful language, novelty/creativity, and argument quality	Yes
Delib.	Self-reflection	Reflection	**E:** Strategic responding → dishonesty; **I:** —	**E:** —; **I:** Playfulness; purpose (safe introspection)	Intrinsic	Paraphrase own and opposing arguments; ask for stance toward each	Yes
Delib.	Decision-making	Plurality; Equality	**E:** Strategic participation → dishonesty; reduced meta-consensus; **I:** Overidentification	**E:** —; **I:** — (only if constrained to reflection/fairness)	Constrained	Use only reflexive prompts and fairness checks; avoid external incentives	High risk
Post-delib.	Generating & validating outputs	Plurality; Authenticity	**E:** —; **I:** —	**E:** —; **I:** —	—	—	—
Post-delib.	Communication	Plurality; Authenticity	**E:** —; **I:** —	**E:** Engagement (last achievement for reviewing output); **I:** Self-reflection on own role → civic culture	Extrinsic/Mixed	Use a "last achievement" for reviewing output; reflection log; efficacy feedback; "what changed" prompts	Yes

Note. This table reports a normative, theory-driven synthesis: risks/affordances and recommendations are conditional design hypotheses grounded in the mismatch analysis and require empirical validation in concrete deployments.

require engagement with multiple viewpoints [50]. This reorients the incentive structure from progression to understanding and helps ensure that plurality is embedded in participants' preparatory experience [60].

As group building is a unique social process, extrinsic reward structures should be avoided entirely or kept invisible at the individual level. Instead, collaborative challenges, shared group goals, or collective achievement markers can foster a sense of community and trust without introducing the status anxiety that leaderboards and individual badges tend to produce. Collaborative reward structures have been shown to promote empathy and mutual understanding in deliberative settings [18,45]

Deliberation: Prioritize Quality Over Volume. During the deliberation phase, the stakes for incentive design are highest because the core epistemic and democratic work is most sensitive to distortion by misaligned rewards. The overarching recommendation is to subordinate all activity-based metrics to quality-oriented ones. In problem definition, designers should avoid displaying prior contributions or endorsement counts before a participant has submitted their own framing, thereby preventing anchoring and herding effects that narrow the range of perspectives considered [53,62]. Where gamification is used, it should reward the introduction of novel frames or underrepresented perspectives rather than the volume of submissions [50,62].

In open discussion, the framework recommends a mixed orientation that leans toward intrinsic support: mechanisms such as prompts for reciprocal uptake ("Which argument changed your thinking?"), recognition for constructive engagement with opposing views, or badges tied to argument quality as assessed by peer review or moderation can sustain motivation without privileging sheer posting frequency [38]. Crucially, public leaderboards and visible point tallies should be suppressed during discussion modules, as their salience redirects attention from reason-giving to status competition [53]. Empirical studies confirm that competitive elements increase participation but may reduce deliberative quality unless balanced with collaborative features [50,62]

For self-reflection, gamification should be limited to intrinsic-supportive scaffolding—structured journaling prompts, perspective-paraphrasing tasks, or private progress indicators—and must not attach extrinsic rewards to reflection outputs, since doing so risks incentivizing strategic or performative responses rather than genuine introspection [25].

Decision-making represents the highest-risk stage: here, the framework advises caution regarding extrinsic incentive mechanism and restricts design interventions to reflexive prompts (e.g., "Have all perspectives been considered?") and procedural fairness checks. Rewarding convergence, speed, or agreement in this module risks suppressing dissent, distorting preference expression, and ultimately undermining the legitimacy of the collective outcome [4].

Post-deliberation: Legitimacy, Validation, and Learning. In the post-deliberation phase, gamification serves two distinct functions: safeguarding the integrity of outputs and fostering lasting civic learning among participants.

For output validation, the primary risk is that speed- or quantity-based rewards encourage superficial review, allowing errors or misrepresentations in the draft report to go uncontested. Designers should therefore avoid completion badges or time-based incentives for validation tasks and instead reward substantive engagement—for instance, by recognizing participants who flag inconsistencies, propose revisions, or identify perspectives that are underrepresented in the draft [43]. Ensuring that all participants have equal opportunity and encouragement to review the output is critical; visibility protections or anonymized review processes can prevent high-status participants from dominating the validation stage [43].

For the communication of outcomes and meta-reflection, a mixed or lightly extrinsic approach becomes viable again: a "final achievement" for completing the full deliberative cycle, including reviewing the output and submitting a reflection, can provide closure and reinforce participants' sense of efficacy without distorting substantive judgments that have already been made. Meta-reflection modules should prompt participants to articulate how their views evolved, which arguments they found most challenging, and what they learned about other stakeholders' perspectives [25]. These prompts serve a dual purpose: they deepen individual learning and generate valuable process data for organizers seeking to evaluate deliberative quality. Throughout the post-deliberation phase, the guiding concern is that gamification must not retroactively undermine the legitimacy of outcomes reached during deliberation—rewards should reinforce careful stewardship of results rather than incentivize hasty closure.

Cross-Stage Design Principles: Across stages, gamification should be treated as a stage-gated incentive regime: align mechanisms to the specific task and normative priority of each step rather than applying a uniform engagement layer [17,29]. Prefer informational over controlling feedback by avoiding public rank and status cues that trigger social comparison. Reward deliberative behaviors—justification, reciprocal uptake, and synthesis—rather than activity counts. Build in equity safeguards such as contribution caps, diminishing returns, or visibility protections to prevent dominance by high-frequency users. Finally, protect cognitive bandwidth by avoiding speed incentives and reducing reward salience when reflection and careful processing are required [29,33]. As an overarching principle, designers should adopt a "do no harm" default: where the risk of distorting a deliberative stage's normative priorities is high and the evidence for beneficial effects is weak, the appropriate choice is to refrain from gamifying that module rather than to gamify speculatively. In the next section, we translate these design implications into governance responsibilities for public-sector platforms [16]. Given limited empirical evidence on safely gamifying deliberation, and the centrality of diverging normative priorities, we argue for a 'no extrinsic incentives by default' principle unless robust, context-specific evidence of compatibility is available.

6 Policy Implications

If gamification mechanisms reshape participation patterns, attention allocation, and the quality of reason-giving, then incentive design is a governance choice with implications for democratic legitimacy. Gamification choices on public deliberation platforms can have similar effects as procedural rules in offline deliberation and should thus be governed with comparable rigor. We argue that publicly mandated deliberative processes should require a *deliberative impact assessment*: a structured ex-ante evaluation, mapping each proposed mechanism to its deliberative stage, the normative priorities at stake, and anticipated effects on inclusion, equality, authenticity, plurality, and reflection. Such assessments should be documented, published, and revisited at defined intervals. Procurement specifications should further require that gamification mechanisms are explicitly mapped to deliberative modules and their normative priorities, that the mechanism orientation is justified per stage, and that high-risk stages such as decision-making and self-reflection are protected from extrinsic incentives by default. Here, the stage-fit test (Sect. 5) can serve as an operational template for both designers and auditors. Evaluation frameworks for publicly funded platforms should include deliberative quality indicators—argument diversity, reciprocal uptake, distributional equality, evidence of position revision—alongside engagement metrics as engagement volume should never serve as the sole proxy for deliberative success. Finally, platform operators should disclose, in accessible language, which behaviors are rewarded, how contributions are ranked or made differentially visible, and how incentive structures differ across stages, while post-hoc reporting on mechanisms employed, observed effects, and adjustments made should be mandatory to support institutional learning and civil-society scrutiny.

7 Summary

This paper argued that gamification in online deliberation cannot be treated as a uniform engagement layer, because deliberation is a staged process with shifting cognitive demands and normative priorities. Based on a normative stage-by-stage mismatch analysis, we developed a process-sensitive alignment framework that, first, maps deliberative steps to stage-specific priorities, risks, and affordances and, second, derives corresponding mechanism orientations, design recommendations, and suitability judgments. We operationalized this contribution as a stage-fit test that supports platform design, procurement, and evaluation by making incentive choices auditable and by specifying safeguards and rollback triggers when activity rises without deliberative contribution. The central governance implication is that gamification functions as an incentive regime and therefore constitutes a public-sector governance choice with direct consequences for inclusion, equality, authenticity, plurality, and reflection. There are still some limiting factors: the framework has not yet been empirically validated in real-world deliberation platform deployments, and the evidence base draws on heterogeneous domains (e.g., education and crowd sensing) such that effects may not fully generalize to political deliberation contexts. Future work should empirically test the framework in real deliberative deployments, refine stage-specific quality indicators (potentially supported by AI-assisted assessment), and evaluate how specific gamification implementations perform across different deliberative contexts and populations.

Acknowledgments. The research presented in this paper was conducted within The AI4Deliberation project. The authors would like to acknowledge the contribution of all consortium partners to formulate the project.

AI4Deliberation is funded by the European Union under grant agreement No. 101178806. Views and opinions expressed are however those of the authors only and do not necessarily reflect those of the European Union or European Research Executive Agency. Neither the European Union nor the granting authority can be held responsible for them.

Disclosure of Interests. The authors have no competing interests to declare that are relevant to the content of this article.

A Stage Clustering: Technical Details

The raw Participedia export contained 106 method records with 1,112 binary `stage_*` indicators. After removing empty, placeholder, or trivially duplicated entries (32 rows), 314 unique normalised stage descriptions were extracted from the remaining 74 methods. Descriptions were encoded with `Qwen3-Embedding-0.6B`[1] (ℓ_2-normalised, 314×1024). A precomputed cosine-distance matrix $D = 1 - \mathbf{E}\mathbf{E}^{\top}$ served as input to agglomerative clustering. Several algorithms and hyperparameter grids were evaluated (K-Means with $k \in \{6–14\}$; Spectral with $k \in \{8–40\}$; HAC[2] with average, complete, and Ward linkage at varying thresholds). Silhouette scores (cosine metric) were low throughout (0.03–0.10), reflecting high semantic overlap among deliberative-process descriptions rather than poor cluster structure.

The selected configuration uses complete-linkage HAC with cosine affinity and a distance threshold of 0.018, yielding 22 clusters. Complete linkage was preferred because it enforces that *every* within-cluster pair lies within the threshold, producing tighter, semantically homogeneous groups. Each cluster was labelled by its *medoid* (the member with the highest mean cosine similarity to all other members). Representative clusters include *Process interaction and decision-making* ($n=29$), *Collective assessment* ($n=27$), *Moderated discussions with procedural rules* ($n=20$), *Provide information on proposed policies* ($n=15$), and *Deliberation and self-reflection* ($n=11$). Full cluster listings will be available in the replication materials. Replication code will be available on GitHub in the Camera-Ready version.

[1] https://huggingface.co/Qwen/Qwen3-Embedding-0.6B. Selected for top-ranked STS[5] performance on the MTEB leaderboard.

[2] Hierarchical Agglomerative Clustering.

References

1. Adams, S.P., Preez, R.D.: Supporting student engagement through the gamification of learning activities: a design-based research approach. Technol. Knowl. Learn. **27**, 119–138 (2021). https://doi.org/10.1007/s10758-021-09500-x
2. Bächtiger, A., Parkinson, J.: Mapping and Measuring Deliberation: Towards a New Deliberative Quality. Oxford University Press (2019). https://doi.org/10.1093/oso/9780199672196.001.0001
3. Bächtiger, A., Shikano, S., Pedrini, S., Ryser, M.: Measuring Deliberation 2.0: Standards, Discourse Types, and Sequenzialization (2010)
4. Bai, N., Azadi, S., Nourian, P., Roders, A.: Decision-making as a social choice game - gamifying an urban redevelopment process in search for consensus. In:Proceedings of the 38th International Conference on Education and Research in Computer Aided Architectural Design in Europe (eCAADe), vol. 2 (2020). https://doi.org/10.52842/conf.ecaade.2020.2.555
5. Bista, S., Nepal, S., Paris, C., Colineau, N.: Gamification for online communities: a case study for delivering government services. Int. J. Cooper. Inf. Syst. **23** (2014). https://doi.org/10.1142/S0218843014410020
6. Carpentras, D., Fricker, R.H., Helbing, D.: Empowering minorities and everyone in participatory budgeting: an agent-based modelling perspective. Phil. Trans. Series A, Math. Phys. Eng. Sci. **382** (2024). https://doi.org/10.1098/rsta.2024.0090
7. Cohen, J.: Deliberation and democratic legitimacy. In: Matravers, D., Pike, J.E. (eds.) Debates in Contemporary Political Philosophy: An Anthology. Routledge (2002)
8. Dah, J., Hussin, N., Zaini, M.K., Helda, L.I., Ametefe, D.S., Aliu, A.A.: Gamification is not working: why? Games Cult. **20**, 934–957 (2024). https://doi.org/10.1177/15554120241228125
9. Dembinska, M., Montambeault, F.: Deliberation for reconciliation in divided societies. J. Deliberative Democracy **11**(1) (2015). https://doi.org/10.16997/jdd.226
10. Dryzek, J.S.: Deliberative Democracy and Beyond: Liberals, Critics, Contestations. Oxford University Press, Oxford (2002)
11. Elstub, S.: Deliberative pragmatic equilibrium review: a framework for comparing institutional devices and their enactment of deliberative democracy in the UK. Brit. J. Polit. Int. Relat. **16**(3), 386–409 (2014). https://doi.org/10.1111/1467-856X.12000
12. Ercan, S.A., et al.: Guidebook for Deliberative Engagement: Key Features and Practical Insights. Australian Government (2025)
13. Estlund, D., Landemore, H.: The epistemic value of democratic deliberation. In: Bächtiger, A., Dryzek, J.S., Mansbridge, J., Warren, M. (eds.) The Oxford Handbook of Deliberative Democracy. Oxford University Press (2018). https://doi.org/10.1093/oxfordhb/9780198747369.013.26
14. Fishkin, J.S.: When the People Speak: Deliberative Democracy and Public Consultation. Oxford University Press, Oxford, New York (2009)
15. Fishkin, J.S.: Democracy When the People Are Thinking: Revitalizing Our Politics Through Public Deliberation. Oxford University Press (2018)
16. Gao, F.: Advancing gamification research and practice with three underexplored ideas in self-determination Theory. TechTrends **68**, 661–671 (2024). https://doi.org/10.1007/s11528-024-00968-9
17. Gastil, J.: To play is the thing: how game design principles can make online deliberation compelling. Am. Behav. Sci. **67**(8), 1042–1060 (2023). https://doi.org/10.1177/00027642221093592

18. Gastil, J., Broghammer, M.: Linking theories of motivation, game mechanics, and public deliberation to design an online system for participatory budgeting. Polit. Stud. **69**(1), 7–25 (2021). https://doi.org/10.1177/0032321719890815
19. Gutmann, A., Thompson, D.: Reflections on deliberative democracy: when theory meets practice, pp. 900–912. Oxford University Press (2018). https://doi.org/10.1093/oxfordhb/9780198747369.013.44
20. Habermas, J.: Between Facts and Norms: Contributions to a Discourse Theory of Law and Democracy. No. 3, Polity (1996)
21. Hamari, J., Koivisto, J., Sarsa, H.: Does gamification work? – a literature review of empirical studies on gamification. In: 2014 47th Hawaii International Conference on System Sciences, pp. 3025–3034 (2014). https://doi.org/10.1109/HICSS.2014.377
22. Hassan, L.: Governments should play games: towards a framework for the gamification of civic engagement platforms. Simul. Gam. **48**(2), 249–267 (2017). https://doi.org/10.1177/1046878116683581
23. Hassan, L., Hamari, J.: Gameful civic engagement: a review of the literature on gamification of e-participation. Gov. Inf. Q. **37**(3), 101461 (2020). https://doi.org/10.1016/j.giq.2020.101461
24. Huang, R.T., et al.: The impact of gamification in educational settings on student learning outcomes: a meta-analysis. Educ. Tech. Res. Dev. **68**, 1875–1901 (2020). https://doi.org/10.1007/s11423-020-09807-z
25. Hwang, G.J., Huang, H., Chen, H.Y.: Promoting students' civic literacy and positive learning behaviors: a supportive feedback-based decision-making gaming approach. Educ. Inf. Technol. **30**, 189–227 (2024). https://doi.org/10.1007/s10639-024-12839-0
26. Iacovides, I., Cutting, J., Beeston, J., Cecchinato, M.E., Mekler, E.D., Cairns, P.: Close but not too close: distance and relevance in designing games for reflection. Proc. ACM Hum.-Comput. Interact. **6**, 1–24 (2022). https://doi.org/10.1145/3549487
27. Khaleghi, A., Aghaei, Z., Mahdavi, M.: A gamification framework for cognitive assessment and cognitive training: qualitative study. JMIR Ser. Games **9** (2021). https://doi.org/10.2196/21900
28. Koivisto, J., Hamari, J.: The rise of motivational information systems: a review of gamification research. Int. J. Inf. Manag. **45**, 191–210 (2019). https://doi.org/10.1016/j.ijinfomgt.2018.10.013
29. Krath, J., Schürmann, L., von Korfleisch, H.F.O.: Revealing the theoretical basis of gamification: a systematic review and analysis of theory in research on gamification, serious games and game-based learning. Comput. Hum. Behav. **125** (2021). https://doi.org/10.1016/j.chb.2021.106963
30. Lafont, C.: Democracy without shortcuts. Constellations **26**(3), 355–360 (2019). https://doi.org/10.1111/1467-8675.12432
31. Landemore, H.: Deliberation, cognitive diversity, and democratic inclusiveness: an epistemic argument for the random selection of representatives. Synth. Int. J. Epistemol. Methodol. Phil. Sci. **190**(7), 1209–1231 (2013). https://doi.org/10.1007/s11229-012-0062-6
32. Landemore, H.: Open democracy and digital technologies. In: Bernholz, L., Landemore, H., Reich, R. (eds.) Digital Technology and Democratic Theory. University of Chicago Press (2021). https://doi.org/10.7208/chicago/9780226748603.003.0003
33. Leclercq, T., Poncin, I., Hammedi, W., Kullak, A., Hollebeek, L.: When gamification backfires: the impact of perceived justice on online community contri-

butions. J. Mark. Manag. **36**, 550–577 (2020). https://doi.org/10.1080/0267257x.2020.1736604
34. Loh, W.: The gamification of political participation. Moral Phil. Polit. **6**, 261–280 (2019)
35. Mansbridge, J.: Using power/fighting power. Constellations **1**(1), 53–73 (1994). https://doi.org/10.1111/j.1467-8675.1994.tb00004.x
36. McPherson, M., Smith-Lovin, L., Cook, J.M.: Birds of a feather: homophily in social networks. Ann. Rev. Sociol. **27**(1), 415–444 (2001). https://doi.org/10.1146/annurev.soc.27.1.415
37. Mekler, E.D., Brühlmann, F., Tuch, A.N., Opwis, K.: Towards understanding the effects of individual gamification elements on intrinsic motivation and performance. Comput. Hum. Behav. **71**, 525–534 (2017)
38. Moon, J., McNeill, L., Edmonds, C., Banihashem, S.K., Noroozi, O.: Using learning analytics to explore peer learning patterns in asynchronous gamified environments. Int. J. Educ. Technol. High. Educ. **21**, 1–36 (2024). https://doi.org/10.1186/s41239-024-00476-z
39. Mora, A., Riera, D., González-González, C., Arnedo-Moreno, J.: Gamification: a systematic review of design frameworks. J. Comput. High. Educ. **29**, 516–548 (2017). https://doi.org/10.1007/s12528-017-9150-4
40. Mouffe, C.: Deliberative democracy or agonistic pluralism? Soc. Res. **66**(3), 745–758 (1999). jSTOR: 40971349
41. Niemeyer, S., Dryzek, J.S.: The ends of deliberation: meta-consensus and intersubjective rationality as ideal outcomes. Swiss Polit. Sci. Rev. **13**(4), 497–526 (2007). https://doi.org/10.1002/j.1662-6370.2007.tb00087.x
42. Niemeyer, S., Veri, F., Dryzek, J.S., Bächtiger, A.: How deliberation happens: enabling deliberative reason. Am. Polit. Sci. Rev. **118**(1), 345–362 (2024). https://doi.org/10.1017/S0003055423000023
43. Nourian, P., et al.: EquiCity game: a mathematical serious game for participatory design of spatial configurations. Sci. Rep. **14** (2023). https://doi.org/10.1038/s41598-024-61093-4
44. Park, S., Kim, S.: Leaderboard Design Principles to Enhance Learning and Motivation in a Gamified Educational Environment: Development Study - Consensus. https://consensus.app/papers/leaderboard-design-principles-to-enhance-learning-and-park-kim/79f418d31ce7521eaab06cc3a89cd21c/?search_id=Y_6CUL22ROyd5jLAZ3-y6A
45. Patricio, R., Moreira, A., Zurlo, F., Melazzini, M.: Co-creation of new solutions through gamification: a collaborative innovation practice. Creat. Innov. Manag. **29**(1), 146–160 (2020). https://doi.org/10.1111/caim.12356
46. Porto, D., Jesus, G., Ferrari, F., Fabbri, S.: Initiatives and challenges of using gamification in software engineering: a systematic mapping. ArXiv arxiv:2011.07115 (2020). https://doi.org/10.1016/j.jss.2020.110870
47. Resek, C.L., Siniscalchi, P.P., Caetano, B., Oliveira, K.W.R., Paula, M.M.V., Souza, J.: Gamification of citizen participation: a systematic mapping. IEEE Trans. Games **16**, 1–15 (2024)
48. Rodrigues, L., Oliveira, A., Rodrigues, H.: Main gamification concepts: a systematic mapping study. Heliyon **5** (2019). https://doi.org/10.1016/j.heliyon.2019.e01993
49. Ryan, R.M., Deci, E.L.: Self-Determination Theory and the Facilitation of Intrinsic Motivation. Social Development, and Well-Being (2000)
50. Sailer, M., Homner, L.: The gamification of learning: a meta-analysis. Educ. Psychol. Rev. **32**, 77–112 (2019). https://doi.org/10.1007/s10648-019-09498-w

51. Seaborn, K., Fels, D.: Gamification in theory and action: a survey. Int. J. Hum Comput Stud. **74**, 14–31 (2015). https://doi.org/10.1016/j.ijhcs.2014.09.006
52. Shortall, R., Itten, A., van der Meer, M., Murukannaiah, P., Jonker, C.: Inclusion, equality and bias in designing online mass deliberative platforms. ArXiv arxiv:2107.12711 (2021)
53. Shortall, R., Itten, A., Meer, M.V.D., Murukannaiah, P., Jonker, C.: Reason against the machine? Future directions for mass online deliberation. Front. Polit. Sci. **4**, 946589 (2022). https://doi.org/10.3389/fpos.2022.946589
54. Steenbergen, M.R., Bächtiger, A., Spörndli, M., Steiner, J.: Measuring political deliberation: a discourse quality index. Comparat. Eur. Politics **1**(1), 21–48 (2003). https://doi.org/10.1057/palgrave.cep.6110002
55. Tang, M., Liao, H., Wu, X.: A Stackelberg game model for large-scale group decision making based on cooperative incentives. Inf. Fusion **96**, 103–116 (2023). https://doi.org/10.1016/j.inffus.2023.03.013
56. Thiel, S.K., Reisinger, M.R., Röderer, K., Fröhlich, P.: Playing (with) democracy: a review of gamified participation approaches. JeDEM: eJournal eDemocracy Open Gov. **8**, 32–60 (2016). https://doi.org/10.29379/jedem.v8i3.440. https://consensus.app/papers/playing-with-democracy-a-review-of-gamified-participation-thiel-reisinger/edbb41d5e4cb5fe3b2547f03985b95d4/
57. Thomas, N., Baral, R.: Mechanism of gamification: role of flow in the behavioral and emotional pathways of engagement in management education. Int. J. Manag. Educ. (2023). https://doi.org/10.1016/j.ijme.2022.100718
58. Warren, M.E., Pearse, H. (eds.): Designing Deliberative Democracy: The British Columbia Citizens' Assembly. Theories of Institutional Design. Cambridge University Press, Cambridge (2008). https://doi.org/10.1017/CBO9780511491177
59. Young, I.M.: Inclusion and Democracy. Oxford University Press (2002)
60. Zainuddin, Z., Chu, S., Shujahat, M., Perera, C.J.: The impact of gamification on learning and instruction: a systematic review of empirical evidence. Educ. Res. Rev. (2020). https://doi.org/10.1016/j.edurev.2020.100326
61. Zhang, Z., Patrício, R., Carella, G., Zurlo, F.: Supporting a sustainable and engaging online transition for co-design through gamification. Sustainability (2022). https://doi.org/10.3390/su14116716
62. Zimmerling, E., Höllig, C.E., Sandner, P.G., Welpe, I.: Exploring the influence of common game elements on ideation output and motivation. J. Bus. Res. (2018). https://doi.org/10.1016/j.jbusres.2018.02.030

A Reality Check for Digital Government Leadership Research: Empirical Insights into Leadership Recruitment in Public Organizations

Till Becker[1(✉)] and Michael Koddebusch[2]

[1] University of Applied Sciences Emden/Leer, Emden, Germany
till.becker@hs-emden-leer.de
[2] ERCIS, University of Münster, Münster, Germany

Abstract. The digital transformation of public organizations progresses slowly, particularly in Germany. Among many contributing factors, leadership is widely recognized as a key determinant of successful transformation, which has led to the emergence of digital government leadership (DGL) as a distinct research topic. Existing studies predominantly provide normative accounts of what DGL should entail in practice. However, there is little empirical evidence on whether these prescriptions are reflected in the actual practices of public organizations. This gap is problematic because research risks mis-targeting its practitioner audience by reinforcing positions that are already routine or by overlooking persisting deficits. To address this gap, this study employs an LLM-supported job-mining approach analyzing 5,949 job advertisements from German public organizations. The analysis examines (1) the career stages and (2) the digital competences demanded in these positions, operationalized using DigComp 3.0, the latest version of the European Digital Competence Framework. The results show that lower-level career stages exhibit a proportionally higher demand for digital competences than higher-level positions. This pattern suggests that digital competences expected from leadership personnel in the German public sector fall short of the expectations articulated in the scholarly literature on DGL. The study contributes to digital government research by providing an empirical reality check for DGL scholarship. Based on the findings, we argue for expanding the research scope of DGL to increase the exploration and design of leadership development interventions, as well as stronger consideration of how leadership behavior interacts with organizational and systemic boundary conditions that shape digital transformation.

Keywords: Digital Government Leadership · Job-Mining · Digital Transformation · Digital Competence

1 Introduction

Digital government, understood as digital transformation in the public sector, progresses slowly. This is particularly true in Germany, where public organi-

Published by Springer Nature Switzerland AG 2027
D. Duenas-Cid et al. (Eds.): ePart 2026, LNCS 16923, pp. 199–214, 2027.
https://doi.org/10.1007/978-3-032-36676-4_13

zations have been described as "latecomers" to digital transformation [16, p. 63]. Scholars link this to persistent bureaucracy, low adoption of digital public services, and overall inert processes [23].

Explanations for this slow pace are multifaceted, but can be broadly grouped into structural and cultural barriers [35]. In both categories, leadership has emerged as a central factor that can either enable or inhibit organizational adaptation [11,18]. In response, research on digital government leadership (DGL) has gained traction. DGL can be defined as the ability "to exploit existing digital opportunities to solve imminent challenges and to explore new digital opportunities to undertake the core mission of the public organization, while managing for results, leading followers, and aligning with the environment towards a clearly articulated vision" [19, p. 3:15]. This admittedly broad definition illustrates the variety of DGL, spanning strategic and operational demands, as well as technical, socio-technical, interpersonal, and public-sector-specific responsibilities. Given this breadth, prior work has proposed DGL competence frameworks [e.g., 1,21,26], highlighting leadership styles, strategic–operational ambidexterity, and capabilities such as stakeholder management, communication, team and trust building, and change management.

However, much of this literature is normative, specifying what DGL should look like. What is currently absent from the literature is an empirically validated account of how public organizations *actually* understand leadership in the context of digital government. To foster such empirically validated accounts, a research stream [e.g., 2,5,22] has emerged that employs job-mining and emphasizes the value of using job advertisements (job ads) as standardized, publicly accessible research data to investigate the competences currently in demand by organizations. All of these accounts, though, treat the public workforce as a homogeneous category, lacking distinction between operational and leadership personnel. We argue that such a distinction between employee groups is crucial: without evidence on how closely leadership understanding aligns with digital competence demands, we cannot assess whether leadership in the practice of digital government aligns with the prescriptive ideals of DGL scholarship. As a result, research risks mis-targeting its practitioner audience by either reinforcing what is already routine or overlooking under-addressed deficits.

Therefore, we ask: *How do digital competence demands evolve across career stages in German public organizations?*

We understand job-mining as a subset of data mining and as the systematic collection and analysis of job-related information from labor-market sources to understand workforce needs [7]. Here, we apply job-mining as competence assessment [13], in which competence profiles (job ads) are compared against a digital competence model. We compare **5,949** job ads from the German public sector and employ DigComp 3.0 [9], the fifth version of the European Digital Competence Framework, as our competence model. Our sample intentionally includes job ads for both leadership positions and lower-hierarchy employees to make meaningful claims about DGL in relation to non-leading employees. The decision for DigComp 3.0 implies the decision against academic competence models [e.g., 1,21,26]. This decision is based on the assumption that if public organizations use competence models for recruitment, they are more likely to utilize practice-oriented models rather than academic ones that are only published in academic literature. Furthermore, DigComp 3.0 explicitly mentions leadership activities in its description of advanced competence levels, further strengthening its choice as a reference model.

Our contribution is an empirically grounded account of the "as-is" state of digital competence demands in public sector leadership recruitment in Germany, offering a reality check for DGL theory and a basis for more targeted leadership development and research agendas.

2 Research Background

2.1 Leadership for Digital Transformation

Digital transformation in the public sector and beyond is a process of ongoing organizational renewal, often triggered by technological disruptions, that results in a fundamental shift of an organization's identity [33,34]. It goes beyond digitization and digitalization by reshaping culture, relationships, and routines [24]. Because it is so profound, digital transformation is difficult to pursue: it can be stoked by individual resistance (e.g., against adopting new technologies or changing established routines) and hindered by organizational inertia, understood as the persistent difficulty of aligning existing resources and capabilities with a rapidly changing environment [33]. Within this dynamic, leadership can either reinforce such barriers or, when guided by a digital mindset, act as a catalyst for change [35]. While the specific foundations of digital leadership differ across contexts, prior work suggests six recurring pillars: how strategy is formulated, how business models are conceived, how workplaces are designed, how employees are treated, enabled, and developed, how platforms are leveraged, and how the role of IT is understood within the organization [15]. Where leadership builds on these foundations, it is more likely to support successful digital transformation.

Digital Government adds a layer of complexity to this broader leadership view. What distinguishes DGL is its commitment to the "public organization's core mission" [19, p. 3:16], that is, the public value. Consequently, digital transformation, and thus DGL, cannot be reduced to a logic of profit maximization. Instead, leaders must navigate many, and sometimes ambiguous, objectives: increasing operational efficiency while simultaneously improving public services for citizens, including those with limited digital skills, and upholding social values like inclusion, democracy, and transparency [32]. This creates tension: DG leaders are expected to champion technologies and foster a digital mindset among employees, while operating in administrative settings that are characterized by clear hierarchies and rigid structures [19].

Research at the intersection of leadership and digital government remains comparatively nascent but is gaining traction. A substantial share of this emerging work is prescriptive, developing DGL competence frameworks [1,21,26] or role concepts for DG leaders [10]. These contributions outline the capabilities leaders should possess to navigate the context described above. Many proposed capabilities align with DigComp 3.0 [9], which we adopt as a reference framework in this study, such as data literacy and analytical skills [21], communication and collaboration [26], and creativity and cultural intelligence [1]. The most recent contribution is van Roekel et al.'s [25] normative framework for digital transformation leadership in public organizations. It differentiates between a strategic dimension, focusing on vision, collaboration, and legitimacy input, and an operational perspective, which includes human resources, politics, structures, and legitimacy through- and output. Although we find this distinction valuable, we did not adopt it in this study because it propagates "normative goals of public sector leadership" (p. 10), while our study aims to investigate whether such scholarly prescriptions translate into practice. Establishing this "as-is" baseline is critical: without it, DGL research risks investigating for an idealized target state while visibility on the impact of investigations.

2.2 Related Works and Research Context

Job Mining: The use of job ads to determine organizations' competence demands has gained traction in recent years, as it enables the analysis of large-scale datasets and thereby helps overcome the limitations associated with resource-intensive qualitative approaches [22]. For the public sector specifically, some studies investigate job ads primarily through qualitative analyses and complement their insights with additional methods such as interviews [e.g., 6] but most job mining research relies on technology support, including natural language processing [e.g., 3,22] or large language models (LLMs) [e.g., 5]. Despite methodological differences, these approaches share a common logic: they extract textual data from large datasets of publicly available job ads and analyze their alignment with competence frameworks. Although the studies differ in emphasis and scope, their conclusions converge. Most find substantial gaps between the competences suggested in normative frameworks and those reflected in actual job ads, while acknowledging the methodological limitations inherent to job mining.

Prior research has either examined developments over time [22] or provided snapshots of current competence demands [2,3,5]. However, differentiation within the public sector workforce remains limited. In particular, existing studies rarely distinguish between employee groups. Addressing this gap, the present study focuses specifically on the differences between leadership and non-leadership positions and interprets the results through the lens of DGL scholarship.

DigComp 3.0: DigComp 3.0 [9] is the fifth version of the European Digital Competence Framework. Although it is a general one, not specific to digital government or public administration, it has previously been used for research in these domains [e.g., 3,5,22], and thus offers the potential to contribute to a growing research stream and benchmarking. DigComp 3.0 foregrounds five competence areas: (1) **Information Search, Evaluation, and Management** (e.g., articulation of information needs, critical evaluation of relevance and sources; further called *Information Search*), (2) **Communication and Collaboration** (e.g., usage of collaboration technologies, digital identity management), (3) **Content Creation** (e.g., usage of technologies for aggregating and integrating content, computational thinking, programming), (4) **Safety, Wellbeing, and Responsible Use** (e.g., protection of devices and privacy alike, mental wellbeing in digital spaces, awareness of environmental impact), and (5) **Problem Identification and Solving** (e.g., adaption of digital environments to satisfy demands, resolving technical and conceptual problems, usage of technologies to provide innovative solutions). Each area identifies several specific competences across four proficiency levels. This study, however, focuses on the aggregated competence area level to allow for interpretation when analyzing the job ads.

Payment Schemes as a Proxy for Leadership Responsibility: Two formal pay schemes structure careers in German public administration. Civil servants are remunerated under the *Bundesbesoldungsgesetz* ("Federal Pay Act", *Beamte*, A-groups), while public employees are paid according to the *Tarifvertrag für den öffentlichen Dienst* ("Labor Agreement for the Public Service", *Angestellte*, E-groups). *Beamte* are state-appointed civil servants with special duties, strong job security, and state pensions. Public *Angestellte* work under regular labor contracts. Although the two groups may not align perfectly in their progression (i.e., E10 may also align with A9 or A11 responsibilities), the scales offer mostly simultaneous career levels: simple, middle, upper, and higher service. A/E1-5 corresponds to simple service; A/E6-9 to middle service; A/E9-13 to upper service; and A/E13-A16 to higher service. These categories reflect task complexity, qualification requirements, and institutional responsibility rather than leadership per se. Accordingly, formal leadership responsibility is very uncommon in the *simple* service and rather uncommon in the *middle* service, although small team supervision may occur. In the *upper* service, leadership responsibility becomes increasingly likely with each A/E grade, especially at higher ones (i.e., A/E12-13). In the *higher* service (A13+ and E14/E15), holding leadership roles is common. We therefore treat A/E levels as sufficiently reliable proxies for leadership responsibility, without claiming perfect accuracy (since, e.g., highly paid technical roles may not have leadership responsibility). Any classification noise is unlikely to bias our results, given the scale of the dataset we analyze. This proxy is also more robust than keyword-based measures derived from job ads, because phrasing like "open to leading teams" does not necessarily imply actual leadership duties, and the absence of such wording does not imply that no team leadership is involved.

3 Method

3.1 Digital Competence and Leadership Responsibility in the Data

Our research approach is situated in the field of job mining, as outlined in Sect. 2.2. We use a dataset of 5,949 job ads that were downloaded from the job portal of the German Federal Employment Agency [8]. The job ads are filtered from a larger dataset containing approximately 16,000 job ads from organizations from the public sector in Germany posted in February 2025. We apply a filter to the 'Profession' field that limits the job ads to jobs from core public administration, that is, individuals working in public administration for the delivery of public services (clerks, advisors, department heads, etc.), but not from other parts of the German public sector, such as higher education or police officers. We also removed job ads for trainings and internships.

We look at the job ads from two perspectives: First, we want to determine which type of digital competences are requested for the advertised position. Naturally, job ads do not contain a standardized list of digital competences. Thus, we operationalize the five competence areas from DigComp 3.0 (see Sect. 2.2) as categories. Each job ad is analyzed to determine which specific competence areas are required from the candidate. We acknowledge that these job ads typically outline the expected competences for a role, not the actual ones of the potential hire. However, we believe they are the most suitable publicly available data for addressing our research question, therefore accepting this potential limitation.

Second, we need to know to what extent "leadership responsibility" is part of the advertised job. This study incorporates a traditional view of leadership associated with hierarchical advancement, where higher-paid positions are more likely to involve formal leadership responsibilities. Although informal leadership, like team- or peer-based leadership, exists, the data do not support these concepts, so they are outside the scope of this study. Leadership responsibility is neither stated nor quantified explicitly in job ads, nor is it a one-dimensional concept. Therefore, we use the payment scheme of the German public sector, as described in Sect. 2.2, as a proxy for leadership responsibility. This approach has two advantages. First, German public organizations are required to adhere to payment schemes. It is also common practice in public sector job ads to explicitly state the salary level or range that will be offered for a position, unlike in the German private sector, where salary information in job advertisements is uncommon. Second, the two payment schemes under consideration (A and E groups) have detailed scales of 18 and 15 levels, respectively. Thus, there is a rather fine-grained scale for assuming increasing leadership responsibility. Furthermore, the division of the salary levels in the four career stages *simple*, *middle*, *upper*, and *higher service* helps to reconstruct the idea of an increasing demand for leadership responsibility as one ascends the career ladder[1].

3.2 LLM-Supported Knowledge Extraction from Job Ads

The classical approach to extract the previously defined information about digital competences and the level of leadership responsibility from the text body of a job ad would be to perform a qualitative analysis of the text, commonly described as qualitative coding [see, e.g., 20]. Since the development of LLMs has gathered pace after their release in public in 2022, the process of qualitative coding is in the focus of automation attempts. The benefits of an automated

[1] Webpage Bundesagentur für Arbeit (in German).

coding approach include faster coding and reduced fatigue compared to human coders, enabling the coding of large text corpora more quickly. Some LLM-based approaches require additional training [e.g., 36], and others utilize regular pre-trained LLMs [e.g. 30]. We focus on the latter approach in this work, as we do not have labeled training data.

Although the use of LLMs in coding increases automation, the entire coding process is still far from fully automated. Than et al. [31] claim generative LLMs hold significant promise for transforming qualitative coding in social science research. By integrating LLMs into the qualitative research workflow, researchers can leverage the strengths of both human insight and machine learning, while maintaining transparency and reproducibility. Barros et al. [4] find that LLMs are effective in automating coding processes and that they offer the potential for extremely high efficiency gains. Hayes [17] emphasizes that LLMs accelerate traditional qualitative tasks (e.g., coding, theme identification) and are able to connect empirical data to theoretical frameworks.

We develop a codebook and a system prompt[2] that instructs the LLM to read the text body of a job ad and to identify parts referring to digital competence; in a next step, the LLM assigns such parts to any of the five DigComp 3.0 competence areas, an approach that aligns with previous works [e.g., 5].

Because we observe that LLMs are prone to assigning a category even when in doubt, we introduce an additional confidence variable. The LLM is requested to include a level of certainty among 0 (if the identification of a digital competence is only based on inference), 1 (if the LLM assumes that digital tools are used with this competence), and 2 (if there is a specific reference to digital technologies or concepts) along with the categorization. After the coding, we discard all assessments with confidence levels 0 and 1 in order to exclude questionable competence assignments from the analysis. Levels 0 and 1 can rather be regarded as a "honeypot" for possible hallucinations by the LLM.

To further strengthen the accuracy of our results, we apply an ensemble strategy. Ensembles of different methods have often been used in Machine Learning to improve the reliability of the results [12]. Schoenegger et al. [28] have identified ensemble LLMs as a promising avenue for AI-assisted decision-making. Their results underscore the potential of aggregating diverse LLM predictions to achieve robust and reliable forecasts. In our case, we let three different models perform the digital competence classification and apply a *majority-vote ensemble*: only if at least two models find at least one occurrence of a competence area in a job ad, this competence area is assigned to the job ad. We opt for this type of voting, because the only alternatives with three models would be either also accepting minority votes or a full consensus. From our perspective, a minority vote would weaken the effort to reduce hallucinations, whereas a full consensus might discard possible categories too early. We select OPENAI-GPT-OSS-120B, MISTRAL-LARGE-3-675B-INSTRUCT-2512, and QWEN3-30B-A3B-INSTRUCT-2507 as three different, recent, open weight models.

[2] codebooks and prompts available for download at https://doi.org/10.5281/zenodo.20664624.

The salary groups are also identified with the help of an LLM. We design a codebook and a system prompt for identifying the groups. Due to the explicit mentioning of the salary groups in the text, a single run with OPENAI-GPT-OSS-120B is sufficient. If multiple salary groups are found (e.g., a range of groups or a position that potentially can be filled by a public employee or a civil servant), we record the highest level found for each career track (A and E).

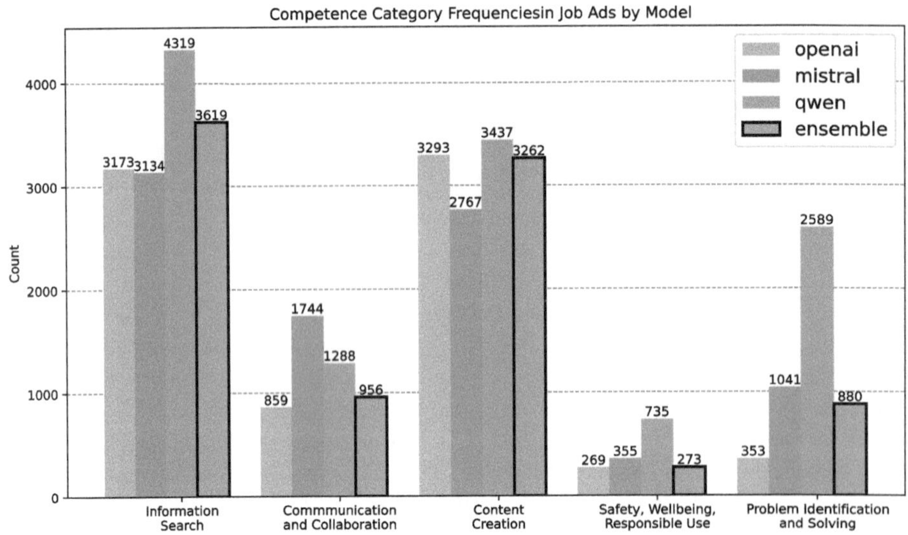

Fig. 1. The LLMs identify similar frequencies of digital competences for each DigComp 3.0 competence area. The *ensemble* count represents the combined judgment of the models, determined by a majority vote. The distribution of digital competences among the job ads is in accordance with previous results.

4 Results

4.1 Distribution of Competence Areas and Career Stages

The 5,949 job ads were categorized into the five competence areas defined in DigComp 3.0. Figure 1 shows the categorization results. The count comprises all job ads in which the LLMs assigned at least one text fragment to a competence area. The *ensemble* count is the result of a *majority-vote ensemble*, which only assigns a competence area to a job ad if at least two of the individual LLMs have identified this competence area with a high confidence level. Generally, the three models exhibit a high degree of agreement in their judgments. An exception to this observation is the behavior of QWEN in the Competence Area 5 (Problem Identification and Solving), where it identifies more than twice the number of job ads as the subsequent model, MISTRAL. However, this divergent number is

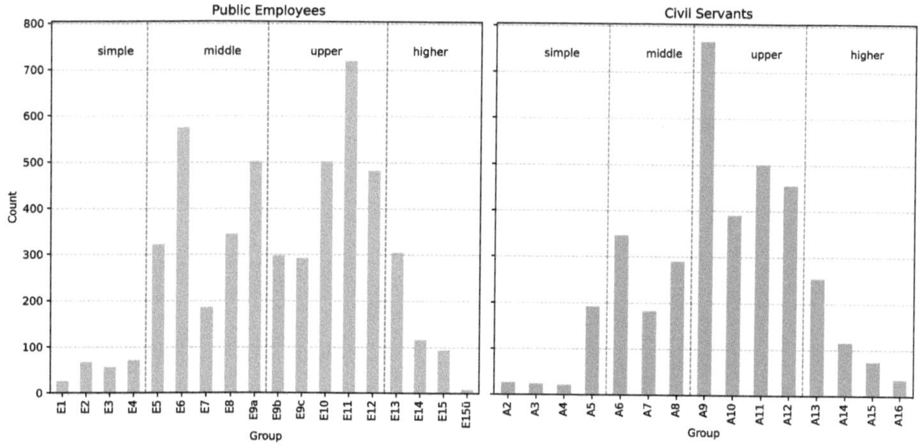

Fig. 2. The two histograms show the distribution of the offered salary levels found in the job ads. We distinguish between offers for public employees and civil servants. A job ad can offer a career in both tracks, which allows that a job ad is counted in both categories. The vertical dotted lines separate the salary levels in the career stages *simple*, *middle*, *upper*, and *higher*.

corrected by the other two models when the ensemble technique is applied. The distribution of competence areas among the job ads shows a similar pattern as previous research [5], which applied the DigComp 2.2 framework: competences in the areas (1) *Information Search* and (3) *Content Creation* are frequently requested, whereas the other areas (2) *Communication and Collaboration*, (4) *Safety, Wellbeing, and Responsible Use* and (5) *Problem Identification and Solving* remain underrepresented.

We use German public sector payment schemes as a proxy for leadership responsibility (see Sect. 2.2). Figure 2 illustrates frequency distributions of the "E" (public employees) and "A" (civil servants) groups found in the job ads. Although there is a frequency concentration in the center part in both career tracks (*middle* and *upper* service), all four career stages are sufficiently represented. Only the "E15Ü" payment level was mentioned in fewer than 20 job ads. Thus, we eliminated this level from the analyses. Some job ads offer career options in both tracks, which means that the same job ad, along with the digital competences assigned to it, can be counted towards both tracks.

4.2 Competence Areas Across Career Stages

Next, we analyze the proportion of job ads at each salary level that require the five areas of DigComp 3.0. We aim to determine whether the demand for digital competence is evenly distributed across different career levels, or if some levels exhibit particularly high or low demand. Figures 3a and 3b clearly show a distinct progression of the demand for digital competences. However, the distribution of demand is not equal across the five competence areas.

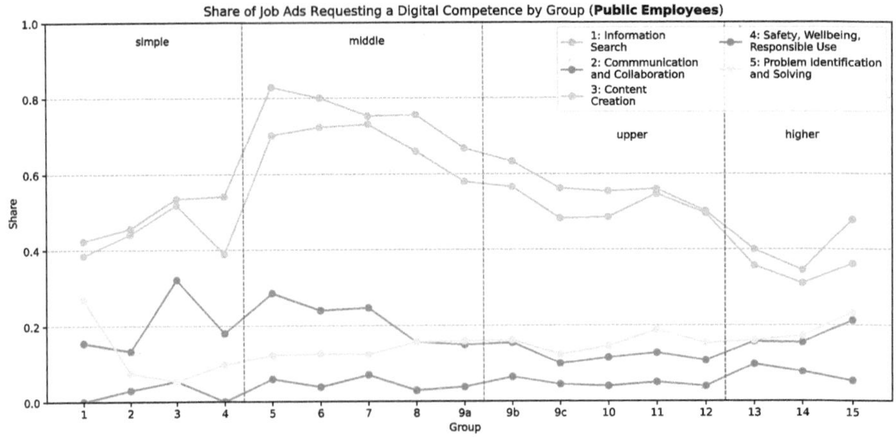

(a) Share of job ads for **public employees** requesting digital competences, by salary level.

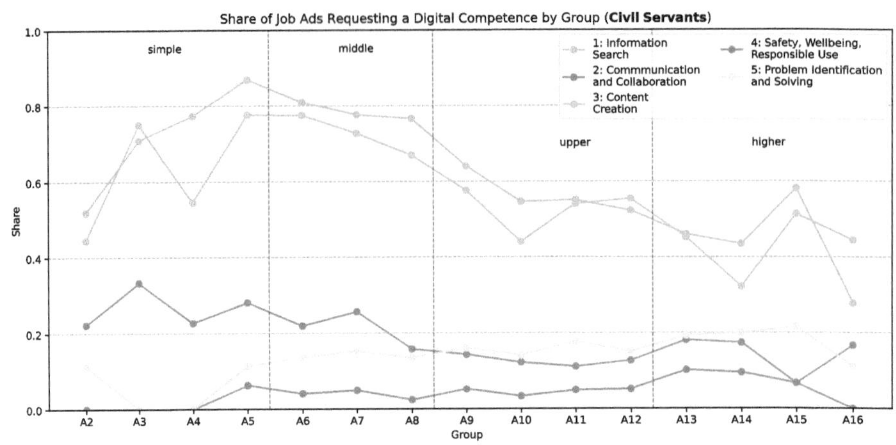

(b) Share of job ads for **civil servants** requesting digital competences, by salary level.

Fig. 3. Digital competence demands vary by salary level: Areas 1–2 peak at *middle*, with area 2 high in *simple*; areas 4–5 show minimal but rising demand.

The most frequently demanded areas 1 (*Information Search, Evaluation, and Management* and 3 (*Content Creation*) are moderately requested at lower salary levels. The demand peaks in the lower middle part and then continuously declines towards the highest salary levels. Area 2 (*Communication and Collaboration*) shows a similar pattern.

However, due to the comparatively low demand for this area, the pattern is not as evident. Although the two remaining areas 4 (*Safety, Wellbeing, and Responsible Use*) and 5 (*Problem Identification and Solving*) tend to increase in demand with higher salary levels, the overall demand remains marginal below 10% and 20%, respectively.

Table 1 underlines these findings by merging the demands of all salary levels to the corresponding career stages. The numbers indicate that areas 1–3 are mainly present in the lower half of the stages (*simple* and *middle*), whereas areas 4 and 5, if demanded at all, can be found in the higher stages (*upper* and *higher*).

Table 1. Share of job ads demanding digital competences, grouped by career stages and career tracks. The coloring is normalized from lowest to highest values for every group of four values *simple*, *middle*, *upper*, and *higher* per digital competence area and career track.

	Public Employees				Civil Servants			
	simple	middle	upper	higher	simple	middle	upper	higher
(1) Information Search	0.50	0.76	0.56	0.40	0.81	0.79	0.58	0.46
(2) Comm. and Coll.	0.20	0.21	0.12	0.17	0.28	0.21	0.13	0.16
(3) Content Creation	0.44	0.67	0.52	0.35	0.72	0.73	0.54	0.43
(4) Safety, Wellb., Resp. Use	0.02	0.04	0.05	0.09	0.05	0.04	0.05	0.09
(5) Problem Id. and Solv.	0.10	0.14	0.16	0.18	0.09	0.14	0.16	0.19

5 Discussion

5.1 Implications for Digital Government Leadership Research

In this article, we investigate how digital competence demands evolve across career stages in German public organizations. This question comes up because most discussions about DGL have been theoretical, with little real-world evidence showing whether practical needs match those theories. Through an LLM-supported job mining analysis, we examined 5,949 job ads from the German public sector. Our findings suggest that the demands vary according to the specific area to which they belong, yet remain at an overall low level. Additionally, we found that lower-level career stages exhibit a proportionally higher demand for digital competences compared to higher career stages.

Our study can be positioned at the frontier of existing research in two ways. First, our findings confirm prior studies indicating a generally low presence of digital competences in the broader workforce. Although the precise results differ across studies [e.g., 3,5,22], our findings suggest that previous scholarly insights have not yet translated into substantially increased demand for digital competences in recruitment within the public sector.

Second, our results provide a new perspective. There is broad agreement that leadership is important in digital transformation initiatives [e.g., 11,18,19]. Assuming, however, that job ads are fairly representative for an organization's competence demands, the data indicate that digital competences are not strongly required of those leaders. As introduced in Sect. 1, DGL spans strategic and operational demands along with (socio-)technical and interpersonal responsibilities [19]. These demands and responsibilities are distributed across all competence areas of DigComp 3.0. However, in reality, competence areas two, four, and five remain low across career stages (between 0%–30%). Areas one and three are more frequently demanded at lower career stages but decline sharply with increasing salary levels. Although there appears to be a slight upward trend in the highest pay levels, we caution against overinterpreting these insights given the relatively small number of job ads in these categories (see Fig. 2).

What do these findings imply for research on DGL? From an empirical perspective, we emphasized in Sect. 1 the broad range of leadership responsibilities in digital transformation, including exploring and exploiting digital opportunities [19]. It is reasonable to assume that these require competences from all five DigComp 3.0 areas, such as staying informed (Area 1), advising others (Area 2), leading content creation (Area 3), guiding strategy design (Area 4), or assessing competence needs (Area 5). Our results do not show these competences are consistently required when recruiting leadership personnel. Thus, we conclude that the digital competences demanded of leadership in the German public sector fall short of scholarly DGL prescriptions.

From a research-stream perspective, our findings suggest that DGL scholarship should broaden its scope. While prescriptive frameworks [e.g., 1,21,26] are valuable for defining competence requirements, it appears equally important to (1) increase awareness of the importance of digital competences in leadership positions and (2) investigate not only leadership itself but also leadership development. In doing so, we align with prior leadership research [14], which emphasizes that scholars should move beyond asking **what** leadership is and also examine **how** leadership can be fostered. This is particularly relevant in the public sector, where digital transformation efforts may conflict with hierarchical structures and organizational rigidities. From this perspective, we conclude that DGL scholarship must broaden its scope, moving beyond prescription towards the exploration and design of leadership development interventions.

While our study cannot fully explain the observed effects, existing literature and contextual understanding allow for some justified assumptions. The German public sector – just as in many other national contexts – is a historically grown construct. It is characterized by a strong legalistic tradition, where adherence to legal and procedural correctness takes precedence over performance and results [16]. This tradition is deeply rooted in Weberian Bureaucracy, emphasizing "a hierarchically structured, professional, rule-bound, impersonal, meritocratic, and disciplined body of public servants" [27]. Leaders are expected to enforce these principles, focusing on accuracy, reliability and accountability in service provision. Implementing principles of digital leadership, such as platform thinking, viewing IT as an enabler, and new-work-like treatment of employees [15], in such an environment will inevitably lead to frictions. Public organizations typically recruit personnel from an education system that follows the principles of Weberian Bureaucracy rather than those of digital transformation. Over time, these recruits ascend to leadership positions. Now, it is critical to recognize that although DGL scholarship has promoted an understanding of what DGL *should* look like in practice, leaders and organizations *in reality* often exhibit rigid structures [29], resistance to change, and organizational inertia [33], making a nonchalant change in the understanding of what is expected from leadership personnel in digital transformation unlikely.

Thus, we argue that DGL should be treated as a highly dynamic and context-dependent concept, which necessitates DGL scholarship to explore leadership behaviour in relation to organisational and systemic boundary conditions.

5.2 Limitations

Our study provides insights into the demand for digital competences in the German public sector. We acknowledge that dynamics may differ elsewhere, encouraging further research in other regions. The extent to which job ads accurately reflect industry demand remains debatable, as some competences may be implicitly required.

Using career stages and salary as proxies for leadership simplifies a multi-dimensional concept, but remains a valid initial approach, as management positions are traditionally higher-paid. The results are indicative rather than conclusive. Additionally, technical senior roles often command higher salaries without involving leadership tasks, yet likely require greater digital competences. If such roles dominate high-level job ads, the observed demand for digital competences among leaders may be overestimated.

Coding of abstract concepts such as competences in textual data presents challenges. This challenge applies to both LLM-supported and manual coding. We followed established procedures and employed an ensemble strategy to further validate our results.

6 Conclusion

This study investigates how digital competence demands evolve across career stages in German public organizations. Our analysis of job ads reveals that lower-level career stages exhibit a higher demand for digital competences compared to higher career stages, suggesting a misalignment between scholarly prescriptions of DGL and practical recruitment demands. This misalignment has implications for both research and practice. For researchers, it highlights the need to broaden the scope of DGL scholarship and investigate leadership development interventions. For practitioners, it underscores the importance of increasing awareness of digital competences in leadership positions and investing in development interventions. The central implications of this study for DGL research are outlined in Sect. 5.1 Overall, this study provides a valuable empirical baseline for understanding digital competence demands in public sector leadership recruitment and underscores the need for further research in this area.

References

1. Adie, B.U., Tate, M., Valentine, E.: Digital leadership in the public sector: a scoping review and outlook. Int. Rev. Public Admin. **29**(1), 42–58 (2024). https://doi.org/10.1080/12294659.2024.2323847
2. Auth, G., Bensberg, F., Christ, J.P.: Competencies for ai adoption in public administration: a demand-side study based on job postings from germany. dms-der moderne staat-Zeitschrift für Public Policy, Recht und Management, pp. 1–21 (2026)
3. Auth, G., Christ, J., Bensberg, F.: Kompetenzanforderungen zur Digitalisierung der öffentlichen Verwaltung: Eine empirische Analyse auf Basis von Stellenanzeigen. In: 16th International Conference on Wirtschaftsinformatik, pp. 1–16 (2021)
4. Barros, C.F., et al.: Large language model for qualitative research: a systematic mapping study. In: 2025 IEEE/ACM International Workshop on Methodological Issues with Empirical Studies in Software Engineering (WSESE), pp. 48–55 (2025). https://doi.org/10.1109/WSESE66602.2025.00015
5. Becker, T.: Koddebusch, M. In: Search of digital talent: evidence from public sector job advertisements in germany using LLM-enabled competence classification. In: Electronic Participation, pp. 344–359. Springer, Cham (2025)
6. Bellmer, A., Becker, T.: Digitale Kompetenzen in der öffentlichen Verwaltung: Ein Überblick über Anforderungen an Mitarbeitende und bestehende Maßnahmen. In: Wimmer, M.A., Räckers, M., Hünemohr, H. (eds.) 7. Fachtagung Rechts- und Verwaltungsinformatik (RVI 2024): Neue Wege der Zusammenarbeit und Vernetzung für digitale Transformation und Verwaltungsmodernisierung, pp. 189–200. Gesellschaft für Informatik e.V., Bonn (2024). https://doi.org/10.18420/rvi2024-15
7. Bensberg, F., Buscher, G.: Job mining als Analyseinstrument für das human-resource-management. HMD Praxis der Wirtschaftsinformatik **53**(6), 815–827 (2016). https://doi.org/10.1365/s40702-016-0256-3
8. bundesAPI: jobsuche-api (2023). https://github.com/bundesAPI/jobsuche-api. Accessed 10 Mar 2026
9. Cosgrove, J., Cachia, R.: DigComp 3.0: European Digital Competence Framework. Technical report. KJ-01-25-521-EN-N,KJ-01-25-521-EN-C, Publications Office of the European Union, Luxembourg (2025). https://doi.org/10.2760/0001149
10. Dang, D., Bui, Q.N., Pham, S., Vartiainen, T., Pekkola, S.: Government roles and dynamic capabilities in digital transformation projects. In: Proceedings of the 33rd European Conference on Information Systems. Amman, Jordan (2025)
11. Day, D.V., Dragoni, L.: Leadership development: an outcome-oriented review based on time and levels of analyses. Annu. Rev. Organ. Psych. Organ. Behav. **2**(1), 133–156 (2015)
12. Dietterich, T.G.: Ensemble methods in machine learning. In: Multiple Classifier Systems, pp. 1–15. Springer, Heidelberg (2000)
13. Draganidis, F., Mentzas, G.: Competency based management: a review of systems and approaches. Inf. Manage. Comput. Secur. **14**(1), 51–64 (2006)

14. Eberl, J.K., Zimmer, M.P., Drews, P.: Digital leadership routines: understanding the role of artifacts in digital leadership development. Inf. Organ. **35**(4), 100599 (2025). https://doi.org/10.1016/j.infoandorg.2025.100599
15. El Sawy, O., Kræmmergaard, P., Amsinck, H., Lerbech Vinther, A.: How LEGO built the foundations and enterprise capabilities for digital leadership. Strateg. Inf. Manage. **15**(2), 141–166 (2016)
16. Hammerschmid, G., Oprisor, A.: German public administration: incremental reform and a difficult terrain for management ideas and instruments. In: Public Administration Reforms in Europe, pp. 63–72 (2016). https://doi.org/10.4337/9781783475407.00012
17. Hayes, A.S.: "conversing"' with qualitative data: enhancing qualitative research through large language models (llms). Int. J. Qual. Methods **24** (2025). https://doi.org/10.1177/16094069251322346
18. Hossain, S., Fernando, M., Akter, S.: Digital leadership: towards a dynamic managerial capability perspective of artificial intelligence-driven leader capabilities. J. Leadersh. Organ. Stud. **32**(2), 189–208 (2025). https://doi.org/10.1177/15480518251319624
19. Kristensen, K., Andersen, K.N.: C-suite leadership of digital government. Digital Govern. Res. Pract. **4**(1), 1–23 (2023). https://doi.org/10.1145/3580000
20. Kuckartz, U., Radiker, S.: Qualitative content analysis: Methods, practice and software. SAGE Publications Ltd (2023)
21. Kusanke, K., Kendziorra, J., Pilgenroeder, S., Christmann-Schwaab, T., Winkler, T.J.: Building digital leadership in the public sector-A literature review. In: Proceedings of the 31st European Conference on Information Systems, pp. 1–17. Kristiansand (2023)
22. Mankevich, V., Magnusson, J., Svahn, F.: The great wave: the increasing demand for digital competence within the public sector. Inf. Polity **28**(3), 411–434 (2023)
23. Mergel, I.: Digital transformation of the German state. In: Kuhlmann, S., Proeller, I., Schimanke, D., Ziekow, J. (eds.) Public Administration in Germany, pp. 331–355. Springer, Cham (2021)
24. Mergel, I., Edelmann, N., Haug, N.: Defining digital transformation: results from expert interviews. Gov. Inf. Q. **36**(4), 101385 (2019). https://doi.org/10.1016/j.giq.2019.06.002
25. van Roekel, H., Branderhorst, M., Tummers, L., Meijer, A.: Digital transformation leadership: a public value-centered measurement scale. Gov. Inf. Q. **42**(4), 102091 (2025). https://doi.org/10.1016/j.giq.2025.102091
26. Roman, A.V., Van Wart, M., Wang, X., Liu, C., Kim, S., McCarthy, A.: Defining E-leadership as competence in ICT-mediated communications: an exploratory assessment. Public Adm. Rev. **79**(6), 853–866 (2018). https://doi.org/10.1111/puar.12980
27. Sager, F., Rosser, C.: Weberian bureaucracy (2021). https://doi.org/10.1093/acrefore/9780190228637.013.166
28. Schoenegger, P., Tuminauskaite, I., Park, P.S., Bastos, R.V.S., Tetlock, P.E.: Wisdom of the silicon crowd: Llm ensemble prediction capabilities rival human crowd accuracy. Sci. Adv. **10**(45), eadp1528 (2024). https://doi.org/10.1126/sciadv.adp1528
29. Selten, F., Klievink, B.: Organizing public sector ai adoption: navigating between separation and integration. Gov. Inf. Q. **41**(1), 101885 (2024)
30. Tai, R.H., et al.: An examination of the use of large language models to aid analysis of textual data. Int. J. Qual. Methods **23** (2024). https://doi.org/10.1177/16094069241231168

31. Than, N., Fan, L., Law, T., Nelson, L.K., McCall, L.: Updating "the future of coding": qualitative coding with generative large language models. Sociol. Methods Res. **54**(3), 849–888 (2025). https://doi.org/10.1177/00491241251339188
32. Twizeyimana, J.D., Andersson, A.: The public value of E-Government - a literature review. Gov. Inf. Q. **36**(2), 167–178 (2019). https://doi.org/10.1016/j.giq.2019.01.001
33. Vial, G.: Understanding digital transformation: a review and a research agenda. J. Strateg. Inf. Syst. **28**(2), 118–144 (2019). https://doi.org/10.1016/j.jsis.2019.01.003
34. Wessel, L., Baiyere, A., Ologeanu-Taddei, R., Cha, J., Blegind Jensen, T.: Unpacking the difference between digital transformation and IT-enabled organizational transformation. J. Assoc. Inf. Syst. **22**(1), 102–129 (2021). https://doi.org/10.17705/1jais.00655
35. Wilson, C., Mergel, I.: Overcoming barriers to digital government: mapping the strategies of digital champions. Gov. Inf. Q. **39**(2), 101681 (2022)
36. Yadav, A., MC, S.P.: Classifying ai vs. human content: integrating bert and linguistic features for enhanced classification. Oper. Res. Forum **6**(2), 77 (2025). https://doi.org/10.1007/s43069-025-00486-1

Contextualizing Coordination in the Dutch AI Oversight Network

Maurus Enbergs(✉) and Roel I.J. Dobbe

Delft University of Technology, Jaffalaan 5, 2628BX Delft, The Netherlands
m.e.enbergs-1@tudelft.nl

Abstract. The Oversight of Artificial Intelligence is emerging as a complex challenge. To tackle this complexity, regulators are increasingly required to collaborate on AI oversight in networks. Yet, little is known about how the coordination of AI oversight must be organized in these arrangements. This paper identifies factors that influence the coordination of AI oversight in regulatory networks, focusing on the Dutch Samenwerkingsplatform Digitale Toezichthouders (SDT). We conducted semi–structured expert interviews with members of AP, DCA (Directie Coördinatie Algoritmes), who are tasked with facilitating collaboration and coordination of AI Oversight in the SDT network. We identify overarching themes that influence coordination in three interlocking core AI oversight activities: risk monitoring; joint evaluation of AI systems; and joint advice for legal compliance. Our findings show that AI oversight has *wicked* characteristics whose resolution is dependent on the critical functions of *knowledge integration*. To strengthen *knowledge integration*, we recommend that regulators work towards system-level assessment methods for AI systems that readily allow them to combine their different perspectives. We believe that such assessment practices should embrace the principles of *knowledge co-production*. Our descriptive approach helps inform coordination of AI oversight in the Netherlands and elsewhere.

Keywords: AI Oversight · Network Governance · Coordination · Collaboration

1 Introduction

The ongoing proliferation of AI in society has revealed a persistent challenge in coordinating effective AI oversight. In the context of pressing societal challenges, AI has the potential to amplify underlying "wicked problems" [2]. These problems are characterized by social pluralism, institutional complexity, and uncertainty [2]. AI Oversight has inherited this same *wickedness*. Firstly, since AI-related harms are often *unstructured*, identifying their causes and effects is difficult. Secondly, AI systems frequently *cut across* regulatory boundaries and policy domains, entangling conflicting values. Finally, AI Oversight is a *relentless* problem [24], that cannot be solved once and for all; it must be continuously addressed. These factors highlight the persistent gap between the emerging

Published by Springer Nature Switzerland AG 2027
D. Duenas-Cid et al. (Eds.): ePart 2026, LNCS 16923, pp. 215–230, 2027.
https://doi.org/10.1007/978-3-032-36676-4_14

harms produced by AI in society and regulators' capacity to oversee or prevent them effectively.

In the Dutch context, the Data Protection Authority *(Autoriteit Persoonsgegevens, AP)* and the Authority for Digital Infrastructure *Rijksinspectie Digitale Infrastructure, (RDI)* have responded to this challenge by jointly advocating for a hybrid approach to algorithmic oversight. Together with other relevant regulators AP, and RDI have set up a platform of digital supervisory authorities the *Samenwerkingsplatform Digitale Toezichthouders*, (SDT), which is a *purpose-oriented network*. Such networks are used to address large-scale social challenges [21]. Research shows that networks can deal with complexity [20] and *'wicked problems'* [5]. Raab points out that networks allow *"organizations to jointly analyze complex problems from different perspectives and can create overall synergies between them"* [21]. This suggests that purpose-oriented networks may be a viable solution to the complex problem of AI oversight.

However, while the Dutch approach offers many advantages, it also poses new challenges. It is currently unclear how best to design the coordination of core AI oversight tasks in networks. Coordination among network actors is "complex"and involves multiple interactions serving different goals. These have political, technical, legal, financial, organizational, and ethical dimensions.

Insurgent research has not addressed these challenges. They remain topics rarely discussed in detail, yet poorly coordinated network structures can increase complexity [21], inadvertently hindering effective AI oversight. Thus, it is essential to contextualize the factors influencing coordination of core oversight activities, specifically (1) **AI Risk Monitoring**; (2) **Joint AI System Assessments**; and (3) **Joint advice on Legal Compliance**.

This paper addresses this gap by investigating the coordination of these activities in the Dutch AI oversight network, SDT. Through thematic analysis, we present an overview of factors influencing coordination for these activities. We conducted semi-structured interviews with experts from the *Directie Coördinatie Algoritmes* (AP DCA) who facilitate collaboration and coordination in SDT. Our findings show AI oversight shares *wicked* characteristics. Their resolution depends on the critical functions of *knowledge integration*. To strengthen this, we recommend that regulators work towards systems-level assessment methods for AI systems, which allow the integration of different 'perspectives'. Finally, we believe arrangements for *knowledge co-production* between researchers and practitioners can help address emerging challenges in AI oversight. In doing so, the paper contributes to the field of algorithmic governance, drawing on concepts from both public administration and Science and Technology Studies.

2 Background

In Sect. 2.1, we introduce relevant background on AI oversight in the Netherlands, including an introduction to the three core, shared oversight activities that form the primary context of our empirical work. In Sect. 2.2, we introduce purpose-oriented networks as a central paradigm used in the Dutch approach to

AI oversight. We conclude with a brief introduction to collaboration and coordination in networks.

2.1 AI Oversight in The Netherlands

AP and RDI have jointly published an advisory report outlining how AI oversight could be structured in the Netherlands. The proposal adopts a *hybrid approach*, recommending *AP DCA* and RDI as co-national market regulators. Oversight responsibilities would also be assigned to sector-specific regulators.

Both RDI and AP DCA emphasized the importance of the *collaborative platform for digital supervisory authorities*, the **SDT**. The SDT comprises three chambers organized around central topics in digital regulation. The first two chambers discuss general and DSA-related subjects. The third chamber is the *AI Chamber*. It is chaired by AP DCA. Currently, the chamber has eight members. This includes sectoral regulators.

Members of the *'AI Chamber'* collaborate to align oversight and enforcement of AI systems across sectors. To achieve this, they must tackle three core, shared activities of AI Oversight: (1) AI Risk Monitoring; (2) Joint AI System Assessments; and (3) Joint Advice on Legal Compliance. These shared tasks provide the backdrop for our analysis of coordination.

AI Risk Monitoring. Within the SDT network, AI risk monitoring involves building a shared, cross-sector risk picture of algorithmic and AI systems deployed in Dutch society. This process begins with each regulator collecting signals from its own sector or area of expertise and sharing these signals within SDT structures like the *'AI Chamber'*. Signals may include incident reports, complaints, insights from recent developments, and more. The findings are discussed to identify emerging patterns, long-term trends, and risks that require monitoring. Critical steps involve jointly gathering incident signals, exchanging knowledge, aligning risk interpretations, and validating conclusions together. Many risks only become apparent when information and knowledge from multiple regulators are combined; thus, sharing, interpreting, and prioritizing risks necessitates coordination among SDT members.

Joint AI System Assessments. Joint AI system assessments occur when an AI system or model affects multiple supervisory domains or poses risks that require a coordinated response. Currently, there is little practical experience on how to conduct such joint oversight assessment projects. SDT members aim to follow these steps: (1) identify a system or system class requiring investigation, often through the shared risk monitoring cycle; (2) agree on a joint assessment project plan, addressing questions such as which questions to examine, what expertise is needed, and who is the primary investigator; (3) collect evidence and conduct evaluations; (4) jointly synthesize the results into a shared understanding and report on the system, its risks, and impacts on fundamental rights and other relevant regulations. Collaboration and coordination are essential, because each regulator brings different legal mandates, technical expertise, and sectoral knowledge. These assessments ensure that AI systems that cut across consumer law, digital infrastructure, and personal data protection are evaluated holistically rather than through fragmented oversight.

Joint Advice on Legal Compliance. The SDT network collaborates on issuing joint legal or compliance guidance to organizations deploying AI systems, ensuring consistency across regulatory domains. This activity involves several coordinated steps: agreeing on the question to address; gathering relevant legal requirements from each regulator's mandate (e.g., privacy law, consumer protection, media regulation); interpreting how these frameworks apply to the AI technology in question; and drafting unified guidance. Collaboration and coordination are essential to ensure that the guidance reflects a coherent regulatory stance and prevents conflicting interpretations among supervisory authorities.

2.2 Collaboration and Coordination in Purpose-Oriented Networks

As the above-described activities highlight, AI Oversight is a complex problem that cannot be resolved by a single actor. Because AI systems are frequently embedded in complex sociotechnical systems whose applications span across regulatory boundaries, AI oversight has inherited their *wickedness*: requiring oversight organizations to seek *collaborative network arrangements* [13,17]. We adopt Alford's and Head's definition of *wicked problems* as complex problems that *"vary in the extent of their wickedness"*, along dimensions such as their cognitive complexity or diversity of opinions among actors or institutions involved [1].

Networks are commonly believed to deal well with complexity [20], enabling actors to pool resources, capacities, and information [21]. They enable organizations to jointly analyze complex problems from different perspectives, allowing them to share risks and create overall synergies with each other [21].

The SDT is primarily a *purpose-oriented network*. These are *"bounded, self-referencing collectives comprised of actors who consciously affiliate to the collective around some shared purpose"* [16]. Since division of labor is inherent to all networks, all networks face collaboration and coordination challenges [11].

Researchers have classified various types of horizontal relationships between actors in networks in different ways [4]. The three most common classifications in the literature are cooperation, coordination, and collaboration [4,11].

Brown and Keast refer to *cooperation* in networks as arrangements where actors have largely informal working relationships without a clearly established common goal [4]. They define *coordination* in networks as working towards alignment or organization of people, tasks, and specialized interventions to achieve a common goal. In this way, coordination is the organization of interdependent parts of a network into a relationship to produce a connected *"whole"* [4].

Lastly, *collaboration* is described as the more integrated form of joint action in which actors – in addition to coordination – also create shared meaning and assume collective responsibility for outcomes [4].

Brown and Keast view the terms as a spectrum, where cooperation is the least integrated form of shared action and collaboration is the most integrated [4]. Network studies have shown that successful network management typically requires a *mix* of all three relationship forms [12].

This study interprets the shared AI oversight activities in the SDT network, as described in Sect. 2.1, as primarily *coordination* challenges. Since each shared activity targets a specific, established goal, i.e., risk monitoring. Nonetheless, we acknowledge that *collaboration* among network actors is required to set up and maintain these shared activities and the network at large Goal-setting and responsibility-sharing are relevant additional dimensions to study in the context of understanding AI oversight networks.

Coordination is not a monolithic activity or responsibility; rather, it is goal- and task-specific. In the context of AI Oversight in SDT, it is insufficient to describe it as a general occurrence. Coordination must be contextualized to the specific oversight activities to identify the relevant factors that influence it.

3 Method

The goal of our research project is to provide a contextualized description of the collaboration and coordination challenges in AI Oversight as they are perceived by actors working to structure joint AI Oversight in the Netherlands. Consequently, conducted semi-structured expert interviews, a qualitative research approach, which fits well with the exploratory nature of our research objective [19].

3.1 Semi-structured Expert Interviews

We conducted interviews with experts from AP DCA. Our focus was on AP DCA, since it is the leading organization in regards to structuring collaboration and coordination in the SDT network. Moreover, AP, as the potential future AI market regulator, watchdog for fundamental rights, and GDPR regulator, will have the broadest mandate on AI oversight. Additionally, it currently has the highest levels of maturity regarding practical AI oversight experience.

We took great care to conduct a *problem-* and *expertise-centered selection process* rather than status-oriented procedure [23]. Consequently, we followed Crouch and McKenzie's logic and opted for a small samples size enabling us to engage closely with participants [8]. We therefore limited our sample size to four participants, to foster the kind of sustained interaction and analytical depth required to generate meaningful insights into this complex topic. With only a few participants, each interview had substantial weight in shaping the study's conceptual development. Yet, as Crouch and McKenzie argue, in exploratory, conceptual research, representativeness and sample size are not central criteria for methodological adequacy. Instead, qualitative inquiry is concerned with uncovering the dynamic qualities of a situation rather than establishing statistical patterns or proportionality of relationships [8]. Nonetheless, this means that our insights cannot be statistically extended beyond the context we examined.

We focused on sampling *experience* embedded within a highly specialized work environment. In line with this rationale, each of the four experts selected have been and are actively involved in architecting and facilitating the core oversight activities within SDT. They all hold senior positions at AP DCA, with backgrounds spanning legal, forensic, and artificial intelligence. However, we recognize that semi-structured interviews are inherently relational. The perceptions expressed by the interviewees may therefore be influenced by the researcher's behaviour during the interview. In addition, the small number of participants means that certain meaningful variants or perspectives may remain absent.

We conducted five interviews. The interviews were predominantly conducted in person, lasting between 70 and 90 min. Regardless, all interviews were recorded and transcribed via Microsoft Teams. Thereinafter, the transcripts were check, corrected, and anonymized, using the recording of the interview as reference. All interviews were conducted in English.

The interviews followed an interview protocol. The interview protocol was uploaded to a repository A. We discussed with participants their practical experiences and perspectives on AI oversight, focusing on how they describe dilemmas surrounding their day-to-day work. The conversations explored recent and planned tasks, such as incident reporting, high-risk AI system assessment, and consultation. A central theme was how coordination and collaboration take place across different actors, including regulators and external stakeholders, alongside the challenges they face—such as resource constraints, information exchange, and legal uncertainty. The interviews also examined how participants conceptualize effective AI oversight and what capabilities, infrastructure, and research would be needed to improve it.

3.2 Analysis

Using Atlas.ti, we followed a reflexive inductive thematic analysis outlined by Friese et al. shown in Fig. 1 [10].

The first author transcribed and anonymized the interviews, allowing an initial familiarization with the data. The dataset consists of approximately 37,156 words. The transcripts are in English. We then generated initial codes by systematically labeling meaningful segments of text, creating new codes when needed. We analyzed individual paragraphs, typically consisting of between 2–6 sentences. We searched for themes by grouping related codes and examining their conceptual coherence. We then reviewed our themes over the full data set. This included comparing coded segments, writing analytic memos, and revisiting earlier phases to ensure coherence. The analysis concluded with defining and naming the themes and assembling them into a narrative that fit our research objective.

We recognize that the chosen approach limits our study by the dependence on the researcher's interpretive skills. Moreover, due to the fluctuating dynamics in the domain of AI policy, we cannot guarantee that the themes we have identified remain static. They are likely to be under flux. Consequently, the findings of our exploratory study remain provisional and speculative.

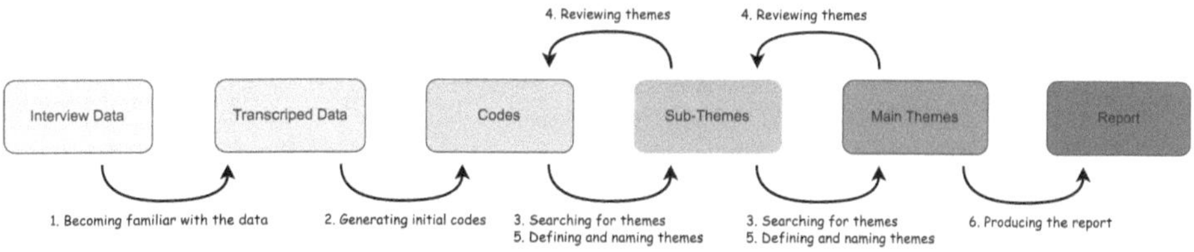

Fig. 1. Thematic Analysis approach according to Friese et al. [10]

4 Result

Our analysis focused on the three core AI oversight activities of AI risk monitoring, joint AI system assessment, and joint advice on legal compliance, as introduced in Sect. 2.1. We organized the factors perceived by experts to influence coordination of AI Oversight into three major themes. The first theme describes factors influencing *Information & Knowledge* sharing in the three AI Oversight activities. The second discusses how *Legal & Mandates* issues shape coordination. Lastly, *Cultural & Organizational*, synthesizes how organizational complexity and cultural diversity influence coordination (Fig. 2).

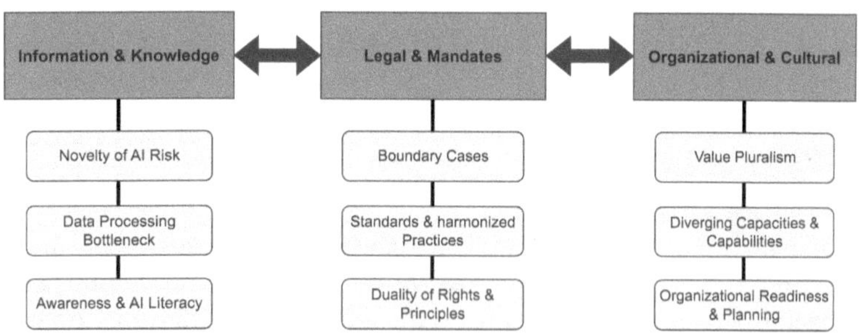

Fig. 2. Theme mapping and interdependancies between different sub-themes

4.1 Information and Knowledge

Information and knowledge transfer play a crucial role in all three AI Oversight activities. The participants highlight the immense challenges this poses. In the following, we will discuss factors influencing information and knowledge sharing, arranged into subthemes, as they are perceived by actors.

Novelty of AI Risks and Harms. A recurring theme in the interviews is the novelty of AI risks and harms. The interviewees perceive three factors contributing to this. First, identifying and understanding the causes and effects of AI risks and harms is challenging. Experts attribute this *unstructuredness* primarily to the lack of data on the effects of novel AI applications. In many cases, a particular application is too new to gather reliable knowledge of its effects. They also attribute part of the blame to the underlying complexity of AI techniques, which complicates tracing harms to a definitive source in an AI system's design or operation. Second, the novelty of AI risks is driven by the *relentless* nature of AI development. This results in information and knowledge about risks quickly becoming outdated. New AI technologies are rapidly evolving and being applied to an expanding range of applications. Third, in many applications, the AI tool's role is *hidden* or not apparent to the public and regulators.

Due to fast-changing circumstances, participants emphasized the importance of staying informed about current developments by monitoring news, recent AI incident reports, and emerging AI applications. Conversely, experts warned that the novelty of certain AI risks can alter perceptions of risks posed by simpler algorithmic models. They noted that risk is not proportional to the complexity or "capability" of AI systems, as recent cases in the Netherlands further substantiate [18]. Rather, the context of the system or use case surrounding the AI tool often determines its potential to create harms [22]. Focusing on *"the madness of the day"*, as one participant put it, can lead to missing long-term trends. Thus, it is crucial to balance current information with established knowledge.

To tackle this conundrum, interviewees emphasized the importance of adopting a *learning first strategy*. They believe that integrating diverse perspectives to analyze AI risks and harms can strengthen the transformation from data to information to knowledge. However, operationalizing this transformation by combining diverse information and knowledge sources presents a significant and under-explored challenge.

Data Processing as a Bottleneck. Regulators are faced with information scarcity created by a bottleneck in data processing. Data processing in AI oversight is extremely challenging because data is immensely diverse, unstructured, and must be gathered from different data sources. Contrasting this with established oversight domains:

"For the financial sector, you are almost literally drowning in data. [...] because the data is extremely structured and organized [...]. Here, early monitoring is actually more early searching because you need to be a detective and [...] you need to think a lot about if something may be of interest to look at [...]."

The quote hints that this is, predominantly, not a data availability issue, but rather related to data being *hidden* or *semantically*, *syntactically*, and *practically* hard to process and share. The following segment further underlines this:

"Take into account [that with] the formal procedures through the GDPR you get to a lot of information and data that is there, but it's very labor intensive [to process it]. [...] So that part is extremely difficult, and the same goes for finding that intelligence [...] through third parties."

Experts say that the semantic and syntactic diversity of source data makes standardized information transfer and presentation difficult. Information is most often embedded in a rich context. Consequently, the *practical knowledge* that analysts hold gains more importance in interpreting information and identifying context-specific AI risks; they need to be *"detectives"*.

The experts also points towards the difficulty of receiving and sharing information with *third parties*. The emphasis that information and knowledge sharing is a *nested* challenge within the SDT network. Information needs to be shared within organization, with other organizations, and with actors outside of the network (e.g. Legal Advise, EU AI Office). The resulting *chain* of sharing and processing information requires interoperability between its links. Yet, the nature of the information, i.e. its diversity, limits the extent to which it can be shared effectively. Experts say that the divergence in practice-based knowledge makes sharing information more difficult because it is likely to carry different meanings and serve different uses for different regulators. Aligning semantic and syntactic standards for information sharing under these conditions is challenging, but is seen as important for information sharing and subsequent analysis.

According to the experts, the nature of the rapidly evolving landscape of AI is also likely to increase the degree of differentiation of partial information and knowledge between different regulators. This factor further strengthens the need for data, information, and knowledge exchange.

Awareness and AI Literacy. The experts also stress the importance of organizational *awareness* and *AI literacy* as a prerequisite to information and knowledge sharing. The previously discussed *"chain"* of information hinges on the awareness of actors in the "chain"about what information needs to be shared and with whom. Experts have described instances where internal actors failed to forward important risk signals to the AI risk monitoring team, either due to a lack of awareness of their relevance or because of conflicting formal procedures. In one example a complaint filed with another department hinted at an underlying AI psychosis. Yet, the information was not treated as such, as the complained was written using legal language. The lacking awareness in this case can be attributed to a mismatch between the practical knowledge that the gatekeeper held and the information that they were confronted with. As a result, experts also reiterated the importance of practicing and teaching AI Literacy. They argue that it has the potential to inform substantive practice across the regulatory landscape, enabling actors to align their practices with AI oversight needs.

The participants stress that regulators must assume that relevant bits of information will likely reach other parts of their organization first. Consequently, it is imperative that the oversight bodies strengthen the AI literacy of their entire workforce to ensure relevant information is flagged. Raising awareness through increasing the visibility and outreach of designated AI oversight teams can help.

4.2 Legal and Mandates

The existing and coming legal provisions and mandates shape the way in which AI Oversight will be conducted. While well-designed legal mandates can improve coordination of AI oversight, experts fear that they have the potential to also inhibit it. Hereinafter, we will highlight the themes influencing legal mandates.

Boundary Cases and Accountability Gaps. Across all interviews, participants have reaffirmed that the need to collaborate with other oversight organizations is born out of the fact that AI *cuts across existing regulatory boundaries* and thus requires multiple regulators to apply their expertise and mandated authority. *Boundary cases* are cases that are situated at the edge between two or more regulators, making it unclear who is empowered to act.

This *cross-cutting characteristic of AI* creates several challenges. First, according to the interviewees, ambiguity in boundary cases can lead to *"Many Hands"problems*. Which refers to a type of dilemma a forum faces in attributing blame for an outcome caused by multiple actors [7]. Experts fear that it is not only unclear, *who is to blame?* – the fundamental question of the *"Many Hands"* problem – but also *who composes the forum?* that attributes the blame. Regulators may have different opinions about who should be taking the lead in certain cases. This could be exacerbated by public pressure.[1] Consequently, it may be difficult for them to agree on a course of action and attribute blame.

Second, experts fear that in *boundary cases*, advice from regulators to AI deployers may differ, leading to inconsistency and catch-22 s. In the literature, such issues are often referred to as *"Multi-accountability disorder* [14]. Experts stress that regulators must coordinate their advice to prevent conflicting instructions.

Finally, boundary cases are likely to exacerbate the importance of other factors. According to interviewees, knowledge sharing is most critical when AI oversight requires several organizations to combine their expertise in a single project.

The interviewees frequently cited examples of *boundary cases* to illustrate the resulting *necessity* for regulators to coordinate. In the eyes of the experts they are seen as the most difficult cases to resolve, yet they also offer a valuable learning opportunity. Joint AI assessments may offer a practice-based learning approach to surface accountability gaps in boundary cases and learn more about AI oversight. Pilot projects could help reduce future friction in collaboration and coordination.

Standards and Harmonized Practices. Standardization and harmonized practices are vital to the success of coordination in AI oversight. However, many of these standards are currently delayed, or offer too little guidance for how these should be applied. Our interviews show that *substantive practices*, which are specific procedures or frameworks that operationalize the application of a standard, principle, or norm, are deemed pivotal to inform and maintain the

[1] *"And then of course big scandal supervisors have to respond. And then yeah, who's in charge? [...] It's meant to show that the likely first response would be a lot of supervisory authorities just looking at the other ones and waiting for somebody to take the initiative because it's not as clear cut [...]."*.

robustness of standard application. Grounding standards in substantive practices can help improve their applicability and generalizability. Yet, due to the novelty of many AI applications and the complexity of AI Act compliance, substantive practices are often missing or their adoption has been slow. That said, there is hope that emerging substantive practices, such as the fundamental rights impact assessment framework (FRIA), if done well and consistently in the context of AI system design, may yield valuable information for AI oversight :

"To treat FRIA as part of your component specs or your product development, could be very successful, depending on how it will be implemented and assuming it won't just be a checklist, but actually something more substantial. [...]"

Aside from the direct impact substantive practices have on improving AI systems, incidents, and risk assessments, they will also yield a high-quality data source that regulators can rely on in the future.

Finally, experts also point out an arising issue between *informal and formal practices.* On the one side, regulators need to be able to adapt to a changing AI landscape and use the *practical knowledge* of analysts to identify risks and assess AI systems. On the other hand, formal practices are required to ensure *legal certainty* and *consistency.*[2] The tension is that informal practices help networks remain agile in a rapidly changing environment but explainability to judges demands structured, principled risk analysis and case selection.

Duality of Rights and Principles. The experts see legislation as both barrier as well as enabler of coordination. On the on hand, legislation such as the GDPR, limits what oversight organizations can share. On the other, it equips specific regulators with a mandate to investigate aspects of algorithmic systems.

European AI law is seen as challenging to implement, but also as a simplifier that can lift some of the barriers to coordination of AI Oversight currently created by legislation. Experts have stressed that under the *principles of sincere cooperation*, regulators can already exchange valuable information. They also state that the AI Act explicitly enables cooperation between AI regulators.

4.3 Organization and Culture

Regulators have diverging organizational structures and cultures. This has a significant impact on the coordination of AI Oversight in the SDT network. The experts stress that these dynamics must be understood and proactively managed to ensure that the oversight activities can be carried out effectively.

Value Pluralism. Value pluralism –fueled by differences in goals, culture, and regulatory perspectives– can create *friction* among SDT members:

"Everyone finds its own perspective to be the focal point of [oversight]. Some see AI as technical. Others as consumers and markets. Others see AI as personal data and fundamental rights. So you started to see friction emerge."

---

2 *"in the end we need to be able to explain to a judge, why we choose to investigate this AI and not another".*

The experts shared that in the past, the SDT had been struggling with agreeing on a shared agenda, due to the different positions held by the participating regulators. However, after the actors realized that their work was very complementary to each other and effective oversight required a joint approach, the conflicts started to be resolved:

"It took long to figure out: 'Wait, but we are very complementary to each other. We don't want to do each other's work. So we should just work together and use each other because we have to. There's no other way to do this.' And I think that was the point that when we started to speak each other's language."

The quote highlights that the experts believe AI oversight should be a collaborative action of the different regulators, that combines their different expertise and perspectives.

Diverging Structure, Capacities and Capabilities. This theme is composed of three sub-factors, each focused on organizational characteristics of the regulators. Practitioners emphasized the importance these characteristics play in shaping the form of coordination that an organizations can engage in.

First, the varying structural arrangements that regulators use in their organizations impact the way collaboration and coordination in the SDT is conducted. Some actors have a centralized approach to AI while others have dispersed AI topics over their entire organization based on synergies with preexisting expertise. This introduces complexity in conducting joint oversight tasks.

Second, this also has a direct impact on the network's ability to effectively combine the capacities of individual organizations. One of the key benefits of purpose-oriented networks is the pooling of resources and capacities. However, this can also be a source of friction and conflict between network members [21]. Capacity is also a primary *constraint* for oversight, influencing *what* will be investigated and *how*. The participants are aware of the importance of capacity constraints and how these vary across organizations. They recognize that in comparison to smaller, sectoral oversight bodies, the core SDT team, having AI-specific capacity, has to do the *"heavy lifting"*. They stress enabling this through cross-authority expert pools hinges on funding and formal agreements. Without these, the expert-pool model may stall, reinforcing dependence on the biggest regulators rather than institutionalized shared capacity.

Finally, interviewees believe that a transdisciplinary pool of experts is crucial for analyzing problems collaboratively from various perspectives. Regulators recognize that AI Oversight *necessitates* them to combine their expertise and produce shared knowledge. However, combining different perspectives and capabilities is not easy and requires shared planning, interoperability and capacity.

Organizational Readiness and Planning. From the interviews it has become apparent that organizational readiness varies among network members. Some are quite aware to the newly emergent AI risks while others have only performed minimal steps to get ready for upcoming AI legislation. This is especially true for sectoral regulators who either, do not have the capacity to build sufficient AI expertise or lack awareness about emerging risks in their sector.

The SDT is coordinating AI oversight across different time frames. Short-term projects are primarily focused on building cooperative processes in SDT, informing the public as well as engage with policymakers. On a time horizon of five years, the SDT is looking to build an information exchange infrastructure, that builds a robust foundation for coordinating oversight activities. In the longer term, the SDT actively engages in *policy pre-figuration*, working together with the OECD and European regulators to shape policy and standardization.

The experts have stressed the importance of following a *learning first strategy*. They are aware that AI oversight requires them to adapt to emerging information and remain open to changing conditions.

5 Discussion

Our results highlight the *wickedness* of AI oversight. Experts perceive the SDT, a network, to be a suitable mechanism for tackling this problem. Generally, this stands is supported by literature [20,21]. However, as Weber and Khademian point out, networks typically evoke a *knowledge challenge* [24]. Knowledge challenges arise because knowledge from different perspectives is needed to tackle wicked problems, but combining this knowledge proves challenging [24]. As Weber and Khademian point out, sending, receiving, and integrating information, are critical tasks, yet they depend on the *semantic, syntactic, and practical knowledge* that actors hold. Our results show, that experts perceive these factors to be divergent between the members in SDT. Yet, they expressed the view that the effectiveness of AI oversight in the SDT network is dependent on the ability of actors to send, receive, and integrate knowledge. However, Practical knowledge recognizes that "what people and organizations know is deeply embedded in what they practice" [24]. Yet, this makes it notoriously difficult to share [24]. As Carlile points out, gathering practice-based knowledge requires tremendous investment, and changing it is costly. Hence, it is commonly referred to as *"hard won"*, and actors are often reluctant to change it [6,24]. The interviewees showed awareness of this and emphasized the importance of continuous engagement between SDT members to overcome the *"hard won"* nature of practical-based knowledge.

Consequently, we believe that strengthening transdisciplinary *knowledge integration* across regulators is pivotal for resolving the "*wickedness*" of AI oversight. Whether it be the novelty of AI risks, data-processing bottlenecks, or emerging *boundary cases*, experts highlight that knowledge integration can help to make sense of all these strings. Following Meijer & Ettlinger, we treat knowledge integration as mutual learning [15]. Mutual learning is conceptualized as a process of exchanging, generating, and integrating existing or newly developed knowledge among distinct organizations and entities [15].

In our view, focusing on knowledge integration in joint AI system assessment projects is of particular use. Joint projects can simultaneously inform and improve all three core oversight activities. For this purpose we encourage the development of systems-level assessment methodologies that enable regulators to combine their perspectives into a holistic system assessment. In our view, such methods are particularly needed to tackle boundary cases. This recommendation concurs with the calls that researchers have made previously [9,22] and with the needs expressed by participants during our interviews. In our future work, we will explore the necessary requirements and evaluate solutions to this challenge.

We believe that such assessment methods should embrace the principles of *co-production*, seeking stronger engagement with the scientific community. Notably, elements of *co-production* already exist in practice. As SDT and AP DCA are engaging with researchers on AI Oversight-related topics. Our research also exemplifies this. *Co-production* has been cited as a valuable approach to tackle wicked problems [3]. We suggest that it can help to inform the core oversight activities, providing a rigorous methodology for evaluation and novel research on the causes and effects of emerging AI risks and harms.

Our study is exploratory, based on a small number of expert interviews in one national context. Moreover, the dynamics persisting in the domain of AI oversight are in flux. AI oversight is a complex adaptive system. Our findings should therefore be interpreted as practice-informed hypotheses and suggestions, not prescriptions. Our main contribution remains descriptive, by contextualizing coordination for the three core AI oversight activities.

6 Conclusion

Our results illustrate the *wickedness* of AI oversight. Experts have described AI risks as *unstructured and relentless*, AI systems *cut across* established regulatory domains, and regulators face *value pluralism*. Our analysis highlights the central role that *knowledge integration* must take to combine these perspectives. The interviews have shown that practitioners have embraced this insight and are actively pursuing a *learning first* strategy. We view *joint oversight projects* as an ideal practice-based learning ground. They have the capacity to inform all three core activities simultaneously, as knowledge about specific AI risks can be gathered, friction in joint AI system assessments can be overcome, and legal certainty can be established.

However, so far, few practical tools and methods have been developed to support joint AI oversight assessments between regulators. Such approaches are required to enable the combination of different perspectives into a single AI system assessment. Consequently, we support the call for systems-level assessment methodologies for AI systems [22]. We believe that such assessment practices should embrace the principles of *knowledge co-production.*

Declaration on Generative AI. During the preparation of this work, the author(s) used **Writefull** to: Grammar and spelling check, Paraphrase, and reword. **Microsoft Copilot (GPT-5.2)** was used to brainstorm ideas, aid your understanding of concepts, and to plan your writing. After using this tool/service, the author(s) reviewed and edited the content as needed and take(s) full responsibility for the publication's content.

A Repository

https://data.4tu.nl/private_datasets/
6JyiMu0B88hrg6qw5krlqJwmUI5DcZgXJogiN-GjgxI Under 'data' click on the file you want to review.

References

1. Alford, J., Head, B.W.: Wicked and less wicked problems: a typology and a contingency framework. Policy Soc. **36**(3), 397–413 (2017). https://doi.org/10.1080/14494035.2017.1361634
2. Andrews, L.: Public administration, public leadership and the construction of public value in the age of the algorithm and 'big data'. Public Admin. **97**(2), 296–310 (2019). https://doi.org/10.1111/padm.12534
3. Bandola-Gill, J., Arthur, M., Leng, R.I.: What is co-production? Conceptualising and understanding co-production of knowledge and policy across different theoretical perspectives. Evid. Policy **19**(2), 275–298 (2023)
4. Brown, K., Keast, R.: Citizen-Government engagement: community connection through networked arrangements. Asian J. Public Admin. **25**(1), 107–131 (2003). https://doi.org/10.1080/02598272.2003.10800411. Publisher: Routledge _eprint
5. van Bueren, E., Klijn, E.H., Koppenjan, J.: Dealing with wicked problems in networks: analyzing an environmental debate from a network perspective. J. Public Admin. Res. Theory **13**, 193–212 (2003). https://doi.org/10.1093/jopart/mug017
6. Carlile, P.R.: A pragmatic view of knowledge and boundaries: boundary objects in new product development. Organ. Sci. **13**(4), 442–455 (2002). https://www.jstor.org/stable/3085976. publisher: INFORMS
7. Cooper, A.F., Moss, E., Laufer, B., Nissenbaum, H.: Accountability in an algorithmic society: relationality, responsibility, and robustness in machine learning. In: Proceedings of the 2022 ACM Conference on Fairness, Accountability, and Transparency. FAccT 2022, pp. 864–876. Association for Computing Machinery, New York (2022). https://doi.org/10.1145/3531146.3533150
8. Crouch, M., McKenzie, H.: The logic of small samples in interview-based qualitative research. Soc. Sci. Inf. **45**(4), 483–499 (2006). https://doi.org/10.1177/0539018406069584
9. Dobbe, R.: AI safety is stuck in technical terms – a system safety response to the international AI safety report (2025). https://doi.org/10.48550/arXiv.2503.04743, http://arxiv.org/abs/2503.04743, arXiv:2503.04743 [cs]
10. Friese, S., Soratto, J., Pires de Pires, D.: Carrying out a computer-aided thematic content analysis with ATLAS.ti. IWMI Work. Pap. **18** (2018)
11. Gulati, R., Wohlgezogen, F., Zhelyazkov, P.: The Two Facets of Collaboration: Cooperation and Coordination in Strategic Alliances (2012)
12. Keast, R., Brown, K., Mandell, M.: Getting the right mix: unpacking integration meanings and strategies. Int. Public Manag. J. **10**(1), 9–33 (2007). https://doi.org/10.1080/10967490601185716
13. Keast, R., Mandell, M.: The collaborative push: moving beyond rhetoric and gaining evidence. J. Manage. Gover. **18** (2014). https://doi.org/10.1007/s10997-012-9234-5
14. Koppell, J.G.: Pathologies of accountability: ICANN and the challenge of "multiple accountabilities disorder." Public Adm. Rev. **65**(1), 94–108 (2005)

15. Meijer, A.J., Ettlinger, K.: Transdisciplinary public administration research: developing and testing a model for transdisciplinary knowledge integration in the public sector. Perspect. Public Manage. Gover. **8**(2), 106–120 (2025). https://doi.org/10.1093/ppmgov/gvaf009
16. Nowell, B., Milward, H.B.: Apples to apples: a taxonomy of networks in public management and policy. Elem. Public Nonprofit Admin. (2022)
17. Nowell, B.L., Kenis, P.: Purpose-oriented networks: the architecture of complexity. Perspect. Public Manage. Gover. **2**(3), 169–173 (2019). https://doi.org/10.1093/ppmgov/gvz012
18. Pascoe, R.: Probation service used error-ridden algorithms to assess risks (2026). https://www.dutchnews.nl/2026/02/probation-service-used-error-ridden-algorithms-to-assess-risks/
19. Pin, C.: Semi-structured Interview (2023). https://scienceetbiencommun.pressbooks.pub/pubpolevaluation/chapter/semi-directive-interview/. book Title: Policy Evaluation: Methods and Approaches Publisher: Quebec City: Éditions science et bien commun
20. Powell, W.: Neither market nor hierarchy: network forms of organization. Res. Organ. Behav. **12**, 295–336 (1990)
21. Raab, J.: Organizing in an Age of Complexity: Reflections on When and How (Not) to Use Organizational Networks (2024)
22. Rismani, S., Shelby, R., Davis, L., Rostamzadeh, N., Moon, A.: Measuring what matters: connecting ai ethics evaluations to system attributes, hazards, and harms (2025). https://doi.org/10.48550/arXiv.2510.10339, http://arxiv.org/abs/2510.10339, arXiv:2510.10339 [cs]
23. Von Soest, C.: Why do we speak to experts? reviving the strength of the expert interview method. Perspect. Polit. **21**(1), 277–287 (2023)
24. Weber, E.P., Khademian, A.M.: Wicked problems, knowledge challenges, and collaborative capacity builders in network settings. Public Adm. Rev. **68**(2), 334–349 (2008). https://doi.org/10.1111/j.1540-6210.2007.00866.x

Political Genres in Social Media

A Study of the 2025 Norwegian Parliamentary Election

Marius Rohde Johannessen(✉)

School of Business, University of South-Eastern Norway, Raveien 215, 3184 Borre, Norway
Marius.johannessen@usn.no

Abstract. In this paper I examine how Norwegian political parties used social media in the 2025 parliamentary election campaign, with particular focus on Facebook, as Facebook remains the largest social medium for political communication. Using genre theory as analytical lens, and the public sphere as theoretical perspective, I examine how political campaigning in social media has evolved from earlier elections to the current (2025) election and discuss if Norwegian political communication is moving more towards an affective public sphere and thereby adopting the rhetorical devices of populism.

Keywords: eParticipation · social media · online campaigning · social networking systems · genre theory · Norway

1 Introduction

In this paper, I examine the Norwegian 2025 parliamentary election in social media, as part of a trend where the Internet and Social media platforms such as Facebook, Instagram, X/Twitter, Snapchat, and TikTok have become important arenas for election campaigning. Some of the first research articles appeared after Howard Dean's 2004 presidential election campaign [1], and during the Obama campaign of 2007–08, social media and data analytics were seen as central elements in his successful campaign [2–4]. In 2015, the European Parliament think tank stated that social media was becoming increasingly important for political participation, and in the 2020's social media has become key infrastructure for political parties and citizens' political participation [5].

eParticipation research has been central in analysing political social media use, for example through studies of the Italian Five Star movement [6, 7], parliamentary elections in Norway [8, 9], local governments' use of Facebook [10], social media's affordances for eParticipation [11], and the impact of cybersecurity and education on eParticipation [12]. In Norway, online campaigning has been on the agenda since the early 2000's, and social media became increasingly important as part of the campaigning toolbox from 2009 and onwards [9].

Early analyses of social media were positive, emphasising deliberation, transparency and the potential for a massive deployment of Habermas' ideals for a deliberative public

Published by Springer Nature Switzerland AG 2027
D. Duenas-Cid et al. (Eds.): ePart 2026, LNCS 16923, pp. 231–246, 2027.
https://doi.org/10.1007/978-3-032-36676-4_15

sphere, exemplified by Obama's 20008 "hope" campaign [3]. However, later research has been less optimistic, citing 2016's Brexit and US election and pointing to issues such as increasing polarisation, emotional mobilisation and misinformation as emerging trends, suppressing early optimism [13–15].

Given this transition, I contrast the public sphere of Habermas, where social media is seen as an arena for deliberation, with Papacharissi's [16] idea of affective publics, which emphasise emotion, affection and feeling over rational deliberation. I apply genre theory as analytical lens to examine how parties and the electorate communicate in social media during the 2025 election. There is a growing body of longitudinal research on political use of social media. Examples include Yoo [17], who has examined how Twitter contributes to agenda-setting in the context of US immigration over several months. Other studies present an 8-year study of US polarization on Twitter/X between 2009–2016 [18], analysis of social media use in US elections between 2008 and 2024 [19], and the growth in the use of Instagram for political communication in Europe between 2012 and 2018 [20].

These studies support the affective hypothesis. Yoo [[17], p.4482] finds that social media influencers can affect public attitudes, and that "compelling and persuasive tweets…can attract news media coverage". Larsson [20] concludes that populist parties have higher engagement rates, Garimella & Weber and Mittal et al. both conclude that social media is drifting towards "emotionally driven narratives… [which] has intensified political polarisation" [[19], p.183]. Much of the existing research is based in the US or at the EU level or concerns specific issues such as US immigration. My study thus contributes to this growing body of time-series research by adding another country to the mix and examining how political communication in Norwegian parliamentary elections has evolved from 2009 until 2025, with the aim of examining if the affective/polarisation trend is evident also in Norway. Hence my research questions for this paper are:

RQ1: How have Norwegian political parties' Facebook communication practices evolved from 2009 to 2025 in terms of genres, issues and use of affect?

RQ2: To what extent do parties' Facebook pages function as arenas for affective publics versus rational-deliberative public-sphere interaction?

The rest of the paper is structured as follows: Sect. 2 presents related theory on social media campaigning, the public sphere, affective publics and populism. Section 3 presents the research approach of the study. Section 4 presents the findings as descriptive statistics of followers, an overview of advertising in social media and an analysis of the genres used in 2025, compared to previous elections. Section 5 discusses if political campaigning has become a sort of affective public sphere, and Sect. 6 presents my conclusions.

2 Related Research

In this section I present related research for the study. First, I introduce background and context from social media campaigning research, then I move on to discussing my theoretical lens: the public sphere vs affective publics and populism, and finally I present my analytical lens of genre theory and briefly discuss social media affordances to link the analytical and theoretical lenses.

2.1 eParticipation and Social Media Campaigning

The eParticipation field, defined as "a participatory process that is made possible by modern information and communication technologies and that allows stakeholders to actively participate in public decision-making processes" [21], has done extensive research on social media, politics and political campaigning. Understanding eParticipation initiatives involves understanding the domain (technology) being used, which stakeholders are involved, how the platform is designed and interaction occurs, what level of engagement to target, different ways of engaging actors (mobilising, bombarding, exposing, publicising) as well as power-relations [22]. Sæbø et al. [23] have a similar approach, focusing on actors performing participatory activities in a given context, which leads to different democratic outcomes. This socio-technical analytical focus on both actors, technology and context provides eParticipation with a broad toolbox for understanding political communication in social media.

In 2013, a study of European eParticipation initiatives emphasized the growing user base of social media, and the potential of social media to attract a broader audience and younger age groups [24], quoting earlier research on the promising nature of "the social Internet" to transform participation [25–27]. In the following years, several eParticipation studies researched various aspects of social media for participation. We saw a wide selection of case studies, for example on how social media was being used to involve citizens in community infrastructure discussions [28], empowering immigrants in decision-making [29] and local government initiatives in developing countries [30]. Other scholars focused on power relations between stakeholders, and implications for the public sphere ideal [31], the impact of trust and openness [32] and more recently in our troubled times also how to use social media in crisis management [33].

Using social media for political campaigning has also been extensively researched in eParticipation. Searching the digital government reference library (DGRL) for "social media", "election" and "campaign*" gives 489 results. While this paper is not intended as a comprehensive review of campaigning in social media, a quick glance at the titles of the DGRL papers imply that social media has become an important part of the campaigning infrastructure in many countries [34], that social media is used for promotion, mobilisation, fundraising and measurement of electoral performance [35, 36], and that youth seems to be an important demographic when studying social media [37]. The DGRL also shows that there are case studies on campaigning from a wide range of countries from all continents. The US, UK, Brazil, Spain and Germany are frequently mentioned, while other European, south American, African and Asian countries appear in one or a few papers.

2.2 The Public Sphere, Affective Publics and Populism

Early research on campaigning in social media focused on democratising effects, opportunities for extended participation and youth engagement. While this is still present, later research has moved towards issues such as affection, emotional language, and the effects of populist rhetoric.

Campaigning is a distinct form of participation [38], where the objective is to get citizens to vote for you. Even so, Norwegian democracy is rooted in the participatory liberal democratic model through the constitution article 100, which states that "the authorities of the state shall create conditions that facilitate open and enlightened public discourse". In the liberal-democratic model, representative democracy depends on open, reasoned public debate—both in election campaigns and in the periods between elections [39]. The liberal-democratic model calls for a Habermasian style public sphere, which is recognised by universal accessibility for all citizens, freedom of expression and assembly, rational-critical debate oriented toward reason-giving, argumentation and facts, matters of common concern and issues of general interest over sectional or private interest, bracketing of social inequalities and autonomy from state and market power, where arguments should ultimately lead to the formation of a common public opinion that informs decision-makers [40, 41].

In contrast to the public sphere, Papacharissi presents the concept of affective publics, which he believes better describes the debates, opinions and tone of online communication. Affect is not the same as emotion. It is rather the intensity with which we experience emotion, and the state between listening to something and labelling our reaction as a particular emotion [42]. Affective publics are assembled through sentiment and shared affects (anger, indignation, hope) towards an issue and operate in networked media platform ecologies with constant streams of updates, comments and reaction. They base communication on emotive statements, memes, hashtags and storytelling, and use narrative practices such as personal stories, strong framing of events and relating issues to everyday life. The effects of affective publics are often more symbolic and tend to shape meaning and agendas more than directly affect institutions. While the term can seem negative when contrasted to the rational public sphere, affective publics can channel both solidarity, empowerment, conflict and fragmentation, depending on the issue, tone and sentiment [16, 43]. Habermas himself went far in agreeing with this move away from the public sphere in his 2023 book on the "new structural transformation of the public sphere", where he says that digitalisation has altered the media landscape controlling public discourse, and that the tabula rasa of social media, combined with social media algorithms, fragmentation of audiences and missing deliberative skills on the side of authors, are moving both political parties and citizens towards a more affective form of communication [44].

Social media, with its algorithmic presentation of content meant to invoke emotion, lends itself well to the formation of affective publics [42], and the combination of network effects, engagement metrics, sharing and algorithms promoting content we engage emotionally with can easily lead to populism [45]. In recent years, we have seen a rise in political communication setting up an "elite vs people" narrative, turning people with different opinions and viewpoints into "the enemy" using emotionally charged language, showing distrust towards established procedures and institutions, and often presenting simplified solutions to complex problems [46]. Affective publics and populism are intertwined, but one does not necessarily lead to the other. However, the affordances of social media, such as algorithmically controlled feeds, shares, comments and visuals, do currently seem to reward populist rhetoric over rational debate [47]. It is worth noting that Habermas does not blame the technology as such but instead is critical towards both the business model of social media and the political elites who have failed to deliver on their part of what he calls the social contract between government and governed [44].

Hence, the current situation is perhaps best understood as a complex sociotechnical mix of technological affordances, sociopolitical context and business logic.

2.3 Genre Theory

Studying political campa can be done using a wide range of theoretical and methodological approaches. I have chosen genre theory as my theoretical lens because I want to examine how political parties and citizens communicate. Genre theory can be applied to classify communication practices and has been a valuable tool for studying online democratic practices and political communication[48]. A genre is defined as "a conventional category of discourse based in large-scale typification of rhetorical action" [49] and are recognized by sharing similar form (physical and linguistic features, tone and style) and content (themes and topics), as well as the functionality of the communication channel when the genre is enacted in a digital medium [50, 51].

Genres can be defined by analysing their form, function, and content using the 5W1H framework. This approach considers six aspects of communication [50]:

- **Where** identifies the setting or context in which communication occurs.
- **Why** explains the purpose or intention of the genre.
- **When** specifies the timing or temporal context of communication.
- **Who** describes the participants involved, including both sender and receiver.
- **What** refers to the content or subject matter being conveyed.
- **How** outlines the technical or material requirements for delivering the genre.

The collection of genres used within a particular community can be understood as a *genre system* [52]. Such a system reveals the community's communicative practices, which evolve in response to shared norms, situational demands, time constraints, and the affordances of available media. Genre theory treats genres as typified social actions and crystallise around recurrent communicative situations [50] such as political speeches or manifestos, and can thus help explain how the enactment of genres seeks strategic outcomes such as mobilizing, legitimizing, attacking opponents etc. Focusing on communication genres and communicative practices, rather than on the technologies through which messages are transmitted, makes it possible to trace how communication changes and develops over time [50]. When we also account for the technological functionalities of the medium in which a genre is realised, we can better grasp the interplay between social and technical factors, leading to a deeper understanding of communication as situated in a specific medium [51].

3 Research Approach

The objective of this paper is to examine the genre system found in social media, particularly Facebook, during the 2025 Norwegian parliamentary election, compare this with findings from previous elections and to discuss if and how the political parties' Facebook use contribute to the realization of the constitutions article 100 on reasoned debate, or if it is moving more in an affective populist direction.

Data collection: Data for this study has been collected over five periods: The Norwegian parliamentary elections of 2009, 2013, 2017, 2021 and 2025. Data was collected in the period from June to election day in September for each election. Data for the 2009 study was collected through semi-structured interviews with representatives from the seven political parties that were represented in the parliament before the election (Socialist Left, Labour, Centre Party, Liberals, Christian people's party, Conservatives and the Progress Party). In the following elections, follow-up interviews were done electronically, using e-mail or online meetings to confirm findings from 2009.

Further, social media content (posts, comments and interactions from the pages of the political parties represented in Parliament) during the main campaign period in June to election day in September, has been archived and analysed using Nvivo and Tableau software. For the 2021 election, Facebook's researcher tool CrowdTangle was used for data collection and initial analysis. In 2025, CrowdTangle was no longer available and downloading content from Facebook has been made all but impossible, so for the latest election I had to do the analysis by observing the Facebook pages and taking manual notes, aided by Google Gemini AI[1]. In addition, statistics from Likealyzer.com, the European Social Survey, and the polling company TNS Gallup has been used to examine trust in media and politics. In 2025, I also downloaded advertising data from Meta and Snapchat. As this is a time-series study situated in my own country, and I am a politically interested person, I also apply Netnographic techniques [53] between elections, where I observe and sometimes participate in the political discourse in the groups and pages of Norwegian political parties, and collect field notes about interesting observations, examples of communicative practices and general observations on how political discourse is evolving.

Data analysis: The combination of interviews and content analysis made it possible to compare what informants say with what we can observe happening. This is used to map the genre system in social media political communication. For this study, only Facebook data has been analysed since Facebook remains by far the most used channel in Norwegian politics. The genre systems have been analysed using the 5W1H method presented in the section on genre theory. Of the around 7000 posts collected, a selection has been coded until saturation (no new genres emerging from further study). When no new genres were identified, the remainders of the posts were quickly scanned to see which genre category they matched. Due to space limitations, the findings are presented using the "form/function/content" constructs, as presented in [54].

4 Findings

4.1 The Past: Evolution of Social Media Genres 2009–2021

In the interviews conducted in 2009, the political parties identified three primary objectives for their use of social media in political communication: fostering dialogue with citizens, encouraging citizen contributions, and promoting involvement in party activities. When asked in 2013 whether these objectives had changed, party representatives indicated that they remained largely consistent. In 2017, and in 2021 and 2025, the parties

[1] Gemini search protocol can be accessed here: https://figshare.com/s/72499fab4f8c9c249550.

continued to regard these aims as their overarching goals; however, several respondents emphasized that their approaches had become more sophisticated, with more detailed strategies and differentiated objectives tailored to specific platforms. In 2021, respondents added that they had implemented a multi-channel approach, with different objectives for different types of social media, for example using Instagram live video for dialogue and contribution.

Note that not all the parties responded in 2021 and 2025, but those who did respond verified that the objectives remained the same. The political parties' objectives for social media use are *Dialogue,* involving engaging citizens in debates on political issues, but encouraging and inviting comments. *Contribution,* seeking knowledge about citizen concerns through Q/A sessions and inviting voters to share their stories. *Involvement,* as fund-raising, getting people to volunteer or mobilise for some kind of action. Finally, *agenda-setting* was a new objective added in 2021, as parties emphasized that they saw informing citizens about party policy and reaching out in a fragmented media landscape had become more important. Hence, this objective concerns setting the agenda for political debate, inform and educate citizens through pointed messages designed for attention and engagement.

Table 1 shows how genres have evolved from 2009–2021. A more comprehensive description of the genres can be accessed in earlier papers from 2010, 2014 and 2022. Over the years, genres have appeared, evolved into new genres, become more or less popular, or disappeared altogether. The genre system has become more fine-grained over time, as more resources have been put into social media, and we also see that Habermasian deliberation has largely been replaced by more affective genres, and that parties have moved from dialogue and involvement towards more one-way communication, exemplified by genres such as slogan, Society & context and we want to/we have. The major change came between 2013 and 2017. In 2013 we saw a lot of interaction, two-way communication and discussions both between parties and citizens and between citizens in the comments. In 2017, the parties took a big step towards more one-way and affective communication, and the same trend was apparent in 2021.

4.2 The Present: Social Media Status in 2025

Followers, Votes and Channel Choice.

With respect to channel use, blogging held a prominent position in 2009 but had nearly disappeared by 2013. During this period, Facebook emerged as the dominant platform, and some parties also began experimenting with Instagram. One party noted that responsibility for social media communication had shifted from communications departments to marketing units, suggesting a more substantial transformation in objectives than was otherwise acknowledged. In 2021, respondents said Facebook was still important, but that the audience was becoming older, and other social media channels were becoming more important, especially for reaching younger voters. Parties began experimenting with TikTok, and YouTube, Twitter (later X) were also used. In 2025, several parties were also active on BlueSky. While it has seemed that Facebook would become less important, as it was seen as an old man's social medium, the numbers show that Facebook is still by far the largest medium also in 2025, at least when it comes to the individual parties and how many followers they have (Table 2):

Table 1. Genre system evolution 2009–2021. Colour coding: Yellow: Evolved or less common. Red: disappeared. Green: Increasing in popularity.

Genre	2009	2013	2017	2021
Policy comments	Comments from citizens on approved party policy	Present to a lesser degree, movement towards support/-non-support and disgruntlement.	Present, but marginal	
Call for action	Parties call for volunteers for specific issues	Still present, and more frequently seen in 2013	Evolved to contribute	
Q&A	Questions from citizens, often unanswered in 2009	Still present, and parties tried to engage more with citizens.	Disappeared as users can only comment on posts, not write their own.	
Appeals to party	Citizens asking what the party intends to do on a specific issue	Evolved into support/non-support and debate	No longer present	
Greeting	Greetings and wishes on politicians' birthdays etc.	Still present		
Personal accounts	Politicians ask citizens to tell their stories on selected issues	Evolved and split between several genres (call for action, debate, policy comments)	No longer present	
Video responses	Citizens or politicians use video instead of text. Back and forth exchange	Disappeared		
Debate	N/A	Rational, Habermasian debate.	Some examples, but very little rational debate	
Support/non-support	N/A	Citizens show support, or lack of support, to parties and party policies.	Remains a popular genre	
Disgruntlement	N/A	Sarcastic comments about the party, unpleasant comments about the party and its politicians	Still present, and growing in popularity	
Link	N/A	Parties link to news articles and other sources. Often accompanied by a short statement or question	Still present	Evolved into Slogan and society&context genres
We want to	N/A			Statements about what the party wants to do if they receive enough votes.
We have	N/A			Ruling parties present their achievements from the past election period.
Slogan	N/A			Related to "we want to", but with a more poster-like form. Example: "we are the best choice for young people!"
Contribute	N/A			Politicians ask voters to participate register for updates, become members, sign up for something.
Society & context	N/A			Links and updates on current affairs, tied in with party policy. Ex: pride, global warming.
Experiments	N/A			Testing new formats: podcast, livestream, video document of "a day in the life" of a supporting citizen.

Table 2. Social media followers, 2025

	Facebook	Instagram	X	Bluesky	TikTok
Progress	186.000	52.200	17.000	170	8.700
Conservative	128.000	50.500	49.000	411	16.800
Liberal D	39.000	13.600	14.800	148	-
Christian	30.000	12.000	20.500	-	-
Centre	56.000	11.000	21.900	246	17
Greens	69.000	36.200	39.600	2.600	5.200
Labour	143.000	42.900	59.400	487	16.600
Socialist	67.000	38.900	-	1.200	16.300
Red	95.000	41.200	-	1.000	23.300

It is also interesting to examine the growth and popularity of social media compared to traditional media and examine the (missing) correlation between social media engagement and votes received in the election. In 2025, 65% of Norwegian read newspapers (on- or offline) on any given day, while 94% use the Internet. Youth are much less frequent consumers of traditional news media and get much of their political and societal news from social media influencers [55]. If we compare the number of votes the part received in the elections with the number of followers they have on social media, the result varies between elections, but correlation is generally low. In 2025, the progress party and Labour received far more votes than their share of social media followers would indicate, while the Greens and the far left Red party receive fewer votes than their share of social media followers would indicate. The far-right "Norway Democrats" are also interesting to examine. If social media followers determined the outcome of the election, they would have 4% and 7 seats in parliament, while they received 0,7% of the vote and no representatives. The five remaining parties were well correlated between social media followers and voters, and if you look at the left and right coalition partners as blocks, the blocks would receive almost the same number of seats in parliament if the election was decided based on social media followers.

Political Advertising

Political advertising on social media has been a debated and heated issue in Norwegian political discourse. Political advertising by parties is prohibited in traditional media such as TV and newspapers, but as the law predates social media, social media advertising is legal, and the parties have spent a lot of money on these channels.

On Snapchat, total spending in the category «society and politics was EUR780.000 in 2025, and the Snapchat data set reveals 723 ads from political parties. The Conservatives, Greens, liberal democrats, Red, socialist left and Labour all had campaigns here, while Progress, Centre and the Christian democrats did not advertise on Snapchat. From the openly available advertising data, it seems that segmentation was low. Most ads had some kind of geographic segmentation, and most were targeted at 18 +, with some examples of more age-bracketing. Very few ads were targeted at specific interests.

For Meta (Facebook and Instagram), I could not access spending for 2025 alone but aggregated for 2021–2025 the political parties in Norway spent a total of EUR 2,5 million on advertising. Focusing on the election period from June to September, the most viewed ads had a total of 45 million views. The right-wing progress party was particularly successful, as many people would share and engage with their political ads. This led to 15 of their paid posts reaching more than 1 million viewers.

The parties have different approaches to advertising. The more they spend, the more they seem to rely on low-volume high reach mass communication, trying to appeal to as many voters as possible with the same message. Others, the liberal party in particular, go for a high volume, low-budget micro-targeting of different segments. The parties at the low-end spending generally use more targeted, niche-based ads to maximize results. In terms of communication genres and message, Labour and the conservatives focus on the party identity and vision, in line with their traditional role as the "responsible" political party on the left and right side of the spectrum. Socialist left, some of the liberal's and much of the progress party's advertising is critique-based, attacking current policies, or even other parties in their own coalition.

Spending 2,5 million euros over a five-year period and reaching up to one million views for a single ad certainly indicates that the political parties should at least discuss the issue of political advertising further. Labour, the progress party and the conservatives spent the most money and are also the three parties who receive the most votes in elections. On top of the spending by the parties themselves, both sides of the political spectrum have NGOs and online news outlets that support the different sides and spend money on furthering left- or right leaning policies, indicating that money is becoming increasingly important also in Norwegian elections.

Genres in the 2025 Election

The genre system in the 2025 election continued its evolution, but the overall impression is that the trend seen in 2021 is continuing, and social media genres are maturing. Genre evolution in 2025 is more a matter of careful experimentation, refinement and adaptation, and less about grand experiments or massive communicative changes. For example, we now see a fragmentation and positioning within the two blocks, where coalition partners will attack each other, albeit with a more nuanced language, on areas where the party disagrees with the coalition.

All genres from 2021 were present, at least to some extent. Using a third-party web scraper, was able to harvest 713 posts from the parties from August and September, and mapped the genres as shown below. This is not a complete data set, as it does not include all content, nor was I able to scrape comments. Still, it is sufficient to verify the existence and popularity of genres (Table 3).

Table 3. Genres in 2025

Genres (2021)	2025 count	Description of use
We want to	71	"we want/wish/will … because …", often combined with concrete policy suggestions from the party

(*continued*)

Table 3. (*continued*)

Genres (2021)	2025 count	Description of use
We have	52	Used by governing parties to show achievements in the previous period
Slogan	158	Very common, especially for mobilising and in the last days of the election. Short, idealist statements, and vote-for-us messages
Personal accounts	60	Many posts presenting top candidates in different districts, or national frontrunners
Society & Context	91	Framing of different issues and topics (climate, economy, crime, immigration) connected to party policy and how to solve the issue
Non-support	33	Attacks on ruling parties from the opposition. Common framing includes using words such as crisis, wrong direction, or concrete single issues
Contribute	3	Parties ask citizens to become a member, support us, volunteer
Experiments	24	Use of podcasting, live broadcast or personal video messages
Greetings	5	Salutations, thank you-messages, marking of special days. Not very common
Disgruntlement/Debate/Policy comments	N/A	These genres are more related to comments than posts, so were not possible to count. But disgruntlement was very present in the manual data analysis, while debate and policy comments only occur here and there

There were also some new genres appearing in 2025, although these are arguably variations or extensions of existing genres. First off is *election mobilisation*, which complements the contribute genre by being more focused on general mobilisation to vote, and which also includes an education "how to vote" component. This seems to be targeted towards young people voting for the first time. The second emerging genre is *crisis issue linkage*. This is likely inspired by the current geopolitical situation, with emphasis on defence and crisis management, in some cases with a whiff of populist rhetoric. The genre is recognised by using current crises (climate, Ukraine, safety, immigration) to create a sense of urgency. Opposition parties use the genre to attack the current government, and the current government use it with "steady course"-arguments. The genre is also recognised by a linking of different and often unrelated cases into a thematic "package", such as claiming we need to cut immigration and foreign aid to pay for better schools. The third and final new genre in 2025 is the *micro-narrative.* We have seen an increase in short form, TikTok-style videos, probably in an attempt to adapt to current algorithms. These videos are 20–60 s, with a clear "hook" or tagline and a slogan-like message. While the first new genre appeals to the rational/practical mind (how to vote, and why you should), the second two appeal more to our emotions. These are summarised in Table 4.

Table 4. Emerging genres in 2025 election

Genre	Genre characteristics	Recognised by
Election mobilisation	*Why*: create a sense of urgency in voting (Get people to the urn now!") *When*: Close to election day. *What*: short, mobilising statements. *How*: Strong encouragement, use of exclamation	Sprint/urgency, exclamation. Last few days before election: examples: Last chance to vote! / election day! Share if you vote for X /
crisis issue linkage	*Why*: Create emotional pressure related to crises. Combine issues to thematic packages *What*: social diagnosis ("X + Y means Z. we need to…") *How*: Combination of facts and emotive language	Opposition: describing how the country is in/moving towards crisis in the economy, crime, immigration, climate change, defence etc. Government: describing how steady course is the best in troubling times
Micro-narrative	*Why*: Increase reach through vertically oriented video / reels. *How*: Short video with text-overlays, music, rapid cuts, emotional language and slogans	Short TikTok-inspired videos, appeal to emotion (common sense, stupidity, anti-elitism)

5 Discussion and Conclusion

In this paper I have examined the 2025 Norwegian parliamentary election in social media, and how communication genres have evolved from 2009 and up to the latest election, with emphasis on Facebook. This time-series analysis shows that campaigning is increasingly moving towards one-way communication, and that the parties are increasingly concerned with adapting to social media algorithms, with short, slogan-like messages, viral videos, and messages of what the party wants to achieve. Communication, both from parties and in the comments, is increasingly moving away from the ideals of a Habermasian rational public sphere towards Papacharissi's affective publics and arguably becoming more populist and more adapted to social media algorithms. Two of three new genres directly appeal to our emotions, and the tone of existing genres is also becoming increasingly emotive, with an increase in exclamation, us vs them or either/or-rhetoric, attacks on other parties or elites, crisis-inducing messages, and few questions or discussions. This is most likely a consequence of both social and technical issues. We have seen a rise in populist and emotive rhetoric in recent years, and at the same time social media algorithms seem to reward messages in this style, while more fact-based posts or longer rational discussions are not as widely distributed by social media algorithms. The long-term consequences of this remain to be seen, but there is at least some cause for pessimism to the future of rational and deliberative democracy.

This paper has several weaknesses that call for future research. First, as I only have been able to access a subset of posts in one social medium, there should be more research on broader data sets, covering longer time periods and other social media channels. Also, I have only focused on the political parties, while there are indications that individuals, both politicians and various influencers, are becoming more important when we decide how to cast our vote, and we should examine this, especially for younger generations who report lower use of traditional media. The role of advertising, lobbying organisations and NGOs should also be examined. Finally, more research is needed on the interplay between technological affordances and societal structures, echoing Habermas' 2023 criticism of how politicians are not delivering what citizens need. Perhaps increased affection is there for more reasons than simply the algorithm?

References

1. Kreiss, D.: Developing the "good citizen": Digital Artifacts, Peer networks, and Formal Organization During the 2003–2004 Howard Dean Campaign. J. Inform. Tech. Polit. **6**, 281–297 (2009)
2. Lutz, M.: The Social Pulpit: Barack Obama's social media toolkit. Edelman, Chicago (2009)
3. Borah, P.: Political Facebook use: Campaign strategies used in 2008 and 2012 presidential elections. J. Inform. Tech. Polit. **13**, 326–338 (2016)
4. Tsvetkova, N., Rushchin, D., Shiryaev, B., Yarygin, G., Tsvetkov, I.: Sprawling in Cyberspace: Barack Obama's Legacy in Public Diplomacy and Strategic Communication. J. Polit. Mark. 1–13 (2020)
5. Vaccari, C., Valeriani, A.: Outside the Bubble: Social Media and Political Participation in Western Democracies. Oxford University Press, New York (2021)
6. Braccini, A.M., Federici, T., Saebø, Ø.: Tensions in Online Communities: The Case of a Mass Size eParticipation Initiative. In: International Conference on Electronic Participa-tion (ePart), pp. 149–160. , St. Petersburg (2017)
7. Sæbø, Ø., Federici, T., Braccini, A.M.: From protest to parliament: The role of digital technologies within an online community. J. Inf. Technol. Teach. Cases **10**, 35–45 (2020)
8. Johannessen, M.R.: Genres of Participation in Social Networking Systems: A Study of the 2017 Norwegian Parliamentary Election BT - Electronic Participation. In: Edelmann, N., Parycek, P., Misuraca, G., Panagiotopoulos, P., Charalabidis, Y., Virkar, S. (eds.) Lecture Notes in Computer science, pp. 64–75. Springer International Publishing, Cham (2018)
9. Johannessen, M.R.: Genres of Participation in Social Networking Systems: A Study of the 2021 Norwegian Parliamentary Election. Lecture Notes in Computer Science (including sub-series Lecture Notes in Artificial Intelligence and Lecture Notes in Bioinformatics). 13392 LNCS, pp. 124–136 (2022)
10. Hofmann, S., Beverungen, D., Räckers, M., Becker, J.: What makes local governments' online communications successful? Insights from a multi-method analysis of Facebook. Gov. Inf. Q. **30**, 387–396 (2013). https://doi.org/10.1016/j.giq.2013.05.013

11. Dini, A.A., Sæbo, Ø., Wahid, F.: Affordances and effects of introducing social media within eParticipation—Findings from government-initiated Indonesian project. Electron. J. Inf. Syst. Dev. Countries. **84**, e12035 (2018)
12. Ahangama, S.: Relating social media diffusion, education level and cybersecurity protection mechanisms to e-participation initiatives: insights from a cross-country analysis. Inf. Syst. Front. **25**, 1695–1711 (2023)
13. Brändle, V.K., Galpin, C., Trenz, H.J.: Brexit as 'politics of division': social media campaigning after the referendum. Soc. Mov. Stud. **21**, 234–253 (2022)
14. Grover, P., Kar, A.K., Dwivedi, Y.K., Janssen, M.: Polarization and acculturation in US Election 2016 outcomes – Can twitter analytics predict changes in voting preferences. Technol. Forecast. Soc. Change. **145**, 438–460 (2019)
15. Grahn, H., et al D.: Mapping research on disinformation and misinformation across the Nordic countries An integrative review. Nordicom Rev. **46** (2025). https://doi.org/10.2478/nor-2025-0015
16. Papacharissi, Z.: Affective publics and structures of storytelling: sentiment, events and mediality. 19, 307–324 (2015). https://doi.org/10.1080/1369118X.2015.1109697
17. Yoo, J.J.: A Multivariate Time-Series Analysis of the Agenda-Setting Effects Among News Media, Twitter Elites, and Twitter Public in the Context of the U.S. Immigration Issue. Int. J. Commun. 18, 22–22 (2024)
18. Garimella, V.R.K., Weber, I.: A Long-Term Analysis of Polarization on Twitter. In: Proceedings of the Eleventh International AAAI Conference on Web and Social Media (ICWSM 2017) (2017)
19. Mittal, P., Mittal, S., Singhmar, A.: the evolution of social media influence in u.s. presidential elections: a longitudinal analysis from 2008 to 2024. International Journal of Management, Public Policy and Research. 4, 176–184 (2025). https://doi.org/10.55829/SKPJCX43
20. Olof Larsson, A.: The rise of Instagram as a tool for political communication: A longitudinal study of European political parties and their followers. New Media Soc. **25**, 2744–2762 (2021). https://doi.org/10.1177/14614448211034158
21. Wirtz, B.W., Daiser, P., Binkowska, B.: E-participation: A Strategic Framework. Int. J. Public Adm. **41**, 1–12 (2018)
22. Ritta Coelho, T., Cunha, M.A., Pozzebon, M.: EParticipation practices and mechanisms of influence: An investigation of public policymaking. Gov. Inf. Q. **39**, 101667 (2022)
23. Sæbø, Ø., Rose, J., Skiftenes Flak, L.: The shape of eParticipation: Characterizing an emerging research area. Gov. Inf. Q. **25**, 400–428 (2008)
24. Tambouris, E., Smith, S., Macintosh, A., Panopoulou, E., Dalakiouridou, E., Tarabanis, K., Millard, J.: eParticipation in Europe: Current State and Practical Recommendations. In: egovernment success around the world. pp. 341–357. IGI Global Scientific Publishing (2013). https://doi.org/10.4018/978-1-4666-4173-0.CH017
25. Pasek, J., More, E., Romer, D.: Realizing the Social Internet? Online Social Networking meets Offline Civic Engagement. J. Inform. Tech. Polit. **6**, 197–215 (2009)
26. Sæbø, Ø., Rose, J., Nyvang, T.: The Role of Social Networking Services in eParticipation Electronic Participation. Presented at the (2009)
27. Boyd, danah m, Ellison, N.B.: Social Network Sites: Definition, History, and Scholarship. J. Comput.-Mediated Commun. **13**, 210–230 (2007)
28. Bagui, L., Bytheway, Prof.A.: Exploring eParticipation in the city of Cape Town. J. Community Inf. **9** (2013)
29. Laanpere, M., Tammsaar, K., Sousa, S.: A case study on using social media for e-participation: design of Initiative Mapper Web service. In: ACM International Conference Proceeding Series, pp. 289–292 (2011). https://doi.org/10.1145/2072069.2072119

30. Rugeiyamu, R., Kumburu, B., Ndunguru, G.: Social media as platform for citizen participation in local government decision-making in Tanzania. Transforming Gov. People, Process Policy 1–24 (2025)
31. Johannessen, M.R., Sæbø, Ø., Flak, L.S.: Social media as public sphere: a stakeholder perspective. Transforming Gov. People Process Policy **10** (2016)
32. Abdulkareem, A.K., Abdulkareem, Z.J., Ishola, A.A., Bello, M.L., Oladimeji, K.A.: The influence of openness of public organizations and social media use on e-participation: the mediating effect of trust in e-government. Int. Rev. Public Adm. **27**, 281–296 (2022). https://doi.org/10.1080/12294659.2022.2136054
33. Garrote-Quintana, A., Urquía-Grande, E., Delgado Jalón, M.L.: Trending topics in E-government, E-participation, and social media in the crisis era. Adm. Soc. **57**, 254–280 (2025)
34. Taras, D., Davis, R.: Electoral Campaigns, Media, and the New World of Digital Politics. Electoral Campaigns Media New World Digital Polit. 1–324 (2022)
35. Baumgartner, J.C., Towner, T.L.: The Internet and the 2020 Campaign, pp. 1–285 (2021). https://doi.org/10.5040/9781978733633
36. Fitzpatrick, J., von Nostitz, F.C.: Reaching the voters: parties' use of Google Ads in the 2021 German Federal election. Media Commun. **12**, 8543 (2024)
37. Brough, M.: Youth Power in Precarious Times. Duke University Press (2020)
38. Medaglia, R.: EParticipation research: Moving characterization forward (2006–2011). Gov. Inf. Q. **29**, 346–360 (2012)
39. Ferree, M.M., Gamson, W.A., Gerhards, J., Rucht, D.: Four models of the public sphere in modern democracies (2002).https://doi.org/10.1023/A:1016284431021
40. Habermas, J.: The Public Sphere. In: Seidman, S. (ed.) Jürgen Habermas on Society and Politics: A reader. Beacon Press, Boston (1989)
41. Jürgen Habermas (Stanford Encyclopedia of Philosophy). https://plato.stanford.edu/entries/habermas/, Accessed 05 Mar 2026
42. Boler, M., Davis, E.: Affective politics of digital media : propaganda by other means. Routledge, Taylor & Francis Group, New York (2021)
43. Papacharissi, Z.: Affective Publics: Sentiment, Technology, and Politics. Oxford University Press, Oxford (2014)
44. Habermas, J.: A New Structural Transformation of the Public Sphere and Deliberative Politics. Polity Press, Cambridge (2023)
45. Olaniran, B., Williams, I.: social media effects: Hijacking Democracy and civility in civic engagement. Platforms, protests, and the challenge of networked democracy. Part F798, **77** (2020). https://doi.org/10.1007/978-3-030-36525-7_5
46. Webber, J.: Understanding Populism. Soc. Leg. Stud. **32**, 849–876 (2023)
47. Schramm, P., Molin, J., Schramm, P., Molin, J.: Feeling the Truth: emotions and digital truth-making within right-wing populist networks on Twitter/X. Ethnol. Eur. **53** (2023). https://doi.org/10.16995/EE.8922
48. Päivärinta, T., Sæbø, Ø.: The genre system lens on e-democracy. Scandinavian J. Inf. Syst. **20** (2008)
49. Miller, C.R.: Genre as social action. Q. J. Speech **70**, 151–167 (1984)
50. Yates, J., Orlikowski, W.J.: Genres of Organizational Communication: A Structurational Approach to Studying Communication and Media. Acad. Manag. Rev. **17**, 299–326 (1992)
51. Shepherd, M., Watters, C.: The evolution of cybergenres. In: System Sciences, 1998., Proceedings of the Thirty-First Hawaii International Conference, Vol. 2 pp. 97–109 2 (1998)
52. Yates, J., Orlikowski, W.: Genre Systems: Structuring Interaction through Communicative Norms. J. Bus. Commun. **39**, 13–35 (2002)
53. Kozinets, R. V: Netnography - Doing Ethnographic Research Online. SAGE, Los Angeles (2010)

54. Orlikowski, W.J., Yates, J.: Genre Repetoire: The Structuring of Communicative Practices in Organizations. Adm. Sci. Q. **39**, 541–574 (1994)
55. Medietilsynet: Feed, fornuft og følelser: ungdoms refleksjoner om politikk, medier og påvirkning | Medietilsynet. https://www.medietilsynet.no/fakta/rapporter/kritisk-medieforstaelse/feed-fornuft-og-folelser-ungdoms-refleksjoner-om-politikk-medier-og-pavirkning/, Accessed 19 Mar 2026

The Success Equation of Data Portability: Centering on Technical Standardization and Trust-Based Infrastructure

Yunam Choi(✉), Hun-Yeong Kwon, and Sang-Pil Yoon

Korea University, 145, Anam-ro, Seongbuk-gu, Seoul, South Korea
{chldbska4834,khy0}@korea.ac.kr

Abstract. Data portability, initially established to guarantee individuals' self-determination over personal information, has evolved into a critical mechanism for fostering a fair data economy. However, its practical implementation often faces significant challenges due to technical barriers and conflicting stakeholder interests. This study analyzes the success and failure factors of data portability across various global cases—including UK Open Banking, Korea's MyData service, and US vehicle data access laws(Right to Repair)—to identify the essential requirements for institutionalizing these rights. Based on these analyses, we propose the 'Data Portability Success Index (S_{DP})' equation, which quantifies success through two primary variables: the reduction in switching costs via technical standardization (ΔC_s) and the ratio of trusted intermediaries (R_{ti}) within the infrastructure. Our findings indicate that technical standardization and trust-based governance are more decisive for the effectiveness of data portability than mere legal codification. Consequently, this paper suggests that the government must fulfill dual roles as a 'Technical Standard Architect' and a 'Trust Governance Builder' to minimize market uncertainty and promote a robust data ecosystem. This research provides a theoretical and quantitative framework for future policy prioritization and empirical analysis across diverse industries.

Keywords: Data Portability · Industry Data · Data Access · GDPR · Data Intermediation · Data Governance · Success Equation

1 Introduction

Governments worldwide have been implementing legal and policy frameworks to guarantee data disclosure, specifically as a strategic response to the growing concerns over data monopolies by large corporations. The Right to Data Portability (RTDP), as initially introduced in the 2012 draft of the European Commission's General Data Protection Regulation (GDPR) [1], has since been codified and specified through various national legislations. This right is intrinsically aligned with the national strategic direction of ensuring equitable data access and usage rights for all stakeholders—including Data Holders, Data Recipients, and Data

Published by Springer Nature Switzerland AG 2027
D. Duenas-Cid et al. (Eds.): ePart 2026, LNCS 16923, pp. 247–261, 2027.
https://doi.org/10.1007/978-3-032-36676-4_16

Subjects—while facilitating the fair distribution of data value and safeguarding the fundamental rights of individuals.

In practice, however, the implementation of these rights often faces significant challenges, as the protection of data subjects' rights frequently clashes with the conflicting interests among various stakeholders. Theoretically, the RTDP is defined as the entitlement of a data subject to receive personal data concerning them from one controller (Data Holder) and transmit it to another (Data Recipient), or to receive it directly. Early scholarship on the legal origins of this right primarily focused on Article 20 of the EU GDPR. De Hert et al. (2018) [2] emphasized that the RTDP transcends the traditional 'Right of Access' and serves as a pivotal instrument for realizing substantive informational self-determination in the digital ecosystem. They analyzed this right as having an 'emancipatory' character, empowering individuals by enhancing data sovereignty and reducing platform dependency. As evidenced by global legal guides such as DLA Piper (2025) [3], the RTDP is proliferating across approximately 160 jurisdictions worldwide, including the United States (CCPA/CPRA) and Asia, following the global spread of data protection frameworks. Prevailing literature converges on the consensus that the RTDP is not a mere duplication of data but an active right that enables individuals to generate value through data reuse.

Furthermore, from an economic perspective, the RTDP has been discussed as a policy intervention to mitigate monopolistic tendencies in digital markets. Studies by Graef (2015) [4] and Kerber (2020) [5] have highlighted the 'Lock-in Effect' resulting from data silos maintained by dominant platforms. If users cannot port their data to competing services, they encounter high Switching Costs, forcing continued reliance on incumbent platforms despite potentially lower service quality. Reports from the OECD (2021) [6] suggest that the RTDP can significantly lower these switching costs, promote multi-homing, and provide equitable competitive opportunities for market entrants and startups. Conversely, some scholars caution that the RTDP may impose excessive technical and financial burdens on large-scale platforms [7], or increase security risks, thereby potentially triggering a 'Privacy-Innovation Paradox' [8] that stifles innovation.

For the RTDP to function effectively in a practical sense, interoperability—the condition where data is interpretable by the receiving party—is a prerequisite. Borgogno & Colangelo (2019) [9] have critiqued that without robust technical standards, the RTDP remains a hollow right. They argue that the efficacy of data portability is determined by the degree of API (Application Programming Interface) standardization. Recent technical research focuses on the integration of Semantic Interoperability and Privacy-Enhancing Technologies (PETs). Key to this is the design of standard ontologies that preserve data integrity during transfer and encryption technologies that ensure security in transit. Particularly in high-dimensional industrial data, such as vehicular data, Zech (2016) [10] argues that granting 'Access Rights' rather than physical data transfer is more advantageous for maintaining data control and preventing security breaches.

In summary, existing literature has sufficiently demonstrated the legal legitimacy, economic necessity, and technical feasibility of the Right to Data Portability. However, most studies have focused narrowly on economic gains or technical solutions within the transmission process. There remains a significant scholarly gap in analyzing the specific factors that determine the success or failure of RTDP implementation across different industries.

Therefore, building upon the theoretical foundations established by prior research, this study aims to identify the 'Success Equation (S_{DP})' and critical factors necessary for the substantive implementation of the RTDP and to derive the corresponding role of government. To achieve this, the study first analyzes implementation cases across various types of industrial data. Second, it identifies the industry-specific factors that lead to either successful implementation or inevitable failure. Finally, based on these findings, This study identifies and compares the relative importance of these factors and establishes the necessary framework for the government's role in this ecosystem.

2 Theoretical Framework of the Right to Data Portability

2.1 Background and Global Proliferation of the Right to Data Portability

The "Right to Data Portability" was first formally introduced in Article 18 of the European Commission's draft GDPR (COM(2012) 11 final) on January 25, 2012 [11]. Its primary objective was to enhance data mobility between social networks and cloud services, thereby preventing "Vendor Lock-in" [12]. This framework was built on the premise that enabling user data mobility by granting data subjects control over their information would foster a more dynamic digital market. Subsequently, on March 23, 2012, the WP29 Opinion 01/2012 evaluated the introduction of this right positively, noting its potential to empower data subjects, prevent service provider lock-in, promote competition, and support the free flow of data and innovation within the digital economy [13]. This conceptual evolution culminated in the final adoption of Article 20 in the GDPR on May 4, 2016. The publication of the "Guidelines on the Right to Data Portability" on October 27, 2017 [14], further served to clarify several legal ambiguities surrounding the practical application of this right [15].

Under GDPR Article 20, the Right to Data Portability is specifically codified to apply only to personal data provided by the data subject to a controller. It guarantees the right to receive such data in a structured, commonly used, and machine-readable format, or to have it transmitted directly to another controller [16]. While the 2017 WP29 Guidelines resolved uncertainties by expanding the scope of "provided by the data subject" to include elements such as search history and raw IoT data, they excluded derived data and emphasized the technical feasibility of direct transmission between controllers [17]. Notably, the European Union has since enacted the "Data Act" to further catalyze the utilization of data. This legislation aims to facilitate access to industrial data—which had previously been managed exclusively—promote switching between data processing service providers, and encourage investment in data value creation.

Since its inception in the EU GDPR, the Right to Data Portability has profoundly influenced the enforcement of laws and policies globally, evolving with consistent core concepts. In Singapore, the Personal Data Protection Act (PDPA) mandates that organizations transmit data to a receiving organization in a commonly used, machine-readable electronic format upon a data porting request, aiming to enhance user autonomy and facilitate service switching [18]. India's Digital Personal Data Protection Act (DPDP), enacted in 2023, is also interpreted to recognize the right of a data principal to obtain their data in a structured, commonly used format and transfer it to other data fiduciaries [19]. Similarly, South Korea is actively fostering the so-called "MyData Industry" based on the right to request data transmission under its Personal Information Protection Act [20]. Furthermore, Australia addresses the concept of Accredited Data Recipients under Sect. 56AB of the Treasury Laws Amendment (CDR) Act 2018 [21], and the United States explicitly enshrines the concept of data portability in the CCPA Sect. 1798.100 (as amended by CPRA, 2018).

2.2 The Legal and Technical Nature of Data Portability

The fundamental essence of the Right to Data Portability (RTDP) lies in guaranteeing the "Right to Informational Self-Determination"—a core facet of data subjects' rights. Through the RTDP, individuals must be empowered to determine to whom, for what purpose, and in what manner their data is utilized. Crucially, the nature of this right transcends mere mitigation of vendor lock-in; it serves as a robust mechanism for safeguarding the data subject's sovereignty. This necessitates a comprehensive consideration of a multi-stakeholder ecosystem that extends beyond the data subject and the Data Holder to include the Data Recipient. This is because the data subject can explicitly mandate the holder to transmit specific data, in a designated format, to a particular recipient at a chosen time and through a specific method. In this regard, when considering the attributes of its practical implementation, the RTDP expands from a simple individual entitlement into a collective and societal dimension [22].

Ultimately, the relationship between the Data Holder and the Data Recipient must be addressed, which directly leads to issues of economic viability. If the RTDP is approached strictly from the perspective of safeguarding an individual's rights, the obligation to transmit data upon a subject's request may be perceived solely as a regulatory and financial burden by the Data Holder. Consequently, the RTDP should be viewed not merely as a matter of rights protection but as a strategic instrument that exerts significant influence on industrial innovation and market competition [4].

Furthermore, in the practical execution of this right, the standardization of the data provision environment emerges as a critical success factor. In the absence of standardization, the resulting friction in data transmission may inadvertently allow large-scale vendors to maintain unilateral control, thereby hindering the activation of the broader data ecosystem [23]. Therefore, while the RTDP is a derivative right for the substantive realization of informational self-determination, its implementation requires a balanced consideration of the holder-recipient relationship, the necessity of standardization for a fair ecosystem, and the economic interests of all stakeholders. In this sense, RTDP is intended to enable innovation and competition across the data ecosystem. Existing literature can be synthesized into three analytical dimensions: (1) legal empowerment, (2) market competition and switching costs, and (3) technical interoperability and governance. This study integrates these dimensions into a unified analytical framework centered on technical standardization and trust governance.

3 Case Studies and Factor Analysis of Data Portability

3.1 Research Design and Comparative Method

This study adopts a comparative case study design to analyze the key success factors for the effective implementation of the Right to Data Portability (RTDP). In particular, rather than relying on random sampling, this study employs a purposive case selection approach. This choice is based on the premise that, given the variation in institutionalization and implementation of RTDP across industries and countries, it is necessary to select cases that exhibit both representativeness and distinctiveness in order to enable a meaningful comparison of success and failure factors.

Case selection was guided by three primary criteria. First, the existence of an explicit legal framework governing data portability or data transfer rights. Second, the presence of operational technical infrastructure, such as standardized APIs, authentication systems, and data intermediary mechanisms. Third, the availability of observable implementation outcomes. Based on these criteria, this study selects the United Kingdom's Open Banking and Korea's MyData initiatives in the financial sector, the United States' Right to Repair framework in the mobility sector, and the Belgium's Athumi model as an example of a public data intermediary system. These cases were deemed suitable for comparative analysis as they differ across dimensions such as industry domain, regulatory maturity, level of technical standardization, and implementation performance.

The comparative analysis is conducted through both within-case and cross-case comparisons. The analytical procedure consists of case identification, variable extraction, cross-case comparison, and factor synthesis. Each case is coded according to a common analytical template, with key dimensions including legal basis, level of standardization, intermediary structure, security and trust governance, and implementation outcomes. Through this approach, the study aims to assess the success of RTDP not merely in terms of the existence of legal provisions, but in relation to the degree of integration between technical standardization and trust-based infrastructure.

3.2 Implementation Cases of the Right to Data Portability

Implementation cases of the Right to Data Portability (RTDP) vary significantly across different nations and industries, depending on the types of data involved.

Financial Sector: Open Banking and MyData

The UK's Open Banking initiative originated from the Competition and Markets Authority's (CMA) "Retail Banking Market Investigation Order" in 2017 [24]. Driven by the recognition of insufficient competition in the retail banking market and high barriers for customers to switch banks, the CMA mandated the "CMA9"—the nine largest banks—to develop and open standardized APIs and establish an independent implementation body. Simultaneously, through the Payment Services Regulations (PSR) 2017, the EU's PSD2 was transposed into domestic law [25], introducing Third-party Payment service Providers (TPPs). TPPs are entities that can access bank account information to provide payment services without holding the accounts themselves, falling into two categories: (1) Account Information Service Providers (AISP) and (2) Payment Initiation Service Providers (PISP). AISPs aggregate a user's scattered account information to provide an integrated view via mobile apps, while PISPs initiate fund transfers from a consumer's account to a recipient's account on their behalf. With the enforcement of Regulatory Technical Standards (RTS) in September 2019, which regulate technical aspects such as customer authentication and data transmission methods, the RTDP was materialized not as a mere "file download" but as a right to move and reuse financial data in real-time through standardized APIs and authorized intermediaries [26]. This resulted in an ecosystem involving hundreds of TPPs and millions of users, effectively promoting competition, reducing card fees, and improving financial inclusion.

Similarly, South Korea established a legal foundation for financial "MyData" following the 2020 amendment of the "Three Data Laws," which introduced the right to request data transmission and the "MyData Service Provider" (Credit Information Management Industry) license [27]. Under this system, fully implemented in 2021, once a data subject provides consent, they gain the right to integrate and transfer their personal credit information—scattered across various financial institutions—into a structured electronic format. The Credit Information Use and Protection Act and its subordinate regulations define the scope of data, standard formats, and the responsibilities of both data holders and recipients. Furthermore, the Financial Services Commission (FSC) ensures security by allowing only entities with sufficient capital and robust security systems to provide MyData services through a licensing process. Korea has notably evolved into a "Super App" model, integrating open banking infrastructure with services ranging from wire transfers and payments to insurance and credit management within a single platform.

Mobility Sector: Challenges in Vehicle Data Access

In contrast to these success stories, the mobility sector in the United States illustrates significant challenges. The 2020 Massachusetts Vehicle Data Access Law (Right to Repair) sought to strengthen independent repairers' access to data by mandating that vehicle manufacturers provide data, including telematics systems. This demand arose because, while users could previously acquire vehicle data via physical On-Board Diagnostics (OBD) ports, modern vehicles now transmit remote diagnostics and telematics directly to manufacturers' servers.

However, the Alliance for Automotive Innovation (AAI) argued that developing a secure open platform for vehicle data is technically unfeasible. Furthermore, in June 2023, the National Highway Traffic Safety Administration (NHTSA) halted data sharing due to hacking risks and remote control vulnerabilities [28].

The NHTSA's stance highlights the practical conflict between data access rights and Public Safety. The agency argued that the Massachusetts law was preempted by the federal Safety Act, warning that granting remote open access to telematics could allow hackers to remotely control critical safety functions such as steering or braking. However, in August 2023, the NHTSA partially pivoted, reaffirming its support for the "Right to Repair" [29]. As a technical compromise, it proposed data access via short-range wireless communication (e.g., Bluetooth) instead of long-range remote access. This approach ensures repairers' access rights while mitigating remote hacking risks by limiting data proximity.

Public Sector: Data Intermediary Services in the EU

The EU has institutionalized data transmission through the establishment of data intermediaries. Under Article 10 of the EU Data Governance Act (DGA, 2022/868), "Data Intermediation Services" (DISPs) act as trusted third parties [30]. Registered DISPs are authorized to use the EU "Trust Mark," signifying they meet the DGA's requirements for neutrality and security [31].

A representative case is Athumi (Vlaams Datanutsbedrijf) in Belgium. Established by the Flemish government in 2022 and launched under the "Athumi" brand in 2024, it is a public-sector company focused on realizing the RTDP under GDPR Article 12 [32]. The Flemish government oversees Athumi's finances and advisory board through government commissioners, representing a public investment to catalyze the data economy [33]. Athumi operates a "Data Vault" platform where citizens, upon consent, can collect, store, and share data from public or private institutions. For example, in collaboration with the HR provider Randstad, citizens can grant employers secure access to verify degree data stored in their personal "Pods," significantly streamlining administrative recruitment processes [34].

3.3 Analysis of Critical Success and Failure Factors

Based on the empirical cases discussed in Sect. 3.1, the most decisive factors determining the success or failure of the Right to Data Portability (RTDP) are 'Technical Standardization' and 'Trust-based Transmission Infrastructure (Table 1).'

Technical Standardization: The Catalyst for Interoperability

The success of the UK's Open Banking and South Korea's MyData initiatives stems from their concretization of the RTDP into a real-time data mobility infrastructure, centered around standardized API specifications and dedicated management bodies. In the UK, the CMA mandated the adoption of common APIs for the CMA9 banks, while the OBIE (now OBL) provided functional and security standards along with conformance testing. This established a de

Table 1. Comparative Cases of Data Portability Implementation

Case	Legal Basis	Level of Standardization	Trusted Intermediary	Security & Governance	Outcome
UK Open Banking	PSD2/CMA	Very High (common APIs, OBIE oversight)	Strong (licensed TPPs)	Very High (RTS, authentication standards)	Success (large-scale adoption)
Korea MyData	PIPA Amend-ment/MyData Act	Very High (JSON, OAuth 2.0 APIs)	Strong (licensed operators under FSC)	Very High (strict licensing & security)	Success (super-app ecosystem)
US Vehicle Data	Massachusetts Right to Repair	Low (lack of unified standards)	Weak (no centralized intermediary)	Fragmented (federal safety conflicts, hacking risks)	Delayed/Partial compromise (Bluetooth access)
Belgium Athumi	EU Data Governance Act (DGA)	Medium–High (Solid Pods, metadata standards)	Strong (EU Trust Mark, public neutrality)	High (government oversight, certified security)	Emerging (early-stage expansion)

facto industrial standard that surpassed the baseline requirements of PSD2 and RTS [35]. Similarly, South Korea, led by the Financial Services Commission and the Financial Security Institute, established granular specifications—such as MyData Standard APIs based on JSON and OAuth 2.0, sector-specific data APIs, and authentication/support APIs—effectively institutionalizing the RTDP as a 'Mandatory API-based Data Transmission' [36]. Consequently, the RTDP evolved beyond mere file downloads into a mechanism that fosters platform competition, allowing hundreds of service providers to participate in a large-scale, interoperable framework with pre-agreed security protocols and formats. This standardized channel enabled Third-party Providers (TPPs) and MyData operators to collect and integrate data at a lower cost, directly enhancing consumer welfare.

The Standardization Gap: Conflict Between Innovation and Safety

Conversely, the case of the Massachusetts Vehicle Data Access Law (Right to Repair) demonstrates how the absence of standardized security protocols acts as a significant barrier to exercising the RTDP. Notably, there was a lack of a centralized standards body and consensus on technical specifications—specifically regarding communication methods, security models, and datasets [37]. In this vacuum, vehicle manufacturers and the AAI argued that the required technology was either non-existent or presented severe cybersecurity risks. The NHTSA also intervened, citing conflicts with the federal Safety Act and warning against the risks of long-range remote hacking [38]. Although the NHTSA and state authorities eventually sought a compromise through short-range wireless access (e.g., Bluetooth), this remained a fragmented, ad-hoc implementation at the device level. It failed to evolve into a universal, guaranteed framework like those seen in the financial sector, revealing a structural tension between data portability and public safety norms.

Trust-Based Infrastructure: Neutral Intermediaries

To address these issues, the EU Data Governance Act (DGA) designed a public infrastructure for data transmission and sharing by defining "Data Intermediary Services" as independent legal entities and granting them an EU-recognized Trust Mark upon meeting neutrality and security requirements [34]. Within this framework, Belgium's Athumi functions as a "public data standards authority," utilizing personal data vaults and Solid-based Pod technology to enable citizens to store and transmit data in structured formats. Athumi has also developed metadata and settlement platforms to ensure that employment and administrative services utilize this data in an interoperable manner. By having a certified, neutral entity manage data formats and security standards, the technical gap between data holders and recipients is bridged, and a robust trust mechanism is established.

Conclusion of Factor Analysis

In conclusion, RTDP becomes effective not merely through legal recognition, but through technical implementation and governance. While the financial sector achieved success through robust standard APIs, the vehicle data sector continues to struggle due to uncertainties in standards and security. To resolve these conflicts, the intervention of governance bodies—similar to Athumi—that can establish data standards and guarantee neutrality is essential. Future RTDP policies must advance beyond simple declarations of rights toward the construction of certified intermediary infrastructures and data standard guidelines that reflect sector-specific characteristics. Across the comparative cases, two recurring patterns emerge. First, successful implementation is associated with lower switching costs enabled by standardized APIs. Second, implementation stability increases where trusted intermediary governance structures are institutionalized. These recurring dimensions form the basis of the S_{DP} model proposed in the following section.

3.4 Formulation of Equations for Validating Success Factors and the Role of Government

To quantitatively measure the success of the Right to Data Portability (RTDP), it is essential to conduct a correlation analysis using the reduction in 'Switching Costs' and the 'Proportion of Trusted Intermediaries' as independent variables. The S_{DP} equation is presented as a conceptual framework and is supplemented with an illustrative pilot application using sectoral indicators to demonstrate how the model can be operationalized. The present study does not claim statistical validation; instead, it specifies the measurement logic required for future empirical testing. This relationship can be modeled into a Success Equation (S_{DP}) as follows:

The Data Portability Success Index (S_{DP}) is expressed as a weighted sum of the reduction in switching costs (ΔC_s) driven by technical standardization and the ratio of trusted intermediaries (R_{ti}) within the infrastructure.

$$S_{DP} = w_1 \cdot \Delta C_s + w_2 \cdot R_{ti} + \epsilon$$

Definition of Variables:

- S_{DP} (Data Portability Success Index): The dependent variable representing the overall success of RTDP implementation.
- ΔC_s (Reduction in Switching Cost): Independent variable 1, representing the rate of decrease in switching costs.
- R_{ti} (Ratio of Trusted Intermediaries): Independent variable 2, representing the proportion of trust-based intermediaries.
- w_1, w_2: Weighting factors indicating the relative importance of each variable ($w_1 + w_2 = 1$).
- ϵ: Other exogenous variables and the error term.

Detailed Variable Definitions and Measurement Methodologies

(1) Reduction in Switching Cost (ΔC_s)
The ΔC_s measures the cost differential before and after technical standardization. As standardization matures, ΔC_s increases, directly leading to an upward trend in S_{DP}.

$$\Delta C_s = \frac{C_{pre} - C_{post}}{C_{pre}}$$

- C_{pre}: The average cost (or time) required for data collection and cleansing prior to the introduction of standard APIs.
- C_{post}: The data acquisition cost following the adoption of standard APIs (e.g., JSON, OAuth 2.0).

(2) Ratio of Trusted Intermediaries (R_{ti})
R_{ti} represents the proportion of data transmission paths that utilize 'trust-based infrastructure' certified by the government or public authorities. As psychological and physical safety nets are established—allowing users to transmit data without fear of security breaches—R_{ti} increases, thereby stimulating the exercise of data portability rights.

$$R_{ti} = \frac{N_{certified}}{N_{total}}$$

- $N_{certified}$: The number of data transmissions via certified third-party intermediaries or secure transmission segments.
- N_{total}: The total number of attempted data transmissions.

Policy Objectives and the Role of Government. The primary policy goal derived from this equation is to maximize S_{DP} by effectively managing two strategic levers:

- Technical Standard Architect: The government must eliminate technical barriers and increase ΔC_s by disseminating robust standard API guidelines.
- Trust Governance Builder: The government must resolve market uncertainties and increase R_{ti} through certification systems for authorized intermediaries and investment in secure infrastructure.

This formalized model can serve as a quantitative tool in future research to calculate the specific weights (w_1, w_2) by inputting empirical data from various industries, ultimately aiding in the prioritization of policy interventions. To maximize the two core variables of the equation, the government must move beyond the role of a regulator and act as a 'Market Maker' and 'Technical Standard Architect'.

First, the government must serve as a 'Technical Standard Architect' to minimize switching costs (ΔC_s). As seen in South Korea's MyData case, leaving standardization to individual firms often leads to delays in interoperability due to conflicting interests. Therefore, the government should directly present industry-specific standard API guidelines (e.g., JSON, OAuth 2.0) and provide incentives to compliant firms to remove technical barriers. This lowers the entry barrier for small-scale service providers, securing a diverse data economy ecosystem.

Second, the government must focus on increasing the ratio of trusted intermediaries (R_{ti}) as a 'Trust Governance Builder'. The conflict between security risks and public safety seen in the US vehicle data case can trigger social distrust in data portability. To address this, the government should foster neutral, certified data intermediaries—similar to Belgium's Athumi—or implement strict certification systems to grant 'Trust Marks'. Such physical and psychological safety nets allow data subjects to exercise their rights without fear of security breaches, effectively increasing the R_{ti} index.

Third, the government must act as an 'Economic Interest Mediator.' If data portability is perceived as a one-sided cost burden for data holders, it risks becoming a mere formality.

Consequently, the government should design reasonable fee structures for data transmission or recognize investments made in value creation, thereby fostering a win-win economic ecosystem for both data holders and recipients. In conclusion, the government must be an active mediator capable of simultaneously controlling the two levers of technical standardization (ΔC_s) and the trust-based infrastructure (R_{ti}) that constitute the S_{DP} equation. This proactive role is the key to ensuring that data portability does not remain a declarative right but functions as a practical mechanism for achieving both industrial innovation and data sovereignty. Although the S_{DP} model is conceptual at this stage, the comparative findings suggest that cases with higher levels of standardization and trusted intermediary governance exhibit stronger implementation outcomes (Table 2).

Table 2. Expected S_{DP} Outcomes by Switching Cost Reduction and Trusted Intermediary Ratio

Case	ΔC_s	R_{ti}	Expected S_{DP}
UK Open Banking	High	High	High
South Korea MyData	High	High	High
US Vehicle Data	Low	Low	Low
Belgium Athumi	High	Middle	Middle

4 Conclusion and Future Research Directions

This study began with the premise that the Right to Data Portability (RTDP), rooted in the protection of a data subject's right to informational self-determination, must transcend a mere legal declaration to be effectively implemented in a real-world market environment. To this end, the study analyzed critical success factors and proposed a formulated model.

The analysis confirms that 'Technical Standardization' and 'Trust-based Transmission Infrastructure' are the most decisive factors for the successful settlement of the RTDP. The cases of UK Open Banking and South Korean Financial MyData succeeded in revitalizing the data ecosystem by drastically reducing switching costs through standardized API systems (JSON, OAuth 2.0, etc.) and dedicated management bodies. In contrast, the US Vehicle Data Access Law (Right to Repair) faced delays in implementation due to the absence of standardized security protocols and a centralized standards organization, encountering significant barriers in public safety and security risks.

The Data Portability Success Index (S_{DP}) equation derived from this comparative analysis quantitatively suggests that the substantive efficacy of the RTDP stems from technical codification and trust governance rather than legal codification. Notably, as seen in the case of Belgium's Athumi, the presence of a neutral and secure 'Data Intermediary' acts as a core mechanism to bridge the technical gap between data holders and recipients and to eliminate market uncertainty.

While this research is significant in theoretically identifying success factors and modeling them through empirical cases, it also presents tasks for future research. Specifically, subsequent empirical data analysis is required to calculate the weights (w_1, w_2, etc.) of the proposed success equation (S_{DP}). The model should be further refined by incorporating actual data transmission volumes across various industries (Finance, Healthcare, Mobility, etc.) and measuring the magnitude of cost reductions before and after the introduction of standardization.

Also, This study has several limitations. First, the comparative analysis is based on a limited number of cases across selected sectors, which may constrain the generalizability of the findings. Second, the proposed S_{DP} equation remains a conceptual and exploratory framework rather than a statistically validated model. Although illustrative sectoral indicators are presented, the study does not include large-scale empirical testing or longitudinal data analysis. Third, sector-specific institutional and technological differences may influence implementation outcomes beyond the variables examined in this study. Future research should therefore expand the number of cases and incorporate empirical datasets to validate the relative weights and causal relationships within the S_{DP} framework.

In conclusion, the Right to Data Portability is a pivotal instrument that goes beyond protecting individual rights to foster a fair ecosystem for the data economy. Moving forward, the government must concentrate its policy efforts on establishing sector-specific standardization strategies and reliable intermediary infrastructures to ensure that the RTDP leads to substantive industrial innovation.

References

1. European Commission: Commission proposes a comprehensive reform of data protection rules to increase users' control of their data and to cut costs for businesses. Press Release IP/12/46 (2012). https://ec.europa.eu/commission/presscorner/detail/en/ip_12_46. Accessed 01 Mar 2026
2. De Hert, P., Papakonstantinou, V., Malgieri, G., Beslay, L., Sanchez, I.: The right to data portability in the GDPR: towards user-centric interoperability. Comput. Law Secur. Rev. **34**(2), 193–203 (2018). https://doi.org/10.1016/j.clsr.2017.11.003
3. DLA Piper: Data Protection Laws of the World (2026). https://www.dlapiperdataprotection.com/. Accessed 01 Mar 2026
4. Graef, I.: Assessing data access issues in online platforms. Telecommun. Policy **39**(5), 375–387 (2015). https://doi.org/10.1016/j.telpol.2014.12.006
5. Kerber, W.: Data sharing in IoT ecosystems and competition law: the example of connected cars. J. Compet. Law Econ. **15**(4), 381–426 (2019). https://doi.org/10.1093/joclec/nhz018
6. OECD: Data portability, interoperability and competition. OECD Competition Committee Background Note (2021). https://www.oecd.org/content/dam/oecd/en/publications/reports/2021/10/data-portability-interoperability-and-competition_f09a402e/73a083a9-en.pdf. Accessed 01 Mar 2026
7. Blind, K., Niebel, C.: The impact of the EU general data protection regulation on innovation in firms. SSRN Electron. J. (2022). https://doi.org/10.2139/ssrn.4257740
8. Galasso, A., Schankerman, M.: Patents and cumulative innovation: causal evidence from the courts. Q. J. Econ. **130**(1), 317–369 (2015). https://doi.org/10.1093/qje/qju029
9. Borgogno, O., Colangelo, G.: Data sharing and interoperability: fostering innovation and competition through APIs. Comput. Law Secur. Rev. **35**(2), 2–4 (2019). https://doi.org/10.1016/j.clsr.2019.03.008
10. Zech, H.: Building a European data economy. Int. Rev. Intellect. Prop. Compet. Law **48**(5), 501–503 (2017). https://doi.org/10.1007/s40319-017-0610-1
11. European Commission: Commission proposes a comprehensive reform of data protection rules to increase users' trust and help maximise the growth of the Digital Single Market. Press Release IP/12/46 (2012). https://ec.europa.eu/commission/presscorner/detail/en/ip_12_46. Accessed 01 Mar 2026
12. Reus, J., Bilderbeek, N.: Data portability in the EU: an obscure data subject right? IAPP News (2022). https://iapp.org/news/a/data-portability-in-the-eu-an-obscure-data-subject-right. Accessed 01 Mar 2026
13. Article 29 Data Protection Working Party: Opinion 01/2012 on the data protection reform proposals, WP 191 (2012). https://ec.europa.eu/justice/article-29/documentation/opinion-recommendation/files/2012/wp191_en.pdf. Accessed 01 Mar 2026
14. Article 29 Data Protection Working Party: Guidelines on the right to "data portability", wp242rev.01 (2017). https://ec.europa.eu/newsroom/article29/items/611233/en. Accessed 01 Mar 2026
15. Diker Vanberg, A.: The right to data portability in the GDPR: what lessons can be learned from the EU experience? J. Internet Law **21**(7), 1, 11–20 (2018). https://business.cch.com/ald/JIL_0118_SPVanberg.pdf. Accessed 01 Mar 2026
16. European Parliament and Council of the European Union: Regulation (EU) 2016/679 of the European Parliament and of the Council of 27 April 2016 on

the protection of natural persons with regard to the processing of personal data and on the free movement of such data, and repealing Directive 95/46/EC (General Data Protection Regulation). Official Journal of the European Union, L 119/1 (2016). https://gdpr-info.eu/art-20-gdpr/. Accessed 01 Mar 2026
17. Meyer, D.: European Commission, experts uneasy over WP29 data portability interpretation. IAPP News (2017). https://iapp.org/news/a/european-commission-experts-uneasy-over-wp29-data-portability-interpretation-1. Accessed 01 Mar 2026
18. Personal Data Protection Commission Singapore: The Distinction between Organisations and Data Intermediaries and Why It Matters (2023), https://www.pdpc.gov.sg/the-distinction-between-organisations-and-data-intermediaries-and-why-it-matters, last accessed 2026/03/01
19. Dalei, P.: The digital personal data protection Act, 2023: A legal analysis in light of global data protection standards. International Journal of Law 11(3), 127–131 (2025), https://www.lawjournals.org/assets/archives/2025/vol11issue3/11064.pdf, last accessed 2026/03/01
20. Korea Credit Information Services: MyData Comprehensive Portal. https://www.mydatacenter.or.kr:3441/. Accessed 01 Mar 2026
21. Australian Government Treasury: Treasury Laws Amendment (Consumer Data Right) Bill 2022: Explanatory Memorandum. Technical report (2022). https://treasury.gov.au/sites/default/files/2022-09/c2022-315575-em.pdf. Accessed 14 Mar 2026
22. Chao, Y., Xu, M., Tamò-Larrieux, A., Kollnig, K.: Data portability strategies in the EU: moving beyond individual rights. Comput. Law Secur. Rev. **57**, 105973 (2025). https://doi.org/10.1016/j.clsr.2025.106135
23. Kranz, J., Kuebler-Wachendorff, S., Syrmoudis, E., Lütge, C., Sunyaev, A.: Data portability. Bus. Inf. Syst. Eng. **65**(5), 603–612 (2023). https://doi.org/10.1007/s12599-023-00815-w
24. HM Treasury, CMA, FCA, PSR: Joint statement by HM Treasury, the CMA, the FCA and the PSR on the future of Open Banking (2022). https://www.gov.uk/government/publications/joint-statement-by-hm-treasury-the-cma-the-fca-and-the-psr-on-the-future-of-open-banking/joint-statement-by-hm-treasury-the-cma-the-fca-and-the-psr-on-the-future-of-open-banking. Accessed 01 Mar 2026
25. PSR, HM Treasury, FCA, CMA: Joint statement from HM Treasury, the PSR, FCA and CMA on the future oversight of Open Banking (2022). https://www.psr.org.uk/news-and-updates/latest-news/news/joint-statement-from-hm-treasury-the-psr-fca-and-cma-on-the-future-oversight-of-open-banking/. Accessed 01 Mar 2026
26. Babina, T., et al.: Customer data access and fintech entry: early evidence from open banking. Bank of England Staff Working Paper No. 1,059 (2024). https://www.bankofengland.co.uk/-/media/boe/files/working-paper/2024/customer-data-access-and-fintech-entry-early-evidence-from-open-banking.pdf. Accessed 01 Mar 2026
27. Kim, J.B.: A Comparative Legal Study on Financial Law Systems. Korea Institute of Finance, Research Report for Financial Services Commission (2020). https://better.fsc.go.kr/fsc_new/upload/2021/01/04/%ED%95%B4%EC%99%B8%EA%B8%88%EC%9C%B5%EB%B2%95%EC%B2%B4%EA%B3%84_%EC%97%B0%EA%B5%AC%EC%9A%A9%EC%97%AD%EB%B3%B4%EA%B3%A0%EC%84%9C.pdf. Accessed 01 Mar 2026

28. Shepardson, D.: Carmakers will not immediately comply with Massachusetts data repair law. Reuters, September 22 (2023). https://www.reuters.com/business/autos-transportation/carmakers-will-not-immediately-comply-with-massachusetts-data-repair-law-2023-09-22/. Accessed 05 Mar 2026
29. Kolodziej, K.: NHTSA Letter Regarding Massachusetts Data Access Law. National Highway Traffic Safety Administration (NHTSA) (2023). https://www.autocare.org/docs/default-source/government-affairs/2023-08-22-nhtsa-letter-regarding-ma-data-access-law.pdf. Accessed 05 Mar 2026
30. European Commission: EU register of data intermediation services. Shaping Europe's digital future (2026), https://digital-strategy.ec.europa.eu/en/policies/data-intermediary-services. Accessed 01 Mar 2026
31. European Commission: EU register of recognised data altruism organisations. Shaping Europe's digital future (2026). https://digital-strategy.ec.europa.eu/en/policies/data-altruism-organisations. Accessed 01 Mar 2026
32. Athumi: Government commissioners represent the Flemish government. https://athumi.be/en/about-us/government-commissioners. Accessed 01 Mar 2026
33. Athumi: Solid: vault technology for consumers and businesses (2025). https://athumi.be/en/technologies/solid. Accessed 01 Mar 2026
34. Inrupt: Flanders Government strengthens a trusted data economy with Inrupt's Enterprise Solid Server (2025). https://www.inrupt.com/case-study/flanders-strengthens-trusted-data-economy, Accessed 01 Mar 2026
35. Open Banking Limited: Open Banking Read/Write API Specifications. https://openbankinguk.github.io/read-write-api-site2/. Accessed 14 Mar 2026
36. Korea Data Agency: Financial MyData Standard API Specification: Guidelines. https://developers.mydatakorea.org/mdtb/apg/dgi/bas/FSAG0101. Accessed 14 Mar 2026
37. Huetter, J.: U.S. regulators put the brakes on Massachusetts vehicle data access law. Aftermarket Matters (2023). https://www.aftermarketmatters.com/national-news/u-s-regulators-put-the-brakes-on-massachusetts-vehicle-data-access-law/. Accessed 14 Mar 2026
38. Bachman, K.L., Fischer, J.L., Kim, S.Y.: NHTSA Sides with Automakers, Informs Court that Massachusetts Data Access Law Preempted by Federal Law. Seyfarth Shaw LLP (2023). https://www.seyfarth.com/news-insights/nhtsa-sides-with-automakers-informs-court-that-massachusetts-data-access-law-preempted-by-federal-law.html. Accessed 14 Mar 2026

From Strategy to Execution: A Framework to Assess Alignment Across Urban Planning Instruments—Evidence from Brazil

Matheus Cruz[1](✉), Flavia Bernardini[1], Vanessa Nunes[2], and Carlos Alberto Malcher Bastos[3]

[1] Institute of Computing, Fluminense Federal University (UFF), Niterói, RJ, Brazil
mmcruz@id.uff.br, fcbernardini@ic.uff.br
[2] Institute of Technology and Leadership (Inteli), São Paulo, SP, Brazil
[3] Engineering School, Fluminense Federal University (UFF), Niterói, RJ, Brazil

Abstract. Although public planning instruments are widely institutionalized, they still lack systematic mechanisms capable of objectively assessing their completeness and alignment. As a result, plans often exhibit gaps and inconsistencies that hinder the translation of strategic intentions into effective actions. This article addresses this limitation by proposing the UPICEF-A framework, designed to assess the completeness and alignment among instruments such as Strategic Planning, the Goal Plan, the Multi-Year Plan, and budgetary instruments.

Methodologically, the study combines a conceptual analysis of planning instruments with the development of an evaluation model that structures objectives, goals, initiatives, and indicators into verifiable relationships. The framework enables the identification of gaps, overlaps, and misalignments, as well as the assessment of the completeness of planning artifacts.

Results indicate that the consistency observed in formal documents does not fully hold when analyzed relationally, revealing weaknesses in the articulation between strategic and operational levels. By providing an objective mechanism for evaluating these relationships, UPICEF-A contributes to shifting public planning analysis from a purely formal perspective to a structured approach focused on completeness and alignment, supporting the development of more coherent and effective planning processes.

Keywords: Public Planning · Alignment Assessment · Completeness Evaluation · Planning Instruments · UPICEF-A Framework

1 Introduction

The increasing complexity of contemporary urban management, driven by accelerated urbanization processes, digital transformation, and environmental pressures, has highlighted the need for more integrated, coherent, and data-driven

Published by Springer Nature Switzerland AG 2027
D. Duenas-Cid et al. (Eds.): ePart 2026, LNCS 16923, pp. 262–277, 2027.
https://doi.org/10.1007/978-3-032-36676-4_17

planning instruments [1,14]. In this context, urban planning plays a central role as a coordinating mechanism between governance, sustainability, and evidence-based decision-making. However, recent studies indicate that, despite advances in technologies such as Big Data and digital platforms, structural challenges related to institutional fragmentation and low integration among planning instruments still persist [7,12]. In many cases, different instruments coexist without clear articulation, compromising the coherence of public policies and the effectiveness of governmental actions.

Although cities formally adopt instruments such as Strategic Planning, Goals Plans, Multi-Year Plans (PPA), Budget Guidelines Laws (LDO), and Annual Budget Laws (LOA), these documents often operate in a disconnected manner, especially when considering different temporal horizons. Recent literature on urban governance highlights that the lack of alignment between strategic planning and operational instruments constitutes one of the main limitations to the effective implementation of public policies [2,9]. As a result, long-term strategies are not always adequately translated into concrete actions, while short-term decisions may occur without proper connection to strategic guidelines. In this scenario, this study addresses the following research problem: how to evaluate, in a structured, measurable, and replicable way, the degree of alignment between urban planning instruments and their ability to translate long-term strategies into concrete actions over time?

To address this problem, this study is guided by the following research questions: **RQ1)** how to identify and structure semantic equivalence relationships between components of different planning instruments? **RQ2)** how to quantitatively measure the degree of alignment between these instruments based on content comparison? and **RQ3)** how to interpret these results in order to assess the extent to which shorter-term instruments execute strategic planning and to which they are previously planned within it? Based on these questions, this article proposes an extension of the *Urban Planning Instruments Completeness Evaluation Framework* (UPICEF), expanding its scope beyond structural completeness by incorporating semantic analysis and comparative content analysis. The study adopts the *Design Science Research* (DSR) paradigm, in which the proposed artifact is developed, applied, and evaluated in a real-world context, operationalized through a Comparative Content Analysis (CCA) at the instance level.

UPICEF-A extends the original Urban Planning Instruments Completeness Evaluation Framework (UPICEF). While the original framework assessed only structural completeness, UPICEF-A incorporates Comparative Content Analysis (CCA), semantic equivalence relationships, and the constructs of alignment, execution, and forecast to evaluate relationships across planning instruments.

The instantiation of the framework in the City of São Paulo—selected due to its high administrative complexity, diversity of planning instruments, and transparency in data availability—enabled the identification of clear and relevant results. In general, it was observed that planning instruments present a strong level of formal alignment among themselves, indicating that strategic guidelines are broadly reflected in other documents. However, this alignment is not equally observed from the perspective of execution, revealing that a significant portion of what is planned is not uniformly materialized in operational instruments. These findings suggest that, although there is coherence in planning formulation, important challenges remain in translating strategies into concrete actions, particularly at levels closer to implementation. Furthermore, the analysis highlighted the complexity involved in comparing planning instruments, reinforcing the need for more structured and scalable approaches to support such evaluations.

This study contributes to both literature and practice by: (i) proposing an extended framework to analyze alignment across urban planning instruments, integrating structural and semantic perspectives; (ii) introducing a semantic analysis approach capable of identifying relationships between planning documents beyond the mere presence of components; (iii) operationalizing the framework through a real-world case study, demonstrating its applicability and analytical value; (iv) providing empirical evidence on alignment and misalignment patterns between strategic and operational planning; and (v) identifying the computational complexity of the analysis, highlighting the need for automation using Natural Language Processing techniques. It is important to emphasize that this study does not aim to support the creation of planning instruments, but rather to provide a method for evaluating them in a qualitative, quantitative, and systematic manner. Although applied in the Brazilian context, the framework was designed with adaptation mechanisms that enable its application in different cities and countries.

Finally, this article is structured as follows. Section 2 presents the theoretical background and conceptual foundations. Section 3 describes the research method based on the Design Science Research (DSR) paradigm. Section 4 introduces the UPICEF-A framework, including its main artifacts and application process. Section 5 presents its instantiation and evaluation through a case study in the City of São Paulo. Section 6 answers the research questions based on the empirical findings. Finally, Sect. 7 concludes the paper by discussing the main contributions, limitations, and future research directions.

2 Background and Related Work

This section presents the theoretical foundations underlying the proposed framework. First, a literature review was conducted to identify studies related to urban planning instruments and their relationships across temporal horizons. Subsequently, the analytical constructs adopted in this study are introduced, providing the conceptual basis for assessing structural completeness and alignment among planning instruments.

2.1 Literature Review and Planning Instruments

A literature review was conducted to identify studies addressing urban planning instruments and mechanisms for evaluating consistency among planning documents. Searches were performed in Scopus, Web of Science, IEEE Xplore, ACM Digital Library, and ScienceDirect using combinations of terms related to urban planning, strategic planning, alignment, and evaluation.

The search string included expressions such as:

("urban planning" OR "urban governance" OR "municipal planning") AND ("framework" OR "model" OR "method" OR "tool") AND ("evaluation" OR "assessment" OR "analysis") AND ("planning instruments" OR "strategic plan" OR "public planning")

The review revealed that previous studies have predominantly focused on individual planning instruments or specific planning levels. Although several studies discuss strategic planning, urban governance, and policy implementation, limited attention has been devoted to the relationships established among instruments operating across different temporal horizons and to mechanisms capable of assessing their coherence. This gap motivated the development of the proposed framework.

Urban planning relies on a set of instruments that operate across different temporal horizons and support the translation of strategic objectives into programs, actions, and budgets. These instruments provide mechanisms for organizing public interventions and coordinating governmental priorities. However, they are often developed independently, which may lead to fragmentation and inconsistencies between planning levels.

Previous studies classify planning instruments according to their purpose and temporal horizon. Strategic instruments define long-term visions and objectives, programmatic instruments translate them into initiatives and targets, and budgetary instruments support operational execution through resource allocation. Based on the literature and the Brazilian planning framework, the instruments analyzed in this study were grouped into strategic, programmatic, and budgetary categories, as summarized in Table 1. This classification provides the conceptual basis for analyzing relationships across planning levels.

2.2 Structural Completeness and Alignment of Planning Instruments

Previous studies have emphasized the importance of evaluating the internal consistency of planning instruments. One approach consists of assessing their structural completeness, understood as the degree to which a document contains the minimum set of components expected for its purpose. This perspective enables the identification of structural gaps and supports comparisons among planning documents.

Although structural completeness contributes to evaluating isolated instruments, it does not necessarily ensure coherence among documents operating at different temporal horizons. Planning effectiveness depends not only on the existence of individual components, but also on the consistency of objectives, initiatives, actions, and indicators across planning levels.

Table 1. Classification of Urban Planning Instruments

Planning Instrument	Nature	Time Horizon
Strategic Plan (PE)	Strategic	Long-term
Master Plan (PD)	Strategic	Long-term
Goals Plan (PM)	Programmatic	Medium-term
Multi-Year Plan (PPA)	Programmatic	Medium-term
Budget Guidelines Law (LDO)	Budgetary	Short-term
Annual Budget Law (LOA)	Budgetary	Short-term

Consequently, assessing relationships among planning instruments requires complementing structural analysis with mechanisms capable of identifying semantic correspondences and evaluating how strategic intentions are translated into operational actions. This broader perspective motivates the analytical constructs adopted in this study and the extension of the original UPICEF framework.

2.3 Analytical Constructs

Assessing urban planning instruments requires more than verifying the existence of planning documents. Previous work introduced the concept of completeness as the degree to which a planning instrument contains the expected minimum components, enabling the identification of structural gaps and comparisons among documents. However, structural completeness alone does not capture whether the content of one instrument is consistently reflected across planning instruments, motivating the adoption of complementary analytical constructs.

To address this limitation, this study adopts four analytical constructs: completeness, alignment, execution, and forecast. Completeness assesses the presence of the minimum structural components expected for each planning instrument. Alignment measures semantic correspondence among planning elements at the same planning level. Execution evaluates how higher-level objectives are translated into lower-level instruments, whereas forecast assesses how operational actions and indicators are anticipated in higher-level planning documents.

Figure 1 summarizes the relationships among planning instruments across temporal horizons. Internal relationships represent alignment among instruments belonging to the same planning level, while relationships between groups represent execution and forecast constructs.

Together, these constructs provide the conceptual foundation of UPICEF-A by integrating structural completeness with semantic relationships among planning instruments, enabling a comprehensive assessment of coherence across strategic, programmatic, and budgetary levels.

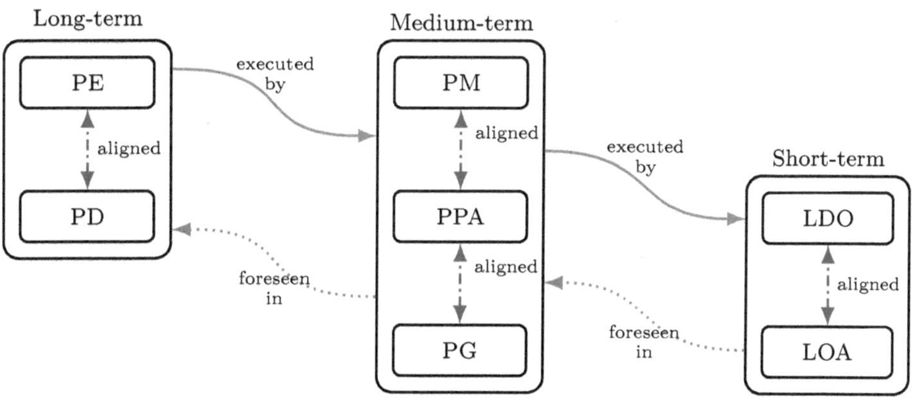

Fig. 1. Relationship constructs between groups of planning instruments across temporal horizons. Internal relationships represent alignment (blue dash-dot lines), while relationships between planning levels represent execution (green solid lines) and forecast (orange dotted lines). *Source: Elaborated by the authors.* (Color figure online)

3 Research Method

This study follows the Design Science Research (DSR) paradigm, which focuses on the design, development, and evaluation of artifacts intended to solve real-world problems while generating prescriptive knowledge [4,8]. DSR is particularly suitable for addressing complex organizational challenges because it combines theoretical foundations with practical applications. Following this paradigm, the present study proposes and evaluates an artifact aimed at assessing relationships among urban planning instruments operating across different temporal horizons.

The research was conducted iteratively through six main activities. Initially, a literature review and exploratory analysis were performed to identify gaps and establish the theoretical foundations of the study. Subsequently, interviews with specialists and an analysis of the Brazilian legal framework supported the identification and classification of planning instruments according to their nature and temporal horizons. Based on these inputs, the framework was designed and refined using insights obtained from the literature, expert knowledge, and empirical observations.

The resulting artifact, named UPICEF-A, extends the original Urban Planning Instruments Completeness Evaluation Framework (UPICEF) by incorporating semantic alignment mechanisms and the analytical constructs of execution and forecast relationships among planning instruments operating across different temporal horizons. Although presented sequentially, these activities were performed iteratively, enabling continuous refinement throughout the research process.

According to DSR principles, artifacts should be evaluated in realistic environments to assess their utility and applicability. Therefore, the proposed framework was instantiated through a case study in the City of São Paulo, selected due to its urban relevance, administrative complexity, and the availability of planning documents. The evaluation focused on the framework's ability to assess both the structural completeness of individual instruments and the relationships established among strategic, programmatic, and budgetary documents. In particular, the study investigated semantic correspondences across planning levels and analyzed how strategic intentions are translated into operational actions.

By instantiating the framework in a real-world context, the study aims to demonstrate its analytical utility and operational feasibility while identifying opportunities for future refinements.

4 The UPICEF-A Framework

Building upon the methodological foundations described in the previous section, this study proposes an extension of the Urban Planning Instruments Completeness Evaluation Framework (UPICEF), referred to as UPICEF-A. While the original framework focused on structural completeness, this extension incorporates mechanisms to assess alignment among planning instruments operating across different temporal horizons. The framework is supported by two complementary artifacts: the Minimum Components Model (MCM) and the Comparative Content Analysis (CCA), the latter operationalized through the Document Relationship Matrix.

4.1 Minimum Components Model (MCM)

The *Minimum Components Model* (MCM) defines the minimum structural elements expected for each planning instrument analyzed by the framework. Rather than prescribing a rigid format, the MCM establishes a baseline set of components that enables the assessment of structural completeness. Its development was based on four complementary sources: a literature review on urban planning and strategic management, an analysis of the Brazilian legal framework, an empirical examination of planning documents from Brazilian capitals, and interviews with public planning specialists.

The integration of these sources resulted in a consolidated set of minimum components for each planning instrument. These components were organized into the MCM artifact, which serves as a reference for completeness assessment during framework application. Table 2 presents the MCM for *Strategic Planning*, including its components, definitions, and theoretical or institutional foundations.

Although Table 2 presents only the Strategic Planning model, the same procedure was applied to the remaining instruments, including the Goals Plan (PM), Multi-Year Plan (PPA), Budget Guidelines Law (LDO), and Annual Budget Law (LOA). The complete models are available in the public repository, allowing consultation, replication, and adaptation to different institutional contexts.

Building upon the structural definition provided by the MCM, the next step consists of identifying semantic relationships among planning components.

Table 2. Strategic Planning (SP)

STRATEGIC PLANNING (SP)		
Definition: Expresses the city's vision for the future and establishes long-term guidelines to guide public policies and investments [3].		
Purpose: Align short- and medium-term decisions with long-term strategic direction, promoting continuity across administrations and a focus on results [6,10].		
Minimum Requirements	**Concept**	**Foundation**
Challenges	Structural problems/opportunities	[10]
Mission	Institutional purpose	[3]
Vision	Desired future state	[3]
Objectives	Broad guiding results	[6]
Goals	Time-bound quantification of objectives	[5]
Strategies	Paths and choices to achieve goals	[3]
Implementation Plan	High-level execution roadmap	[11]
Indicators	Monitoring and evaluation measures	[13]
PlanningClassification	Political	
Classification byTime Horizon	Long-term (10–20 years)	
Classification byLevel	Strategic	
Minimum number of components: 8		

4.2 Comparative Content Analysis (CCA)

Building upon the structural definition provided by the Minimum Components Model (MCM), the Comparative Content Analysis (CCA) extends the framework from assessing *what* planning instruments contain to examining *how* they relate to one another. While the MCM focuses on document structure, the CCA identifies semantic correspondences across instruments, enabling the analysis of alignment and coherence among different temporal horizons.

The CCA relies on a systematic comparison of planning components based on their meaning, purpose, and role within the planning process. Thus, components with different labels may be considered equivalent when they perform similar functions, such as translating strategic objectives, operationalizing actions, or supporting monitoring and evaluation. This perspective allows the framework to move beyond structural analysis and capture how planning is articulated across instruments.

To ensure consistency and reduce subjectivity, the identification of semantic relationships follows four steps: (i) conceptual analysis of components; (ii) cross-comparison between components from different instruments; (iii) application of semantic equivalence criteria; and (iv) consolidation and validation through expert interviews. All protocols, criteria, and matrices are publicly available[1], ensuring transparency and reproducibility.

[1] Available at: https://github.com/MatheusMoreiraCruz/UPICEF-Framework.

4.3 Framework Application Process

Following the definition of components (MCM) and their semantic relationships (CCA), the application of UPICEF-A is operationalized through a structured sequence of analytical activities. These steps integrate both artifacts to enable a systematic evaluation of planning instruments in real-world contexts (Fig. 2).

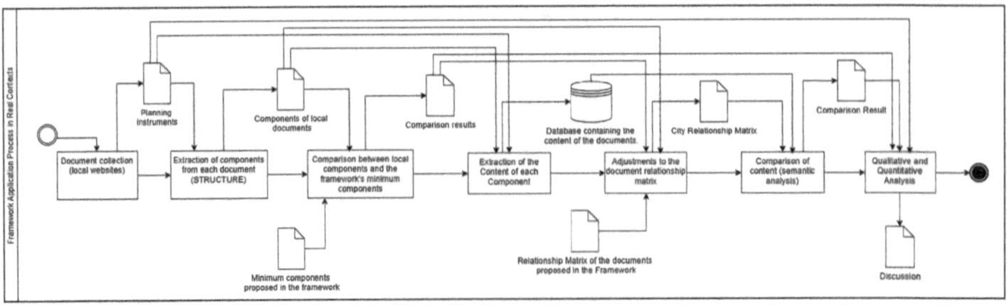

Fig. 2. Framework application process in real-world scenarios

The process begins with the collection of planning documents from official institutional sources, followed by the extraction of their internal structure, identifying the components that organize each document. These components are then compared against the Minimum Components Model (MCM), allowing the assessment of structural completeness. Next, the textual content associated with each component is extracted and organized into a dataset to support semantic analysis. An essential step is the *adjustment of the Document Relationship Matrix*, which enables the framework to be tailored to different institutional contexts. This ensures that the predefined relationships accurately reflect the characteristics of the analyzed instruments.

Once the relationships are defined, the content of comparable components is analyzed to identify alignment, execution, and forecast relationships across documents.

5 Framework Application: Case Study of the City of São Paulo

The framework was applied through a case study in the City of São Paulo, selected due to its urban relevance, administrative complexity, and maturity of planning instruments. As the largest city in Brazil, São Paulo provides a comprehensive and publicly available set of planning documents, making it a suitable environment for instantiating and evaluating the proposed artifact. Within the Design Science Research (DSR) paradigm, real-world instantiation is a key evaluation mechanism. Thus, the case study aims to demonstrate the framework's analytical utility and operational feasibility while also revealing limitations and opportunities for improvement. The analyzed documents include:

Strategic Planning (Municipal Agenda 2030), the Goals Plan (PM), the Multi-Year Plan (PPA), the Budget Guidelines Law (LDO), and the Annual Budget Law (LOA) (Figs. 3, 4, 5, 6 and 7).

Fig. 3. Municipal Agenda 2030 (Strategic Planning)

Fig. 4. Goals Plan 2021–2024

Fig. 5. Multi-Year Plan (PPA) 2022–2025

Fig. 6. Budget Guidelines Law (LDO) 2024

Fig. 7. Annual Budget Law (LOA) 2024

The application followed the process defined in the previous section. First, documents were collected and their structural components extracted. These components were then compared against the Minimum Components Model (MCM) to assess completeness. Results reveal structural gaps, particularly the absence of mission and vision in the Strategic Planning document. This is explained by the adoption of Agenda 2030 as the strategic instrument, which emphasizes SDGs rather than traditional planning elements. Table 3 summarizes the completeness assessment.

Aggregated results for the remaining instruments (Table 3) indicate high completeness levels in operational instruments, contrasting with gaps in strategic planning. This suggests that the existence of formal instruments does not necessarily ensure structural coherence across planning levels.

After assessing structural completeness, the next step consists of analyzing semantic relationships among planning instruments.

Table 3. Completeness assessment summary for São Paulo.

Document	Present/Expected	Completeness (%)
Agenda 2030	6/8	75.0
Plano de Metas	5/5	100.0
PPA	5/6	83.3
LDO	2/2	100.0
LOA	2/2	100.0
City Average	–	**91.7**

5.1 Content Comparison and Semantic Analysis

Content comparison and semantic analysis represent the main advancement of this approach over traditional methods. While structural analysis verifies the presence of components, this stage evaluates the degree of semantic alignment between planning instruments. The procedure starts by selecting a source component (e.g., an SDG from Strategic Planning) and comparing it only with target components previously defined as semantically equivalent in the relationship matrix. For each pair of equivalent components (C_A, C_B), the analysis is conducted at the content (instance) level. Each element of C_A is compared against all elements of C_B, generating all possible combinations.

Table 4. Sample of correspondences between SDGs (SP) and Objectives (GP)—direction SP→GP.

SDG (SP)	Correspondence in Objectives (GP)
No poverty	Reduce poverty and expand access to rights for vulnerable populations.
Zero hunger and sustainable agriculture	Improve food and nutritional security, supporting urban agriculture.
Health and well-being	*No correspondence.*
Quality education	Ensure access to quality education.
Gender equality	Promote diversity and equal opportunities.
Affordable and clean energy	Expand public lighting with energy efficiency.

This results in a many-to-many relationship, with a maximum of $m \times n$ comparisons, where m and n are the number of elements in the source and target components, respectively. In the São Paulo case, this process was performed manually, without computational support. More than one million potential comparisons were identified for Strategic Planning alone, highlighting both the complexity of the task and the limitations of manual approaches. Given the large number of possible combinations, only representative examples are presented in this article.

Table 5. Sample of correspondences between Objectives (GP) and SDGs (SP)—direction GP→SP.

Objective (GP)	Correspondence in SDGs (SP)
Expand public lighting with energy efficiency.	Affordable and clean energy
Promote diversity and equal opportunities.	Gender equality
Promote international cooperation and innovation.	Industry, innovation and infrastructure
Expand access to housing and land regularization.	No poverty
Promote economic development and decent work.	Decent work and economic growth
Reduce poverty and expand access to rights.	No poverty

To illustrate the procedure, Tables 4 and 5 present sample correspondences between SDGs and Goals Plan objectives in both directions. Results show that alignment is not symmetric: operational objectives often reflect strategic guidelines, but the reverse is less consistent.

The correspondence percentage of a source component relative to a target component is defined as: number of correspondences (c)/total elements in source (m) × 100.

This metric quantifies alignment by measuring how many source elements have semantic correspondence. Table 6 presents the results. While some components exhibit strong alignment, others—particularly indicators and actions—reveal significant gaps, highlighting challenges in translating strategy into operations.

Figure 8 visualizes these relationships at the component level, while Table 7 summarizes the aggregated results. In general, forecast relationships are stronger than execution relationships, suggesting that operational instruments tend to reflect higher-level planning elements, whereas execution relationships are more heterogeneous, particularly for the LDO.

Building on these results, the framework evaluates alignment through the constructs of *execution* and *forecast*. Figure 9 shows that 51.2% of Strategic Planning is executed by the Goals Plan, 58.6% by the PPA, and 17.2% by the LDO, indicating varying levels of translation from strategic intentions into operational actions.

Forecast analysis (Fig. 10) reveals that 94.36% of the GP, 95.30% of the PPA, and 85.70% of the LDO are grounded in Strategic Planning, indicating strong coherence across planning levels. Together, these results suggest that formal alignment among planning documents does not necessarily translate into effective implementation.

Table 6. Correspondence percentages by document pair and components (São Paulo).

Pair (direction)	Compared components	m	c	%
SP → GP	SDG × Objectives	17	17	100%
	Goals × Goals	135	37	27.4%
	Indicators × Indicators	545	222	40.7%
	Actions × Initiatives	655	240	36.6%
SP → PPA	Goals × Linked Goals	135	81	60.0%
	SDG × Objectives	17	17	100%
	Indicators × Indicators	545	84	15.4%
SP → LDO	Goals × Targets	135	23	17.0%
GP → SP	Objectives × SDG	27	27	100%
	Goals × Goals	77	75	97.4%
	Indicators × Indicators	84	74	88.1%
	Initiatives × Actions	366	344	94.0%
PPA → SP	Goals × Goals	83	82	98.8%
	Objectives × SDG	31	27	87.1%
	Indicators × Indicators	477	84	17.6%
LDO → SP	Targets × Goals	49	42	85.7%

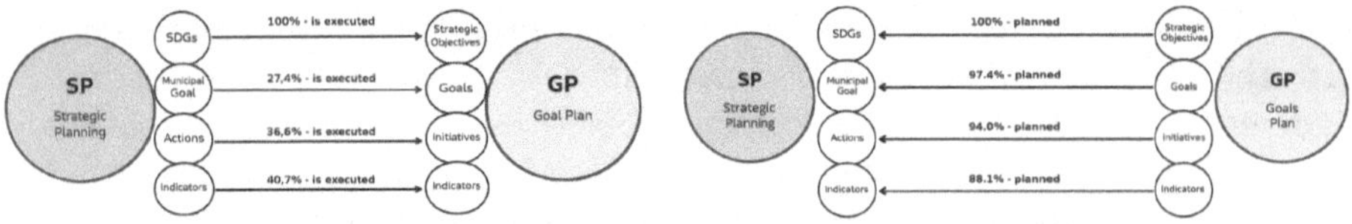

(a) SP → GP (b) GP → SP

Fig. 8. Component-level correspondences between SP and GP.

Table 7. Average percentage of relationships between documents (São Paulo).

Document Pair	Direction	Average
PE ↔ PM	PE → PM	51.2% – executed
	PM → PE	94.9% – planned
PE ↔ PPA	PE → PPA	58.5% – executed
	PPA → PE	67.8% – planned
PE ↔ LDO	PE → LDO	17.0% – executed
	LDO → PE	85.7% – planned

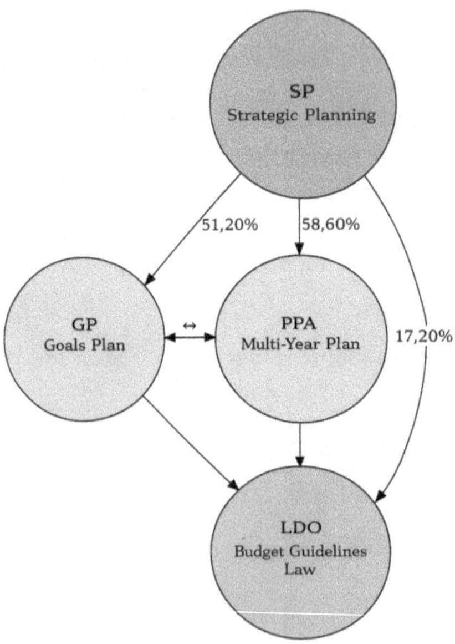

Fig. 9. Execution alignment between SP and GP

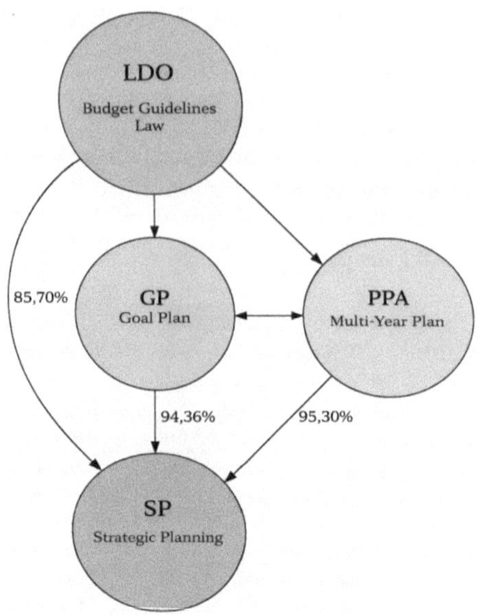

Fig. 10. Planning alignment across instruments

Minor deviations—especially in LDO—are expected due to non-strategic expenditures. Overall, the framework demonstrates its ability to identify alignment patterns not captured by conventional analyses. The high manual effort required also highlights the need for future automation using Natural Language Processing. From a practical perspective, these results provide decision-makers with a structured approach to diagnosing misalignment between planning and budgeting, supporting more coherent and data-driven urban governance.

6 Answering the Research Questions

This section answers the research questions that guided the study based on the empirical findings obtained from the São Paulo case study.

RQ1: How to identify and structure semantic equivalence relationships between components of different planning instruments?

Semantic equivalence relationships were identified through Comparative Content Analysis (CCA), supported by the Document Relationship Matrix. By considering the meaning and function of components rather than their labels, the framework establishes structured and replicable correspondences among planning instruments.

RQ2: How to quantitatively measure the degree of alignment between planning instruments based on content comparison?

Alignment was measured by comparing the contents of semantically equivalent components. The proposed metric, based on the ratio between identified correspondences and the total number of elements in the source component, enabled an objective quantification of alignment. Results showed stronger correspondence for objectives and goals than for indicators and actions, revealing limitations in the operational translation of strategies.

RQ3: How to interpret the results to assess execution and forecast relationships across planning instruments?

Results were interpreted using the constructs of *execution* and *forecast*, derived from average correspondence percentages between instruments. While medium- and short-term instruments exhibited high forecast levels, execution was more heterogeneous, especially in budgetary instruments. This asymmetry suggests that formal alignment does not necessarily ensure coherence in implementation.

Overall, the findings demonstrate that UPICEF-A is capable of identifying semantic relationships, quantifying alignment, and revealing patterns of coherence and misalignment across planning levels, supporting the evaluation and improvement of urban governance.

7 Conclusion

This study demonstrated that formal consistency among public planning instruments does not necessarily translate into coherent relationships across planning levels. The application of the UPICEF-A framework revealed significant gaps, asymmetries, and inconsistencies in the articulation between strategic, programmatic, and budgetary instruments, highlighting limitations in the translation of strategic intentions into operational actions.

The main contribution of this work is the introduction of UPICEF-A, a framework that transforms a traditionally qualitative problem into a structured, measurable, and verifiable analysis. By combining structural completeness with semantic alignment, the framework provides a robust and replicable approach for assessing relationships among planning instruments. The semantic equivalence criteria were supported by both theoretical foundations and expert validation, strengthening the practical applicability of the proposed model.

The framework supports researchers, public managers, and urban planners by enabling evidence-based diagnostics and promoting greater coherence between planning and implementation. As a limitation, the framework was evaluated in a specific context, which may influence the observed patterns. Future work should investigate its application in different administrative settings and explore automated techniques for semantic analysis, improving scalability and enabling continuous assessments. Overall, UPICEF-A provides a systematic and replicable approach for evaluating planning coherence, offering a practical contribution to digital government, urban governance, and evidence-based public planning.

Author contributions. Author contributions: Matheus Cruz (conceptualization, formal analysis, investigation, methodology, validation, writing – original draft, review & editing); Flavia Bernardini (conceptualization, formal analysis, methodology, supervision, validation, review & editing); Vanessa Nunes and Carlos A. Malcher Bastos (conceptualization, formal analysis, methodology, validation).

Funding Information. The authors thank the National Council for Scientific and Technological Development (CNPq - Brazil) for financial support through process n. 441713/2023-8 and 303916/2025-7.

Data Availibility Statement. All data, materials, and resources supporting this study are publicly available in the project repository: https://github.com/MatheusMoreiraCruz/UPICEF-Framework. The repository includes the framework documentation, and related artifacts necessary to reproduce and understand the results.

Disclosure of Interests. There is no conflict of interest.

References

1. Alesaily, Z.: The role of urban planning in designing future cities. Sustain. Futures (2025)
2. Bahmanfar, H.: City strategies, public services and strategic digital city. EPSIR (2026)
3. Bryson, J.M.: Strategic Planning for Public and Nonprofit Organizations, 5th edn. Jossey-Bass, San Francisco (2018)
4. Hevner, A., March, S., Park, J., Ram, S.: Design science in information systems research. MIS Q. **28**(1), 75–105 (2004)
5. Inter-American: Development Bank: Planning and managing for development results (2020)

6. OECD: OECD principles on urban policy (2019)
7. Ogunkan, D.: Big data applications in sustainable urban infrastructure. Sustain. Cities (2025)
8. Peffers, K., Tuunanen, T., Rothenberger, M., Chatterjee, S.: A design science research methodology for information systems research. J. Manag. Inf. Syst. **24**(3), 45–77 (2007)
9. Saenz de Tejada, C.: Urban planning, design and management approaches. Urban Plan. J. (2024)
10. UN-Habitat: World cities report 2020: The value of sustainable urbanization (2020)
11. United Nations Development Programme: Strategic planning guidance note (2016)
12. Varzeshi, S.: Integrating smart city technologies and urban resilience. Smart Cities (2025)
13. World Bank: Results-based monitoring and evaluation for urban development (2017)
14. Zhang, Y.: Metacity: data-driven sustainable development of complex urban systems. Nat. Commun. (2025)

Digital Sovereignty in AI-Enabled Public Sector Transformation

Gideon Mekonnen Jonathan(✉), Shengnan Han, and Erik Perjons

Department of Computer and Systems Sciences (DSV), Stockholm University, Borgarfjordsgatan 12, 164 55 Kista, Sweden
{gideon,shengnan,perjons}@dsv.su.se

Abstract. Digital sovereignty—the capacity of states to exercise control over digital infrastructures, data, platforms, and technological development—has become a critical concern in the era of Artificial Intelligence (AI)-driven digital transformation. As governments increasingly rely on global cloud providers, proprietary AI systems, and transnational data ecosystems, questions of autonomy, accountability, and long-term strategic capability intensify. Existing literature, however, offers limited empirical insight into how digital sovereignty is understood and operationalised in the Global South, particularly within AI-enabled public sector contexts. This study addresses this gap by developing and empirically grounding a multi-dimensional framework of digital sovereignty in Kenya's public sector. Employing a qualitative case study approach, the research draws on fourteen semi-structured interviews with government, policy, and industry stakeholders. The findings reveal that digital sovereignty is negotiated across operational, strategic, and epistemic dimensions, shaped by infrastructural dependency, fragmented governance, and trade-offs between efficiency and control, with implications for trust and service delivery.

Keywords: Digital Sovereignty · Public Sector · Digital Transformation · Artificial Intelligence (AI) · Responsible AI · AI Platforms

1 Introduction

1.1 Background

The phenomenon of digital sovereignty—defined as "*the capacity of a state or society to exercise autonomous control over its digital infrastructures, data, technologies, and governance systems, while managing dependencies within global digital ecosystems and safeguarding economic, political, and societal interests*" [1–3]—has attracted increasing attention among researchers and practitioners as the rapid diffusion of emerging technologies such as Artificial Intelligence (AI), Big Data, and cloud computing reshapes digital infrastructures and governance models [4–6]. This growing prominence reflects concerns over technological

D. Duenas-Cid et al. (Eds.): ePart 2026, LNCS 16923, pp. 278–294, 2027.
https://doi.org/10.1007/978-3-032-36676-4_18

dependence, control over data, and the concentration of power within global platform ecosystems [7,8]. As a result, digital sovereignty has become a critical lens for understanding how states and organisations navigate autonomy, governance, and strategic capability in an increasingly data-driven world [1,2].

Although the concept lacks a universally agreed definition, it is generally understood as an extension of traditional state sovereignty into the digital domain, encompassing issues of autonomy, governance, and control over digital ecosystems [1,4,9]. Early debates framed digital sovereignty primarily in terms of national security and protection against external surveillance. More recent scholarship, however, highlights its broader significance as a geostrategic and socio-economic construct, involving power relations, technological dependence, and the governance of data and digital platforms [4,10]. There is a growing consensus that digital sovereignty encompasses not only state control but also the capacity of societies to shape technological development, promote domestic innovation, and safeguard citizens' rights within digital environments [7]. This broader perspective positions digital sovereignty at the intersection of economic independence, technological capability, and democratic governance. Related notions, such as strategic autonomy, further emphasise the importance of enabling states to make independent decisions regarding their digital futures [2]. At the same time, some scholars question whether sovereignty, traditionally grounded in territorial authority, can be meaningfully applied to cyberspace, given its inherently borderless and networked character [11]. These tensions are particularly evident in the Global South, where digital sovereignty is shaped by historical and structural inequalities. The concept of data colonialism suggests that global digital platforms extract value from local data while limiting local control and benefit [7]. Similarly, concerns about digital neo-colonialism highlight how reliance on foreign infrastructure and platforms may reproduce asymmetrical power relations [12]. Consequently, in these contexts, digital sovereignty is not solely about control, but also about negotiating participation in global digital ecosystems while safeguarding national interests.

Within Africa, and particularly in Kenya, digital transformation has accelerated considerably over the past decade. According to previous studies (e.g., [9]) Kenya is widely recognised as a regional leader in digital innovation, with initiatives such as mobile money, e-government platforms, and emerging AI applications contributing to socio-economic development. Recent policy developments, including the Kenya National AI Strategy (2025–2030) and the Data Protection Act (2019), demonstrate a growing commitment to establishing governance frameworks for digital technologies. These developments reflect an increasing recognition of the need to balance innovation with control over data, infrastructure, and digital services. However, Kenya's digital ecosystem remains deeply embedded within global technological networks. Public sector organisations frequently depend on foreign cloud providers, proprietary AI platforms, and external data infrastructures, raising critical questions regarding technological dependence, data governance, and algorithmic transparency. In this context, digital sovereignty emerges as a dynamic and contested concept, shaped by the interplay between global integration and national autonomy.

These dynamics mirror broader patterns observed in the Global South, where digital transformation is often accompanied by structural dependencies on external technologies and platforms [6,7]. At the same time, efforts to strengthen local capabilities and regulatory frameworks reflect an emerging push towards greater strategic autonomy and governance capacity [2,8]. Consequently, Kenya represents a critical case for examining how digital sovereignty is negotiated in practice within rapidly evolving and globally interconnected digital ecosystems.

1.2 Research Problem

Despite the growing importance of digital sovereignty, existing research has predominantly focused on major geopolitical actors such as the United States, China, and the European Union [9,11,13,14]. These studies typically examine how these powers assert influence over digital infrastructures, regulate platforms, and govern data flows. While such contributions are valuable, this focus tends to marginalise the experiences of countries in the Global South, where the challenges and priorities of digital sovereignty differ considerably [6,8]. Within the relatively limited body of research on developing countries, attention has often centred on large-scale digital infrastructure initiatives, such as India's digital public infrastructure, including Aadhaar and UPI. Although these cases demonstrate significant potential for technological innovation, they do not fully capture the complex socio-technical and political dynamics of digital sovereignty, particularly in contexts characterised by resource constraints and external dependencies. Moreover, much of the existing literature adopts a macro-level perspective, focusing on policy frameworks and national strategies, while overlooking how digital sovereignty is perceived, experienced, and operationalised by key stakeholders within organisations [15].

In the African context, research remains relatively limited and fragmented. Existing studies have highlighted issues such as data governance, digital dependency, and regulatory development, yet there is a lack of integrated frameworks that connect these dimensions to digital transformation outcomes [6,8]. In Kenya specifically, although policy initiatives and digital innovation have been widely documented, there is limited empirical research examining how public sector actors navigate the tensions between efficiency, innovation, and sovereignty in their adoption of AI platforms. This gap is particularly significant given the increasing role of AI in public administration. AI systems, especially those delivered through global platforms, introduce new challenges related to algorithmic opacity, data control, and institutional capacity [16]. Without a clear understanding of how digital sovereignty operates in practice, public sector organisations risk adopting technologies that may undermine their autonomy, compromise citizen trust, or constrain long-term strategic capabilities [6,7].

1.3 Research Aim and Question

The aim of this study is to develop and empirically ground a conceptual framework of digital sovereignty in the context of AI-enabled public sector transformation in Kenya. In particular, the study seeks to identify the key dimensions of digital sovereignty relevant to AI adoption in the public sector, examine how these dimensions influence organisational decision-making, governance practices, and transformation outcomes, and explore the strategies and trade-offs employed by stakeholders as they navigate the tensions between global technological integration and national autonomy.

By focusing on Kenyan public sector as a single case, the study contributes to a more nuanced understanding of digital sovereignty in the Global South, where it is shaped not only by policy ambitions but also by structural constraints and evolving institutional capacities. To guide this study, the following research question is formulated: *What are the key dimensions, challenges, and strategic approaches shaping digital sovereignty in the adoption of AI platforms within Kenya's public sector?*

2 Related Work

2.1 Digital Sovereignty as a Multi-level Construct

Digital sovereignty has emerged as a central yet contested concept in contemporary digital governance, reflecting shifting power dynamics in increasingly data-driven and platform-based societies. While broadly understood as the capacity of states or political entities to exercise authority over digital infrastructures, data, and technologies, the concept lacks a universally accepted definition and continues to evolve across academic and policy debates [1,8].

Early conceptualisations framed digital sovereignty primarily as an extension of traditional state sovereignty into cyberspace, emphasising control, autonomy, and territorial authority. However, such perspectives have been challenged by the inherently borderless and networked nature of digital systems, which complicates the application of territorially bounded governance models [11]. More recent scholarship reconceptualises digital sovereignty as a relational and contested construct, shaped by interactions among states, corporations, and other actors competing for control over digital ecosystems [3,6,8]. From this perspective, digital sovereignty is not a fixed condition but a dynamic and context-dependent process, often mobilised as a political and strategic narrative. It reflects broader geopolitical tensions, particularly as states seek to reduce dependency on external technological infrastructures while maintaining participation in global digital networks [15]. At the same time, sovereignty claims are increasingly linked to economic competitiveness, innovation capacity, and the ability to define technological standards [6].

To capture this complexity, digital sovereignty can be conceptualised as a multi-level construct encompassing operational, strategic, and epistemic dimensions [1,3]. ***Operational sovereignty*** refers to control over the day-to-day functioning of digital infrastructures, including cloud systems, data storage, and communication networks. This dimension highlights the importance of infrastructural capacity and resilience, particularly in contexts where reliance on external providers may constrain autonomy. ***Strategic sovereignty*** relates to the ability to shape long-term technological trajectories through investment, innovation, and policy frameworks. Increasingly, digital sovereignty is tied to the development of "critical" or "key enabling" technologies such as artificial intelligence, semiconductors, and quantum computing, which are seen as foundational for economic and geopolitical competitiveness [6,17]. ***Epistemic sovereignty*** captures control over knowledge production, data governance, and algorithmic systems. As digital systems increasingly mediate decision-making, the ability to understand, design, and govern these systems becomes a critical source of power. This dimension is closely linked to concerns about data extraction, algorithmic opacity, and the concentration of knowledge within global technology firms [7]. However, digital sovereignty does not imply complete autonomy or isolation. Instead, it reflects a balancing act between independence and interdependence, in which states seek to manage dependencies while remaining integrated into global technological ecosystems. This tension is particularly evident in discussions of strategic autonomy, which emphasise the capacity to act independently while maintaining openness and cooperation [2].

2.2 AI Platforms as Infrastructural Dependencies

The rapid diffusion of artificial intelligence has transformed digital infrastructures into platform-based ecosystems, where core functionalities are increasingly delivered through cloud services, APIs, and proprietary models. These AI platforms operate as critical infrastructures, shaping how organisations access, process, and utilise data [6]. From a digital sovereignty perspective, this shift introduces new forms of dependency. As organisations adopt AI systems provided by global technology firms, control over key components of the digital stack—including data storage, processing capabilities, and algorithmic models—often resides outside national or organisational boundaries. This creates what has been described as structural dependency, where technological reliance constrains autonomy and decision-making capacity [6].

A key feature of AI platforms is the emergence of lock-in effects, whereby organisations become dependent on specific providers due to proprietary standards, data integration mechanisms, and ecosystem compatibility. Such lock-in limits the ability to switch providers or develop alternative solutions, reinforcing asymmetrical power relations between platform providers and users [6]. In addition, AI platforms are characterised by algorithmic opacity, often described as "black-box" systems. Limited transparency restricts the ability of organisations, particularly in the public sector, to fully understand or audit algorithmic decision-making processes. This raises significant governance challenges, as accountability mechanisms become difficult to enforce in the absence of access to underlying models and data [16]. These dynamics reflect a broader reconfiguration of power in digital ecosystems. Rather than residing solely with states, governance authority is increasingly distributed across a range of actors, including multinational technology firms that design and operate digital infrastructures. This aligns with arguments that digital sovereignty must be understood in terms of multi-actor governance, rather than state-centric control [6,8].

At the same time, efforts to reduce dependency through policies aimed at technological autonomy may introduce new challenges. The literature on technological protectionism suggests that restricting international collaboration or supply chains can hinder innovation, particularly in fields characterised by global knowledge networks and distributed production systems [18,19]. This highlights a central paradox: achieving sovereignty may require openness, while attempts to enforce autonomy may undermine innovation capacity. Furthermore, emerging technologies such as quantum computing illustrate the uncertainties associated with sovereignty-driven innovation strategies. While often framed as critical for future competitiveness, their practical applications remain uncertain, and their development is deeply embedded in global research and industrial ecosystems [16,20]. This raises important questions about the feasibility of achieving sovereignty through technological leadership alone.

2.3 Responsible AI Meets Sovereignty

The increasing prominence of Responsible AI reflects growing concern with the ethical, legal, and societal implications of AI systems. Frameworks emphasising principles such as transparency, fairness, and accountability have become central to both academic and policy discussions [21–23]. However, these frameworks often overlook a critical prerequisite: the existence of sufficient digital sovereignty to operationalise these principles. Without adequate control over data, infrastructure, and algorithmic systems, the practical implementation of Responsible AI remains constrained, particularly in contexts characterised by platform dependency and asymmetrical power relations [6,7,24].

Transparency requires access to algorithmic systems and decision-making processes. In platform-based environments characterised by proprietary models, such access is often limited, constraining organisations' ability to audit or explain AI systems [16]. Without epistemic control, transparency remains aspirational rather than actionable. Fairness depends on control over data. Bias in AI systems is intrinsically linked to the data used for training and validation. Where data governance is externalised or controlled by third-party providers, organisations may lack the ability to ensure that systems operate in a fair and contextually appropriate manner [7]. Accountability requires institutional authority and enforcement capacity. While regulatory frameworks such as the GDPR and emerging AI regulations demonstrate strong normative ambitions, their effectiveness depends on the ability to enforce compliance within complex and transnational digital ecosystems [6]. This is particularly challenging where key technological components are controlled by actors beyond national jurisdiction [8].

These interdependencies suggest that Responsible AI and digital sovereignty are mutually constitutive. Responsible AI frameworks require sovereignty-related capabilities—such as control over data, infrastructure, and algorithms—to be effectively implemented [6,7]. Conversely, sovereignty without ethical governance risks legitimising exclusionary or protectionist approaches that may undermine trust and societal welfare [8]. Moreover, the relationship between sovereignty and innovation introduces additional complexity. While digital sovereignty is often framed as a means to foster innovation and competitiveness, there is limited empirical evidence that sovereignty-driven policies necessarily lead to improved societal outcomes. Indeed, tensions between openness, innovation, and control remain a central challenge in digital governance [2,6].

2.4 Integrating Digital Sovereignty, AI Platforms, and Governance

The extant literature on digital sovereignty provides us with three key insights. First, digital sovereignty is a multi-dimensional and relational construct, shaped by interactions between technological infrastructures, governance systems, and geopolitical dynamics [1,6,8]. Second, AI platforms introduce new forms of dependency, reconfiguring power and constraining organisational autonomy [6]. Third, Responsible AI frameworks emphasise normative governance principles, but often lack the operational foundations required for implementation [6,7]. However, scholars argue that important gaps remain. To start with, conceptually, existing research often treats sovereignty as either a political discourse or a technical capability, without fully integrating these perspectives [3]. Besides, empirically, there is limited understanding of how digital sovereignty is operationalised in practice, particularly within public sector organisations. Finally, the relationship between sovereignty, innovation, and governance remains underexplored, especially in contexts outside major geopolitical actors [2]. This study addresses these gaps by developing and empirically grounding a multi-dimensional framework of digital sovereignty in AI-enabled public sector transformation, focusing on how sovereignty is enacted, negotiated, and contested in practice within the Kenyan context. Figure 1 depicts the conceptualisation of digital sovereignty and how it is operationalised in practice.

Based on the synthesis of prior studies presented above, we conceptualise digital sovereignty as a multi-level construct comprising operational, strategic, and epistemic dimensions, embedded within AI-driven digital infrastructures. AI platforms introduce structural dependencies—such as lock-in effects and governance opacity—that shape the extent to which sovereignty can be exercised. These dependencies influence the capacity of public sector organisations to develop responsible AI capabilities, including transparency, fairness, and accountability. In turn, the level of digital sovereignty achieved affects key transformation outcomes, such as AI adoption success, citizen trust, and policy compliance. The model therefore positions digital sovereignty as a mediating construct between infrastructural dependencies and public sector outcomes.

3 Research Methodology

This study adopts a qualitative research approach to explore how digital sovereignty is conceptualised and operationalised in the context of AI adop-

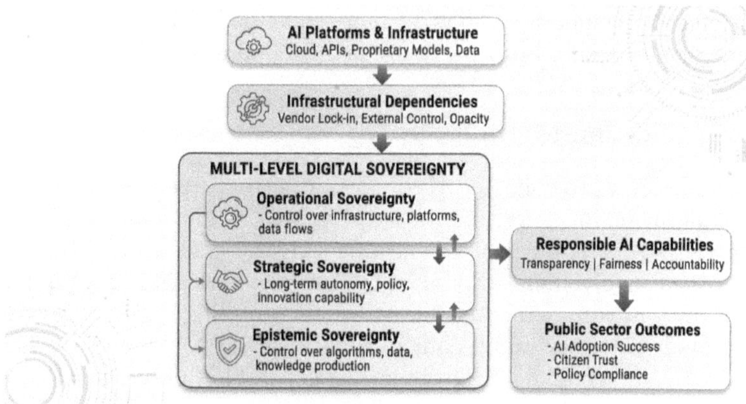

Fig. 1. Multi-Level Digital Sovereignty in AI-Enabled Public Sector Transformation.

tion within Kenya's public sector. A qualitative design is particularly well-suited to investigating complex socio-technical phenomena such as digital sovereignty, where meanings, perceptions, and organisational practices are central [25]. It enables an in-depth understanding of how different stakeholders interpret and navigate issues related to control, autonomy, and governance within digital ecosystems.

3.1 Research Strategy

A single-country qualitative case study strategy was employed, focusing on Kenya as a critical and information-rich case. Case study research is appropriate when the objective is to examine a contemporary phenomenon within its real-life context, particularly when the boundaries between the phenomenon and its context are not clearly defined [26]. In this study, digital sovereignty is closely embedded within Kenya's evolving digital transformation landscape, making a case study approach especially appropriate.

Kenya was selected for several reasons. It is widely recognised as a leading digital innovation hub in Africa, characterised by advanced digital public services and an increasing adoption of AI technologies. In addition, the country has recently introduced key policy frameworks, including the National AI Strategy (2025–2030) and the Data Protection Act (2019), which reflect a growing emphasis on digital governance and sovereignty. At the same time, Kenya demonstrates a notable tension between technological ambition and structural dependence on global platforms, making it a particularly suitable case for examining how digital sovereignty is negotiated in practice.

The case study design enables the integration of perspectives from multiple stakeholder groups involved in digital transformation, including government actors, public-sector implementers, policy experts, and private-sector actors.

This multi-actor perspective offers a more holistic understanding of digital sovereignty, encompassing both its strategic and operational dimensions.

3.2 Data Collection Method

Data for this study were collected through semi-structured interviews, enabling in-depth exploration of participants' experiences while ensuring consistency across interviews [27]. A purposive sampling strategy was used to select respondents with direct involvement in AI adoption, digital governance, and public sector transformation, complemented by snowball sampling to identify additional key informants in advisory and technical roles. In total, fourteen interviews were conducted with stakeholders from government, public sector implementation, policy and research, and private sector organisations, capturing both strategic and operational perspectives on digital sovereignty. The respondents represented core institutions shaping Kenya's digital ecosystem, including national regulatory bodies, implementing agencies, county governments, research institutions, and technology firms. Their roles spanned policy formulation, system implementation, advisory functions, and platform provision, offering a comprehensive view of both governance and technical dimensions. This diversity enabled the study to capture cross-sectoral dynamics and contrasting perspectives on digital sovereignty.

The interviews, lasting 45–75 min, were conducted virtually via Zoom, recorded with consent, and transcribed for analysis. The interview guide was aligned with the study's conceptual framework and focused on AI platform use, infrastructural dependencies, data governance, algorithmic transparency, and institutional arrangements, as well as the trade-offs between efficiency, innovation, and sovereignty. This approach ensured the collection of rich, relevant data to inform the development and refinement of the digital sovereignty framework.

3.3 Data Analysis Method

The data were analysed using thematic analysis, a widely used qualitative method for identifying, analysing, and interpreting patterns within qualitative data [28]. This method was selected for its flexibility and suitability for generating theory-informed insights while remaining grounded in empirical material [29]. The analysis followed the six-phase approach outlined by Braun and Clarke [28], namely familiarisation with the data, generating initial codes, searching for themes, reviewing themes, defining and naming themes, and producing the report.

To enhance analytical rigour, the study employed an iterative coding process in which emerging insights were continuously compared across interviews. This facilitated the identification of cross-cutting patterns and variations among stakeholder groups. In addition, the analysis involved mapping the empirically derived themes onto the proposed conceptual framework, enabling the refinement of theoretical constructs in light of the observed data.

4 Results

This section presents the empirical findings structured in line with the conceptual model of multi-level digital sovereignty—operational, epistemic, and strategic—and its relationship with AI platform dependencies, responsible AI capabilities, and public sector outcomes. Our analysis indicates that digital sovereignty in Kenya is uneven, negotiated, and shaped by infrastructural dependency, governance fragmentation, and strategic trade-offs.

4.1 AI Platforms and Infrastructural Dependencies

Across all stakeholder groups, AI adoption in Kenya's public sector is embedded within platform-based ecosystems dominated by global providers. The reliance on external infrastructure is both enabling and constraining. The Technical Lead for a platform explained that the system "relies almost entirely on a global cloud provider" to meet scalability and reliability requirements. Similarly, one Director of Digital Transformation noted that government systems depend on "cloud services from a global hyperscaler", particularly for AI experimentation and analytics. According to the interviewees, this dependency manifests in three key ways. First, infrastructural reliance is widespread, as confirmed by one Head of Enterprise Architecture, who observed that many ministries and agencies in the country still prefer global providers due to their advanced capabilities. Second, lock-in effects emerge, with migration to local infrastructure described as "massive and risky". Third, governance opacity is prevalent, particularly in proprietary systems; one Chief Officer in a County reported having "very little control" over a traffic management system supplied by a foreign vendor. At the same time, there are emerging efforts to reduce dependency. The Head of Enterprise Architecture highlighted initiatives such as the G-Cloud and an AI sandbox based on open-source models, while another respondent emphasised growing concern over "the concentration of critical digital infrastructure in a few foreign-owned entities".

4.2 Operational Sovereignty: Control over Infrastructure and Data

Operational sovereignty—control over infrastructure and data flows—emerges as partial and uneven across institutions.

Our interviewees argue that, at the national level, there is a clear policy direction towards localisation. For instance, one Director of Digital Transformation at a Ministry stated that "critical government data should be stored within Kenya", reflecting a strategic commitment to local control. Similarly, the Head of Enterprise Architecture confirmed that the G-Cloud stores data within Kenya and that policies are being developed to mandate local storage for sensitive data. However, in practice, hybrid infrastructures seem to dominate. Government data is distributed across local servers, global cloud platforms, and legacy systems. A Senior Data Protection Officer, however, highlighted increasing concerns around cross-border data transfers, particularly where "data subjects' rights may not be enforceable". On the other hand, at the county level, operational sovereignty varies significantly. One Chief Officer in a County described a hybrid model with limited visibility over externally hosted systems, while the Director of ICT emphasised a deliberate "local-first approach" to retain control over data and avoid vendor lock-in. This contrast illustrates how resource constraints and strategic priorities shape outcomes of sovereignty.

Overall, our analysis suggests that operational sovereignty is aspirational but constrained, with growing policy attention but incomplete implementation.

4.3 Epistemic Sovereignty: Algorithmic Control and Transparency

Epistemic sovereignty—control over algorithms and knowledge systems—emerges as one of the most critical gaps in the Kenyan public sector.

Multiple respondents highlighted limited visibility into AI systems. For instance, the Director of Digital Transformation at a Ministry described this as "a major gap", noting that for many systems, the government relies on "vendor assurances" rather than direct understanding. Similarly, the Technical Lead of a platform explained that while APIs are used, "we don't know what's happening inside", particularly for fraud detection models. This opacity creates governance challenges. The Senior Data Protection Officer further noted that organisations often "cannot explain how their AI systems work", which complicates the enforcement of data protection rights. The Programme Lead at a digital rights NGO further highlighted the consequences, citing cases in which citizens are affected by automated decisions "with no explanation and no recourse". In contrast, locally developed systems provide greater epistemic control. The Director of ICT in a County emphasised that open-source models allow them to "modify" and "audit" algorithms, while the Co-Founder of a Kenyan AI startup noted that transparency and explainability are key competitive advantages in sovereignty-sensitive contexts.

These findings reveal a clear divide between high-performance proprietary systems with limited transparency and locally controlled systems with greater auditability but lower capability.

4.4 Strategic Sovereignty: Policy Direction and Institutional Capacity

Strategic sovereignty—long-term autonomy through policy and institutional capacity—is emerging but fragmented.

Our interviewees agree that, at the policy level, there is strong recognition of sovereignty challenges. For instance, the AI Policy Advisor to a Ministry highlighted that reliance on global platforms raises risks related to "data sovereignty, vendor lock-in, and lack of transparency", which are central concerns in the national AI strategy. However, institutional arrangements remain dispersed across multiple actors. The Director of Digital Transformation at the same Ministry described AI governance as "fragmented", with no single coordinating body, while the Head of Enterprise Architecture noted that many agencies procure AI systems independently, limiting oversight at a national level.

Respondents from regulatory actors also emphasised gaps. For instance, one Director described the governance landscape as "patchy", and the Senior Data Protection Officer noted that existing frameworks do not fully address issues such as bias, fairness, and transparency. However, our analysis also suggests that efforts to address these gaps are underway. Proposals for a national AI council and new governance frameworks indicate movement towards greater coordination, but current arrangements remain insufficient for comprehensive oversight.

4.5 Trade-Offs: Efficiency Versus Control

A central theme across all interviews is the tension between efficiency and control, which directly shapes outcomes of digital sovereignty.

Our respondents agree that global platforms are consistently associated with efficiency, scalability, and advanced capabilities. The Director of Digital Transformation at a Ministry described this as "the central tension", noting that while global platforms provide efficiency, "control is essential for sovereignty and trust". One interesting finding of this study was that different actors navigate this trade-off differently. The Chief Officer in a County argues that "efficiency often wins" due to service delivery pressures, albeit at the cost of "loss of control and long-term lock-in". In contrast, the Director of ICT in another County explicitly prioritised control, arguing that sovereignty is a "long-term investment" even if it reduces efficiency. A common approach, however, is to adopt hybrid models. The AI Policy Advisor described a "risk-based framework" where control is prioritised for sensitive applications and efficiency for others. Similarly, the Managing Director of a data science firm recommended starting with control and selectively integrating global capabilities.

These findings suggest that digital sovereignty is not binary but context-dependent and negotiated through trade-offs.

4.6 Outcomes: Trust, Service Delivery, and Innovation

The level of digital sovereignty achieved has direct implications for key public sector outcomes.

Our respondents recognise that citizen trust is closely linked to perceptions of control. As the Director of Digital Transformation at a Ministry puts it, "if citizens do not trust that their data is safe, they will not use digital services". Similarly, the Research Fellow at a National University highlighted that sovereignty over data and algorithms is essential for maintaining trust.

Respondents have also recognised the service-delivery benefits of AI-enabled efficiency, but it is vulnerable to sovereignty risks. For instance, a national Coordinator of an online public service noted that AI improves service delivery but must not compromise accountability, while the Technical Lead of a platform highlighted concerns about introducing risks alongside efficiency gains. On a related note, our analysis suggests that innovation is shaped by the balance between global capability and local capacity. The Head of Enterprise Architecture stressed that "innovation cannot be outsourced", while the Co-Founder of an AI startup demonstrated how local innovation can address sovereignty concerns by combining performance with transparency.

Overall, the findings suggest that digital sovereignty acts as a mediating factor, shaping how infrastructural dependencies translate into outcomes in trust, service delivery, and innovation.

5 Discussion

This study set out to examine how digital sovereignty shapes public sector AI strategies, addressing the research question: *How does digital sovereignty shape public sector AI strategies in the context of Kenya?* By integrating the empirical findings with existing literature, the study advances a multi-level understanding of digital sovereignty and demonstrates how it operates as a mediating construct between AI platform dependencies and public sector outcomes.

First, the findings reinforce the conceptualisation of digital sovereignty as a multi-level construct comprising operational, strategic, and epistemic dimensions. Consistent with prior studies, sovereignty is not a fixed attribute but a relational and context-dependent process shaped by interactions among state actors, private platforms, and institutional arrangements [1,8]. Empirically, the Kenyan case shows that these dimensions are unevenly developed. Operational sovereignty is partially realised through initiatives such as local data infrastructure, while epistemic sovereignty remains limited due to reliance on proprietary AI systems. Strategic sovereignty, although recognised at the policy level, is constrained by fragmented governance and institutional capacity gaps. This uneven development supports the argument that digital sovereignty is best understood as a dynamic equilibrium between capability, dependency, and governance, rather than as absolute control.

Second, the study highlights the central role of AI platforms as infrastructural dependencies, extending prior work on platform power and digital dependency [6]. The findings confirm that AI platforms function as critical infrastructures, introducing lock-in effects, governance opacity, and reliance on external providers. This aligns with concerns regarding the limited transparency of AI systems and the challenges this poses for accountability [16]. However, the empirical evidence also nuances this perspective by showing that such dependencies are not passively accepted but are actively negotiated through hybrid strategies, where organisations selectively combine global platforms with local solutions. This supports the view that sovereignty is exercised not through isolation but through managed interdependence [2].

Third, the findings extend the literature on Responsible AI by demonstrating that its implementation depends on digital sovereignty. While prior research emphasises principles such as transparency, fairness, and accountability [21], this study shows that these principles cannot be operationalised without sufficient control over data, algorithms, and infrastructure. The lack of epistemic sovereignty, in particular, limits public institutions' ability to audit or explain AI systems, reinforcing arguments about the implications of data dependency and extraction [7]. At the same time, the Kenyan case illustrates that efforts to enhance sovereignty—such as prioritising local solutions—may introduce trade-offs with efficiency and performance, highlighting the tension between ethical governance and technological capability.

Fourth, the study furthers our understanding of the emerging debate on sovereignty–innovation dynamics. While policy discourse often assumes that increased sovereignty leads to greater innovation and competitiveness, the findings suggest a more complex relationship. On the one hand, global platforms enable rapid innovation by providing access to advanced tools and infrastructure. On the other hand, local initiatives foster long-term capability building and contextual relevance. This duality reflects the paradox identified in the literature, where efforts to reduce dependency may constrain innovation if they limit access to global ecosystems [18]. Our case demonstrates that innovation and sovereignty are not inherently aligned but must be balanced through strategic choices.

Finally, the study empirically demonstrates that digital sovereignty acts as a mediating mechanism linking infrastructural dependencies to public sector outcomes. Variations in sovereignty levels influence key outcomes, including citizen trust, service delivery, and innovation capacity. For example, limited control over data and algorithms undermines trust, while dependence on global platforms enhances efficiency but introduces long-term risks. This confirms that digital sovereignty is not merely a normative or policy concept but a practical determinant of digital transformation outcomes.

6 Concluding Remarks

This study makes several contributions to both research and practice by advancing a multi-dimensional understanding of digital sovereignty in the context of AI-enabled public sector transformation. Theoretically, it integrates insights from digital sovereignty, platform governance, and responsible AI literature to conceptualise sovereignty as a multi-level construct encompassing operational, strategic, and epistemic dimensions. By positioning digital sovereignty as a mediating mechanism between AI platform dependencies and public sector outcomes, the study extends existing research that has often treated these domains in isolation. Empirically, it contributes to the limited body of work on digital sovereignty in the Global South by providing context-rich evidence from Kenya, demonstrating how sovereignty is enacted and negotiated in practice. From a practical perspective, the findings offer actionable insights for policymakers and practitioners, highlighting the need for coordinated governance frameworks, investment in local capabilities, and context-sensitive strategies that balance efficiency with control.

The study underscores that achieving digital sovereignty is not about technological isolation but about strategically managing dependencies while strengthening institutional and technical capacities.

Despite these contributions, the study has several limitations. First, it adopts a qualitative case study approach focused on a single country, which limits the generalisability of the findings to other contexts. While Kenya provides a critical and information-rich case, the dynamics of digital sovereignty may differ in countries with varying institutional capacities, regulatory environments, and levels of technological development. Second, the study relies primarily on interview data, which reflects stakeholders' perceptions and experiences and may be subject to bias or incomplete knowledge, particularly in areas involving proprietary technologies or sensitive governance issues. Third, the analysis focuses on public sector actors and selected private and policy stakeholders, potentially overlooking perspectives from other important actors such as citizens or smaller technology providers. Finally, the study captures a snapshot of an evolving landscape in which policies, technologies, and institutional arrangements are rapidly changing.

Future research can build on this study in several ways. Comparative studies across multiple countries in Africa or other regions of the Global South would help to identify common patterns and contextual differences in how digital sovereignty is conceptualised and operationalised. Quantitative or mixed-method approaches could further validate and refine the proposed conceptual framework, examining the relationships between sovereignty dimensions and transformation outcomes at scale. In addition, future research could explore the role of specific technologies—such as large language models, digital public infrastructure, or federated data systems—in shaping sovereignty dynamics. Longitudinal studies would be particularly valuable in understanding how digital sovereignty evolves over time as institutional capacities develop and dependencies shift. Finally, there is a need for deeper investigation into the societal implications of digital sovereignty, including its impact on citizen rights, inclusion, and democratic governance, particularly in contexts where balancing control and openness presents ongoing challenges.

References

1. Pohle, J., Nanni, R., Santaniello, M.: Unthinking digital sovereignty: a critical reflection on origins, objectives, and practices. Policy Internet **16**(4), 666–671 (2025)
2. Moerel, L., Timmers, P.: Reflections on digital sovereignty. EU Cyber Direct, Research in Focus Series (2021)
3. Santaniello, M.: Attributes of digital sovereignty: a conceptual framework. Geopolitics **31**(2), 788–809 (2026)
4. Glasze, G., Cattaruzza, A., Douzet, F., et al.: Contested spatialities of digital sovereignty. Geopolitics **28**(2), 919–958 (2022)
5. Mekonnen, G., Reychav, I.: Digital sovereignty in the global south: autonomy in a connected world. Procedia Comput. Sci. **278**, 598–605 (2026)

6. Broeders, D., Cristiano, F., Kaminska, M.: In search of digital sovereignty and strategic autonomy: normative power Europe to the test of its geopolitical ambitions. JCMS: J. Common Market Stud. **61**(5), 1261–1280 (2023)
7. Couldry, N., Mejias, U.A.: Data colonialism: rethinking big data's relation to the contemporary subject. Television New Media **20**(4), 336–349 (2019)
8. Couture, S., Toupin, S.: What does the notion of "sovereignty" mean when referring to the digital? New Media Soc. **21**(10), 2305–2322 (2019)
9. Jonathan, G.M.: Charting the crossroads of digital sovereignty and digital transformation. In: International Conference on Electronic Government and the Information Systems Perspective. Springer, Cham (2025)
10. Pierucci, F.: Sovereignty in the digital era: rethinking territoriality and governance in cyberspace. Digit. Soc. **4**(27), 1–19 (2025)
11. Mueller, M.L.: Against sovereignty in cyberspace. Int. Stud. Rev. **22**(4), 779–801 (2020)
12. Gravett, W.: Digital neo-colonialism: the Chinese model of internet sovereignty in Africa. Afr. Hum. Rights Law J. **20**(1), 125–146 (2020)
13. Budnitsky, S., Jia, L.: Branding Internet sovereignty: digital media and the Chinese-Russian cyberalliance. Eur. J. Cult. Stud. **21**(5), 594–613 (2018)
14. Thumfart, J.: The norm development of digital sovereignty between China, Russia, the EU and the US: From the late 1990s to the Covid-crisis 2020/21 as catalytic event. In: Enforcing Rights in a Changing World. Computers Privacy Data Protection (CPDP), vol. 14, pp. 1–44 (2022)
15. Barrinha, A., Christou, G.: Speaking sovereignty: the EU in the cyber domain. Eur. Secur. **31**(3), 356–376 (2022)
16. Hoofnagle, C.J., Garfinkel, S.L.: Law and Policy for the Quantum Age. Cambridge University Press, Cambridge (2022)
17. Edler, J., Blind, K., Kroll, H., Schubert, T.: Technology sovereignty as an emerging frame for innovation policy: defining rationales, ends and means. Res. Policy **52**(6), 104765 (2023)
18. Shelley-Egan, C., Vermaas, P.: European technological protectionism and the risk of moral isolationism: the case of quantum technology development. J. Responsible Technol. **18**, 100084 (2024)
19. Lee, K., Malerba, F., Primi, A.: The fourth industrial revolution, changing global value chains and industrial upgrading in emerging economies. J. Econ. Policy Reform **23**(4), 359–370 (2020)
20. Preskill, J.: Quantum computing in the NISQ era and beyond. Quantum **2**, 79 (2018)
21. Roumate, F.: AI and information: the global landscape of AI ethics and regulation. In: AI, Information. and Global Dynamics: Shaping the Future of Geopolitical Landscape, pp. 157–170. Springer, Cham (2026)
22. Ammah, L.N.A., Lütge, C., Kriebitz, A., Ramkissoon, L.: AI4People–an ethical framework for a good AI society: the Ghana (Ga) perspective. J. Inf. Commun. Ethics Soc. **22**(4), 453–465 (2024)
23. Mittelstadt, B.D., Allo, P., Taddeo, M., Wachter, S., Floridi, L.: The ethics of algorithms: mapping the debate. Big Data Soc. **3**(2), 2053951716679679 (2016)
24. Kalluri, P.: Don't ask if AI is good or fair, ask how it shifts power. Nature **583**(7815), 169 (2020)
25. Denscombe, M.: The Good Research Guide: For Small-Scale Social Research Projects, 5th edn. Open University Press, Maidenhead (2014)
26. Yin, R.K.: Case Study Research and Applications: Design and Methods, 6th edn. Sage, Thousand Oaks (2017)

27. Bryman, A.: Social Research Methods, 5th edn. Oxford University Press, Oxford (2018)
28. Braun, V., Clarke, V.: Using thematic analysis in psychology. Qual. Res. Psychol. **3**(2), 77–101 (2006)
29. Eisenhardt, K.M.: Building theories from case study research. Acad. Manag. Rev. **14**(4), 532–550 (1989)

A Systematic Literature Review on Current Research Trends in Legal Information Retrieval and Legal Databases

Diogo Sasdelli(✉) and Gregor Eibl

Department for E-Governance and Administration, University for Continuing Education Krems, Dr.-Karl-Dorrek-Straße 30, 3500 Krems an der Donau, Austria
{diogo.sasdelli,gregor.eibl}@donau-uni.ac.at

Abstract. Legal information retrieval (LIR) has become a core capability in contemporary legal practice and a central topic in AI and Law. This paper synthesises LIR and legal database research published between 2020 and 2025. Within a PRISMA-guided systematic literature review, we analyse 162 peer-reviewed papers and code them along six dimensions: (i) geographic distribution of research activity, (ii) technical versus non-technical orientation, (iii) legal database focus, (iv) focus on AI methods, (v) employed technologies and methods, and (vi) application areas and objectives. Results indicate that LIR research during the period is concentrated in a limited set of countries and splits into technical and non-technical contributions. Across the corpus, recurring technology families and application areas cluster around retrieval tasks, inference-oriented tasks (e.g., entailment and question answering), and data structuring and enrichment. Linking technologies to application contexts enables us to derive a four-step process model of LIR: (1) structuring, (2) finding, (3) entailment, (4) case analysis. We argue that this process model enables a better conceptualisation and classification of LIR research. Furthermore, given the increasing embedment of advanced LIR in jurists' daily routines, we argue for complementing technology-centric research with empirical studies of contemporary practice, focussing on both system designers/managers and end users in situated legal workflows.

Keywords: Legal databases · legal information systems · legal information retrieval (LIR) · legal artificial intelligence (AI) · Process Model of Legal Information Retrieval (PMLIR)

1 Introduction

Employing computing technology to manage legal information by means of *legal databases* or *legal information systems* is one of the oldest and most well-established research areas at the intersection between law and technology. While first legal databases such as Canada's *QUIC-LAW* (1972) [35], USA's Lexis

Published by Springer Nature Switzerland AG 2027
D. Duenas-Cid et al. (Eds.): ePart 2026, LNCS 16923, pp. 295–306, 2027.
https://doi.org/10.1007/978-3-032-36676-4_19

(1973) [24], Europe's CELEX (1981) [31] and Germany's JURIS (1984) [39] were first made publicly available over the 1970s and 1980s, the research efforts that made their development possible are considerably older [13,37]. As early as in the 1940s, Louis Kelso already urged for a "technological revolution" to cope with the post-war explosion of statutes, regulations, and judicial opinions [21], what Greek-German jurist Spiros Simitis would later call the "information crisis" of law [38]. Likewise, topics related to legal information systems and legal information retrieval (LIR) were intensively discussed in early legal informatics research (cf., e.g., [2,20]). Since then, legal databases have become an essential part of a legal practitioner's everyday life, and further advancements in the wide field of information retrieval, including the usage of *generative artificial intelligence*, have been enabling the development of ever more advanced legal databases.

In this context, the short paper at hand reports on a PRISMA-based [33] systematic literature review performed in order to identify and analyse current research trends in LIR and legal databases. The paper's main contributions are:

- An overview of the current research landscape on LIR and legal databases, including aspects such as *geographic distribution*, *technologies* and *research topics*, with a particular focus on the usage of AI technologies.
- The proposal of a novel, so-called *Process Model of Legal Information Retrieval*, which can be used to better conceptualise further research on LIR and legal databases.

This topic is relevant for digital government research, because these technologies play a fundamental role in shaping the AI-driven transformation of the legal domain, especially with respect to citizens' access to legal information and to the overall digital transformation of the judicial system. The rest of this paper is structured as follows. Section 2 first briefly discusses related work providing overviews of the research and practice landscape in the field of LIR. Section 3 then outlines the research methodology underlying the literature review. Results are presented in Sect. 4 and discussed in Sect. 5, which also introduces the novel process model. Section 6 concludes and refers to future work.

2 Related Work

Despite the importance and long tradition of research on legal databases, systematic analyses of the respective research landscape such as the one presented here are fairly rare. In particular, there are, to our knowledge, no other contemporary studies performing a comprehensive analysis of the overall research landscape on LIR for specific applications in legal databases. Hence, the study presented in the paper at hand fills an important research gap.

The most similar study to the one presented here was conducted by *D'Alotto et al.* [11], who recently performed a mapping study analysing literature on the usage of machine learning, natural language processing techniques and ontologies to improve LIR. Their study included articles in English and Spanish, published between 2010 and 2024, and included in the IEEE Xplore Digital Library, the ACM Digital Library, and the scientific literature databases *ResearchGate*, *Science Direct*, *ArXiv*, *Springer* and *Sedici*. While their initial scope was relatively wide, their final inclusion criteria were rather strict, only considering papers that fully addressed the aspects they were interested in analysing, i.e., papers that clearly stated (1) the respective *research motivations*, (2) the *AI techniques applied*, (3) the main *problems, barriers and setbacks* encountered, (4) the *datasets and evaluation metrics* employed, and (5) the *key contributions and future directions* identified. This led to only 34 papers being fully analysed. Furthermore, the queries employed were relatively general, resulting in many of the papers ultimately analysed having a focus outside of the area of LIR in a more strict sense: for example, five of the 34 papers they analysed focussed on the prediction of outcomes of legal cases.

Feng et al. [12], in their turn, present a survey of the state of the art on legal case retrieval. However, beyond focussing only on the specific topic of legal case retrieval, their literature survey does not seem to follow any systematic review methodology, focussing instead on selected aspects of legal case retrieval and on papers presented in established venues such as the *COLIEE* competitions.

Finally, Jahnel [17] presented a survey on AI tools for supporting legal research by legal practitioners, albeit with a focus on Austrian law. His survey considers openly available legal databases such as *Lexis 360*, *LinDa*, *RDB*, *RIDA*, as well as tools like *AI:socciate*, *Genjus KI*, and *RechtGPT*. Unlike the review presented here, Jahnel's survey focusses on specific databases and tools, rather than on current research on legal information retrieval.

3 Methodology

The survey presented here followed a systematic literature review methodology based on the PRISMA (*Preferred Reporting Items for Systematic reviews and Meta-Analyses*) statement [33]. The survey sought to include all scientific publications related to legal databases and LIR which were published since 2020. Documents were retrieved from the databases *Scopus* and *Web of Science*, using the queries shown in Table 1. These databases were chosen because they include publication venues within the specific legal informatics community (e.g., ICAIL and JURIX conferences, or the journal *Artificial Intelligence and Law*) and publication venues for the wider field of information retrieval (e.g. annual meetings of the *Forum for Information Retrieval Evaluation – FIRE*).

A total of 198 documents were retrieved from *Scopus* and 273 documents were retrieved from *Web of Science*, although the latter were reduced to only 74 after the exclusions of unrelated research areas (e.g., medicine, surgery). Thus, a total of 272 documents were identified and selected for screening of titles and abstracts. This screening led to the exclusion of 110 documents, which were either duplicates or documents on subjects outside of the scope of the survey (for example, any paper investigating some specific legal topic and explicitly mentioning the words "legal database" in its abstract would be retrieved through the queries used). Thus, after screening, a total of 162 documents were included

Table 1. Databases and Queries

Database	Query
Scopus	`TITLE-ABS-KEY("Legal Database*" OR "Legal Information Retrieval") AND PUBYEAR > 2019 AND PUBYEAR < 2026 AND (EXCLUDE (SUBJAREA , "MEDI")) AND (EXCLUDE (DOCTYPE , "cr")) AND (LIMIT-TO (LANGUAGE , "German") OR LIMIT-TO (LANGUAGE , "English"))`
Web of Science	`"legal database*" (topic) or "legal information retrieval" (topic)`

in the survey.[1] An overview of the literature gathering procedure is given in the flow diagram in Fig. 1.

The 162 documents included in the survey (110 complete papers and 52 records for which only titles, abstracts, and keywords were available) were carefully analysed and assessed with respect to the following criteria:

(i) *Geographic Distribution*: country or countries whose legal systems are considered, if any.
(ii) *Tech/Non-Tech*: whether the paper has a technical focus (e.g., developing or enhancing information retrieval methods, ontologies, LLMs) or not (e.g., policy-oriented or discussing issues related to legal education etc.).

Papers with a non-technical focus were further analysed individually to assess their respective subjects. Papers with a technical focus were also assessed with respect to the following criteria:

(iii) (Specific) *Database-Focus*: whether the paper has a focus on legal databases or not (e.g., LIR on individual documents).
(iv) *AI-Focus*: whether the paper focusses on AI-methods like machine learning or language models.
(v) *Technology/Methods*: which specific technologies or methods were employed (e.g. retrieval-augmented generation).
(vi) *Applications*: the concrete applications or objectives pursued with the technologies or methods employed.

The criteria were jointly defined by all authors; the assessment was performed by one of the authors. Difficult cases were discussed and a solution was agreed upon by all authors. The criteria *Technology/Methods* and *Applications* were assessed by *inductive coding*, i.e., without pre-defined categories.

[1] A list of all included documents (including their evaluation with respect to the employed criteria) is available under https://door.donau-uni.ac.at/o:6293.

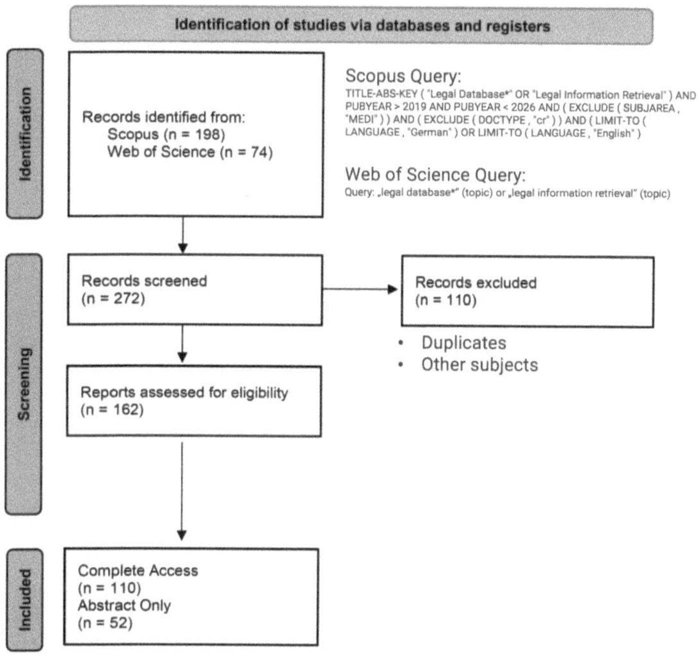

Fig. 1. Adapted PRISMA flow diagram

4 Results

4.1 Geographic Distribution

The countries whose legal systems were most often considered in the papers surveyed were India (15), Brazil (14), the United Kingdom (8) and Vietnam (8), followed by Austria (5), Germany (5), Greece (5), the Netherlands (5) and the USA (5). 58 papers were not associated with any country (24 of which were subject to access restrictions).

4.2 Tech/Non-tech

Of the 162 surveyed papers, a total of 26 had a *non-technical* focus. The main venue of these papers was the journal *Legal Information Management* (9). These papers investigated a wide array of topics ranging from the development of user-centric metrics for evaluating legal databases (cf., e.g., [41]) to issues surrounding legal education (cf., e.g., [1]), legal visualisation (cf., e.g., [7]), up to foundational analysis of the connection between digitalisation and legacies of colonialism [16]. The remaining 136 papers had a technical focus (97 with full access). The main venues for these papers were JURIX proceedings (9) and the journal *Artificial Intelligence and Law* (8). Many papers were also related to specific events, in particular to the *Competition on Legal Information Extraction and Entailment* – COLIEE (14). Noteworthy is also the workshop series *Legal Information Retrieval meets Artificial Intelligence* – LIRAI (3). Tech-papers were also assessed on the criteria discussed below.

4.3 (Specific) Database-Focus

Of the 136 surveyed tech-papers, 88 had a specific focus on legal databases, e.g., discussed technologies for improving LIR within a set of legal documents such as statutes or cases (cf., e.g., [15,18,23,43]). The remaining 48 papers either focussed on LIR within individual documents (e.g., within a single statute [50]), gave overviews of specific events (e.g., of COLIEEs [14] or LIRAIs [42]) or discussed a variety of other topics such as (automated) annotation methods [30,48], drafting contracts [3], or matching videos to transcripts of court hearings [36]. Despite lacking a specific focus on databases, these papers are relevant for the survey at hand because they discuss LIR technologies and methods that could readily be implemented for improving legal databases.

4.4 AI-Focus

Among tech-papers, 114 had a focus on AI-methods such as machine learning (cf., e.g., [49]) and/or the application of LLMs (cf., e.g., [27]); of these, 69 also had a specific focus on legal databases (cf. Sect. 4.3, above). 22 tech-papers did not deal primarily with AI, focussing instead on topics such as methods for legal knowledge representation (e.g., ontologies/knowledge graphs [28] or standards like Akoma Ntoso [5,34]), and on bibliometric [45] or other statistical analyses of legal documents (often called *jurimetrics*) [29].

4.5 Technology/Methods and Applications

The usage of an *inductive coding* method for the last two criteria (*Technology/Methods* and *Applications*) was deemed necessary, because the current LIR research landscape is relatively diverse, encompassing a variety of technologies and methods which, in turn, can be employed for manifold applications. This, however, made the establishment of useful classifications and the identification of coherent trends rather challenging. For example, 52 of the surveyed papers discussed some kind of (large) language model, with 29 papers analysing one or more variants of *bidirectional encoder representations from transformers* (BERT); cf., e.g., [26]. This diversity within one type of technology/method is further potentiated when considering the various applications for which these models were employed, e.g., recognising textual similarity between documents [25], to preprocess search queries [10] or to solve legal question answering tasks [27]. The search for a coherent method for grouping these diverse technologies and applications motivated the development of the novel process model of LIR, as discussed below in Sect. 5.

5 Discussion

5.1 General Research Landscape

The results on the *geographic distribution* criterion generally correspond to the expectation that legal systems of larger and/or higher income countries are more often analysed in research papers. Some interesting outliers are *Vietnam* and smaller European countries such as Austria/Greece (all five papers were published by researchers from the same research groups working together) and the Netherlands. We conjecture these outliers could be explained by digitalisation policies (for Vietnam, cf., e.g., [32]) and by relatively strong research communities in fields such as *E-Governance* and *Legal Informatics* in Austria/Greece and the Netherlands respectively.

The analysis of publication venues reveals a relatively clear division between technical and non-technical discussions, with rather few authors publishing papers of both categories (examples are *C. Alexopoulos* [5,41] and *G. Wiggers* [44,46]). Furthermore, despite a clear majority of papers dealing with topics related to AI, there is still a considerable amount of research on older topics within information retrieval in general, especially concerning knowledge representation methods. In this context, it seems particularly promising to investigate how different approaches to data structuring might influence the performance of LIR methods based on AI.

5.2 The Process Model of Legal Information Retrieval

The relatively large diversity of technologies, methods and respective applications within the current LIR research landscape made it difficult to extract meaningful patterns on the basis of our inductive coding approach to the criteria *Technology/Methods* and *Applications*. For this reason, inspired by the traditional COLIEE-tasks [14], we first attempted to group the surveyed papers based on whether they were focussed on information retrieval tasks in a more strict sense or on entailment/question-answering tasks. However, after a first screening, this idea was discarded for two reasons. First, it would exclude a significant number of papers (26) which were focussed on neither of these tasks, but instead discussed ontologies or other knowledge representation methods or presented new datasets (e.g., [47]). Second, even among papers that focussed on one of these tasks, we still verified a significant diversity. For example, even within COLIEE itself, while some papers tackled information retrieval tasks by focussing on more traditional retrieval methods (cf., e.g., [22]), others focussed on improved ways of structuring or pre-processing data or queries (cf., e.g. [4,9]). Hence, it would be arguably more reasonable to group these papers with other papers focussing on *data structuring*, rather than on information retrieval in a more strict sense, i.e., *data finding*. Furthermore, some of the papers surveyed seem to go beyond mere entailment tasks, addressing instead *case analysis* in a more holistic sense. Motivated by these observations, we propose adding two further classes or *steps* to the two classes inspired by the COLIEE-tasks, resulting in a *four-step* Process Model of Legal Information Retrieval (PMLIR), which is visualised in Fig. 2. We call it a *process* model, because each step generally builds on the results of the steps behind it.

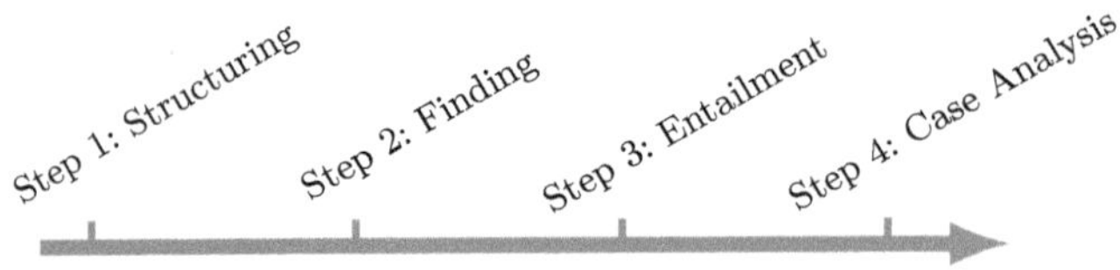

Fig. 2. Process Model of Legal Information Retrieval

The first step is data *structuring*. It encompasses primarily the usage of knowledge representation methods, ontologies/knowledge graphs and standards like Akoma Ntoso. Traditionally, most research has focussed on structuring the datasets containing the information one intends to retrieve; however, one can also enhance the structure of search-queries [8,10,40]. This step also includes research on *automated* data structuring (cf., e.g., [6]), which may include topics such as automated legal text summarisation.

The second step concerns the methods and algorithms employed for actually *finding* the data. These include classic information retrieval and natural language processing methods (e.g., *bag of words*), but also LLM-based methods such as RAG. Due to the large amount of legal documents (and to the large amount of information they contain), the main objective within this step is not merely finding information, but finding the *most relevant* information.

The third step, *entailment*, goes beyond a merely syntactical approach to information retrieval (i.e., finding specific words or sentences or slight variants thereof) to also consider the actual *meaning* contained in these words or sentences. This encompasses tasks such as legal question answering and automated subsumption, in which one must not only find relevant information within a dataset, but also assess it with respect to given questions.

Finally, the fourth step, *case analysis*, takes a more holistic approach, e.g., by attempting to determine which legal frameworks are relevant to a given case and assessing which legal consequences would follow. It involves all steps necessary for legal case analysis, in particular with respect to deciding on possible solution approaches and formulating a concrete solution applicable to the particularities of the specific case.

Following the PMLIR's framework, a paper is to be assigned to one or more steps depending on which contributions the paper provides to the respective steps. Thus, if a paper employs a new knowledge graph approach in order to tackle an entailment task, without contributing to the entailment methods themselves, it is to be assigned to the *structuring* step.

Of the 136 surveyed tech-papers, 43 were assigned to *structuring*, 38 to *finding*, 14 to *entailment* and two to *analysis* (e.g., [19]). 11 were assigned to both *finding and entailment*, nine to both *structuring and finding* and four to both *structuring and entailment.* One paper was assigned to *structuring, finding and entailment.* The remaining 14 papers did not fit in the classification for being literature reviews or having a *jurimetrics* focus. These results further corroborate the view that more classic approaches related to data structuring (e.g., knowledge representation methods) and data finding (e.g., relevance ranking algorithms) remain highly relevant, especially when combined with AI-methods. Of course, due to the employed search queries, these numbers only refer to research papers that are explicitly situated within research on LIR and legal databases. It remains possible that a substantial part of research on legal entailment or case analysis is developed outside this specific scope; such papers have not been considered in this study.

6 Conclusion and Future Work

80 years after Kelso's [21] plea for the usage of computing technology for managing legal information, legal databases have become a central element of a legal practitioner's everyday life. As research on information retrieval techniques advances, legal databases continue to evolve. Complementing the overview of the current research landscape provided here, future work should focus on analysing the current practice of legal databases; namely both from the perspective of the entities managing such systems (e.g., with respect to which new technologies are being implemented within legal information systems), and from the perspective of their users, e.g., citizens or legal practitioners. Overall, our proposed PMLIR could prove a valuable tool for better classifying, conceptualising and analysing future developments within LIR and legal databases.

References

1. Alabi, R.: Legal research without barriers: free legal databases as alternatives for law students. Int. J. Leg. Inf. **53**(2), 148–154 (2025). https://doi.org/10.1017/jli.2025.10065
2. Allen, L.E., Brooks, R.B.S., James, P.A.: Storage and retrieval of legal information: possibilities of automation. MULL: Mod. Uses Logic Law **2**(2), 68–84 (1960). http://www.jstor.org/stable/29760833
3. Amazou, Y., Tayalati, F., Mensouri, H., Azmani, A., Azmani, M.: Accurate AI assistance in contract law using retrieval-augmented generation to advance legal technology. Int. J. Adv. Comput. Sci. Appl. **16**(2), 1141–1150 (2025). https://doi.org/10.14569/IJACSA.2025.01602113
4. Askari, A., Verberne, S.: Combining lexical and neural retrieval with longformer-based summarization for effective case law retrieval, vol. 2950, pp. 162–170 (2021). https://ceur-ws.org/Vol-2950/paper-02.pdf
5. Avgerinos Loutsaris, M., Lachana, Z., Alexopoulos, C., Charalabidis, Y.: Legal text processing: combing two legal ontological approaches through text mining. In: Proceedings of the 22nd Annual International Conference on Digital Government Research, dg.o 2021, pp. 522–532. Association for Computing Machinery, New York (2021). https://doi.org/10.1145/3463677.3463730

6. Belhoucine, K., Mourchid, M., Mbarki, S., Mouloudi, A.: A bottom-up approach for Moroccan legal ontology learning from Arabic texts. Commun. Comput. Inf. Sci. **1389**, 230–242 (2021). https://doi.org/10.1007/978-3-030-70629-6_20
7. Brunschwig, C.R.: Die Rechtsbilddatenbank der Abteilung Rechtsvisualisierung an der Universität Zürich: Ein Kurzporträt und Überlegungen zu ihrer Vorbildrolle für Rechtsdatenbanken und moderne Rechtsbilddatenbanken. Jusletter IT, pp. 421–445. (2023). https://doi.org/10.38023/ea0604f6-3442-40e6-814f-2efd7b634273
8. Catacora, J., Casali, A., Deco, C.: Legal information retrieval system with entity-based query expansion: case study in traffic accident litigation; sistema de recuperación de información legal con expansión de la consulta basada en entidades: Caso de estudio en litigios por accidentes de tránsitos. J. Comput. Sci. Technol. (Argentina) **22**(2), 151–163 (2022). https://doi.org/10.24215/16666038.22.e12
9. Donabauer, G., Kruschwitz, U.: A reproducibility study of graph-based legal case retrieval. In: Proceedings of the 48th International ACM SIGIR Conference on Research and Development in Information Retrieval, SIGIR 2025, pp. 3135–3144 (2025). https://doi.org/10.1145/3726302.3730282
10. Dung, D., Pham, V., Tran, H., Phan, M., Huynh, H., Nguyen, H.: Extracting core meaning from legal queries using semantic technologies. In: Proceedings of the 24th International Conference on New Trends in Intelligent Software Methodologies, Tools and Techniques (SoMeT_25), pp. 429–442. IOS Press (2025). https://doi.org/10.3233/FAIA250543
11. D'alotto, J.E., Pons, C.F., Antonelli, L.R.: Artificial intelligence applied in legal information: a systematic mapping study; inteligencia artificial aplicada en información jurídica: Mapeo sistemático de la literatura. J. Comput. Sci. Technol. (Argentina) **25**(1), (2025). https://doi.org/10.24215/16666038.25.e03
12. Feng, Y., Li, C., Ng, V.: Legal case retrieval: a survey of the state of the art. In: Proceedings of the Annual Meeting of the Association for Computational Linguistics, vol. 1, pp. 6472–6485 (2024). https://doi.org/10.18653/v1/2024.acl-long.350
13. Geist, A.: Rechtsdatenbanken und Relevanzsortierung. Österreichische Computer Gesellschaft (2017)
14. Goebel, R., Kano, Y., Kim, M.Y., Rabelo, J., Satoh, K., Yoshioka, M.: Summary of the competition on legal information. Extraction/Entailment (COLIEE) **2023**, 472–480 (2023). https://doi.org/10.1145/3594536.3595176
15. Gomes, T., Ladeira, M.: A new conceptual framework for enhancing legal information retrieval at the Brazilian Superior Court of Justice. In: Proceedings of the 12th International Conference on Management of Digital EcoSystems, MEDES 2020, pp. 26–29. Association for Computing Machinery, New York (2020). https://doi.org/10.1145/3415958.3433087
16. Hoque, S.T.: Neocolonial digitality: analyzing digital legal databases using legal pluralism. Asian J. Law Soc. **10**(3), 516–549 (2023). https://doi.org/10.1017/als.2023.9
17. Jahnel, D.: KI-Einsatz bei der juristischen Recherche. Ein aktueller Überblick, Österreichisches Anwaltsblatt (2025)
18. Jamil, H.M.: Semantic querying of knowledge rich legal digital libraries using prism. In: JURIX 2022: The Thirty-fifth Annual Conference. vol. 362, pp. 63–72 (2022). https://doi.org/10.3233/FAIA220449
19. Kang, X., Qu, L., Soon, L.K., Li, Z., Trakic, A.: Automating IRAC analysis in Malaysian contract law using a semi-structured knowledge base. Artif. Intell. Law (2025). https://doi.org/10.1007/s10506-025-09467-5

20. Kehl, W.B., Horty, J.F., Bacon, C.R.T., Mitchell, D.S.: An information retrieval language for legal studies. Commun. ACM **4**(9), 380–389 (1961). https://doi.org/10.1145/366696.366744
21. Kelso, L.: Does the law need a technological revolution? Rocky Mountain Law Rev. 378–392 (1946)
22. Kim, M.Y., Rabelo, J., Okeke, K., Goebel, R.: Legal information retrieval and entailment based on BM25, transformer and semantic thesaurus methods. Rev. Socionetwork Strategies **16**(1), 157–174 (2022). https://doi.org/10.1007/s12626-022-00103-1
23. Lee, J.: Legal text classification in Korean sexual offense cases: from traditional machine learning to large language models with XAI insights. Artif. Intell. Law (2025). https://doi.org/10.1007/s10506-025-09454-w
24. LexisNexis: About LexisNexis (2026). https://www.lexisnexis.com/en-hk/about-us/overview
25. Mandal, A., Ghosh, K., Ghosh, S., Mandal, S.: Unsupervised approaches for measuring textual similarity between legal court case reports. Artif. Intell. Law **29**(3), 417–451 (2021). https://doi.org/10.1007/s10506-020-09280-2
26. Melo, R., Santos, P.A., Dias, J.: A semantic search system for the supremo tribunal de Justiçca. In: Moniz, N., Vale, Z., Cascalho, J., Silva, C., Sebastião, R. (eds.) Progress in Artificial Intelligence, EPIA 2023. LNCS, vol 14116. Springer, Cham (2023). https://doi.org/10.1007/978-3-031-49011-8_12
27. Moreira, J., da Silva, A.S., de Moura, E.S., Marinho, L.B.: A study on unsupervised question and answer generation for legal information retrieval and precedents understanding. In: Proceedings of the 47th International ACM SIGIR Conference on Research and Development in Information Retrieval, SIGIR 2024, pp. 2865–2869 (2024). https://doi.org/10.1145/3626772.3661354
28. Ngo, H.Q., Nguyen, H.D., Le-Khac, N.A.: Ontology knowledge map approach towards building linked data for Vietnamese legal applications. Vietnam J. Comput. Sci. **11**(2), 323–342 (2024). https://doi.org/10.1142/S2196888824500015
29. Nishikawa-Pacher, A., Hamann, H.: Is every law for everyone? Assessing access to national legislation through official legal databases around the world. Oxf. J. Leg. Stud. **43**(2), 298–321 (2023). https://doi.org/10.1093/ojls/gqac032
30. Nunes, R.O., Spritzer, A.S., Freitas, C.M.D.S., Balreira, D.G.: Out of sesame street: a study of Portuguese legal named entity recognition through in-context learning. In: International Conference on Enterprise Information Systems, ICEIS - Proceedings, vol. 1, pp. 477–489 (2024). https://doi.org/10.5220/0012624700003690
31. Office for Official Publications of the European Communities: 25 years of European law online - The event. Publications Office (2007)
32. Open Development Vietnam: Vietnam digital transformation agenda (2023). https://vietnam.opendevelopmentmekong.net/topics/vietnam-digital-transformation-agenda/
33. Page, M.J., et al.: The PRISMA 2020 statement: an updated guideline for reporting systematic reviews. BMJ **372** (2021). https://doi.org/10.1136/bmj.n71
34. Palmirani, M.: Lexdatafication: Italian legal knowledge modelling in Akoma Ntoso. In: Rodríguez-Doncel, V., Palmirani, M., Araszkiewicz, M., Casanovas, P., Pagallo, U., Sartor, G. (eds.) AICOL/XAILA 2018/2020. LNCS (LNAI), vol. 13048, pp. 31–47. Springer, Cham (2021). https://doi.org/10.1007/978-3-030-89811-3_3
35. Queen's Faculty of Law: History (2026). https://law.queensu.ca/about/history
36. Saadany, H., Orăsan, C., Breslin, C., Barczentewicz, M., Walker, S.: Employing AI for better access to justice: an automatic text-to-video linking tool for UK Supreme

Court hearings. Appl. Sci. (Switzerland) **15**(16) (2025). https://doi.org/10.3390/app15169205
37. Sasdelli, D., Steffes, B., Janisch, S., Krempelmeier, S., Mader, P., Staudegger, E.: Rechtsdatenbanken zwischen Logik und Praxis. In: Thiele, C. (ed.) Am Glasfaserpuls - Festschrift für Dietmar Jahnel, pp. 545–562. Jan Sramek Verlag (2025)
38. Simitis, S.: Informationskrise des Rechts und Datenverarbeitung. C. F. Müller (1970)
39. Strzolka, R.: JURIS: Das juristische Informationssystem in Vergangenheit und Gegenwart. Koeckert (1986)
40. Udayakumar, R., Abbas, H.M., Ramasamy, V., Vanathi, A., Kalyana Sundaram, N., Nailevich, I.R.: Context-aware query interpretation in legal information systems. Indian J. Inf. Sources Serv. **15**(3), 122–130 (2025). https://doi.org/10.51983/ijiss-2025.IJISS.15.3.14
41. Virkar, S., Alexopoulos, C., Tsekeridou, S., Novak, A.S.: A user-centred analysis of decision support requirements in legal informatics. Gov. Inf. Q. **39**(3) (2022). https://doi.org/10.1016/j.giq.2022.101713
42. Wehnert, S., Fiorelli, M., Picca, D., De Luca, E.W., Stellato, A.: LIRAI'24: 2nd Workshop on Legal Information Retrieval Meets Artificial Intelligence, pp. 390–392 (2024). https://doi.org/10.1145/3648188.3675120
43. Westermann, H., Savelka, J., Benyekhlef, K.: Paragraph similarity scoring and fine-tuned BERT for legal information retrieval and entailment. In: Okazaki, N., Yada, K., Satoh, K., Mineshima, K. (eds.) JSAI-isAI 2020. LNCS (LNAI), vol. 12758, pp. 269–285. Springer, Cham (2021). https://doi.org/10.1007/978-3-030-79942-7_18
44. Wiggers, G., Hartz, C.: Lererag: measuring legal relevance in retrieval augmented generation applications. Front. Artif. Intell. Appl. **395**, 404–406 (2024). https://doi.org/10.3233/FAIA241284
45. Wiggers, G., Verberne, S., van Loon, W., Zwenne, G.J.: Bibliometric-enhanced legal information retrieval: combining usage and citations as flavors of impact relevance. J. Am. Soc. Inf. Sci. **74**(8), 1010–1025 (2023). https://doi.org/10.1002/asi.24799
46. Wiggers, G., Zuccon, G.: The task: distinguishing tasks and sessions in legal information retrieval. In: ADCS 2022: Proceedings of the 26th Australasian Document Computing Symposium (2022). https://doi.org/10.1145/3572960.3572983
47. Wrzalik, M., Krechel, D., Aletras, N., Androutsopoulos, I., Barrett, L., Goanta, C.: GerDaLIR: a German dataset for legal information retrieval. In: Preotiuc-Pietro, D. (ed.) Proceedings of the Natural Legal Language Processing Workshop 2021, pp. 123–128. Association for Computational Linguistics, Punta Cana, Dominican Republic (2021). https://doi.org/10.18653/v1/2021.nllp-1.13
48. Zanoli, E., et al.: Annotators-in-the-loop: testing a novel annotation procedure on Italian case law. In: Proceedings of the Annual Meeting of the Association for Computational Linguistics, pp. 118–128 (2023). https://doi.org/10.18653/v1/2023.law-1.12
49. Zhebel, V.V., Devyatkin, D.A., Zubarev, D.V., Sochenkov, I.V.: Approaches to cross-language retrieval of similar legal documents based on machine learning. Sci. Tech. Inf. Process. **50**(5), 494–499 (2023). https://doi.org/10.3103/S0147688223050167
50. Šavelka, J., Ashley, K.D.: Legal information retrieval for understanding statutory terms. Artif. Intell. Law **30**(2), 245–289 (2022). https://doi.org/10.1007/s10506-021-09293-5

Legal Challenges and Technical Solutions for Data Spaces in Complex Value Chains

Diogo Sasdelli[1(✉)], Viola Gallina[2], Arko Steinwender[2], and Thomas Lampoltshammer[1]

[1] Department for E-Governance and Administration, University for Continuing Education Krems, Dr.-Karl-Dorrek-Straße 30, 3500 Krems an der Donau, Austria
{diogo.sasdelli,thomas.lampoltshammer}@donau-uni.ac.at
[2] Fraunhofer Austria Research GmbH, Theresianumgasse 7, 1040 Wien, Austria
{viola.gallina,arko.steinwender}@fraunhofer.at

Abstract. The European regulatory framework for data spaces, established by the Data Governance Act (DGA) and the Data Act (DA), introduces various legal requirements surrounding the development of the data economy across different sectors. Complying with these legal requirements while also ensuring an economically sustainable business model is particularly challenging in sectors such as the electronics and the textile industries, which often involve complex international value chains including several participants of different sizes and market power, many of which may not widely employ technological solutions from the digital age. Focussing on the electronics and textile industries as general examples, the paper at hand analyses the European regulatory framework for data spaces and investigates how state-of-the-art data space technology can provide (at least partial) solutions to emerging legal challenges.

Keywords: Data Spaces · Data Governance Act (DGA) · Data Act (DA) · Electronics · Textile Industry

1 Introduction

The *European Strategy for Data* [6] (henceforth "Data Strategy") is the main policy document outlining the European Union's approach to the data economy. It explicitly establishes the goal of creating "a single European data space – a genuine single market for data, open to data from across the world – where personal as well as non-personal data, including sensitive business data, are secure and businesses also have easy access to an almost infinite [*sic!*] amount of high-quality industrial data, boosting growth and creating value, while minimising the human carbon and environmental footprint." The Data Strategy itself is fairly vague – and, in many passages, even contradictory [17]. In particular, despite the central role of data spaces for the data economy, the strategy contains no clear definition of "data space". Notwithstanding, other documents published by the Commission [7,8,10,11] reveal that the framework envisaged by the EU

Published by Springer Nature Switzerland AG 2027
D. Duenas-Cid et al. (Eds.): ePart 2026, LNCS 16923, pp. 307–319, 2027.
https://doi.org/10.1007/978-3-032-36676-4_20

involves three categories of data spaces at different abstraction levels [17]. First, *individual data spaces*, consisting of decentralised data management mechanisms that enable secure, reliable and trustworthy usage and sharing of data among various stakeholders [11] – this corresponds, *in nuce*, to the usual technical definition of data spaces [3] (cf. below Sect. 2). Second, so-called *common European data spaces*, which should emerge – with direct support from the EU – from the interconnection of individual data spaces within sectors (e.g., health, finance, energy, agriculture etc.) [8]. Third, the *single European data space*, envisioned by the Data Strategy, which would be achieved by interconnecting the various common European data spaces across different sectors [8,17].

An important step in the development of this genuine single market for data in a way consistent with European values involves the establishment of a European regulatory framework for data spaces [6]. Currently, this framework is centred on two European regulations: the *Data Governance Act* (Regulation (EU) 2022/868, henceforth "DGA"), which seeks to enhance the availability of data in general by establishing a framework for promoting (mostly voluntary) sharing and reusing data (e.g., by means of *data altruism*), and the *Data Act* (Regulation (EU) 2023/2854, henceforth "DA"), which sets rules concerning the using and sharing of – and the generation of value from – data [18,20].

The regulatory framework established by the DGA and the DA gives rise to a range of legal challenges, notably for actors in sectors like the electronics and the textile industries, which are characterised by intricate, transnational value chains involving numerous stakeholders with differing scales and bargaining power, many of whom may have yet to adopt digital-age solutions. Focussing on the electronics and textile industries as general examples, the paper at hand analyses the legal challenges emerging from the DGA and DA (Sect. 2) and assesses the extent to which state-of-the-art data space technologies can offer at least partial solutions to each of them (Sect. 3). Finally, Sect. 4 concludes and refers to future work.

2 Legal Challenges

A first, more general legal challenge emerges from the fact that – like the Data Strategy – neither the DGA nor the DA explicitly introduce a proper legal definition of "data space". From a technical perspective, a data space is a form of data management framework involving decentralised storage of heterogeneous data (unlike, e.g., a database, which usually stores semantically homogeneous data centrally) [3]. Data spaces should enable trustworthy data sharing among the so-called data space participants [13,14]. For this, so-called *data space connectors* are employed, i.e., technical solutions managing data exchange among individual actors (the data space participants) in the data space following the so-called *data space protocol* (DSP) [3,14]. To further promote trustworthiness, the identities of the data space participants are managed and verified by an identity provider, e.g., by following the *Verifiable Credentials* (VCs) model from the *World Wide Web Consortium* (W3C) [3,21]. Thus, a so-called *minimal viable data space* consists of at least two participants (one *data provider* and one *data consumer*) and their respective *connectors*, as well as of an *identity provider* [3].

Although no *legal concept* of data space is explicitly introduced in the DGA or in the DA, these acts do contain legal definitions for the main components of a data space as outlined above. Data providers and any further intermediaries entitled to grant access or share data are *data holders* in the sense of Art. 2(8) DGA (or, depending on the context, in the sense of the slightly diverging definition in Art. 2(13) DA). Data consumers, in their turn, correspond to the *data users* in the sense of Art. 2(9) DGA, or, depending on the context, to the *data recipients* in Art. 2(14) DA. Furthermore, the DGA also introduces, in its Art. 2(11), the concept of *data intermediation service* (DIS), which is also adopted in the DA in its Art. 2(10). According to this definition, DIS are services that aim "to establish commercial relationships for data sharing between an undetermined number of data subjects and data holders on the one hand and data users on the other, through technical, legal or other means [...]." Following Art. 10 DGA, it becomes clear that this definition also includes "making available the technical or other means to enable such services". Thus, the two main services involved in data spaces, i.e., those referring to data space connectors and to identity providers, can be subsumed under the definition of DIS (cf. also the discussion in [19], concerning services in the context of the project *Gaia-X*).

Overall, actors providing services related to data space connectors and to identity providers are, thus, to be seen as *data intermediation service providers* (DISP) and have to comply with the respective requirements deriving from the DGA and from the DA. Under particular circumstances, especially within the electronics and textile industries, specific legal challenges might emerge. In the following, general requirements and particular challenges emerging from the DGA are discussed in Sect. 2.1; general requirements and specific challenges emerging from the DA are analysed in Sect. 2.2.

2.1 Legal Challenges in the Data Governance Act

Most of Chapter III DGA (Art. 10–15) focuses on matters concerning the monitoring of DISP by competent authorities. In particular, Art. 11 DGA requires all DISP to actively notify the respective monitoring authorities about their intention of providing DIS. These authorities notify the Commission, which keeps an up-to-date register of all DISP in the EU (Art. 11(10) DGA; for the register, cf. [9]). This strict monitoring framework should promote the trustworthiness of DISP, resonating with the EU's approach to data spaces and the data economy in general [17]. The core legal requirements for DISP are listed in Art. 12 DGA and can be summarised as follows:

- Neutrality and impartiality: Art. 12(a)-(c) and (f) DGA introduce general requirements concerning the impartiality and neutrality of DISP: they may not use the data (or metadata) subject to their service for purposes other than the provision of the respective DIS; they also have to ensure that their service provision is fair, transparent and non-discriminatory.

- Interoperability: Art. 12(d) and (i) DGA require DISP to facilitate data exchange in the format in which data was received from a data subject or a data holder. However, they should also take appropriate measures to ensure interoperability with other DIS, e.g., by using open standards common to the respective sectors in which they operate. Interoperability requirements are further specified in the DA (cf. below Sect. 2.2).
- Additional services: Art. 12(e) DGA allows DISP to offer additional services for the specific purpose of facilitating the exchange of data. Listed examples are temporary storage, curation, conversion, anonymisation, and pseudonymisation.
- Resilience and cybersecurity: Art. 12(g), (j)-(m) DGA establish resilience, cybersecurity, and data protection requirements. DISP have to take measures to prevent fraudulent or abusive practices in relation to parties seeking access through their DIS. They have to adopt adequate technical, legal and organisational measures to prevent unlawful access to – or transfer of – data, immediately informing data holders in the event of a violation. Overall, they have to ensure an appropriate cybersecurity level for data storage, processing and transmission. They also have to facilitate data subjects' exercise of their data protection rights.
- Continuity in case of insolvency: Art. 12(h) requires DISP to ensure a reasonable continuity of their DIS even in the event of their insolvency. This includes guaranteeing to data holders access to their data and the possibility to transfer and retrieve the data, as well as guaranteeing that data subjects can exercise their data protection rights.
- Consent and permission tools: Art. 12(n) requires DISP that use tools for obtaining consent from data subjects (or permission from data holders) to, whenever relevant, specify third country jurisdictions where data use is intended to take place, providing the possibility to both give and withdraw consent or permission for data processing. This resonates with the requirements in Art. 44–50 GDPR, which regulate transfers of personal data to third countries or international organisations.
- Log record: Art. 12(o) DGA requires DISP to maintain a log record of their data intermediation activity.

Many of these requirements are particularly relevant in the context of industries which, like the electronics or the textile industries, involve diverse actors of varying size, economic leverage, and degrees of digitalisation across a complex production and value chain. This strong diversity within the production and value chain stresses, in particular, the need for interoperability: different actors might adopt different strategies for collecting and using data, leading to a plethora of adopted technical and operational standards.

A further reason stressing the importance of interoperability in these industries emerges from the EU's goal of establishing overarching, sector-specific *common European data spaces*. So far, the EU has begun rolling out common European data spaces across various sectors, e.g., agriculture, cultural heritage, energy, green deal, manufacturing, health, etc. [8]. However, due to their complexity, the many actors involved in industries such as the electronics and the textile industries can be expected to participate in various common European data spaces *at the same time*. For example, the value chains of both the electronics and the textile industries are likely to include actors involved, e.g., in the energy, agriculture, green deal, and manufacturing common European data spaces. The importance of interoperability becomes evident when considering that, as shown by the example of the common European Data Space for the health sector (cf. Regulation (EU) 2025/327, the so-called European Health Data Space Regulation), different common European data spaces might introduce specific requirements or adopt specific standards.

Finally, the fact that many actors within the complex value chains of sectors such as the electronics and the textile industries display a relatively low degree of digitalisation and data usage – in particular those primarily involved in the extraction of raw materials and natural resources – could make the provision of additional services (data curation, conversion, storage, etc.) in the sense of Art. 12(e) DGA a necessity. Thus, DISP acting in such industries should be flexible enough to also be able to provide such services. Finally, international production and value chains common to these industries are likely to involve situations falling under the requirement concerning tools for gathering consent and permission for data processing (Art. 12(n) DGA).

2.2 Legal Challenges in the Data Act

While the DGA has a strong focus on governance models aimed at promoting data availability in the EU by facilitating data sharing mechanisms (e.g., DIS, data altruism, the establishment of the common European data spaces), the DA aims at promoting data availability by imposing data sharing obligations on data holders – in the sense of Art. 2(13) DA, which deviates from the definition of data holder in Art. 2(8) DGA – in diverse scenarios, e.g., with respect to so-called *connected products* (Art. 2(5) DA), i.e., in the context of so-called *Internet of Things* (IoT) applications, or in cases of particular public interest. The DA also introduces requirements concerning interoperability standards, safeguards against unlawful access to non-personal data, and instruments facilitating switching between data processing services (in the sense of Art. 2(8) DA). Thus, overall, among the various participants and roles within data spaces, the primary addressees of the DA are not the DISP (i.e., those providing services related to data space connectors and identity providers), but the data holders, especially when they offer data processing services (Art. 2(8) DA) to other data space participants. Depending on the context, these data holders (in the sense of Art. 2(13) DA) might, in the technical sense outlined above in Sect. 2, be either data producers or data consumers. DISP are nonetheless strongly affected by these requirements, either because they act, within the data space, as intermediaries for these data processing services, or because they offer such services themselves as additional services in the sense of Art. 12(e) DGA (cf. above Sect. 2.1), thus acting, at the same time, both as DISP and as data holders.

In this context, the DA introduces the following, potentially challenging requirements:

- *Data availability requirements*: Following Art. 3 DA, data pertaining to connected products (IoT context) or related services has to be, by-default, readily accessible to the respective users (Art. 2(12) DA). If such accessibility cannot be guaranteed, data holders have to make this data available, if relevant and feasible, continuously and in real-time (Art. 4(1) DA). Upon a request by the user, the data holders also have to make this data available to third parties (Art. 5 DA), who are potentially not participants in the respective data space. Likewise, data holders might also be required, in cases of exceptional need, to make data available to public sector bodies, the Commission, the European Central Bank, and Union bodies (Art. 14 *et seq.* DA).
- *Switching between data processing services*: Following Art. 23 *et seq.* DA, providers of data processing services have to take measures to enable customers to switch to another data processing service of the same type by a different provider or to use several such services at the same time. These measures include, e.g., interoperability aspects (Art. 30(3) DA), harmonised standards (Art. 30(5) DA), and open interfaces (Art. 30(2) DA). Furthermore, following Art. 30(4) in conjunction with Art. 26(b) DA, providers of data processing services have to provide the customer with an up-to-date online register with details of all data structures, formats, and standards in which exportable data is available.

Furthermore, the DA introduces, in its Art. 33 *et seq.*, a wide range of interoperability requirements for data processing services within data spaces and among common European data spaces. This includes the adoption of common standards and practices concerning data structures, programming interfaces, etc. In particular, Art. 33(2) DA empowers the Commission to adopt delegated acts specifying essential interoperability requirements.

3 Technical Solutions

Addressing the legal challenges outlined in the DGA and in the DA requires an understanding of the technical architecture that underpins modern data spaces. Contemporary research and initiatives such as the International Data Spaces Association (IDSA), Gaia-X, and the Data Spaces Support Centre (DSSC) have converged on a shared architectural vision [3]. This vision defines data spaces as federated, decentralised, and interoperable ecosystems that enable trusted data exchange across organisational and sectoral boundaries. From a technical standpoint, a data space is not a centralised repository, but a distributed infrastructure connecting independent data ecosystems (cf. above Sect. 2). It enables data providers, consumers, and intermediaries to share, access, and reuse data in accordance with clearly defined policies and governance frameworks. The distinguishing feature of data spaces lies in their ability to preserve data sovereignty –

data remain under the control of their originators even while being shared and reused elsewhere.

As shown in the previous chapter, the DGA and the DA establish a range of obligations for DISP, which are central to the governance and operation of data spaces and need to be reflected within the employed technical building blocks of data spaces. Current technological and organisational frameworks for data spaces already offer solution approaches to complying with each of these requirements, as discussed below:

3.1 Neutrality and Impartiality

As discussed above, the DGA requires that data spaces ensure neutrality and impartiality by acting as trusted intermediaries that strictly limit data use to the provision of the data intermediation service, preventing any secondary use of data or metadata. This is technically enforced by so-called *Policy and Usage Control Mechanisms*, such as *LUCON* or *Open Policy Agent* (OPA). These tools allow data providers to define fine-grained, machine-readable policies that are attached to their data. These policies are then interpreted and enforced by the data connectors (e.g., the *Eclipse Dataspace Components* (EDC)) at the moment of any data request. This provides a technical guarantee that the DISP (or any other participant) cannot access or use the data in a way that violates the provider's explicit rules, thereby enforcing structural separation [4,16].

3.2 Interoperability

The DGA requires DISPs to facilitate the exchange of data in their original format while also taking measures to ensure interoperability with other services, for example, by using open standards. Data space architectures address this requirement on two interoperability levels:

- *Technical interoperability*: data connectors are designed to communicate using a common, open standard like the *Data Space Protocol* (DSP) [14]. This ensures that connectors from different vendors can "talk" to each other.
- *Semantic Interoperability*: To ensure that the meaning of the data is adequately understood, so-called *Semantic Interoperability Layers* are employed. These implement shared vocabularies, ontologies, and data models (e.g., *FIWARE Smart Data Models*, *ECLASS*, or *NGSI-LD*) that act as a "common language" for all participants. [1,2,15].

3.3 Additional Services

The DGA explicitly allows DISPs to offer additional services like temporary storage, curation, conversion, or anonymisation, provided they facilitate the data exchange. The modular architecture of data spaces is the key technical enabler. These additional services (for example, an anonymisation tool) are not built into the core connector but are treated as separate, "pluggable" components. A data provider can then use the *Policy and Usage Control Mechanisms* to grant explicit permission for their data to be sent to this specific conversion or anonymisation service as part of a data-sharing agreement. This modularity allows for flexibility while maintaining data sovereignty.

3.4 Resilience and Cybersecurity

As discussed above, Art. 12(g), (j)-(m) DGA mandates robust resilience and cybersecurity measures to prevent fraudulent practices, unlawful access, and data breaches, requiring DISPs to inform data holders of violations. This is addressed by a combination of building blocks:

- *Identity and Trust Management*: using technologies like W3C's *Verifiable Credentials* (VCs) [12] and *Decentralised Identifiers* (DIDs), one can ensure that only authenticated and verified participants may join the data space.
- *Data connectors* and their embedded *Policy and Usage Control Mechanisms*: these can provide granular access control, encryption, and enforcement of data usage policies at all times.
- *Adopting Federation and Compliance Services*: services like *monitoring* and *logging* can provide the *incident response protocols* needed to detect breaches and inform stakeholders.

3.5 Continuity in Case of Insolvency

The DGA requires DISPs to ensure "reasonable continuity" of their service, allowing data holders to access and retrieve their data even if the DISP becomes insolvent. The foundational principle of *decentralisation* in data spaces constitutes the core of the primary technical solution to this problem. Following this principle, data are not actually stored by the DISP, remaining instead with the data provider. The DISP only provides the intermediation service (the *data connector* and *identity services*). If a DISP (e.g., a connector vendor) becomes insolvent, the data holder still has their data. They can then switch to a different, interoperable DISP, using the same *Data Space Protocol* (DSP), to maintain continuity of service with other participants.

3.6 Consent and Permission Tools

Art. 12(n) DGA requires DISPs to provide tools for obtaining and managing consent from data subjects (or permission from data holders), including specifying third-country jurisdictions and allowing for withdrawal. This is a direct application of the *Policy and Usage Control Mechanisms* already discussed above. Consent and permission are codified as fine-grained, machine-readable rules (e.g., using OPA or *LUCON*). These policies can specify purpose, duration, and even jurisdiction. They are enforced by the data connectors for every transaction, providing a dynamic and auditable way to manage and (just as importantly) withdraw consent or permission in real-time.

3.7 Log Record

Art. 12(o) DGA requires DISPs to maintain a log record of their data intermediation activity. This function is a core part of the data connectors themselves, as well as the overarching *Federation and Compliance Services* (like monitoring and cataloguing). Every request, policy evaluation, and data exchange is logged by the connector. These logs create a secure, tamper-evident audit trail, providing the transparency and accountability required by the DGA.

3.8 Data Availability Requirements

As discussed above, the DA imposes significant data-sharing obligations, particularly for data from "connected products" (IoT). It mandates that data must be readily accessible to the user by default. If direct access isn't possible, the data holder must make this data available, potentially continuously and in real-time. Crucially, upon the user's request, this data must also be shared with third parties, who may or may not be part of the same data space. Generally, data space architecture directly addresses this *by design*. The combination of data connectors and *Policy and Usage Control Mechanisms* creates a secure, standardised, and auditable pipeline for data access. In particular, this is guaranteed by the following aspects:

- *Access for Users & Third Parties*: Instead of building custom APIs for every request, a data holder uses a Data Connector (like EDC) to manage all data sharing. When a user requests their data or nominates a third party, the data holder simply grants them access rights via a machine-readable policy (e.g., using OPA or *LUCON*). The connector then enforces this policy, enabling seamless, secure, and real-time data flow without the data holder losing sovereignty.
- *Federated Management*: Another central element is so-called *federated data management*, in which the IoT data remains with the data holder but is made discoverable and accessible to authorized users and third parties via the data space's common protocols. This is an implementation of the general *decentralisation* principle already discussed above.

3.9 Switching Between Data Processing Services (*portability*)

As discussed above, the DA introduces measures to facilitate switching between providers of DIS. It mandates that providers must remove barriers and assist customers in moving their data. This includes providing open interfaces, adhering to harmonised standards, and maintaining an *up-to-date online register* with details of all data structures, formats, and standards for exportable data. Overall, the fundamental architecture of data spaces is built on the principles of interoperability and federation, which directly address this requirement:

- *Open Interfaces & Standards*: Instead of proprietary APIs, data spaces use data connectors that communicate via the open *Data Space Protocol* (DSP). For data formats, they leverage *Semantic Interoperability Layers*, which were already discussed above. This use of open, common standards is the primary technical enabler for switching, as it ensures that data exported from one provider can be understood and ingested by another.
- *Data Registries*: The DA's requirement for a *register* of data formats is technically implemented by the *Federation and Compliance Services* [5], specifically the cataloguing and discovery mechanisms. These services allow a provider to publish its data models and API specifications in a standardised way, making it transparent for a customer (or a new provider) to understand how to export and migrate data.

3.10 Summary

An overview of all assessed legal requirements in the DGA and the DA and the respective technical solutions is provided in Table 1.

Table 1. Overview of requirements and respective technical solutions

General Requirement	Source	Tech. Solution
Neutrality and Impartiality	DGA	Policy and Usage Control Mechanisms
Interoperability	DA/DGA	Open communications standards like the DSP and semantic interoperability layers
Additional Services	DGA	Modular architecture and Policy and Usage Control Mechanisms
Resilience and Cybersecurity	DGA	VCs and DIDs, Policy and Usage Control Mechanisms, Federation and Compliance Services
Continuity in case of insolvency	DGA	Decentralisation principle and the DSP
Consent and Permissions Tools	DGA	Policy and Usage Control Mechanisms
Log Record	DGA	Part of data connectors' implementation; Federation and Compliance Services
Data Availability	DA	Policy and Usage Control Mechanisms, federated data management
Portability	DA	Nearly *by design* within data spaces' architecture, DSP, Federation and Compliance Services

4 Conclusion and Future Work

The analysis developed above demonstrates that compliance with the European regulatory framework for data spaces requires paying particular attention to their core technical building blocks – including, in particular, *data connectors*, *Policy and Usage Control Mechanisms*, and *Semantic Interoperability Layers*. Adequate implementations offer tangible ways to operationalise diverse legal principles and requirements like neutrality, data availability, and interoperability.

However, this analysis also reveals that technical components alone are insufficient. For example, while policy engines can technically enforce neutrality, they cannot eliminate subtle biases that may emerge in data handling. Similarly, while semantic layers advance interoperability, "perfect" interoperability is hindered by evolving standards and inherent semantic ambiguities. Furthermore, technical solutions face challenges that are not purely technical: cybersecurity must constantly evolve to meet new threats, and the principle of decentralisation cannot, by itself, resolve the legal uncertainties of data ownership during an insolvency event. These challenges are reinforced by the practical realities of integrating legacy devices and the legal exclusion of trade secrets from data-switching mandates.

These limitations underscore the central success factor for embedding data spaces in the data economy: the alignment of legal compliance, technical feasibility, and business usability. This is especially important within sectors characterised by heterogeneous value chains and a high degree of SME involvement. It is therefore imperative that future research and technical developments prioritise enhancing the accessibility and reliability of data and data sharing structures across various sectors – especially in view of the common European data spaces envisaged in the Data Strategy. At the same time, legal scholarship has to contribute to the clarification of the relevant legal requirements and respective liability issues, especially when several legal frameworks are applicable at the same time. For example, besides the DA and DGA, one will likely also have to consider requirements emerging from the GDPR and, when employing digital product passports (DPP), the Ecodesign for Sustainable Products Regulation – Regulation (EU) 2024/1781.

Consequently, new methods and measures for automation must be designed and implemented to enable automated data generation, handling, storage, and exchange. This is necessary not only to reduce human effort and cost but also to manage the technical complexities identified, such as balancing comprehensive logging with privacy concerns or integrating diverse consent tools. Technological innovations, including semantic enrichment, interoperable metadata standards, and federated AI approaches, have the potential to play an essential role in this development.

Generally, it is paramount to ensure that these advanced technologies remain accessible to SMEs. The adoption of low-threshold entry points, standardised compliance tools, and pre-certified service components has been demonstrated to substantially lower barriers to participation. This will enable smaller actors to remain competitive while meeting their regulatory requirements. The sustainable evolution of data spaces depends on establishing trustworthy, high-quality, and automation-ready data infrastructures. This will not only facilitate legal compliance and operational efficiency but will also enable the digital transformation of the sector in line with European values of transparency, sovereignty, and fairness.

Acknowledgments. This work was carried out as part of the project PASSAT (*Digital Product Passport Austria & Beyond*). This bilateral flagship project is funded in Austria by the Federal Ministry for Innovation, Mobility and Infrastructure (BMIMI) and the Austrian Research Promotion Agency (FFG), and in Germany by the Federal Ministry for Economic Affairs and Energy (BMWE) and the German Aerospace Centre – Project Management Agency (DLR-PT).

Disclosure of Interests. The authors have no competing interests to declare that are relevant to the content of this article.

References

1. Ahle, U., Hierro, J.J.: Fiware for data spaces. In: Otto, B., Hompel, M., Wrobel, S. (eds.) Designing Data Spaces: The Ecosystem Approach to Competitive Advantage, pp. 395–418. Springer, Cham (2022)
2. Alexopoulos, K., et al.: An industrial data-spaces framework for resilient manufacturing value chains. Procedia CIRP **116**, 299–304 (2023)
3. Bacco, M., Kocian, A., Chessa, S., Crivello, A., Barsocchi, P.: What are data spaces? Systematic survey and future outlook. Data Brief **57**, 110969 (2024). https://doi.org/10.1016/j.dib.2024.110969
4. Dam, T., Klausner, L.D., Neumaier, S., Priebe, T.: A survey of dataspace connector implementations. In: Proceedings of ITADATA 2023: The 2nd Italian Conference on Big Data and Data Science, Naples, Italy (2023)
5. Data Spaces Support Centre: Federation services (2025). https://dssc.eu/space/BVE2/1071255059/Federation+Services
6. European Commission: Communication from the commission to the European parliament, the council, the European economic and social committee of the regions, a European strategy for data. https://eur-lex.europa.eu/legal-content/EN/TXT/?uri=celex:52020DC0066
7. European Commission: What is a data space? (2023). https://ec.europa.eu/newsroom/lds/items/788753/en
8. European Commission: Common European data spaces (2025). https://digital-strategy.ec.europa.eu/en/policies/data-spaces
9. European Commission: Eu register of data intermediation services (2025). https://digital-strategy.ec.europa.eu/en/policies/data-intermediary-services
10. European Commission: European data strategy: Making the EU a role model for a society empowered by data (2025). https://commission.europa.eu/strategy-and-policy/priorities-2019-2024/europe-fit-digital-age/European-data-strategy_en#a-single-market-for-data
11. European Commission: JRC data spaces knowledge space: Definitions (2025). https://wikis.ec.europa.eu/spaces/jrcdataspaceswiki/pages/57443811/1.3+Definitions

12. Fotiou, N., Siris, V.A., Polyzos, G.C.: Access control for data spaces. In: Proceedings of the 2025 28th Conference on Innovation in Clouds, Internet and Networks (ICIN), pp. 54–58. IEEE (2025)
13. International Organization for Standardization: ISO/IEC DIS 20151: Information technology—cloud computing and distributed platforms—dataspace concepts and characteristics (2025). https://www.iso.org/standard/86589.html
14. Koen, P., et al.: Dataspace protocol 2025-1 (2025). https://github.com/eclipse-dataspace-protocol-base/DataspaceProtocol
15. Martín, L., Lanza, J., González, V., Santana, J.R., Sotres, P., Sánchez, L.: A connector for integrating NGSI-LD data into open data portals. Sensors **24**(5), 1695 (2024)
16. Mboudia, O., Heppner, S., Miny, T., Kleinert, T.: Access control for asset administration shell applications. In: Proceedings of the 2025 IEEE 30th International Conference on Emerging Technologies and Factory Automation (ETFA), pp. 1–7. IEEE (2025)
17. Novak, A.S., Sasdelli, D.: The European approach to data spaces in light of the data governance act. In: Benfer, M., Lampoltshammer, T., Bachlechner, D. (eds.) Intelligent and Sovereign Use of Data in Manufacturing: Examples from the Wood Industry. Springer (2026). [expected]
18. Richter, H.: Looking at the data governance act and beyond: how to better integrate data intermediaries in the market order for data sharing. GRUR Int. **72**(5), 458–470 (2023). https://doi.org/10.1093/grurint/ikad014
19. Schütrumpf, M.: Anbieter von Datenvermittlungsdiensten als neue Intermediäre: Korrelationen zwischen dem Infrastrukturprojekt Gaia-X und dem Data Governance Act. Recht Digital, p. 373 (2023)
20. Specht-Riemenschneider, L.: Der Entwurf des Data Act: Eine Analyse der vorgesehenen Datenzugangsansprüche im Verhältnis B2B, B2C und B2G. Zeitschrift für das Recht der Digitalisierung, Datenwirtschaft und IT (MMR), pp. 809–826 (2022)
21. Sporny, M., Longley, D., Chadwick, D., Herman, I.: Verifiable credentials data model v2.0 (2025). https://www.w3.org/TR/vc-data-model-2.0/

Safety First: A Citizen Perspective on Privacy in Digital Public Administration Services

Luzia Borgmann(✉), Julian Hildebrandt, Luisa Vervier, and Martina Ziefle

Communication Science, RWTH Aachen University, Campus Boulevard 57, Aachen 52074, Germany
{borgmann,hildebrandt,vervier,ziefle}@comm.rwth-aachen.de

Abstract. Public administration services increasingly face challenges stemming from growing populations and demographic change, leading to inefficient processing of requests and subsequently to citizens' frustration with public administration. Digitization and the integration of AI into public administration services have been identified as integral to designing future-oriented public administration services. Especially regarding systems for citizen interaction, citizens' perception of privacy can influence service adoption. To understand citizens' perceptions of privacy in digital public administration systems and to identify their concerns and requirements, results from three qualitative studies concerning digital transformation in public administration were consolidated through a qualitative meta-synthesis (n = 29). The results revealed five aspects of privacy: (1) Differences between human and AI interaction, (2) General privacy perceptions, (3) Factors influencing trust in privacy perception, (4) Concerns, and (5) Requirements. The results showed that citizens focus mainly on technical aspects of the systems, emphasizing data security, prevention of possible harm through data misuse, and transparent information on data use, processing, and storage. While some concerns relate specifically to AI technologies, many reflect broader concerns regarding data processing in public services. Based on these findings, the article offers practical recommendations to enable a service-oriented and privacy-conscious design of trustworthy digital public services.

Keywords: Privacy Perception · Digital Public Services · AI in Public Administration · User-requirements

1 Introduction

Demographic change, especially a growing and aging population, poses unique challenges for public administration. Particularly in cities, this leads to higher workload for public services while public administration increasingly struggles

Published by Springer Nature Switzerland AG 2027
D. Duenas-Cid et al. (Eds.): ePart 2026, LNCS 16923, pp. 320–335, 2027.
https://doi.org/10.1007/978-3-032-36676-4_21

with staff shortages resulting in a significant backlog in requests [1]. Although this challenge is not entirely novel and steps have been taken in the past to improve the process – such as the separation of front and back office [2] – the effects of the problems – such as long waiting times – are becoming increasingly noticeable to the general public and may lead to frustration with the public service [1,3]. Comparisons to other countries and studies on previous German efforts suggest that digitization of public services and the possible introduction of artificial intelligence (AI) for certain use cases are integral to designing future-oriented public administration both in back and front office [1,3,4]. With public administration being a data driven context, citizens' perspective on privacy is an aspect to consider when designing digital public administration services [5], especially regarding the involvement of AI technologies, which pose unique challenges in terms of data processing [4,6]. Using a qualitative approach combining interviews and focus groups, this study addresses how citizens perceive privacy in regard to digital and AI-based public administration services and which concerns and requirements are voiced in this context.

2 Related Work

Understanding citizens' perception of privacy in digital services in public administration is important in order to enable a service-oriented and privacy-conscious design of trustworthy digital public services. This section outlines the current development of digitization of German public administration and describes existing research on privacy, trust, and acceptance of digital systems in this context. It concludes by identifying gaps in the literature that motivate the conducted study.

2.1 Digital Transformation in Public Administration

Public administration is part of public services, meaning services provided to citizens by public organizations. In Germany, most public administration requests are handled through citizen service offices ("Bürgerämter"), which serve as municipal centers for various types of administrative requests [3]. Compared to other European countries such as Estonia or Sweden, Germany shows lower levels of digitization in public administration while facing challenges to provide timely administrative services, especially in cities [1,7]. Over the last decades, Germany has implemented several strategies and regulations to further advance the digitization of public administration [1,8], including additional strategies for the integration of AI into public services, with a focus on preparing Germany for the consequences of demographic change [4]. Literature shows several public administration tasks that can be automated or simplified through the use of AI, such as responding to questions, performing repetitive tasks, and procedural decision making [7]. When looking at citizens' interactions with public administration, AI-based chatbots show potential for increasing effectiveness and efficiency by outsourcing time-intensive but frequent requests and possibly providing more accurate answers to requests based on higher amounts of data [9,10].

2.2 Characteristics of Public Administration

Regarding the digitization of public administration and the integration of AI in such services, some key differences between public administration and market-oriented companies can be considered. One of the main differences between public services and private services is that there is typically only one provider for public services, limiting citizens' freedom of choice. Due to this non-voluntary nature, classical models of technology acceptance may only partially apply to public administration contexts, as citizens cannot freely decide whether or not to use digital public services [11]. As a result, public administration can be argued to be obligated to provide citizens with effective and efficient service, meeting their expectations while also considering the communal budget [5,12].

2.3 Privacy, Trust and Acceptance

With citizens as stakeholders and main user group for digital public services, the citizen perspective plays an important role in ensuring the success of the digitization initiative [9,13]. The acceptance of digital public service systems is connected to citizens' trust in these offers [14], with several factors influencing trust in digital public administration, such as *Trust in Government*, *Trust in Internet*, *Perceived Risk* and *Perceived Privacy and Security* [15]. The influence of trust in AI and trust in government has been widely studied, both in terms of AI offerings in general and AI offerings in public administration [5,7,16–18]. In the following, we explore the factor of privacy and security and its connection to trust and acceptance. When considering privacy as a factor for trust in public services, one must first look at what privacy includes. The concept of privacy has no clear definition and can be considered as an umbrella term [19]. Early definitions of privacy focus on the desire to control access to personal information [20]. With increasingly digital technology, privacy aspects have been expanded to include concerns related to the collection, processing, and distribution of information, as well as invasion of personal information spaces [19]. Especially in the context of AI, privacy is an important factor to consider, as such systems generally work on large datasets. This can lead to ethical questions, for example, when data sharing becomes a condition for service access. Based on this, considerations of costs and benefits from the citizens' perspective are necessary when implementing AI-based services [6]. Empirical studies show that the aspect of privacy in the public administration context is closely interconnected with trust in government and the intention to use public service systems [5]. Furthermore, understanding how and why data is collected reassured citizens and strengthened their intention to trust and use an exemplary system [5]. Other studies showed that the trustworthiness of a system was not significantly influenced by the systems' data sharing and privacy policies but indicated that overall attitude to privacy might influence trust [21]. In general, privacy concerns affect usage intention of digital systems, with some studies indicating that these concerns are more connected to the system provider than the system itself [22]. When designing digital public services, transparency about data collected and used can be treated as a user-requirement [23]. The information required includes the data source, collection, processing purpose, storage, access, and bias, with potential users requiring control over the data and expressing desire for active consent [23]. With regard to AI-based systems, generative chatbots pose further privacy challenges, with the main concerns being data leakage and data rights, stemming from several aspects of AI systems, such as the processing methods and its state as an evolving technology [24].

2.4 Study Motivation and Research Questions

Prior research on digital public administration services has shown that privacy is a key requirement for citizens. Furthermore, privacy-related concerns may inhibit usage intention or lead to distrust in digital public services. However, few studies explicitly examine which aspects citizens refer to when talking about privacy and how these aspects and subsequent concerns and requirements relate to public administration specifically and within the German cultural context. We identified the following research questions to address these topics:

1. What aspects constitute privacy in digital public administration services from a citizen point of view?
2. What privacy concerns do citizens have regarding digital public administration services?
3. What requirements do citizens impose on digital public administration services with regard to privacy?

3 Methodological Approach and Logic of Procedure

Due to the exploratory nature of the research questions, we chose an approach of consolidating privacy aspects from three qualitative studies to achieve triangulation of the empirical data (see Fig. 1). Rather than statistically aggregating results, the study applies a qualitative meta-synthesis, i.e., a secondary qualitative analysis of the three studies' empirical material, in order to identify recurring privacy-related themes across the different study contexts. This section describes the survey and analysis methods in order to achieve the highest possible degree of intersubjective comprehensibility [25].

3.1 Study Design

The studies considered in this research were conducted with german citizens between late 2023 and early 2025. Each study addressed the topic of digital transformation in public administration using semi-structured interview and focus group approaches, with different research foci. Figure 1 provides an overview

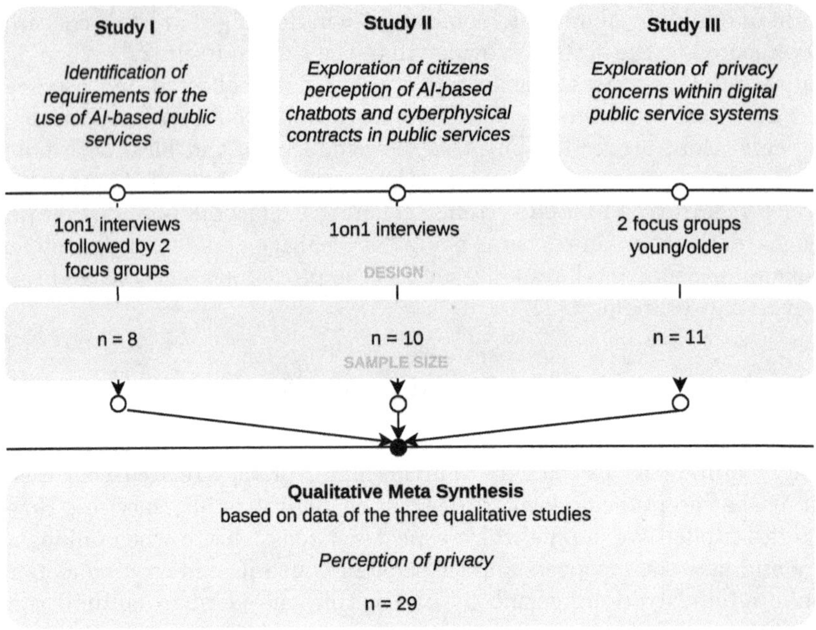

Fig. 1. Study design of the qualitative meta-synthesis integrating three qualitative studies on privacy in digital public services (n = 29).

of the three study designs and illustrates how the empirical material from the individual studies was consolidated in the qualitative meta-synthesis focusing on citizens' perception of privacy (n = 29).

Study I (n = 8) combined a semi-structured interview approach (average duration d = 27 min) with a focus group approach (d = 91 min), aiming to identify citizens' requirements for the use of digital and AI-based public service systems. This allowed us to gain individual insights into participants' concerns and requirements regarding digital public service systems while establishing a common basis and basic concepts, as well as to take advantage of the focus group situation to develop and discuss possible systems and suitable use-cases.

Study II (n = 10) used a semi-structured interview approach (d = 36 min) to explore citizens' perception of AI-based chatbots and cyber-physical contracts in digital public service systems. The major part of the interview focused on chatbots, specifically on data privacy in regard to the use of such systems in public administration. Participants were also asked to compare their privacy perception regarding AI-based services with human services.

Study III (n = 11) consisted of two focus groups (d = 97 min) discussing privacy concerns and requirements regarding digital public service systems, especially concerning the use of AI. The groups were divided into older and younger participants, allowing a comparison between age groups. This approach enabled us to create two different but homogeneous settings in which the participants could discuss matters openly with each other without age-related confounding factors, such as mental models or needs, having too much of an influence.

Although the studies differed in their specific research focus and empirical setting, all three addressed citizens' perceptions of privacy in the context of digital or AI-supported public administration services. This thematic overlap enabled the consolidation of privacy-related findings across studies.

3.2 Qualitative Content Analysis

Audio recordings were made of all studies and then transcribed into reading versions, as the analysis focuses on the verbalized content. All studies were conducted and transcribed in German. Anchor examples were translated as faithfully as possible for the results report. The individual analyses of Studies I and II were each conducted by one researcher, while Study III was analyzed collaboratively [26]. For all studies, coding units ranged from single words to paragraphs, with the requirement that the unit bears meaning [27]. All transcripts were analyzed separately in a deductive-inductive approach following Kuckartz (2014) [28]. Main categories were deductively based on the respective interview structures and subcategories devised inductively. To extract combined perceptions of privacy from all three studies, we then identified all privacy-related codes from the individual analyses through a secondary qualitative analysis. These codes included statements referring to data sharing, data storage, data use, trust in service providers, perceived risks of digital or AI-based processing, and citizens' expectations regarding privacy protection. In an initial step, we combined identical categories. We then used an inductive bottom-up approach to combine similar codes into larger categories. The resulting category system was reviewed by three professional evaluators and unanimously accepted. In cases of ambiguity, passages were revisited in their original study context to ensure consistent interpretation across studies.

3.3 Sample of Qualitative Meta-synthesis

A total of n = 29 participants, aged from 22 to 76 years (M = 42.76, *SD* = 16.84) participated in the studies. Overall, n = 13 participants were female (45%), n = 16 male (55%; rounded). Due to the fact that potential users of digital public administration systems were the target group, only German citizens or persons permanently residing in Germany were interviewed.

4 Results

In this section, the results of the combined qualitative content analysis are described. The structure follows the five main categories, as can be seen in the category system in Fig. 2. The resulting code system consists of five main categories: *Differences between Human and AI Services*, *General Privacy Perception*, *Trust*, *Concerns and Risks*, and *Requirements for Public Service Systems*. Each main category contains up to three levels of subcategories.

MAIN CATEGORY	SUBCATEGORY I	SUBCATEGORY II	OBSERVED IN STUDY
Differences in Human and AI Interaction	Human more Privacy		I & II
	No Difference		I, II & III
	Human as Risk		II & III
General Privacy Perception	Impossible		I & III
	Necessary Data Collection		I & III
	No Concern		I & III
Trust Influencing Privacy Perception	Organization		I, II & III
	Laws and Regulations		I, II & III
	System and Developer		II
Concerns	Technology Turning Against Humans		I & III
	Inappropriate Data Use	Identity Theft	I & III
		Surveillance and Analysis	I & III
	Unwanted Data Distribution	AI Stealing Data	I
		Data Transmission	II & III
		Data Storage	II
		Hackers	I, II & III
		Third Party Access	II
Requirements	Shared Data	Type and Amount	I, II & III
		Sovereignty	II & III
		Security	I & II
	Data Storage	Security	I & II
		Location	I & II
	Data Use	Transparency	I, II & III
		AI Training	I

Fig. 2. Category system resulting from the combined qualitative content analysis across Studies I–III (n = 29).

4.1 Differences in Privacy Between Human and AI Interaction

The main category *Differences in Privacy between Human and AI Interaction* contains participants' different perceptions of privacy depending on whether a public service interaction takes place via a human or an AI-based system. Overall, the results did not provide a clear assessment of which version meets higher privacy requirements. In addition, participants stated that humans are still dependent on (non-explicitly AI-based) digital systems and are therefore vulnerable to IT security issues ("Although data is ultimately always stored somewhere, I trust people much more than a computer system." (Study II; male, 22 years)) or emphasized that analog scenarios are also relevant to data protection, as human administrative officials may discuss sensitive information or leave confidential documents lying around. When participants expressed a preference for interacting with a human employee, they tended to describe it as a feeling rather than giving a rational explanation.

4.2 General Privacy Perception

Participants' general perception of privacy in digital public services could be divided into three subcategories: *Impossible*, *Necessary Data Collection*, and *No Concern*. Some participants considered that any use of data constitutes the risk of an invasion of privacy and that it would be *impossible* to achieve full privacy in digital services, because "whenever something new is developed, there will always be attempts to counteract it." (Study III; female, 62 years). In addition, participants argued that a certain amount of *Data Collection is Necessary* for any automated process and that public administrations already possess the required data anyway. Several participants stated that they did not worry much about privacy online, again referring to the perceived *Impossibility* of privacy in digital services. Without being asked to do so, participants distinguished between the public use of digital services and everyday applications, arguing that public authorities are generally more trustworthy than private providers and that people tend not to worry much about privacy when it comes to the latter. This distinction suggests that participants perceived privacy in public services differently from privacy in everyday digital applications. In addition, general risks to data protection were mainly discussed outside of the public service context and related to everyday technologies.

4.3 Trust Factors Influencing Privacy Perception

Trust in the *Organization* providing the service, *Laws and Regulations*, and the *System and Developer* were summarized in the main category *Trust Influencing Privacy Perception*. Participants mentioned all three aspects as factors influencing their perception of privacy. With regard to the influence of the *Organization*, participants stated that they generally have a high level of trust in the state and their local government. However, some conditions were also mentioned, including that data should not be collected by external companies or, in some cases, by the city itself. Concerning digital public services, participants generally highlighted high levels of trust in German and EU *Laws and Regulations*, as well as in public administration or cities as organizations: "Here in the EU, one already has the impression that at least the state is providing a certain protective shield for citizens, so that companies cannot take complete advantage of the situation" (Study III; male, 34 years). The category *System and its Developer* reflected differing opinions: some participants perceived more privacy in government-run systems, while others preferred specialized technology firms. In the latter case, participants were concerned regarding the security of smaller, government-run systems.

4.4 Concerns

The fourth main category, *Concerns*, captures participants' privacy concerns regarding digital public service systems. Such concerns were *Technology Turning Against Humans*, as well as *Inappropriate Data Use* and *Unwanted Distribution* of data. Many of these concerns referred to digital data processing in public services more broadly, while AI-specific concerns mainly related to training, bias, and autonomous system behavior. The subcategory *Technology Turning Against Humans* mainly included mentions of AI becoming fully autonomous and using personal data acquired through services to harm humans. Participants sometimes referred to science fiction scenarios, partly humorously, when discussing AI becoming uncontrollably autonomous. The subcategory of *Inappropriate Data Use* could be further divided into two potential concerns: *Identity Theft* and *Surveillance and Analysis*. Participants voiced concerns of both AI and humans committing *Identity Theft* based on the provided data, providing examples such as "Then I'll call on behalf of my neighbor and report new trash every day so that he has to pay ten Euros each time" (Study I; male, 55 years). Several interviews discussed *Surveillance and Analysis*, covering both commercial and government uses of citizen data for surveillance purposes. Regarding data analysis and decision-making, participants pointed to risks of bias, particularly in AI-supported decisions, which could disadvantage certain groups in sensitive cases such as asylum applications. Furthermore, the risks of a transparent citizen were discussed, for example, when user profiles are created or when "the identity card is linked to information about how many garbage bags are purchased" (Study I; male, 23 years). Concerns regarding data access by unauthorized parties were summarized in the subcategory of *Unwanted Data Distribution*. One participant voiced concerns about *AI Stealing Data*, expressing that "AI itself is just data theft" (Study I; female, 24 years). Interestingly, this example references AI as a technology in general, not just the use of AI-based systems. In contrast to that, participants' concerns about the security measures for *Data Transmission* ("My main concerns relate to data exchange—the data would have to be transmitted in encrypted form." (Study II; male, 26 years)) and *Data Storage* ("My concerns relate primarily to the storage and subsequent handling of the data." (Study II; male, 44 years)) mainly referred to the use of IT-systems in general. Furthermore, *Hackers* were seen as a potential risk to data provided for digital public services, referencing examples of hacker attacks in public institutions in Germany. Lastly, participants mentioned unwanted *Distribution to Third Parties* in reference to private companies as well as other public institutions.

4.5 Requirements

The main category Requirements encompasses the aspects that citizens wish to see in order to strengthen their sense of privacy. Three subcategories were defined: The *Shared Data*, the *Data Storage* and the *Data Use*. Requirements such as the *Type and Amount* of data shared, the *Sovereignty* the users retain over their data and the *Security Measures* during the sharing process are covered by the subcategory *Shared Data*. Regarding the *Type and Amount* of data shared, participants demanded that only data *Relevant to the Process* should be shared. For the type of data participants considered relevant in terms of privacy, they focused mainly on *Identity Determining Data*, such as information for identity cards. In summary, participants demanded that only data strictly necessary for the process should be used. *Sovereignty* over shared data included *Options to Edit* data, an *Opt-out Option* for data processing, and *Privacy Settings*. Participants referenced opt-out options for cookies and mentioned that "[...] one can always decide not to visit the website[...]" (Study III; male, 34 years) when comparing public services to private services. In terms of *Security*, participants emphasized the necessity of *Encryption* in data storage and transfer and called for appropriate methods of *Verification* as soon as privacy-relevant data were used. The subcategory *Data Storage* contains requirements for storage *Security* and *Location*. Regarding storage security, participants called for *Non-Permanent* data storage ("safety measures and permanent deletion of data are the top priority." (Study II; female, 35 years)) and expected measures to protect the stored data against *Misuse* and external access. One participant requested that a storage server "should provide a security architecture in which my data is treated just as trade secrets are and protected with modern encryption methods" (Study II; male, 39 years). Participants mentioned that they wished for data to be stored on *Local Servers* or at least on *European Servers*, showing distrust and less perceived privacy when data is stored on non-European servers hosted by private companies. The subcategory *Data Use* includes the aspects *Transparency* and *AI Training*. In general, participants expected *Transparency* regarding data processing, requesting insights into what data is used and stored, how, and for what purpose. Participants showed concern when talking about data being used for *training* ("the necessary data should be collected, but not used for further training in any way." (Study I; male, 23 years)), suggesting that no personal information should be used in this case. Transparency about data use also included information on whether data is used for *AI Training*, but also a general wish to gain more information on how citizens' data is used and processed. This was summarized by a participant as "Transparency regarding functionality, stored data, and security standards would be crucial" (Study II; male, 26 years).

5 Discussion

The main objective of the qualitative meta-synthesis was to gain insights into potential users' perception of privacy in the public administration context and to identify requirements and perceived risks of digital public administration systems. Our results confirm findings of previous studies that potential users are concerned with data usage of digital systems [23,24]. We could also identify the relationship of trust and privacy in our results, with privacy being argued in the context of trusting a system or provider, which aligns with previous studies suggesting trust being related to the company providing the system. [22].

5.1 Perception of Privacy

Most of the participants across all three studies were very conscious of their understanding of privacy and aspects influencing it. In general, they seemed to have either a resigned or an accepting perspective on privacy, and showed an understanding of the fact that especially public administration services heavily depend on personal data. While this was not always viewed positively, citizens were not generally averse to sharing data; rather, privacy perception depended on the interaction context, the actors involved, and transparent communication on system functionality [23]. In many cases, the general point of reference for positive privacy perception of digital public service was a human interaction, generally in a public service office. Considering this comparison, citizens generally seemed aware that while an interaction on location might be completely non-digital and excluding the use of AI, their shared data was very likely to still be processed and transferred digitally as German public service does already process and store data digitally [3,4]. With privacy perception being linked to the (by participants generally well received) status quo, it could be argued that citizens intuitively identify current privacy regulations and their enactment through human actors as the benchmark for positive privacy perception. However, some participants also seemed to realize certain problems within the current standards and proposed solutions through digital services. In general, citizens seemed to be less interested in whether or not their information was shared with a digital or AI-based system, but rather on the specifics of the information handling, once again strengthening that transparent communication furthers trust and adoption intent [23]. Another aspect that participants were considering when discussing their perception of privacy was the actors involved in the information sharing process. While trust in the system provider and trust in the government have been identified as trust aspects in previous studies [22], our research showed a more nuanced look into actors in which potential users trust. We identified the organization offering the service, the organization deciding on and enforcing the laws and regulations, and the provider and developer of the system as actors. Concerning public service, the municipality is generally seen as the service provider but might also be responsible for regulations and provide the system. However, while the general groups of actors could be identified from the participants comments, the mental model of who these actors are exactly might be individual or even depending on further context, such as the exact service provided by the system. Furthermore, more than one actor in each group might be relevant to citizens' privacy perception. Research also showed how the perception of the individual actors might vary greatly from personal experience, cultural factors, or possibly political opinion [29]. While some participants of our studies viewed the local government as a reliable trust actor, others had less positive opinions and voiced security concerns and doubts of capabilities. In general, the study confirmed that privacy is no singular aspect in a digital public service system, but interconnected with several other aspects of the discussed service and system. This reinforces the status of privacy as an umbrella term [19], uniting various different aspects that influence potential users' trust and adoption intention for a system. Especially considering trust in different actors as a vital aspect of privacy perception, privacy cannot be considered as a fixed value, but is situational, depending on context, prior experience, and personal opinions and values, which is a result not only noticed in regard to digital public service systems, but also towards the public service in general [29].

5.2 Concerns and Requirements

Our research also allowed insights into which concerns and requirements users have for digital public services regarding privacy. Most concerns voiced by the participants were of a technical nature, suggesting that citizens consider digital public service first and foremost as a technological system and referring less to the public service as an institution. This indicates that citizens often frame privacy concerns in digital public services primarily as technological risks rather than as institutional trust issues related to public administration. Concerns such as loss of control over processed data, security concerns, or fear of hackers are not unique to the public service, but instead general concerns related to digital processing of information. This concern about how data is used could be identified in various prior studies [24,30], with our studies participants mainly focusing the concerns of their data being "leaked", as further identified by Ye et al. [24]. Similarly, participants requested the system to prevent harm from various directions, such as external actors (hackers) and internal misuse (surveillance). Participants further commented on harm being done by misuse through other citizens. Said concern was mentioned in reference to a need for authentication, which highlights authentication as an aspect to strengthen the perception of privacy. The potential influence of negative citizen behavior into AI-behavior can also be related to the choice of AI training data. This raises the question of how to ensure that a model trained on citizen interaction does not learn negative or harmful behavior [6]. Furthermore, the concern of surveillance stands out, as it focuses on different types of data than other concerns about data misuse. While participants largely considered their personal identification data when discussing privacy concerns, surveillance could be linked more to behavioral data learned by a system through citizen interactions. As previous research suggests that privacy perception might vary depending on the concerned type of data [30], the concern of surveillance also reinforces a need for careful consideration of type and amount of data collected by a service [6]. In the context of the public service, the difference in privacy perception between identification data and behavioral data might be heightened, as identification data is inherently connected to public administration [3], while the collection of behavioral data could be considered as additional data collection irrelevant to the process. Generally, participants' statements indicate that a "the less the better" approach concerning the processing of data would benefit citizens' privacy perception. This requirement aligns with the GDPR, which regulates that data processing must be purpose-specific and minimal. A further requirement strongly voiced by the studies participants was transparency over data processing, use, and storage. Participants stated that their perception of privacy concerning a service depended on the information provided. Our results aligned with the findings that citizens were interested in information on purpose, functionality and processing, with potential risks being a less sought out information [23]. The GDPR requirements for transparency clarify that information on data processing must not only exist, but also be communicated in a way for users to understand, a requirement that some participants also voiced. Connecting the requirement for transparency with the privacy aspect of involved actors, it can also be noted that the source of the information might be relevant to citizens' perception of privacy, considering the provider as a trust factor [22] and the importance of initial trust for system adoption [31].

6 Practical Recommendations

Our results allow several practical recommendations for the design of digital and AI-based public administration services. Public service workers should be informed about data security standards and privacy regulations, as citizens identified privacy concerns in any context involving digital processing of personal data. Furthermore, only data necessary for the service should be collected, citizens should retain control over their data where possible, and data should be transferred and stored securely under European standards aligning with the GDPR. Participants were skeptical of using personal data for AI training, suggesting that training should not rely on personal or identifiable user data. Finally, transparent information on data processing should be provided in accessible language and at different levels of detail.

7 Limitations and Future Research

Overall, the chosen evaluation approach of method and data triangulation turned out to be suitable for addressing the research questions: Constituent aspects of privacy could be identified just as effectively as concerns and requirements in the area of digital public service. The variation of interview and focus group studies allowed for the examination of several scenarios, resulting in a comprehensive overall picture. The impact of the identified requirements, such as secure authentication and data use, on service adoption can be further evaluated in scenario-based questionnaires or conjoint approaches. We also mainly examined citizens' perception, concerns and requirements in hypothetical scenarios. With increasing implementation of digital and AI-based public services, possible implicit benefits of use or external incentives could be taken into consideration by evaluating post-prototype solutions. Real-life evaluations (such as user tests or field studies) would also allow common conceptions to be examined, such as that privacy regulations make systems more complicated and less user-friendly. Furthermore, the topic of privacy concerns in government-related contexts remains as a potential field for further qualitative research. With trust in the provider as a privacy aspect for digital and AI-based public administration services, evaluation of the relation between privacy, trust, and government could be helpful to address possible user concerns especially considering cultural and regional factors beyond the German context [29].

8 Conclusion

The study examined citizens' perception of privacy in digital public administration services and possible concerns and requirements regarding such systems. Using a qualitative meta-synthesis based on three qualitative studies, citizens' perceptions were explored and categorized, focusing on differences between human and AI interaction, general privacy perceptions, actors influencing trust, concerns, and requirements. The results show that citizens compare perceived privacy in digital and AI-based public services to the status quo of human interactions, while being aware of their data being inherently digital when using a variety of services. Furthermore, trust in different system actors was identified as a privacy relevant factor, namely the service provider, the organization providing and enforcing laws and regulations, and the systems' developer. Citizens' concerns were largely of a technical nature and focused on the possibility of data misuse and potential harms stemming from this misuse, often linked to existing GDPR regulations. Additionally, a need for transparent information on data use, processing, and storage could be identified. Concerning AI specifically, the results suggest that no personal identification data or behavioral data should be used for training purpose. Overall, the results suggest that citizens primarily interpret privacy in digital public services through the lens of data security and technical system reliability rather than through institutional characteristics of public administration. The study therefore contributes a citizen-centered perspective on privacy in digital government and highlights the importance of transparency, data minimization, and secure data practices for the design of trustworthy digital public services.

Acknowledgments. Funded by the Bundesministerium für Forschung, Technologie und Raumfahrt (BMFTR, Federal Ministry of Research, Technology and Space) under P2Broker - Public Private Broker for Secure AI-supported Administration in Municipalities - 16KIS2041 and FENI-X - Flexibility and Efficiency in Production Networks: Determined Reaction to X - 02J23C102.

The authors would like to express their sincere gratitude to Semih Yildirim, Uwe Assmann, Luisa Loreng, and Greta Clemens for their research support and all participants who took part in the interviews and focus groups.

Disclosure of Interests. The authors have no competing interests to declare that are relevant to the content of this article.

References

1. Wangler, L., Botthof, A.: E-governance: digitalisierung und KI in der öffentlichen Verwaltung. In: Wittpahl, V. (eds.) Künstliche Intelligenz, pp. 122–141. Springer, Heidelberg (2019). https://doi.org/10.1007/978-3-662-58042-4_8
2. Klages, H.: Wie lässt sich Bürokratie "unspürbar" machen? Das Bürgeramt als Paradigma. In: Brüggemeier, M. & Lenk, K. (eds.) Bürokratieabbau Im Verwaltungsvollzug, pp. 99–112. Nomos, Baden-Baden (2011). https://doi.org/10.5771/9783845267890-99

3. Kuhlmann, S., Bogumil, J.: The digitalisation of local public services. Evidence from the German case. In: Bergström, T., Franzke, J., Kuhlmann, S. & Wayenberg, E. (eds.) The Future Of Local Self-Government, pp. 101–113. Palgrave Macmillan, Cham (2021). https://doi.org/10.1007/978-3-030-56059-1_8
4. Etscheid, J., Von Lucke, J., Stroh, F.: Künstliche Intelligenz in der öffentlichen Verwaltung: Anwendungsfelder und Szenarien (2020). https://publica.fraunhofer.de/entities/publication/8d3f320b-12d4-4a68-a0f5-d317c285f04c/details
5. Schmager, S., Grøder, C., Parmiggiani, E., Pappas, I., Vassilakopoulou, P.: What do citizens think of AI adoption in public services? Exploratory research on citizen attitudes through a social contract lens. In: Proceedings of the 56th Annual Hawaii International Conference on System Sciences, pp. 4472–4481 (2023). https://doi.org/10.24251/HICSS.2023.544
6. Madan, R., Ashok, M.: AI adoption and diffusion in public administration: a systematic literature review and future research agenda. Gov. Inf. Q. **40**(1), 101774 (2023). https://doi.org/10.1016/j.giq.2022.101774
7. Kaun, A., Larsson, A., Masso, A.: Automating public administration: citizens' attitudes towards automated decision-making across Estonia, Sweden, and Germany. Inf. Commun. Soc. **27**(2), 314–332 (2024). https://doi.org/10.1080/1369118X.2023.2205493
8. Weiss, J.: Digitizing bureaucracy: insights from the digital transformation of Germany's local government sector. In: Urs, N., Špaček, D., & Nõmmik, S. (eds.) Digital Transformation In European Public Services: Complexities, Challenges, And Good Practices, pp. 79–105. Palgrave Macmillan, Cham (2025). https://doi.org/10.1007/978-3-031-81425-9_5
9. Makasi, T., Nili, A., Desouza, K., Tate, M.: Public service values and chatbots in the public sector: reconciling designer efforts and user expectations. In: Proceedings of the 55th Hawaii International Conference on System Sciences, pp. 2234–2343 (2022). https://doi.org/10.24251/HICSS.2022.292
10. Van Noordt, C., Misuraca, G.: Artificial intelligence for the public sector: results of landscaping the use of AI in government across the European Union. Gov. Inf. Q. **39**(3), 101714 (2022). https://doi.org/10.1016/j.giq.2022.101714
11. Renn, O., Zwick, M.: Risiko- und Technikakzeptanz. Springer, Heidelberg (1997)
12. Gesk, T., Leyer, M.: Artificial intelligence in public services: when and why citizens accept its usage. Gov. Inf. Q. **39**(3), 101704 (2022). https://doi.org/10.1016/j.giq.2022.101704
13. Horvath, L., James, O., Banducci, S., Beduschi, A.: Citizens' acceptance of artificial intelligence in public services: evidence from a conjoint experiment about processing permit applications. Gov. Inf. Q. **40**(4), 101876 (2023). https://doi.org/10.1016/j.giq.2023.101876
14. Wirtz, B., Weyerer, J., Sturm, B.: The dark sides of artificial intelligence: an integrated AI governance framework for public administration. Int. J. Public Adm. **43**(9), 818–829 (2020). https://doi.org/10.1080/01900692.2020.1749851
15. Gupta, P., Hooda, A., Jeyaraj, A., Seddon, J., Dwivedi, Y.: Trust, risk, privacy and security in e-government use: insights from a MASEM analysis. Inf. Syst. Front. **27**(3), 1089–1105 (2025). https://doi.org/10.1007/s10796-024-10497-8
16. Djeffal, C.: Künstliche Intelligenz in der öffentlichen Verwaltung (Artificial intelligence in public administration). Berichte Des NEGZ (3), 1–32 (2018). https://ssrn.com/abstract=3289109
17. Petriv, Y., Erlenheim, R., Tsap, V., Pappel, I., Draheim, D.: Designing effective chatbot solutions for the public sector: a case study from Ukraine. In: Chugunov,

A., Khodachek, I., Misnikov, Y., Trutnev, D. (eds.) Electronic Governance And Open Society: Challenges In Eurasia, pp. 320–335. Springer, Cham (2020). https://doi.org/10.1007/978-3-030-39296-3_24
18. Yigitcanlar, T., Li, R., Beeramoole, P., Paz, A.: Artificial intelligence in local government services: Public perceptions from Australia and Hong Kong. Gov. Inf. Q. **40**(3), 101833 (2023). https://doi.org/10.1016/j.giq.2023.101833
19. Solove, D.: A taxonomy of privacy. Univ. Pennsylvania Law Rev. **154**, 477 (2005). https://ssrn.com/abstract=667622
20. Altman, I.: The Environment and Social Behavior: Privacy, Personal Space, Territory, and Crowding. Brooks/Cole, Monterey (1975)
21. Kleizen, B., Van Dooren, W., Verhoest, K., Tan, E.: Do citizens trust trustworthy artificial intelligence? Experimental evidence on the limits of ethical AI measures in government. Gov. Inf. Q. **40**(4), 101834 (2023). https://doi.org/10.1016/j.giq.2023.101834
22. Li, J., Wu, L., Qi, J., Zhang, Y., Wu, Z., Hu, S.: Determinants affecting consumer trust in communication with AI chatbots: the moderating effect of privacy concerns. J. Organ. End User Comput. **35**(1), 1–24 (2023). https://doi.org/10.4018/JOEUC.328089
23. Drobotowicz, K., Kauppinen, M. & Kujala, S. Trustworthy AI services in the public sector: what are citizens saying about it?. In: Dalpiaz, F., Spoletini, P. (eds.) Requirements Engineering: Foundation For Software Quality, pp. 99–115. Springer, Cham (2021). https://doi.org/10.1007/978-3-030-73128-1_7
24. Ye, X., Yan, Y., Li, J., Jiang, B.: Privacy and personal data risk governance for generative artificial intelligence: a Chinese perspective. Telecommun. Policy **48**(10), 102851 (2024). https://doi.org/10.1016/j.telpol.2024.102851
25. Steinke, I.: Gütekriterien qualitativer Forschung. In: Flick, U., von Kardorff, E., Steinke, I. (eds.) Qualitative Forschung. Ein Handbuch. Rowohlt Taschenbuch, Reinbek b. Hamburg (2000)
26. Spangler, P., Liu, J., Hill, C.: Consensual Qualitative Research for Simple Qualitative Data an Introduction to CQR-M (2012)
27. Mayring, P., Fenzl, T.: Qualitative inhaltsanalyse. In: Baur, N., Blasius, J. (eds.) Handbuch Methoden Der Empirischen Sozialforschung, pp. 633–648. Springer, Wiesbaden (2019)
28. Kuckartz, U.: Qualitative Text Analysis: A Guide to Methods, Practice and Using Software. Sage Publications Ltd., London (2014). https://doi.org/10.4135/9781446288719
29. Morin, G., Charbonneau, É., Bromberg, D.: Canadians' gentle shift of opinions on privacy: an assessment of public values. Can. Public Adm. **68**(1), 41–62 (2025). https://doi.org/10.1111/capa.70005
30. Böhm, G., Pfister, H., Ayres-Pereira, V., Tjøstheim, I.: E-privacy concerns: a facet theoretical approach. J. Risk Res. **27**(1), 1–20 (2024). https://doi.org/10.1080/13669877.2023.2288012
31. Aoki, N.: An experimental study of public trust in AI chatbots in the public sector. Gov. Inf. Q. **37**(4), 101490 (2020). https://doi.org/10.1016/j.giq.2020.101490

The Global Stocktake: Exploring the Intersection Between Generative AI and Climate Negotiations

Tove Engvall[1](✉) and Leif Sundberg[1,2]

[1] Mid Sweden University, Holmg. 10, 85170 Sundsvall, Sweden
{tove.engvall,leif.sundberg}@miun.se, leif.sundberg@umu.se
[2] Umeå University, Universitetstorget 4, 901 87 Umeå, Sweden

Abstract. Every five years, a global stocktake (GST) is conducted to evaluate progress towards the goals in the Paris Agreement. The GST involves processing large quantities of information produced by a diverse range of stakeholders. Scholars have suggested exploring the use of Generative AI (GenAI) technologies such as large language models (LLMs) to support the GST. To examine this proposal, we conduct a scoping review to identify challenges in the GST and map these challenges to a framework of GenAI modes of operations, and literature on trustworthy AI. We propose three problem-solution mappings (information challenge, effectiveness, GST outcomes) where GenAI can support the GST process. Then, we present important considerations that need to be addressed when using AI in this context and propose scenario building as a path for future research. By doing so, this paper links literature on climate negotiations and AI, and draws on principles of trustworthy AI to guard against inaccuracy, bias, and inadequate human oversight.

Keywords: Global stocktake · trustworthy AI · climate governance

1 Introduction

Every five years, a global stocktake (GST) is conducted to evaluate progress towards the goals in the Paris Agreement, as well as inform further action to close the gap [1]. The GST has an important role in driving the implementation of the Paris Agreement and climate action among various stakeholders [2].

The global stocktake is conducted in three phases: information collection and synthesis, technical assessment, and political output. A major challenge in the process is to synthesise and analyse an extensive amount of information [3–6], comprising more than 170,000 documents in a variety of formats and languages, produced by diverse stakeholders [4]. The facilitators of the technical assessment dialogues have suggested exploring the use of AI to enable more effective synthesis and use of information [6]. This has been further supported by other scholars, specifically pointing towards the possibilities of using Generative AI (GenAI)-technologies such as large language models

D. Duenas-Cid et al. (Eds.): ePart 2026, LNCS 16923, pp. 336–352, 2027.
https://doi.org/10.1007/978-3-032-36676-4_22

(LLMs) in this context [4]. Against this backdrop, in this paper we conduct a literature review to identify challenges in the global stocktake and propose ways in which GenAI could support the process, along with key considerations for its use. By doing so, we contribute to digital governance literature by proposing tentative use cases for the use of GenAI in climate negotiations, and suggestions for aligning this use with ideals of trustworthy AI.

The first GST was conducted in 2023, and the process is still evolving. It is therefore valuable to review experiences and discuss how it could be enhanced by digital technologies. Digitalisation in climate negotiations has been suggested as both an opportunity to transform climate diplomacy for the better and a risk that could further exacerbate existing trust deficits and power inequalities [7]. It is thus important to understand the objectives of and challenges in the GST process per se, but also contextual challenges. Reaching agreements among countries with very different conditions in a conflict-prone context with multiple tensions and competing stakeholder interests, is a daunting task. Using digital technologies in this process must therefore be sensitive to the complex context of climate negotiations [8]. Here, previous research describes AI as a double-edged sword [9]. GenAI entails a significant increase of resource consumption and negative environmental implications [10], while at the same time having the potential to improve, for instance, policy design and policy evaluations [11]. With transparency being a primary governance mechanism in climate governance, a major challenge is to make sense of the overflow of information [12]. For transparency to have an impact, it is important to focus on the use of information and how it affects governance activities for different stakeholders [13]. AI has the potential to process large amounts of information, and to tailor information to different actors' needs, which could support climate action among stakeholders [14]. Meanwhile, the use of AI to process information comes with its own challenges as the output from, for example, LLMs, is often "black-boxed" in the sense that it is difficult to derive their underlying logic [15]. Thus, the use of AI presents both possibilities and critical challenges. While previous research has focused on how predictive AI technologies, such as machine learning, can be used to monitor climate change [16, 17], there is limited evidence of benefits associated with GenAI in this context [18].

The objective of this paper is to map challenges related to the global stocktake and examine how AI could facilitate the process, while also identifying associated risks and key considerations. This aligns with a growing literature that examines how AI can be applied across contexts while adhering its use to appropriate safeguards and governance mechanisms.

This paper proceeds as follows. Section 2 outlines a framework of human-AI knowledge production and trustworthy AI which will be used as an analytical lens. Design of the literature review is described in Sect. 3, followed by findings in Sect. 4. The findings are discussed in Sect. 5, followed by conclusions, limitations and suggestions for further research in Sect. 6.

2 Conceptual Framework: Modes of human–AI knowledge production and Trustworthy AI

GenAI systems and LLMs excel compared to previous AI systems in one important aspect: they are pre-trained on data that enables them to process a wide range of content "out of the box", without being provided extra context. In addition, they can further enhance content such as images and texts, based on the input (prompt) provided by a user. This is a major difference compared to older AI systems that needed to be trained on specific content to make predictions in narrow tasks. GenAI can thus be used for several diverse purposes to support various modes of knowledge production [19, 20]. To conceptualize these modes, we adapt Sundberg and Holmström's [21] framework, which distinguishes between AI systems' capabilities to navigate content and context.

LLMs are trained on large corpora and acquire knowledge of patterns, grammar, semantics, and associations which enable them to have a degree of contextual knowledge. The statistical and semantic knowledge embedded in the model's weights and generation process underpins output quality. Knowledge produced through human interaction with LLMs thus depends on two elements: what the model already encodes (its training-induced representations) and a dynamic component in which the user supplies content via prompts [21]. Between these components, factors such as memory (e.g., conversation history) and retrieval are important [22]. We further outline these mechanisms in terms of a model's semantic capacity and users' contextual constraints below.

2.1 Semantic Capacity and Contextual Constraint

Where Sundberg & Holmström [21] differentiate between GenAI systems' capabilities to use contextualized knowledge to process user-supplied content to conduct operations such as translation, summarization, classification, and inform dynamic chatbots, we instead distinguish between semantic capacity [23] of a model, and user supplied contextual constraints [24] in these operations.

Here, semantic capacity ranges from low, useful for specific pattern recognition tasks, to high, which enables abstraction, recombination, and reinterpretation. Meanwhile, low contextual constraint facilitates open-ended exploratory operations, while a high contextual constraint tightens operations to specific audiences and formats by reducing variance.

Building on prior distinctions between content and context awareness, we reinterpret GenAI outputs as the result of two underlying mechanisms: semantic capacity and contextual constraint. Within this space, we identify four recurring modes of generative operation: translation, classification, summarization, and amplification. These modes do not represent fixed capabilities of the AI system but emerge from how generative processes are constrained and orchestrated in practice. These mechanisms jointly determine how generative output is shaped, and outcomes depend on how they are orchestrated, as outlined in the following section.

2.2 Generative Operations

Echoing Sundberg & Holmström [21], we distinguish between translation, classification, summarization, and amplification. However, we do not define them as capabilities, but as generative operations that emerge under different configurations of semantic capacity and contextual constraint.

- Translation is the transformation of content across representational systems (e.g., languages, modalities, disciplinary codes) while preserving meaning. It requires high semantic capacity (to preserve meaning across representations) and operates under strong contextual constraints (target language, format, audience, domain)
- Classification means assigning inputs to predefined or emergent categories based on learned semantic regularities. It sits primarily in the low semantic capacity half, because it depends more on discrimination than recombination. It often operates under strong task constraints.
- Summarization builds on compressing and re-representing information while preserving salient meaning. It emerges when high semantic capacity is paired with high contextual constraint (audience, purpose, scope).
- Amplification appears when high semantic capacity is paired with low contextual constraint, enabling expanding, recombining, or extrapolating content to generate novel or enriched representations. Here, GenAI is used for ideation, sensemaking, and creative support.

2.3 Modes of Human-AI Knowledge Production

Figure 1 conceptualizes GenAI outputs as a function of semantic capacity (Y-axis) and contextual constraint (X-axis). Rather than representing fixed system capabilities, classification, summarization, translation, and amplification are depicted as generative operations that arise under different configurations of these mechanisms. High semantic capacity enables abstraction and transformation, while increasing contextual constraint steers generation toward faithful re-representation rather than expansion. It should be noted that the quadrants should not be interpreted as rigid categories, as hybrid tasks exist (e.g., translate and summarize which preserves quantity but changes representation, followed by a reduction of quantity of content), and interactions between multi-agent systems [25].

	High Contextual Constraint	**Low Contextual Constraint**
High Semantic Capacity	Situated problem-solving (summarization)	Exploratory ideation (amplification)
Low Semantic Capacity	Rule-bound assistance (translation)	Generic automation (classification)

Fig. 1. Modes of human–AI knowledge productionTrustworthy AI

To make the framework useful for the purpose at hand, we integrate a discussion regarding the inherent limitations of GenAI and need to create trustworthy AI systems. A well-known limitation in GenAI systems and LLMs is their stochastic, non-deterministic nature. They are prone to produce "hallucinations", factual errors that may sound plausible [26]. These AI systems are also biased towards "western" content as their training data is often English content from various databases and Internet content [27]. Adding to this, misalignment between task demands and generative operations (e.g., models engaging in amplification when classification is the goal) degrades perceived usefulness.

Trustworthy AI (TAI) is a holistic approach that includes not just technology per se, but also the processes and actors designing, developing, and deploying AI systems [28]. TAI has been presented as a particularly important focus area for digital government research [29] and various initiatives have been developed in practice, with the EU framework being one of the most comprehensive [28]. Díaz-Rodríguez et al. [28] emphasize the importance of critically discussing the roles of AI versus humans, and what functions AI is assigned to, and to attain a risk-based approach. In the following, we will raise some aspects related to the requirements for TAI.

Human agency & oversight refers to the ability to maintain human choice and control, and that AI should support, not override, human agency. In complex information processing, it is a challenge to ensure human agency. Mechanisms such as human-in-the-loop, human-on-the-loop, and human-in-command can be applied to safeguard human agency and oversight [28].

Technical robustness and safety refer to objectives to minimize harm and build resilience, but also to ensure accuracy, reliability and reproducibility, and that the system operates as expected [28]. In the GST context, where information informs decision-making and each GST recurs every five years, outputs must remain comparable across a long time span.

Data governance must ensure data quality and integrity, privacy, legitimized access, and data management throughout the AI cycle in a way that mitigates bias, especially of training data sets [28, 30].

Transparency ensures that the relevant stakeholders have access to appropriate information. In a TAI context, this includes aspects of traceability (to track the system's data, development and deployment processes), explainability (making AI operations understandable) and communication (how people are informed about the AI system) [28, 31].

Diversity, non-discrimination and fairness involve the avoidance of unfair bias in AI outputs. It also means ensuring diversity of opinions. Ensuring diversity, representativeness, and completeness in data, diversity of people involved, and applying principles in design, training, and operation of AI systems that prevent and detect discrimination is important [28].

Social and environmental wellbeing addresses the social and environmental impact of AI, including carbon footprint and consumption of natural resources. Measures can be taken to reduce environmental impact, including the use of carbon footprint calculators, selection of necessary data (to avoid superfluous data processing), efficiency measures, and use of renewable energy [28].

Accountability includes both responsibility and accountability for the development, deployment, maintenance and use of AI systems and their outcomes. This is closely linked to fairness and risk management. Auditability enables assessment of algorithms, data and design processes. Accountability establishes liability of decisions derived from AI outputs [28].

3 Method

3.1 Research Context: The Global Stocktake

The Paris Agreement builds on three common goals, which countries make voluntary commitments to contribute to: a) mitigating climate change, b) adapting to the impacts of climate change, and c) making finance flows consistent with a pathway towards low greenhouse gas emissions and climate-resilient development. The Agreement builds on voluntary measures and emphasizes the consideration of equity and common but differentiated responsibilities and capabilities [1]. The Paris Agreement is backed by an Enhanced Transparency Framework (ETF), to promote implementation and build mutual trust. This framework requires countries to regularly report on their emissions, commitments and implementation. Every fifth year, a global stocktake evaluates collective progress towards the goals in the Paris Agreement [1].

The global stocktake (GST) is conducted in three phases: information collection and preparation, technical assessment, and consideration of output. The secretariat prepares synthesis reports in the information collection and preparation phase. The technical assessment should consider the inputs and assess collective progress, through dialogues in an open, inclusive and facilitative manner. The consideration of outputs consists of high-level events where the findings of the technical assessment are presented and discussed. The outputs of the global stocktake should: identify opportunities, challenges and good practices in enhancing action and support for collective progress; summarize key political messages and recommendations to strengthen action and support; and be referenced in a COP decision and/or declaration [32].

The information input to the GST includes a wide range of sources, including government reports under the ETF and the Convention, reports from the Intergovernmental Panel on Climate Change, subsidiary and constituted bodies, UN Agencies and other international organizations, synthesis reports from the secretariat, as well as voluntary submissions from Parties and non-Party stakeholders and UNFCCC observer organizations [32].

3.2 Sampling

This study employs a scoping review to identify challenges in the GST and map these challenges to potential GenAI applications. A scoping review is suitable to create an overview of a body of knowledge, map key characteristics and identify research gaps in the literature [33, 34]. We followed the five-stage process for scoping reviews proposed by [33]: 1. Identifying the research objective; 2. Developing a search strategy including selection of databases, search terms, inclusion/exclusion criteria; 3. Selection of studies; 4. Data extraction and analysis; and 5. Collating, summarising and reporting results.

Research objectives and questions posed in scoping reviews are often broad. As mentioned in the introduction, our aim is to identify challenges and suggestions for the global stocktake, and identify both how GenAI could facilitate the process, as well as risks and key considerations in this context.

Searches for publications were conducted in the databases Google Scholar, Web of Science, and Scopus. Search terms included "global stocktake AND artificial intelligence OR AI"; and "global stocktake AND lessons learned OR experiences". The selection of papers primarily focuses on experiences from the first GST conducted in 2023, while also including some papers published upfront to the GST, containing suggestions for its successful implementation. Only peer-reviewed scientific papers in English were included. We did not limit the selection based on the field of research. After duplicates across the databases were removed, 42 papers were selected for a closer read.

3.3 Data Analysis

Our analysis of the 42 papers is structured around a "problem–solution mapping" [35] in three steps, outlined below.

First, we scanned the sample to identify challenges mentioned in relation to the GST process. As the GST is a fairly new process, first undertaken in 2023, this was mainly inductive thematic work [36], without the use of any previous classifications. The articles were coded with the objective to identify key challenges. The codes were inductively aggregated into three main themes: information challenges, effectiveness, and outcomes of the GST.

Second, sensitized by the GenAI framework in Sect. 2, we refined the identified challenges to match the modes of operation (translation, summarization, classification, and amplification) proposed by Sundberg & Holmström [21]. As these solutions were pre-defined, this step was deductively performed based on the four modes.

Finally, based on literature on trustworthy AI in Sect. 2.4, we highlighted potential challenges and critical issues to explore what type of safeguards needs to be in place when using GenAI in the context of climate negotiations.

4 Findings

In the following section, findings from the literature review are reported. Challenges in the GST are thematized into three main categories: 1) The information challenge; 2) effectiveness, and 3) GST outcome. These challenges are summarized in Table 1 and further outlined in Subsects. 4.1–4.3.

4.1 The Information Challenge

A major challenge in the GST is to synthesise, analyse and make sense of a vast amount of diverse information [3–6].

> "With over 170 thousand pages of documents to be considered (...), the scale of the Stocktake is daunting. Submissions are also highly varied, as they are made by a wide variety of Party and non-Party actors (...), and they can be in any of the United Nations (UN) languages." [4, p. 1]

Table 1. Summary of findings

GST challenge	Findings
Information challenge	The information to be analysed in the GST includes a multitude of sources in different formats and from different stakeholders [5] and comprises over 170,000 pages [4]
Effectiveness	The GST's role to assess progress and drive ambition can be strengthened through a sectoral approach [2, 14, 44–46]
GST outcomes	Ensure that key findings from the technical dialogue are included in the final output [3], focus more on solution and means of implementation [4, 42, 48]. Tensions undermine trust and the ability to reach consensus decisions [48]

The UNFCCC secretariat produces synthesis reports based on the sources of input to the GST. Synthesis reports are also produced from the outcome of the technical dialogues, which draws on consultations with not only governments and scientists, but also other stakeholders including cities, businesses, farmers, indigenous people, civil society, and others [5]. Thus, there are various steps of synthesizing and analysing information, to inform the dialogues and negotiations in the GST process. The amount of information to process entails challenges in making the synthesis transparent, repeatable, and comparable across future GSTs. Sietsma et al. [4] suggest that GenAI can be helpful in identifying key issues in large volumes of text, and making the process more transparent, efficient, and repeatable. The co-facilitators of the technical dialogues expressed major challenges to efficiently manage and synthesise information and raised the question whether AI could support this [6]. Insufficient language translation and capturing vivid discussions were also raised as practical problems [37].

Moreover, the GST process should enable a broad participation of stakeholders, including the possibility to comment on information provided by Parties [38]. The process should be transparent, flexible, fair, and efficient. It should also be a process of learning and meaning making, and give sufficient time for science-policy dialogues [39], and to enable that, the vast amount of information must be effectively organized and analysed to derive relevant insights and clear messages [40].

4.2 Effectiveness

The GST has a double purpose to both assess progress and increase ambition over time [1]. The GST is a review mechanism with the objective to ensure effectiveness of the Paris Agreement. Effectiveness includes capacity for problem-solving, goal-attainment, and behavioural change. However, while many review mechanisms have standards of goals and obligations stated in international agreements which countries are held accountable to, the GST does not have that kind of standard. The GST assesses collective progress and does not focus on individual countries [39]. Several scholars suggest that the technical dialogues could define best practices that illustrate due diligence [37] and benchmarks [41]. Publicly showing performance distribution visualisations and evaluating individual countries can further strengthen accountability. This could also help countries evaluate their own trends and highlight good examples [41]. The GST further has a significant role in collective learning, and shaping a shared understanding on the meaning, measurement, and status of progress towards the goals in the Paris Agreement [39].

International institutions' core functions can be categorized into guidance & signal; rules & standards; transparency & accountability; means of implementation; and knowledge & learning [42]. Regarding the GST specifically, scholars [43] have emphasised its role for agenda setting, driving ambition, ensuring accountability, and providing guidance & signal. The public review and assessment of progress contribute to accountability, and the GST informs the next round of nationally determined contributions (NDCs), in which countries set their goals and commitments. The GST also shapes a global vision of the climate transition (guidance & signalling) and is a forum for setting agenda [43]. Improvements of the effectiveness of the GST can preferably be discussed towards these functions.

Various scholars call for a sectoral approach to the GST [2, 14, 44–46]. The climate transition requires systemic transformation, where each sector (such as transportation, energy, agriculture) requires tailored approaches to its system, including institutions, infrastructure, markets, business models, and social practices [46]. An integrated approach centred around sectors could coordinate mitigation, adaptation, and finance for transformation, and emphasize crosscutting considerations better [44]. A sectoral approach can be more specific on climate action [45], and clarify expectations for each sector, enabling comparison and evaluation of sectoral performance, and policy making [41]. Sectoral decarbonisation roadmaps can provide guidance for international cooperation [46] and strengthen the GST's role to drive ambition and provide guidance and signal [42]. A sectoral approach to finance and support could also more clearly specify needs for support, provided support, and gaps, and thereby enhance effective allocation of means of implementation [42].

4.3 Outcomes of the GST

A major critique of the first GST was that it did not deliver sufficient outcomes, and formulations in the output were weak [44, 47, 48], and diluted compared to the outcome of the technical assessment [37]. Scholars suggest the implementation of mechanisms to track progress of key issues raised in the technical dialogues and efficiently disseminate the findings from the technical assessment among negotiation groups [3].

> "Possible improvements need to ensure a greater alignment between the technical dialogues and the final outcome of the GST, promoting a more balanced outcome between key sectors and ensuring that considerations of climate justice are reflected more prominently in the GSTO." [3, p. 11]

However, the content of the outcome is not primarily a technical problem, but a political one. Climate negotiations entail persistent tensions, related to equity, balancing mitigation and adaptation measures, and finance and responsibilities, also undermining trust [48], and making it challenging to reach consensus. The facilitators of the technical dialogues expressed the need to balance various aspects (such as mitigation and adaptation) in the reports from the technical dialogues, and to use well established language to reach consensus [6]. To strengthen the political outcome of the GST, the outcome of the technical assessment has to be translated into strong political signals that build momentum for ambitious decisions and drive action by various actors [14, 49]. The technical assessment should have a package of different outputs, targeting different actors, to inform national action and international cooperation. This could include both political agreements and practical toolkits, focusing on how to fill the gaps. AI might in this context be helpful to tailor presentation of information to different stakeholders [14].

The GST has further been criticised for being more descriptive than solutions oriented [4], emphasizing goals rather than focusing on implementation [53], and failing to establish resources and mechanisms for implementation [42]. The GST should catalyse action within and outside the UNFCCC process, across a range of actors. To this end, the GST should identify opportunities for action and focus on overcoming the implementation gap [45]. Hermwille et al. [50] propose that the communication of the GST can be tailored to different stakeholders and stages of the NDC cycle (agenda setting, policy formulation, and decision making of the Nationally Determined Contributions) to promote implementation of GST outputs and increase ambition in NDCs [50].

5 Discussion

In the previous section, we identified challenges in the GST process and mapped them to proposed GenAI solutions. In this section, we discuss how these solutions need to be accompanied by safeguards outlined in literature on trustworthy AI. Together, these elements form a problem–solution mapping [35] linking climate governance, GenAI management, and trustworthy AI. This mapping is summarized in Table 2 and further detailed below.

5.1 Problem-Solution Mapping #1: Using Genai to Process Large Information Volumes

The first problem–solution mapping is the most straightforward: LLMs' capacity to process large volumes of text could help stakeholders make sense of the extensive GST documentation. This involves use cases of high contextual constraint, focusing on the content of the GST, to translate between different languages and provide summaries as needed. Related to the framework of GenAI operations, translation and summarization could help to synthesise large volumes of heterogeneous information. Here, *translation* enables the transformation of content across languages and formats, while preserving meaning [21]. This facilitates the interpretation and synthesis of submissions in different languages and formats, thus addressing the *heterogeneity* problem. Although the reporting requirements under the Paris Agreement have specified standard formats, this only covers the government reporting. The GST should in addition to this include a broad variety of submissions from other stakeholders, and those will be in a variety of formats. *Summarization* with GenAI offers compressing and re-representing this information [21]. This addresses the *information volume* problem by enabling synthesising of an

Table 2. GST challenges, mapped to GenAI solutions, and notions of trustworthy AI

GST challenge	GenAI solution(s)	Trustworthy AI
The information challenge	*Translation and summarization* to promote efficient synthesis and analysis to evaluate progress and inform negotiations	Translations and summaries need to be robust while at the same time supporting several types of data and represent diverse perspectives and priorities
Effectiveness	*Classification* to enable the identification of gaps, best practices, needs, to support learning, guide climate action, and promote international cooperation	Classifications need to be made in transparent ways, based on ideas of traceability and explainability
GST outcomes	*Amplification* to create creative solution spaces and support novel ideas beyond polarized positions	AI ideation cannot be reduced to automation: it needs to be aligned with diverse human values, and subject to human oversight

extensive amount of information to something more manageable. A challenge though, is the problem of bias, where AI models tend to perpetuate western perspectives. Translation and summarization are closely related; to be able to summarize heterogeneous information, the AI must be able to translate the content across languages and formats. Here, there is a need to configure GenAI systems to provide accurate representations of the content - technical robustness and safety in terms of trustworthy AI [28] - to ensure a fair representation of topics and enable a range of diverse stakeholders to make use of the materials.

The objective of the GST is to a) evaluate the progress of the goals in the Paris Agreement, and b) inform further action [1]. The information provided to policy makers will affect how they act in climate governance. Clarity, accuracy, and a comprehensive understanding of progress is imperative [48], where gaps are clearly indicated and problems are not concealed, to provide clear guidance on what needs to be done onwards. Considering the political nature of many of the submissions to the GST [4], the content of the GST may be highly contested, and the risk of bias may further escalate tensions and distrust in dialogues and negotiations. Data governance frameworks that ensure transparency, fairness, and mitigate bias [28, 30] must consider these risks. Bias may in this context include political priorities, and what issues are emphasized or left out. On the other hand, the GenAI capacity to synthesise large volumes of information across different languages, is also a possibility to improve fairness and inclusion. The role of AI in shaping environmental understanding in climate policy making has risks [51] and should be critically evaluated. As with any AI system, GenAI and LLMs rely heavily on their training data. Because most models are trained on openly available content, they tend to align with dominant views and risk overlooking edge cases and outliers. It is thus crucial to design models that have broader representation even if it means reducing the size of the training data. Here, it is important to use and adapt principles from data- and AI governance frameworks that support fairness in synthesizing information in the GST context in particular, and global negotiations in general.

5.2 Problem-Solution Mapping #2: Using Genai to Improve Effectiveness

In addition to processing language, GenAI systems also come with a range of analytical capabilities like their predecessors in various natural language processing and machine learning systems. Here, LLMs may support the evaluation of the goals in the Paris Agreement, by identifying gaps, successes, and needs for support. This in turn can inform further priorities on climate action, benchmarking, allocation of resources, international cooperation, and accountability. The objective of the GST is to evaluate collective progress and not assess countries individually [1]. However, scholars have also suggested having sector-specific visions, benchmarks, and visualisation of performance to strengthen accountability and highlight good examples [41]. Previous research has further suggested that it would be beneficial if the GST would show the emissions of each country, region, sector, and greenhouse gas, and what their trends are [12]. Here, GenAI's *classification* capabilities imply that information can be categorized according to predefined or emergent regularities [21]. This could, for instance, facilitate thematization of the contents of submissions according to the different objectives of the GST: review progress towards the goals in the Paris Agreement; identify opportunities, challenges, and best practices, and inform further action [32]. The suggestion for a sectoral approach and benchmarking might also be facilitated by GenAI classification to cluster information according to sectors, and in those sectors identify best practices that can be the foundation for benchmarking, which can drive transformation through positive examples and praise, while also identifying underperformance. GenAI can thus support provision of information in a way that promotes a sectoral approach, and to tailor information to different stakeholders [14], which could more clearly inform stakeholders to develop sector-specific policies, international cooperation [46], market- and finance mechanisms, and civic participation. This would however require a more profound institutional innovation along the use of GenAI. However, transparency alone is not enough to ensure accountability; information provision must be accompanied by institutional mechanisms [52].

As with earlier generations of AI systems, a notable issue here is the lack of transparency - it is challenging to derive how GenAI came to a certain conclusion [28]. Such issues can, at least partially, be solved by measures that force GenAI systems to "explain" their "reasoning" by providing step by step explanations, and sources used [53]. This mode of GenAI operation (classification) has a higher level of impact, compared to synthetization of information, and may have a greater impact as advice to policymakers. It is therefore increasingly important to consider the risk of bias in political priorities, especially given the deeply uneven distribution of the causes and effects of climate change and of agenda-setting power. Here, GenAI could either reinforce existing power dynamics or help level the playing field. Transparency is not a neutral space that informs climate action but is highly contested politically [52]. Moreover, considering the strong resistance to the climate transition and the frequency of disinformation, ensuring the accuracy and integrity of the information is of utmost importance [54], so that the information to policy makers is not skewed or concealing major problems [55]. Data governance frameworks, and AI robustness measures should particularly consider these risks. Information and digital technologies have interpretative, facilitative, and normative roles, influencing understanding, decision making and accountability [55]. The role of GenAI in influencing those aspects should further be critically evaluated and discussed, to not retreat from human responsibility.

5.3 Problem-Solution Mapping #3: Using Genai to Strengthen Gst Outcomes

In the third and final problem-solution mapping, GenAI capabilities operating under low contextual constraints are envisioned as a creative partner that not only operates on a content level but amplifies it by bringing in ideas from training data and other sources, and combining them into e.g., solutions and paths forward for climate negotiations. GenAI's capacity for amplification means that AI can be used to generate novel insights based on the information available, through sensemaking and creative support [21]. In the context of the GST, this could imply using GenAI to assist with creative solutions to climate action challenges, as well as negotiation challenges, thus facilitating coming to agreement. It may also assist in the formation of international cooperation, by identifying common problems or matching needs and solutions. While this is the most novel suggested use of GenAI here, it is imperative that such efforts are accompanied by measures of human oversight to ensure AI is augmenting rather than automating processes. This is particularly important considering the impact AI would have in mediating understanding of the environment and shaping visions in climate policy, potentially leading to an estranged ecological understanding [51].

A major problem of the GST was insufficient outcomes [44, 47, 48], and a persistent problem in climate governance is inadequate action and implementation. Scholars have emphasized the GST's role for collective learning [42], shaping a global vision for the climate transition, providing guidance and signal, and driving ambition [2]. Information provision is important for these tasks, and AI has further been proposed to facilitate the tailoring of outputs to different stakeholders, to drive climate action and strengthen implementation [14]. GenAI could potentially strengthen the focus on creative solutions, indicate potential collaboration (such as matching needs and provision of support), overcome negotiation blocks, and orient the negotiations in general towards possibilities for positive societal transformation rather than barriers and resistance to change. However, considering the risks of misrepresentation [51], inaccuracy and bias, it is critical to ensure AI robustness and human oversight [28], and use AI to augment and not replace human governance. This mode of GenAI operation (amplification) represents an additional level of complexity, compared to the previous ones. This may further entail a greater consumption of electricity and thus environmental footprint. The use of GenAI here is of course partially speculative. Echoing emerging research, there are several scenarios for the role of AI in negotiations and diplomacy, ranging from replacing humans and reducing them to 'data gatherers' to more augmentative uses where AI can facilitate better crisis management, communication, and negotiations [56]. This paper can contribute to a discussion on how to deliberately use GenAI to enhance climate negotiations, while managing its risks.

6 Conclusions, Limitations, and Further Research

The objective of this paper was to map challenges in the global stocktake, and identify both how AI could facilitate the process, as well as risks and key considerations in this context. By analysing a corpus of scientific literature and using a framework of human-AI knowledge production and trustworthy AI, we have outlined potential ways for using GenAI *for* governance, in this case during climate negotiations, as well as explored governance *of* AI in complex contexts.

By using problem-solution mappings, we have identified three key areas in which GenAI could support the GST, as well as key risks and trustworthiness concerns that need to be addressed. However, it is important to understand the limitations of the study, regarding how AI can be applied, as well as the risks addressed. What is discussed in this paper should not be considered as all-encompassing, but a contribution to the discussion on potential implementation of GenAI in the GST.

This paper thus contributes to the literature on AI governance in two main ways. First, the literature review ties together the topics of GenAI and climate negotiations. The proposed problem-solution pairs provide specific examples of how GenAI can support complex negotiation processes such as the GST, thus highlighting areas for *governance with AI*. Second, we have both adapted and applied the GenAI framework proposed by Sundberg & Holmström [26], with specific use cases in a context. In addition to refining the dimensions of the initial framework, we have also tied it to ideas of trustworthy AI by showing the potential tensions that arise when aiming to use GenAI in sustainable ways. By doing so, we have created a foundation for key principles for *governance of AI* in this context.

Further research is needed to a) analyse stakeholder views on possibilities and concerns of using GenAI in the GST; b) review experiences of implementing AI in global governance institutions, in a globally diverse context; and c) develop AI and data governance frameworks that take the specific needs and challenges in this context into consideration. This paper has highlighted some of them, but more comprehensive work on governance frameworks is needed if AI is to be used in this highly complex context. Here, we propose scenario-building as a path forward to generate diverse future trajectories and avoid falling into the traps of "AI solutionism". Such investigations may provide more detailed explorations of what "kind" of GenAI is needed to be useful in these contexts, by examining issues such as diversification of training data, model choice and size, and localized AI initiatives to reduce reliance on "Big Tech".

Disclosure of Interests. The authors have no competing interests to declare that are relevant to the content of this article.

References

1. Paris Agreement to the United Nations Framework Convention on Climate Change (2015). www.unfccc.int
2. Hermwille, L., Siemons, A., Förster, H., Jeffrey, L.: Catalyzing mitigation ambition under the Paris Agreement: elements for an effective Global Stocktake. Clim. Policy **19**(8), 988–1001 (2019)

3. Herrera, O., Alcaraz, O., Sureda, B., Lazzarini, B.: Cherry-picking climate science: how is the GST technical dialogue reflected in the final GST outcome? Clim. Policy, 1–13 (2025)
4. Sietsma, A.J., Groenendijk, R.W., Biesbroek, R.: Progress on climate action: a multilingual machine learning analysis of the global stocktake. Clim. Change **176**(12), 173 (2023)
5. Voigt, C.: The first Global Stocktake under the Paris Agreement: role, functioning and legal implications. Chin. J. Environ. Law **8**(2), 258–270 (2024)
6. Winkler, H., Akhtar, F.: Navigating the technical dialogue of the first global stocktake from process to findings. Nat. Clim. Change 1–7 (2024)
7. Klein R., Harris, K., Bakhtaoui, I., Lager, F., Lindblom, A., Carson, M.: Building climate diplomacy back better: imagining the UNFCCC meetings of tomorrow. S. E. Institute (2021)
8. Engvall, T.S., Flak, L.S., Sæbø, Ø.: The role of digital technologies in global climate negotiations. Gov. Inf. Q. **40**(4), 101867 (2023)
9. Große, C., Sundberg, L.: Generative AI and digital resilience: a research agenda. J. Risk Res. (2025)
10. Carroll, N., Holmström, J., Stahl, B.C., Fabian, N.E.: Navigating the utopia and dystopia perspectives of artificial intelligence. Commun. Assoc. Inf. Syst. (2024)
11. Clutton-Brock, P., Rolnick, D., Donti, P.L., Kaack, L.H.: Climate Change and AI Recommendations for Government Action (2021). https://www.climatechange.ai/press_releases/2021-11-08/release
12. Engvall, T.S., Flak, L.S.: The state of information infrastructure for global climate governance. Transforming Gov. People Process Policy **16**(4), 436–448 (2022)
13. Pauw, W.P., Klein, R.J.: Beyond ambition: increasing the transparency, coherence and implementability of Nationally Determined Contributions. In: Taylor & Francis (2020)
14. Jeudy-Hugo, S., Leon, C.: Towards a successful outcome of the first global stocktake of the Paris Agreement. OECD/IEA Climate Change Expert Group Papers (2023)
15. Tredinnick, L., Laybats, C.: Black-box creativity and generative artificial intelligence. Bus. Inf. Rev. **40**(3), 98–102 (2023)
16. Ukoba, K., Onisuru, O.R., Jen, T.-C., Madyira, D.M., Olatunji, K.O.: Predictive modeling of climate change impacts using Artificial Intelligence: a review for equitable governance and sustainable outcome. Environ. Sci. Pollut. Res. (2025)
17. Inam, S.A.: A review of artificial intelligence for predicting climate driven infectious disease outbreaks to enhance global health resilience. Disc. Public Health (2025)
18. Joshi, K.: The AI Climate Hoax: Behind the Curtain of How Big Tech Greenwashes Impacts (2026)
19. Banh, L., Strobel, G.: Generative artificial intelligence. Electron. Mark. **33**(1), 63 (2023)
20. Schneider, J., Meske, C., Kuss, P.: Foundation Models. Bus. Inf. Syst. Eng. **66**(2), 221–231 (2024)
21. Sundberg, L., Holmström, J.: Innovating by prompting: How to facilitate innovation in the age of generative AI. Bus. Horiz. **67**(5), 561–570 (2024)
22. Klesel, M., Wittmann, H.F.: Retrieval-augmented generation (RAG). Bus. Inf. Syst. Eng. **67**(4), 551–561 (2025)
23. Li, Z., et al.: Rethinking LLM-as-a-Judge: Representation-as-a-Judge with Small Language Models via Semantic Capacity Asymmetry. arXiv preprint arXiv:2601.22588 (2026)
24. Tang, T., Li, J., Zhao, W.X., Wen, J.R.: Context-tuning: Learning contextualized prompts for natural language generation. In Proceedings of the 29th International Conference on Computational Linguistics, pp. 6340–6354 (2022)
25. Allmendinger, S., Bonenberger, L., Endres, K., Fetzer, D., Gimpel, H., Kühl, N.: Multi-agent AI. Electron. Mark. **36**(1), 18 (2026)
26. Maleki, N., Padmanabhan, B., Dutta, K.: AI hallucinations: a misnomer worth clarifying. In: 2024 IEEE conference on artificial intelligence (CAI), pp. 133–138. IEEE (2024)

27. Ghosh, S., Venkit, P. N., Gautam, S., Wilson, S., Caliskan, A.: Do generative AI models output harm while representing non-Western cultures: Evidence from a community-centered approach. In: Proceedings of the AAAI/ACM Conference on AI, Ethics, and Society, Vol. 7, No. 1, pp. 476–489 (2024)
28. Díaz-Rodríguez, N., Del Ser, J., Coeckelbergh, M., de Prado, M.L., Herrera-Viedma, E., Herrera, F.: Connecting the dots in trustworthy Artificial Intelligence: From AI principles, ethics, and key requirements to responsible AI systems and regulation. Inf. Fusion **99**, 101896 (2023)
29. Medaglia, R., Gil-Garcia, J.R., Pardo, T.A.: Artificial intelligence in government: taking stock and moving forward. Soc. Sci. Comput. Rev. **41**(1), 123–140 (2023)
30. Janssen, M., Brous, P., Estevez, E., Barbosa, L.S., Janowski, T.: Data governance: organizing data for trustworthy Artificial Intelligence. Gov. Inf. Q. **37**(3), 101493 (2020)
31. Cantens, T.: How will the state think with ChatGPT? The challenges of generative artificial intelligence for public administrations. AI & Soc. **40**(1), 133–144 (2025)
32. UNFCCC. Report of the Conference of the Parties serving as the meeting of the Parties to the Paris Agreement on the third part of its first session, held in Katowice from 2 to 15 December 2018 (2018)
33. Arksey, H., O'Malley, L.: Scoping studies: towards a methodological framework. Int. J. Soc. Res. Methodol. **8**(1), 19–32 (2005)
34. Peters, M.D., et al.: Scoping reviews: reinforcing and advancing the methodology and application. Syst. Rev. **10**(1), 263 (2021)
35. Lee, J.W., Daly, S.R., Huang-Saad, A., Rodriguez, G., DeVries, Q., Seifert, C.M.: A solution in search of problems: a cognitive tool for solution mapping to promote divergent thinking. J. Eng. Des. **32**(6), 300–321 (2021)
36. Clarke, V., Braun, V.: Thematic analysis. J. Posit. Psychol. **12**(3), 297–298 (2017)
37. Klinsky, S., Holz, C., Nautiyal, S., Maurya, J., Nath, A., Athanasiou, T.: Did the First Global Stocktake Successfully Assess Progress Towards the Paris Goals 'In Light of Equity'? Available at SSRN 5263948 (2026)
38. Obergassel, W., Hermwille, L., Siemons, A., Förster, H.: Success factors for the global stocktake under the Paris Agreement. Wuppertal Institute for Climate, Environment, Energy. Part of the iGST Designing a Robust Stocktake Discussion Series (2019)
39. Milkoreit, M., Haapala, K.: The global stocktake: Design lessons for a new review and ambition mechanism in the international climate regime. Int. Environ. Agreements: Polit. Law Econ. **19**(1), 89–106 (2019)
40. Stringer, L.C., et al.: Ratcheting up effectiveness to improve the Global Stocktake process. One Earth **6**(9), 1069–1073 (2023)
41. Nascimento, L., Posada, E., Jeffery, L.: Submission to the technical phase of the Global Stocktake. New Climate Institute, Report (2023)
42. Obergassel, W., et al.: The Bitter (COP of) Truth (2025)
43. Hermwille, L., Siemons, A., Förster, H., Jeffery, L.: Catalyzing mitigation ambition under the Paris Agreement: elements for an effective Global Stocktake. Clim. Policy **19**(8), 988–1001 (2019)
44. Qi, J.J., et al.: Reflections on the first global stocktake of the Paris Agreement. Earth Syst. Gov. **21**, 100212 (2024)
45. Rajamani, L., Oberthür, S., Guilanpour, K.: Designing a Meaningful Global Stocktake (Center for Climate and Energy Solutions, Issue (2022)
46. van Asselt, H., et al.: Leveraging the Global Stocktake for effective sectoral climate governance. PLOS Clim. **2**(6), e0000229 (2023)
47. Darby, S.J., Hammond, G.P., Wu, J.: Briefing: Stocktaking global warming: the outcomes of the 2023 Dubai Climate Summit (COP28). Proc. Inst. Civil Eng.-Energy **177**(5), 193–204 (2024)

48. Zhou, Y., Pan, J.: Analysis of the process and outcomes of the first Global Stocktake under the Paris agreement. Chin. Journal Popul. Resour. Environ. **23**(2), 143–152 (2025)
49. Jeudy-Hugo, S., Leon, C.: Translating outputs to outcomes under the global stocktake of the Paris Agreement. OECD/IEA Climate Change Expert Group Papers (2022)
50. Hermwille, L., Elsner, C., Obergassel, W.: How can the Global Stocktake be leveraged for enhanced climate action?: Discussion paper (2024)
51. Francisco, M.: Artificial intelligence for environmental security: national, international, human and ecological perspectives. Current Opin. Environ. Sustain. **61**, 101250 (2023)
52. Gupta, A., van Asselt, H.: Transparency in multilateral climate politics: Furthering (or distracting from) accountability? Regul. Gov. **13**(1), 18–34 (2019)
53. Arslan, M., Ghanem, H., Munawar, S., Cruz, C.: A Survey on RAG with LLMs. Procedia Comput. Sci. **246**, 3781–3790 (2024)
54. Engvall, T.S., Flak, L.S., Sæbø, Ø.: Sharing, Cooperation or Collective Action? A Research Agenda for Online Interaction in Digital Global Governance. In: 21st IFIP WG 8.5 International Conference, EGOV 2022, Linköping, Sweden, September, 6–8 (2022)
55. Engvall, T.: Conceptualizing the Roles of Information and Digital Technologies in Digital Global Climate Governance. HICSS, Waikoloa Village, Hawaii, 7–10 January (2025)
56. Manor, I.: From digital diplomacy to AI-driven diplomacies? Communication and the Public, Mapping the potential impact of AI on Diplomacy (2026)

Producing Futures Through Large-Scale Dataset Annotation in Public Authorities

Klara Källström(✉), Marie Eneman, and Jan Ljungberg

University of Gothenburg, Box 200, 405 30 Gothenburg, Sweden
klara.kallstrom@gu.se

Abstract. Data preparation plays a central role in shaping how administrative categories become available for computational use, yet its consequences often remain difficult to trace once algorithmic systems are operational. This paper conceptualises large-scale dataset annotation as infrastructural and anticipatory work through which administrative distinctions are prepared for later algorithmic applications. Drawing on sociotechnical perspectives on data, infrastructure, and anticipatory governance, the paper examines how large-scale annotation practices contribute to different future orientations: future suspicion, future identifiability, and future knowledge and representation. Using illustrative analytical vignettes from public authority contexts, the analysis explores how annotation establishes conditions for later forms of scrutiny, recognition, and knowledge production while separating where classifications are established from where their consequences later emerge. By shifting analytical attention from algorithmic outputs and model performance to annotation, the paper highlights data preparation as a constitutive site of governance through which what becomes visible, actionable, and governable is organised. In doing so, the paper develops a temporal account of how classifications travel across organisational contexts and acquire consequence at a distance from the settings in which they were originally produced. Positioned within critical data and AI studies, the paper contributes by articulating the classificatory and infrastructural work through which the conditions for later algorithmic action are assembled and by examining the implications for accountability related to public authorities' deployment of AI systems.

Keywords: Large-Scale Dataset Annotation · Anticipatory Governance · AI in Public Authorities · Data Infrastructures

1 Introduction

Across public authorities, artificial intelligence (AI) is increasingly incorporated into everyday organisational operations, giving rise to concerns around data practices and governance. As part of broader digital government transformations, AI systems are increasingly positioned as instruments for administrative efficiency and decision support. Current developments such as AI factories at the European level and proposed AI workshops for public administration in Sweden illustrate how public authorities are investing

Published by Springer Nature Switzerland AG 2027
D. Duenas-Cid et al. (Eds.): ePart 2026, LNCS 16923, pp. 353–365, 2027.
https://doi.org/10.1007/978-3-032-36676-4_23

in shared infrastructures for AI development and deployment (European Commission, 2025; Government Offices of Sweden, 2026). At the same time, public and scholarly discussions of these systems frequently emphasise automation and analytical capacity alongside questions of accountability (OECD, 2025). Research on data-intensive systems has shown that much of the work through which algorithmic systems become operational takes place upstream of deployment, where data are prepared, structured, and aligned in organisational settings that shape later system behaviour but are often difficult to trace once systems are in use (e.g., Kitchin, 2017; Monteiro et al., 2013; Parmiggiani et al., 2023).

Yet much less attention has been directed toward the practices through which data become prepared for computational processing. Among these practices, annotation occupies a formative position. Annotation refers here to the situated work through which administrative distinctions, professional judgements, and legally grounded concepts are articulated into data structures that can participate in computational processes, drawing on classificatory logics that shape organisational action over time (Bowker & Star, 1999). Annotation unfolds within existing information infrastructures, engaging with categories that already carry meaning in organisational practice and adapting them for incorporation into data-driven systems. In this paper, however, annotation is not treated only as a way of describing or stabilising existing administrative realities. It is approached as a future-oriented practice through which public authorities help produce the conditions for later intervention, identification, and knowledge production.

Public authorities developing AI capabilities internally often encounter dataset construction as a practical site where organisational knowledge and computational requirements come into relation. While large language models (LLMs) and foundation models are often trained through self-supervised learning processes, classificatory and annotative practices continue to shape downstream processes of retrieval, alignment, and administrative operationalisation (Bommasani et al., 2021; Ouyang et al., 2022). In retrieval-based systems, including retrieval-augmented generation (RAG), such classificatory work may also influence which documents, records, or categories become available for model responses. Across different AI systems, annotation may therefore contribute to distinct orientations for later action. These concepts provide an analytical vocabulary for examining how upstream classificatory work contributes to shaping later forms of organisational action. When categories become embedded within datasets and analytic workflows, their formative work gradually recedes from view while continuing to shape later forms of organisational action. Many of the implications of category consolidation remain only partially visible at the point of data preparation and instead surface later through system operation, organisational uptake, and institutional circulation (Star & Ruhleder, 1996).

Building on this concern, the paper is guided by the following analytical question: *How do large-scale annotation practices in public authorities contribute to the production of particular future orientations by separating where classifications are made from where their consequences and accountabilities later emerge?*

This conceptual paper approaches large-scale annotation practices in public authorities as socio-technical practices that link upstream data work with later forms of intervention, identification, and knowledge production. Drawing on perspectives from data studies, infrastructure studies, and anticipatory governance, it develops a conceptual account of how annotation contributes to shaping organisational action across time. Three analytical vignettes concerning AI-based risk profiling, real-time facial recognition, and large language models are used to illustrate how different forms of annotation contribute to these future orientations.

The paper proceeds as follows. First, it develops a theoretical background around data as raw material, large-scale dataset annotation, annotation as future-making, and the relationship between classification, consequence, and organisational distance. Second, it presents three analytical vignettes from public authority contexts: AI-based risk profiling, real-time facial recognition, and large language models. Finally, the discussion returns to the theoretical framework to examine how annotation practices contribute to particular future orientations while separating sites of classification from sites of consequence and accountability. The paper ends with a conclusion.

2 Theoretical Background

2.1 Data as Raw Material

The Contemporary AI development is often premised on the assumption that data constitute a resource from which models, predictions, and services can be produced. Policy documents, industry reports, and AI strategies frequently describe data as a strategic asset and a prerequisite for innovation. Such formulations encourage an understanding of data as a form of raw material that can be collected, processed, and transformed into value through computational techniques. However, scholarship within science and technology studies and data studies has long questioned the notion of data as naturally occurring resources (Bowker & Star, 1999; Gitelman, 2013; Leonelli, 2019). Rather than existing independently of organisational settings, data are produced through organisational, technical, and social processes that shape what becomes visible, recordable, and available for analysis. Decisions concerning documentation practices, measurement procedures, metadata standards, and data curation all contribute to the production of data and influence their subsequent movement across organisational settings.

From this perspective, data preparation becomes a central part of AI development. Before data can be incorporated into models or infrastructures, they must be organised, structured, interpreted, and rendered computationally operable. Annotation forms part of this broader process. Through annotation, data become aligned with categories, distinctions, and forms of organisational knowledge that shape how they can later participate in computational systems.

2.2 Large-Scale Dataset Annotation

Large-scale dataset annotation occupies a central position within these preparatory processes. In this paper, large-scale dataset annotation is understood broadly as the classificatory, metadata-related, and organisational work through which data are prepared for computational processing (Edwards et al., 2011; Leonelli, 2019). Annotation unfolds within existing information infrastructures and draws upon legal, administrative, and organisational categories already embedded in public authority practice. Public authorities rarely begin with blank datasets. Instead, annotation typically engages records, classifications, and documentation practices that already carry institutional meaning and consequence. Scholarship on data, classification, and infrastructure provides useful resources for analysing these processes. Bowker and Star (1999) show how classifications become embedded in organisational life and continue to shape action over time. Information infrastructure research similarly highlights how standards, routines, and installed bases provide continuity while making later modification increasingly difficult (Monteiro et al., 2013; Star & Ruhleder, 1996). From this perspective, annotation can be understood as an infrastructural practice through which data are prepared for movement across datasets, systems, and organisational settings.

Through this work, annotation establishes categories that can travel across contexts. Distinctions articulated in one organisational setting may later become embedded in datasets, models, and workflows that extend beyond the circumstances in which they were originally produced.

2.3 Annotation as Future-Making

The organisation of data through annotation also has anticipatory dimensions. Annotation prepares data for possible future uses by aligning categories with forms of analysis that have not yet been fully specified. In this sense, annotation participates in governance by contributing to how algorithmic systems are oriented toward future cases without determining their outcomes in advance (Amoore, 2020; Rouvroy & Berns, 2013). As Parmiggiani et al. (2024) argue in their account of data curation as anticipatory generification, infrastructural data work often configures future possibilities before concrete applications are defined. From this perspective, annotation can be understood as a form of anticipatory structuring that shapes how later forms of action become conceivable and actionable.

This anticipatory orientation can also be understood through Amoore's notion of algorithmic propensities. Algorithmic systems shape the conditions under which certain futures become more likely than others. Annotated categories do not fix decisions in advance, but configure fields of possibility that orient attention, prioritisation, and action over time (Amoore, 2020). Annotation thus participates in the gradual formation of tendencies whose effects materialise through subsequent use. In this sense, annotation participates in organising which futures become actionable, visible, and available for intervention, while others remain outside the horizon of organisational attention.

Approached in this way, annotation can be understood as a practice through which different future orientations are assembled. The term future orientations refers to ways in which distinctions established during annotation shape possibilities for later organisational action. Risk profiling establishes conditions for future suspicion through the preparation of cases for possible scrutiny and intervention. Facial recognition establishes conditions for future identifiability through the preparation of individuals for later recognition and matching. Large language model applications participate in assembling future knowledge and representation through the preparation of information, interpretation, and response for later use. Together, these orientations illustrate how annotation participates in organising the conditions under which particular futures become visible, actionable, and available for intervention.

These orientations emerge later, through deployment, interpretation, and organisational use, when classifications prepared upstream become connected to concrete practices of intervention, identification, and representation. Annotation therefore establishes a relationship between where classifications are produced and where their consequences later become visible. This relationship becomes particularly significant in large-scale AI infrastructures where categories circulate across datasets, organisational settings, and future applications.

2.4 Classification, Consequence and Organisational Distance

A central concern of this paper is the relationship between where classifications are established and where their consequences later emerge. Through large-scale annotation, data are prepared for computational processing and organised through distinctions that may subsequently travel across datasets, systems, and organisational settings. As these distinctions circulate, the organisational site in which they were produced may become increasingly separated from the settings in which they later acquire consequence.

This observation builds on scholarship concerning classification and infrastructure. Bowker and Star (1999) show how classifications become embedded in organisational life and continue to shape action beyond the circumstances in which they were originally established. Similarly, infrastructure studies emphasise how standards, routines, and installed bases persist across organisational settings while becoming increasingly difficult to modify (Star & Ruhleder, 1996; Monteiro et al., 2013). From this perspective, the consequences of annotation may not be fully visible at the moment of data preparation. They often emerge later, when classifications are operationalised through systems, organisational routines, and institutional practices.

Understanding annotation therefore requires attention not only to how distinctions are established, but also to how they travel and acquire consequence across time and organisational space. This relationship between classification, consequence, and organisational distance provides an important foundation for the analysis that follows.

3 Producing Futures Through Annotation

The three analytical vignettes presented in this section are used to illustrate the conceptual argument developed in the paper. Together, they support the development of three analytical categories through which annotation contributes to different future orientations: future suspicion, future identifiability, and future knowledge and representation.

The vignettes are not intended as formal case studies or empirical evaluations. Rather, they illustrate how distinctions established during data preparation become connected to later forms of intervention, identification, and representation. Together, they suggest that annotation takes different forms depending on the data, classifications, and organisational settings involved, while also revealing a common dynamic in which classifications are established upstream and become consequential downstream through organisational use.

3.1 Future Suspicion

The first vignette concerns AI-based risk profiling within the Swedish Social Insurance Agency (Eneman, Källström & Ljungberg, 2026). Annotation practices involve administrative cases, control cases, repayment cases, deviations, and risk indicators. Through these practices, distinctions are established between what appears ordinary and what appears worthy of further scrutiny.

Annotation practices in public authority settings typically begin from administrative records produced for purposes other than algorithmic modelling (Star & Ruhleder, 1996; Kallinikos, 2004). These records emerge through routines of case handling and legal reporting and carry categories that already matter within public administration (Ruppert, 2011). When such records are prepared for AI systems, they carry forward existing administrative distinctions, documentation practices, and assumptions about relevance. Annotation brings these categories into relation with computational requirements as decisions are made regarding which attributes should be included, how records should be segmented, and how administrative judgements should be rendered into forms capable of circulating within datasets and models.

Studies of data science practices describe this preparatory work as taking place in organisational 'backrooms', where decisions concerning inclusion, exclusion, and categorisation are made before modelling begins while remaining largely invisible in later accounts of system performance (Parmiggiani et al., 2023). Categories appear familiar because they are continuous with established administrative practice, yet their role within AI systems may differ from the purposes for which they were originally created.

The upstream work therefore consists of preparing administrative records, control cases, repayment cases, deviations, and risk indicators for computational use. Through annotation, distinctions are established regarding what counts as ordinary, anomalous, or requiring further review.

This form of annotation contributes to what can be described as future suspicion. Categories established during data preparation participate in shaping which cases later become visible as potential sites of risk and therefore available for scrutiny, investigation, and control. Suspicion is not simply discovered through algorithmic analysis but assembled through earlier acts of classification, reflecting broader anticipatory forms of governance through which future interventions become possible (Amoore, 2020). Future suspicion refers to the assembly of subjects as potentially suspicious before suspicion itself exists. Through annotation, individuals, cases, and behaviours become positioned within a field of possible future intervention.

Downstream, these classifications become consequential when risk scores, recommendations, or alerts inform organisational attention, case prioritisation, investigation, and intervention. The classifications are established during dataset preparation, whereas their consequences emerge later through organisational action. Future suspicion is therefore produced upstream through annotation but enacted downstream through subsequent administrative practices.

3.2 Future Identifiability

The second vignette concerns real-time facial recognition. Annotation practices involve facial images, biometric characteristics, matching criteria, and identification parameters.

Through these processes, faces become transformed into computationally comparable entities capable of participating in systems of automated recognition (Fussey & Murray, 2025; Eneman et al., 2025).

A second form of annotation can be observed in systems designed to identify individuals. Real-time facial recognition depends upon image datasets prepared through categories such as facial images, biometric characteristics, matching criteria, and identification parameters. Annotation establishes distinctions through which individuals become comparable and recognisable within computational systems (Bowker & Star, 1999; Crawford & Paglen, 2021).

The upstream work consists of preparing biometric data through classifications concerning similarity, correspondence, and matchability. Through these processes, facial features are rendered computationally comparable, enabling later recognition, matching, and identification across datasets and operational settings. This form of annotation contributes to what can be described as future identifiability. The significance of annotation lies not in recognising who someone is today but in creating the conditions under which they may become identifiable tomorrow. Through dataset construction, individuals become positioned as potentially identifiable subjects within future surveillance environments (Amoore, 2020). Future identifiability therefore refers to the capacity for later recognition, matching, and tracking that is assembled through earlier acts of annotation and classification.

Downstream, these classifications become consequential when recognition systems are deployed in operational settings. Identification, tracking, location, and intervention become possible because earlier acts of annotation have already established the conditions through which individuals can later be recognised. Future identifiability refers to the assembly of individuals as potentially identifiable subjects before identification itself takes place.

3.3 Future Knowledge and Representation

The third vignette concerns large language models in public authority settings. Annotation practices involve language data, instructions, preferences, safety rules, retrieval structures, and various forms of alignment work that shape how models generate responses.

When annotation practices prepare language data for algorithmic use, they also participate in shaping the forms of knowledge and representation that later become available through AI systems. This dynamic is particularly visible in large language models. Public-sector deployments of LLMs depend upon upstream processes through which language data are organised, aligned, prioritised, and connected to organisational requirements (Bommasani et al., 2021; Ouyang et al., 2022).

The incorporation of organisational knowledge into algorithmic systems unfolds through situated engagement. Categories acquire meaning as models are trained, outputs are interpreted, and responses are connected to organisational concerns (Suchman, 2007). Annotation therefore contributes not only to model performance but also to shaping what later becomes available as information, interpretation, and representation. This work includes the preparation of language datasets, instructions, preferences, safety rules, retrieval structures, and alignment mechanisms through which models are configured. Through these practices, certain forms of information become prioritised, certain interpretations become available, and certain perspectives become visible through model outputs (Leonelli, 2019; Amoore, 2020).

This form of annotation contributes to what can be described as future knowledge and representation. Unlike risk profiling or facial recognition, the issue is not who becomes suspicious or identifiable. Rather, it concerns how organisational knowledge is prepared for later retrieval, interpretation, and communication through AI systems. Annotation shapes what becomes knowable, representable, and actionable through AI systems. Future knowledge and representation therefore refer to the ways in which earlier annotation practices shape what can later be known, communicated, and acted upon through interactions with AI systems.

Taken together, the three vignettes illustrate how different forms of large-scale annotation link upstream data preparation to downstream organisational consequences. In risk profiling, administrative cases and risk indicators are prepared in ways that establish conditions for future suspicion and later become consequential through scrutiny and intervention. In real-time facial recognition, biometric characteristics and matching criteria establish conditions for future identifiability, later becoming consequential through identification, tracking, and intervention. In large language models, language data and alignment mechanisms participate in assembling future knowledge and representation and later become consequential through information provision and decision support. Across all three examples, distinctions are established upstream during data preparation, while their consequences emerge downstream through organisational use. This separation between classification and consequence provides the foundation for the discussion that follows.

4 Discussion

The vignettes demonstrate that large-scale annotation in public authorities is not a uniform practice. What is annotated, how annotation is organised, and what consequences follow differ substantially across risk profiling, real-time facial recognition, and large language models. At the same time, the examples reveal a common dynamic: distinctions are established upstream during data preparation and become consequential later through organisational use. This separation between where classifications are established and where their consequences emerge provides a useful lens for understanding annotation as a site of governance.

4.1 Future-Making Through Annotation

The vignettes suggest that annotation functions as more than a technical activity concerned with preparing data for computational processing. Across the three examples, annotation contributes to the assembly of distinct future orientations: future suspicion, future identifiability, and future knowledge and representation. These orientations highlight different ways in which distinctions established through annotation become connected to later organisational action. While they differ in their content and effects, they share a common relationship to upstream data preparation through which organisational distinctions are rendered computationally operable.

Building on Bowker and Star's (1999) analysis, classifications can be understood as practical arrangements embedded in organisational life. They do not merely describe phenomena but also organise them by establishing boundaries, categories, and relationships that shape how action becomes possible. Situated judgements become stabilised through their incorporation into standards, routines, metadata structures, and information systems, while the work through which these judgements are produced gradually recedes from view. As classifications persist and circulate across organisational settings, they acquire infrastructural qualities that extend their influence beyond the contexts in which they were originally established (Star & Ruhleder, 1996; Monteiro et al., 2013).

From this perspective, annotation participates in translating organisational distinctions into forms capable of travelling across datasets, systems, and institutional settings. Administrative records become available as indicators of risk, biometric characteristics become available as markers of identifiability, and language data become available as resources for knowledge production and representation. Annotation contributes to organising data and to shaping the conditions under which later practices of scrutiny, identification, and knowledge production become possible. Annotation can therefore be understood as future-making because it helps configure the conditions under which particular forms of action later become recognised, justified, and enacted. In this sense, annotation is political not because it determines future outcomes, but because it participates in organising which futures become visible, actionable, and available for intervention in the first place.

4.2 Separating Classification from Consequence

The analysis suggests that annotation contributes to a separation between where classifications are established and where their consequences later emerge. While classifications are produced upstream during data preparation, annotation, and model development, their effects become visible downstream through organisational use, interpretation, and intervention. This separation is not incidental. It is a structural feature of how large-scale annotation operates within contemporary AI systems.

Large-scale annotation unfolds through established routines of documentation, classification, and case handling, drawing on existing administrative arrangements and adapting them for computational use. As an infrastructural practice, it is shaped by prior categorisations, data formats, and documentation standards that continue to influence how distinctions move across datasets, systems, and workflows over time (Bowker & Star, 1999; Leonelli, 2019). Once incorporated into datasets and models, later modification can become difficult as these arrangements become embedded within the installed base of information infrastructures (Monteiro et al., 2013).

The significance of this process lies not only in the classifications that are established, but also in the distance that emerges between classification and consequence. Organisational judgements are translated into datasets, indicators, matching criteria, and alignment structures during data preparation. Their implications may only become visible later, when classifications are operationalised through risk assessments, biometric identification, or algorithmically generated representations. What appears downstream as a risk score, a facial match, or a model response is therefore not simply the outcome of later computation. It is the manifestation of earlier classificatory decisions through which some futures were made actionable while others remained outside the scope of organisational and computational attention.

Across the three vignettes, this separation takes different forms. In risk profiling, classifications become consequential when previously established indicators inform later assessments of risk and intervention. In facial recognition, classifications become consequential when biometric distinctions support identification and operational action. In large language models, classifications become consequential through the representations and forms of knowledge made available through model outputs. In each case, classifications are established in one setting while their consequences emerge in another.

4.3 Accountability, Consequence, and Temporal Distance

The separation between classification and consequence has important implications for accountability. If annotation contributes to producing the conditions under which later interventions become possible, accountability cannot be understood solely in relation to the moment when consequences become visible. The distinctions that shape later action are often established long before risk assessments, identifications, or model outputs emerge in practice.

This temporal dynamic challenges conventional understandings of accountability within public-authority AI governance. Attention is frequently directed towards systems at the point of deployment or use: whether a decision can be explained, whether an intervention is justified, or whether an output is accurate and lawful. While these concerns remain important, they risk locating accountability primarily at the point where consequences become visible rather than where the conditions that made those consequences possible were established.

As classifications travel across datasets, models, infrastructures, and organisational settings, they continue to shape action while becoming increasingly detached from the circumstances under which they were originally established. Responsibility therefore becomes distributed not simply across actors, but also across time. Accountability cannot be confined to the point of deployment, intervention, or output. It must also encompass the upstream practices through which distinctions are assembled, stabilised, and made available for later action.

4.4 Assembling Futures Through Infrastructure

The dynamics identified in the analysis become particularly significant when annotation is organised through infrastructures that support the circulation and reuse of datasets, models, and classifications across organisational boundaries. Annotation assembles data together with the conditions under which later interventions, identifications, and representations become possible.

Viewed through this lens, the significance of emerging AI infrastructures lies not simply in enabling AI development, but in organising the circulation of classifications through which future interventions become possible. As annotation becomes embedded within shared datasets, interoperable systems, and reusable AI capabilities, distinctions established in one organisational setting become increasingly capable of travelling beyond the contexts in which they were originally produced. Risk indicators, biometric distinctions, and language-model alignment mechanisms may circulate beyond the settings in which they were first established, extending the organisational reach of earlier classificatory work over time.

The implications for accountability are substantial. As classifications become embedded within infrastructures, the temporal and organisational distances identified throughout the analysis do not disappear; they become infrastructural. The relationship between where classifications are established, where consequences emerge, and where accountability is sought may become increasingly difficult to trace. Understanding annotation as a constitutive site of governance therefore requires moving analytical attention upstream towards the classificatory work through which later forms of intervention, identification, and representation are made possible.

5 Conclusion

This paper has examined how large-scale annotation practices in public authorities participate in shaping later forms of intervention, identification, and knowledge production. More specifically, it has shown how annotation separates where classifications are established from where their consequences and accountabilities later emerge. Through three vignettes, the analysis illustrated how risk profiling establishes conditions for future suspicion, facial recognition for future identifiability, and language-model applications for future knowledge and representation. Across these examples, classifications are established upstream during data preparation, while their consequences emerge downstream through organisational use, intervention, and interpretation.

By shifting attention from algorithmic outputs and model performance to annotation, the paper shows how data preparation can function as a constitutive site of governance. The analysis approaches annotation as more than a technical stage of data preparation and shows how it helps establish the conditions under which later interventions, identifications, and representations become possible. Annotation does not merely organise data. It participates in organising what becomes visible, actionable, and governable through AI systems. In doing so, the paper contributes to critical data and AI studies by directing analytical attention towards the classificatory and infrastructural work through which the conditions for later algorithmic action are assembled.

The analysis also shows how annotation introduces a temporal and organisational separation between classification, consequence, and accountability. Classifications established during data preparation continue to shape what becomes visible, actionable, and governable long after the circumstances of their original construction have receded from view. As classifications travel across datasets, models, infrastructures, and organisational settings, the relationship between where distinctions are established and where their consequences emerge becomes increasingly difficult to trace.

The analysis further suggests that accountability cannot be confined to moments of deployment, intervention, or system use. If these future orientations are assembled through upstream practices of annotation, then questions of transparency, accountability, and democratic oversight must also extend to the classificatory work through which they are made possible.

As public authorities increasingly invest in shared AI infrastructures, the significance of annotation extends beyond the preparation of individual datasets or systems. Annotation becomes part of the infrastructural arrangements through which classifications circulate, become actionable, and acquire consequence across organisational settings. Understanding how futures are assembled through annotation is therefore important for maintaining accountability, legitimacy, and democratic oversight of AI in public authorities, and for analysing how governance increasingly operates before interventions become visible in practice.

References

Amoore, L.: Cloud Ethics: Algorithms and the Attributes of Ourselves and Others. Duke University Press (2020)

Bommasani, R., et al.: On the opportunities and risks of foundation models. arXiv:2108.07258 (2021). https://doi.org/10.48550/arXiv.2108.07258

Bowker, G.C., Star, S.L.: Sorting Things Out: Classification and its Consequences. MIT Press (1999)

Crawford, K., Paglen, T.: Excavating AI: the politics of images in machine learning training sets. AI Soc. **36**(4), 1105–1116 (2021). https://doi.org/10.1007/s00146-021-01162-8

Edwards, P.N., Mayernik, M.S., Batcheller, A.L., Bowker, G.C., Borgman, C.L.: Science friction: data, metadata, and collaboration. Soc. Stud. Sci. **41**(5), 667–690 (2011). https://doi.org/10.1177/0306312711413314

Eneman, M., Ljungberg, J., Miranda, D., Urquhart, L., Webster, W.: The ontological shift in surveillance: revisiting the "surveillant assemblage" in the age of facial recognition. Surveill. Soc. **23**(4), 498–504 (2025). https://ojs.library.queensu.ca/index.php/surveillance-and-society/index

Eneman, M., Källström, K., Ljungberg, J.: The opacity of seeing: algorithmic welfare surveillance and the politics of visibility. In: UK Academy for Information Systems Conference Proceedings, Sheffield, UK (2026)

European Commission. AI factories – Shaping Europe's digital future. European Commission (2025). https://digital-strategy.ec.europa.eu/en/policies/ai-factories

Fussey, P., Murray, D.: Facial Recognition Surveillance: Policing and Human Rights in the Age of Artificial Intelligence. Oxford University Press (2025)

Gitelman, L.: Raw data is an oxymoron. In: Gitelman, L. (ed.) "Raw Data" is an Oxymoron, pp. 1–14. MIT Press (2013)

Government Offices of Sweden. Assignment to the Swedish Social Insurance Agency and the Swedish Tax Agency to establish an AI workshop for public administration [Uppdrag till Försäkringskassan och Skatteverket att etablera en AI-verkstad för den offentliga förvaltningen (Diarienummer Fi2026/00018)]. Government Offices of Sweden (2026). https://www.regeringen.se/regeringsuppdrag/2026/01/uppdrag-till-forsakringskassan-och-skatteverket-att-etablera-en-ai-verkstad-for-den-offentliga-forvaltningen

Kallinikos, J.: Deconstructing information packages: organizational and behavioural implications of ERP systems. Inf. Technol. People **17**(1), 8–30 (2004). https://doi.org/10.1108/09593840410522152

Kitchin, R.: Thinking critically about and researching algorithms. Inf. Commun. Soc. **20**(1), 14–29 (2017). https://doi.org/10.1080/1369118X.2016.1154087

Källström, K., Eneman, M., Ljungberg, J.: Anticipatory governance and the temporalities of dataset construction in AI factories. In: Proceedings of the 17th Scandinavian Conference on Information Systems (SCIS), Iceland (2026)

Leonelli, S.: Data-Centric Biology: A Philosophical Study. University of Chicago Press (2019)

Monteiro, E., Pollock, N., Hanseth, O., Williams, R.: From artefacts to infrastructures. Comput. Supported Coop. Work (CSCW) **22**(4–6), 575–607 (2013). https://doi.org/10.1007/s10606-012-9167-1

Parmiggiani, E., Monteiro, E., Storni, C., Østerlie, T.: In the backrooms of data science: Exploring the work practices behind data-driven innovation. Comput. Supported Coop. Work (CSCW) **32**(1–3), 1–33 (2023). https://doi.org/10.17705/1jais.00718

Parmiggiani, E., Amagyei, N.K., Kollerud, S.K.S.: Data curation as anticipatory generification in data infrastructure. Eur. J. Inf. Syst. **33**(5), 748–767 (2024). https://doi.org/10.1080/0960085X.2023.2232333

OECD. Governing with Artificial Intelligence: The State of Play and Way Forward in Core Government Functions. OECD Publishing (2025). https://doi.org/10.1787/795de142-en

Ouyang, L., et al.: Training language models to follow instructions with human feedback. arXiv (2022). https://doi.org/10.48550/arXiv.2203.02155

Rouvroy, A., Berns, T.: Algorithmic governmentality and prospects of emancipation: disparateness as a precondition for individuation through relationships? Réseaux **177**(1), 163–196 (2013). https://doi.org/10.3917/res.177.0163

Ruppert, E.: Population objects: interpassive subjects. Sociology **45**(2), 218–233 (2011). https://doi.org/10.1177/0038038510394027

Star, S.L., Ruhleder, K.: Steps toward an ecology of infrastructure: design and access for large information spaces. Inf. Syst. Res. **7**(1), 111–134 (1996). https://doi.org/10.1287/isre.7.1.111

Suchman, L.A.: Human–Machine Reconfigurations: Plans and Situated Actions. Cambridge University Press (2007)

Role-Based Analysis of Public Administration in National AI Strategies Under Evolving AI Governance

Isabell Peters(✉) and Christian Weidner

Technical University of Applied Sciences Wildau, Wildau, Germany
isabell.peters@th-wildau.de

Abstract. Artificial intelligence (AI) is rapidly reshaping public administration by influencing how governments design services, manage data, and exercise regulatory authority. National AI strategies have become essential tools for governments to articulate their ambitions and define their role in governing AI, yet the policy landscape has evolved significantly since their adoption. This paper examines how national AI strategies define the role of public administration in AI governance. Drawing from concepts in AI governance, digital government, and policy design, we propose a role-based analytical framework that conceptualizes public administration as AI adopter, regulator, and orchestrator of AI ecosystems. We apply the framework in a qualitative comparative policy analysis of national AI strategies in Germany, the Netherlands, and the United Kingdom. Using a deductive-inductive qualitative content analysis, we reconstruct country-specific role profiles and compare strategic emphases across the three cases. Our findings reveal that all three strategies foreground the adopter role, with extensive public-sector use cases and capacity-building measures, while systematic scaling beyond pilots is made explicit only in the UK's "scan, pilot, scale" approach. Regulatory framings are more prominent and tightly coupled to EU-level developments in Germany and the Netherlands, whereas the UK emphasizes innovation and competitiveness with fewer detailed legal commitments. Orchestration arrangements likewise diverge, ranging from partnership-oriented ecosystem coordination in the Netherlands and the UK to more hierarchical steering in Germany. We argue that a role-based lens helps to uncover how early strategic choices position public administrations within emerging AI ecosystems and to identify misalignments between these framings and the evolving European AI governance landscape, thereby providing an empirically grounded lens to assess whether existing strategies remain fit for purpose.

Keywords: Artificial Intelligence · Public Administration · Strategy · Comparative policy analysis · Qualitative content analysis

D. Duenas-Cid et al. (Eds.): ePart 2026, LNCS 16923, pp. 366–381, 2027.
https://doi.org/10.1007/978-3-032-36676-4_24

1 Introduction

Artificial intelligence (AI) has moved from experimental pilots to a key enabling technology in public administration, shaping how governments design services, manage data and exercise authority. Even several years after their adoption, national AI strategies remain key reference points for how states justify, steer and institutionalize the use of AI in the public sector. At the same time, the landscape around them has changed profoundly: developments such as the EU's AI Act, rapid advances in generative AI, and the emergence of new AI governance bodies have increased the pressure on governments to clarify the role of public administration not only as a user of AI, but also as a regulator and ecosystem coordinator [4,6,11,14]. Analyzing these strategies today thus serves a dual purpose: It reveals how early strategic choices continue to frame current governance debates, and it provides a critical lens on whether existing role conceptions of public administration are still adequate for contemporary AI developments.

National governments increasingly rely on AI to transform public administration, yet they differ markedly in how they envision the state's role in governing and using AI [6]. In particular, national AI strategies vary in the extent to which they frame public administrations as a proactive adopter of AI, as a regulator of AI systems, and as an orchestrator of wider AI ecosystems. This paper investigates these role conceptions through a comparative policy analysis of the national AI strategies of Germany, the Netherlands and the United Kingdom with a specific focus on their public-sector dimension.

The three cases offer a theoretically and empirically interesting contrast. Germany has adopted a federal "Artificial Intelligence Strategy" [17] and an updated 2020 version [18], that combine strategic objectives with an extensive catalogue of measures, but without a separate, government-wide AI action plan for public administration. The Netherlands follows an integrated approach, as its "Strategic Action Plan for Artificial Intelligence" functions both as the central national strategy and as an action-oriented agenda [19], including public-private coordination and measures for the public sector. An updated "Government-wide vision on Generative AI" was published in 2024 [20]. The United Kingdom, by contrast, has replaced its 2021 "National AI Strategy" with completely revised AI Action Plans, which specify how AI should be deployed and governed across government [2,3].

Against this backdrop, the paper asks:

To what extent do national AI strategies frame public administration as an adopter, regulator, and orchestrator of AI ecosystems?

Drawing on concepts from AI governance, digital government and policy design, the study develops a role-based analytical framework and applies it in a qualitative comparative document analysis of the three national strategies. The contribution is twofold: empirically it maps and contrasts how three European governments position their administrations in emerging AI ecosystems; conceptually, it refines a roles-based lens for analyzing AI governance in the public sector that can inform both future comparative research and the design of national AI strategies.

2 Theoretical Framework

National AI strategies serve as a central policy instrument through which governments articulate their vision for integrating AI into public administration. This paper applies a role-based analytical framework drawn from AI governance, digital government, and policy design literature. The framework positions public administration along three core roles and serves as a set of sensitizing concepts guiding the systematic reconstruction of strategic framings while remaining open to inductive refinement.

2.1 AI Governance and Digital Government

Recent scholarship on AI governance emphasizes that states play multiple roles in AI governance beyond mere regulation [5,10,11,14]. While early work primarily focused on controlling AI-related risks through top-down regulation [9], newer contributions highlight that governments simultaneously act as:

- **Adopters:** deploying AI internally to improve efficiency, service delivery, and decision-making [5,14];
- **Regulators:** establishing rules, oversight structures, and normative principles to ensure responsible AI use [5,16];
- **Orchestrators:** facilitating public-private AI ecosystems through partnerships, infrastructure, and coordination [1,5].

Digital government literature complements this by focusing on public sector innovation. Janowski outlines four maturity stages of digital government [8]. Margetts and Dorobantu introduce the concept of the "algorithmic state", where public administration acts as platform operator, leveraging AI for predictive services while managing risks like bias and opacity. These works underscore adoption (like AI-driven process automation) and orchestration (e.g., data spaces as public goods), framing public administration as both technological modernizer and ecosystem enabler [12].

This perspective provides a comprehensive lens for understanding how governments balance innovation promotion with risk mitigation, particularly in the public sector where AI intersects with service delivery, data stewardship, and democratic accountability [9,14]. In the context of this study, these roles directly inform the coding framework used to analyze national strategies.

2.2 Policy Design and Instrument Choice

Policy design theory provides tools to analyze how strategies translate roles into instruments. Howlett et al. classify policy tools as authority, treasure, nodality, and organization [7], revealing how AI strategies mix substantive and procedural tools, such as sandboxes. Thierer advocates "permissionless innovation", where new technologies should be permitted by default absent compelling evidence of serious harm [21]. This framework positions governments as enablers rather than precautionary gatekeepers, supporting hybrid orchestration tools that balance innovation with public-sector oversight in AI strategies. [15].

2.3 Operationalizing Governmental Roles in AI Strategies

Existing work on national AI and digitalization strategies highlights that governments assume multiple roles in relation to AI. Hjaltalin and Sigurdarson show that recent AI strategies predominantly frame the state as an enabler and facilitator of AI ecosystems, while giving comparatively less detailed attention to the internal use of AI in public administration [6]. Madan and Ashok's systematic review similarly documents the importance of governments as adopters, regulators and ecosystem builders, although these roles remain largely background categories rather than an explicit analytical framework [11]. OECD reports on "governing with AI" further emphasize that governments increasingly combine internal AI use, regulation, and ecosystem steering, but typically discuss these dimensions separately rather than through an integrated role-based lens [14].

Table 1. Coding Scheme.

	DimensionSubdimension	Definition
	Adopter	*Internal AI use for efficiency/service transformation*
A1	Objectives	Strategic goals for AI in administration (e.g., efficiency, personalization).
A2	Use Cases	Concrete applications (e.g., chatbots, predictive analytics).
A3	Capacity Building	Skills, training, organizational change.
A4	Implementation Logic	Pathways like scan-pilot-scale.
	Regulator	*Oversight and norm-setting for trustworthy AI*
B1	Regulatory References	Laws or binding standards (e.g., EU AI Act).
B2	Oversight Structures	Institutions (e.g., AI safety institutes).
B3	Normative Principles	Public values (like transparency, non-discrimination).
B4	Assurance Mechanisms	Tools like audits, sandboxes.
	Orchestrator	*Ecosystem coordination and enablement*
C1	Infrastructure	Data/compute provision (e.g., sovereign clouds).
C2	Partnerships	Public-private coalitions.
C3	Ecosystem Facilitation	Funding or hubs for innovation.
C4	Multi-Level Coordination	Integration into international bi- or multilateral agreements.

Building on these contributions and the theoretical synthesis developed above, our study advances a more explicit and comparably fine-grained role framework that disaggregates public administration into adopter, regulator and orchestrator roles and operationalizes them through the subdimensions A1–A4, B1–B4 and C1–C4 in Table 1. This compact set of categories is particularly well suited to national AI strategies, as it captures both the internal use of AI in public administration and the external governance and ecosystem functions attributed to the state. The resulting codebook, accompanied by definitions, inclusion and exclusion criteria, and anchor examples, forms the basis for the qualitative document analysis and is available in the Supplementary Materials.[1]

3 Methods

This study employs a qualitative comparative policy analysis to address the research question. It systematically examines how national AI strategies frame public administration in the three specified roles, drawing on principles of qualitative content analysis adapted for policy documents. The methodological approach is deductive-inductive [13], with a theory-derived coding framework operationalizing the roles of AI adopter, regulator, and orchestrator.

3.1 Case and Document Selection

The analysis focuses on three European countries selected for their substantive AI strategies and varying institutional traditions: Germany (federal system), the Netherlands (consensus-oriented), and the United Kingdom (unitary with strong central coordination). The primary corpus consists of core national AI strategy documents which are displayed in the following Table 2.

Table 2. AI strategy documents of three European countries.

	Primary Documents	Year	Rationale
DE	Artificial Intelligence Strategy; 2020 Update	2018; 2020	Core strategic framework with measures for public administration
NL	Strategic Action Plan for AI; The government-wide vision on Generative AI	2019; 2024	Integrated strategy-action document addressing public sector use and governance digitalization
UK	AI Opportunities Action Plan: Independent Report; Governmental Response	2025	Implementation-focused recommendations paired with declarations of intent

Our document corpus deliberately focuses on core national AI strategy documents and their most recent AI-specific updates: the German AI Strategy and its 2020 update, the Dutch Strategic Action Plan for AI and the 2024 government-wide vision on generative AI, and the UK AI Opportunities Action Plan (independent report and government response). We acknowledge that each country has produced additional, partly overlapping documents on digitalization and AI (such as UK white papers or broader national digital strategies), but these were only consulted selectively because they either extend beyond AI-specific

[1] Supplementary Materials: https://doi.org/10.5281/zenodo.18980276.

governance or have a separate, domain-specific scope and thus would reduce cross-country comparability. This strategic focus on core AI strategies is therefore a conscious scope decision, which we explicitly treat as a limitation of our sample in the methodological reflections in Sect. 5.5.

3.2 Analytical Framework and Coding Procedure

A role-based coding scheme operationalizes the research question across three main categories, each with subdimensions derived from AI governance and digital government literature, cf. Table 1.

This study employed a multi-step coding procedure to ensure rigour and transparency. The primary coder first deductively coded all national AI strategy documents using the theory-derived framework of public administration as AI adopter, regulator, and orchestrator, further differentiating subcodes where necessary and documenting ambiguous passages and borderline cases. The two coders then discussed these borderline cases at a conceptual level to clarify coding rationales and refine category definitions. Subsequently, the secondary coder independently reviewed all critical segments and applied her own coding decisions. Finally, both coders jointly examined all divergent codings and remaining borderline cases; some discrepancies were resolved through consensual reinterpretation, while others were retained as analytically meaningful tensions. Details of the coding process, decision rules, and documentation of borderline cases and unresolved discrepancies are provided, see Supplementary Materials.

3.3 Data Analysis and Validity

First, the analysis yields country-specific role profiles, followed by cross-case comparison to identify patterns. For example, the UK emphasizes adoption orchestration, while Germany has a stronger regulatory focus. The robustness of the analysis is supported by transparent codebook documentation (see Table 1), verbatim examples for each category, intercoder discussion and agreements (see Supplementary Materials), and an explicit discussion of limitations. One limitation is the focus on *policy-as-written* rather than implementation. For the UK and German case, we used the English- and German-language originals, as well as the official English-language version of the Dutch strategies. The sample design of three cases with six strategy documents prioritizes analytical depth over breadth. Together, these measures provide systematic and replicable insights into strategic role framing.

4 Findings

The analysis along the predefined main categories adopter, regulator, and orchestrator provides an overall impression of what goals national AI strategies aim to achieve. This enables a comparative analysis of the various government tendencies and characteristics, identifying systematic overrepresentation or gaps in the individual strategies.

4.1 Adopter

Across all three strategies, public administration is framed as a central domain for AI adoption, but with different degrees of ambition and specificity. The German and Dutch documents present the broadest range of public-sector use cases, from chatbots for citizen communication and automation in procurement to data analytics for climate action, often described with comparatively high technical detail. Capacity-building measures span individual skills and organizational capabilities, with Germany focusing on continuing education and experimentation environments, the Netherlands emphasizing Science, Technology, Engineering, and Mathematics (STEM) and digital literacy, and the United Kingdom prioritizing the attraction and immigration of AI talent. At the same time, all three strategies link public-sector capabilities indirectly to wider ecosystem investments in labor markets, academia, citizens, and industry. The implementation logic is most explicit in the UK's "scan, pilot, scale" approach, which casts government as an early adopter and scaling pioneer, while Germany and the Netherlands largely remain at the levels of preparation and piloting, with only limited discussion of systematic scaling as a governmental responsibility. Building on this overall pattern, the role of public administration as an AI adopter can be unpacked along four dimensions: strategic objectives, use cases, capacity building, and implementation logic.

Objectives. Across the three strategies, objectives for AI use in public administration range from task-centered improvements (efficiency, productivity, quality) to organization-wide modernization and broader macro-level goals such as sustainability, security, and social well-being. Germany and the Netherlands place particular emphasis on relieving routine workloads and modernizing administrative processes, whereas the UK links public-sector AI adoption more explicitly to competitiveness and national security. This mixture of micro-, meso- and macro-level objectives underlines that AI is framed not only as a technical tool, but as a lever for wider public-sector and societal transformation.

In addition to the positively framed objectives, such as improving accessibility of services or reducing psychological stress in challenging work situations, negative aspects and risks are also highlighted to the reader, particularly in the German and Dutch strategies.

Use Cases. The strategies also differ in how concretely they specify public-sector use cases. Germany and the Netherlands provide the richest and most technically detailed catalog of administrative applications, including chatbots for citizen interaction, automation of procurement, and data analytics for climate and sustainability policies. The more recent documents in particular focus closely on the potential of generative AI and large language models. In all three countries, however, public safety and defense organizations stand out as particularly prominent adopters, with examples such as data mining for law enforcement, predictive analytics for disaster management, and autonomous systems in defense.

Capacity Building. Capacity-building measures span both individual skills and organizational capabilities. Germany emphasizes continuing education for civil servants and the creation of experimentation environments such as labs and sandboxes, while the Netherlands highlights STEM skills, digital and media literacy and broader investments in citizens' and workers' competencies. The UK, by contrast, places stronger emphasis on attracting and recruiting AI talent from abroad, situating public-sector capabilities within a broader strategy to make the country an attractive AI location. All strategies underline the importance of lifelong learning for civil servants and citizens.

Implementation Logic. Finally, the implementation logic of AI adoption in public administration is articulated with different levels of specificity. Germany and the Netherlands mainly describe preparatory activities (such as data-sharing frameworks and legal clarification) as well as piloting in controlled environments, with only limited and implicit reference to systematic scaling into routine operations. In this regard, German strategies focus heavily on evaluations and selective support for so-called *Leuchttürme* (flagship projects). The Netherlands also wants to test and utilize products in a responsible manner and experiment in "safe government environments" [20]. The United Kingdom stands out with a more explicit "scan, pilot, scale" approach, in which government is portrayed as an early adopter and scaling pioneer that accelerates procurement, standardize AI systems and rapidly moves from prototypes to everyday use in administrative work. Figure 1 provides an overview of the identified phases, showing the British approach of scaling up alongside the progression from piloting to utilization.

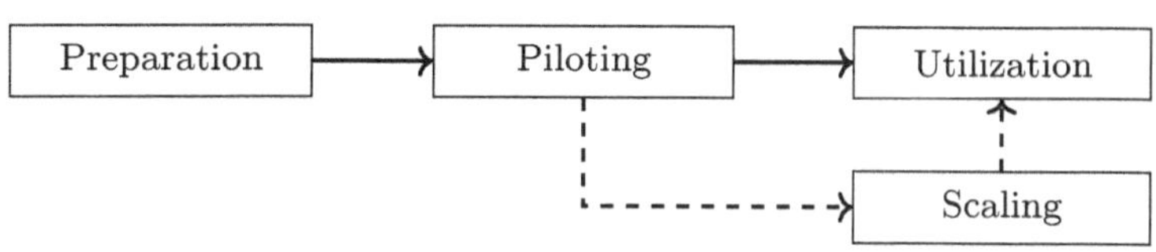

Fig. 1. Implementation logic

4.2 Regulator

Across the three strategies, public administration is clearly expected to help govern AI risks, but the regulator role is articulated with varying depth and emphasis. While Germany and the Netherlands anchor their approaches strongly in existing and emerging European frameworks, the UK adopts a more competitiveness-oriented stance with comparatively fewer explicit regulatory commitments.

Regulatory References. All three strategies refer to adapting or extending legal frameworks to address AI, but they do so with different scopes and levels of specificity. Germany and the Netherlands explicitly reference data protection, copyright, procurement, and digitalization laws as well as broader commitments such as the UN Sustainable Development Goals and, in the 2024 Dutch document, alignment with the EU AI Act. With regard to administrative digitization laws, the strategies reviewed place particular emphasis on data sharing, public access to government information as well as specific considerations regarding the use of data by police and law enforcement. By contrast, the UK strategy contains fewer detailed legal references and instead stresses maintaining or surpassing the EU in areas such as the text and data mining regime, reflecting a more deregulatory, competitiveness-driven framing.

Oversight Structures. The strategies also recognize the need for oversight structures to monitor AI risks, yet concrete institutional designs are unevenly developed. All three acknowledge threats to national security, data protection, competition, controllability and public values, but the Netherlands goes furthest in specifying national and multinational supervisory actors, including European and international bodies, and in sharing oversight responsibilities with the private sector. Both Germany and the UK mention oversight and monitoring more generically. Additionally, Germany highlights the role of works councils in company-level oversight.

Normative Principles. Normative principles provide an important reference frame for regulatory activity across the three cases. The reliability and security of AI systems are key elements in this regard. Openness is a central theme, encompassing open data policies and FAIR principles, which are intended to provide a more efficient framework for government and business at the administrative level. Furthermore, the principle of openness is also intended to help achieve algorithmic transparency, thus efforts to prevent "black box" systems, alongside ethical principles such as non-discrimination, transparency to prevent misuse, and protection of public values.

In all cases, public values refer primarily to democracy, the rule of law, and an open market. While the strategies of the continental European cases place specific emphasis on the principles of consumer protection and the protection of the European single market, the Anglo-Saxon perspective prioritizes international competitiveness and innovation-friendliness. Sustainability considerations, including "sustainability by design", feature particularly prominently in the German and Dutch strategies, underscoring that AI in the public sector is expected to support not only efficiency but also environmental and social goals. The issue of accessibility for marginalized groups is also addressed, as this contributes significantly to democratic values.

Assurance Mechanisms. Assurance mechanisms are presented as concrete instruments that translate abstract principles into governable practice, but with different emphases across the three strategies. Germany focuses on product safety and occupational health and safety, disclosure duties for AI systems to prevent misuse and discrimination, and the development of evaluation criteria for safe and reliable systems. The Netherlands prioritizes audits of AI use by government and public–private partnerships, monitoring of AI in the public sector and the administration of justice, and ensuring compliance with legal frameworks, supported by EU-accredited certification schemes. The UK links assurances more directly to innovation capacity and regulatory performance, highlighting public spending on AI capabilities, attention to the effectiveness of AI regulation, and the development of safe and reliable systems and consumer protection. Against this background, Table 3 summarizes assurance mechanisms mobilized under the regulator role.

Table 3. Mechanisms that apply specifically to public administration are printed in bold.

Mechanism	Purpose
Setting up Audits	**Usage of AI by government and PPPs** (NL)
	Public spending on AI capabilities (UK)
	Development of safe and reliable AI systems (DE; NL)
	Benchmarking occupational safety and health (DE)
Certification system *accredited by the EU*	**Usage of AI in the administration of justice** (NL)
Disclosure of AI systems	Prevent misuse and discrimination (DE)
Assurance Tools/ Standard Procedures	**Procurement and implementation of AI systems in Public Administration** (DE)
	Development of safe and reliable AI systems, also regarding consumer protection (DE; UK)
Monitoring	**Usage of AI in the Public Sector** (NL)
	AI regulation effectiveness (UK)
	AI talent availability, capabilities, and demand (DE)
	Availability of computing power (NL)
	AI compliance with legal frameworks (NL)
Develop Evaluation Criteria	Development of safe and reliable AI systems (DE)
Involve experts	Consider legal framework when scaling up pilots (DE)

4.3 Orchestrator

Infrastructure. Infrastructure plays a crucial role in enabling the development and deployment of advanced digital technologies. Data sharing is emphasized in every case, particularly in the German and Dutch context, to facilitate innovation and data-driven applications. In addition, increasing network capabilities is highlighted as a key prerequisite for supporting data-intensive services and ensuring reliable connectivity. Another important aspect concerns the promotion of cooperation and "co-opetition" through interoperability, which can be achieved by establishing common technical standards—an approach that is also strongly emphasized in Germany. Finally, the allocation of sufficient computing power represents a central infrastructural measure. Recent strategic initiatives in the Netherlands underscore the importance of expanding access to computing resources, in some cases in conjunction with measures to ensure an adequate energy supply, as is being discussed in detail in the UK in the context of small modular reactors.

Partnerships. Partnerships serve the public sector in direct exchange or contractual relationships to share knowledge, test technologies, enable innovation, or identify needs. This requires a certain degree of reciprocity on behalf of both partners. The strategies primarily involve the private sector and academic institutions, but also civil society with social partnerships as a German specialty, as well as intergovernmental relations as outlined in the Dutch strategy. Horizontal partnerships appear in all countries, and can be grouped into three ideal-typical forms:

- **Formal cooperation:** such as contractual public-private partnerships (PPP) that are considered particularly in the Netherlands and the UK when private actors support or co-deliver public tasks.
- **Semi-formal collaboration:** including structured access for selected companies to security or critical-infrastructure communities in the UK and participation in institutionalized round tables or working groups in Germany with similar formats sketched in the Dutch strategy.
- **Informal exchange:** such as participation in events, dialogues, and community-building activities, which all three countries use to share experiences, identify needs, and surface potential AI use cases.

These partnerships serve various purposes. They often involve the exchange of knowledge and experience such as integrating different perspectives to build data ecosystems and ICT standardization in Germany or consolidating AI skills within the Dutch public sector. Additionally, partnerships can effectively bind companies and their capacities to build innovation clusters, such as the AI growth zones presented in the British AI Action Plans, which offer enhanced access to power and support. Partnerships can also encourage companies to deliver public tasks in the Netherlands. The Netherlands and the UK consider PPP instruments more often, especially for public tasks. In the British context, the strategies highlight the governmental role as guarantor of competitive international companies and products.

Ecosystem Facilitation. In contrast to direct public–private partnerships, where the public and private sectors engage in a horizontal relationship, the government may also assume the role of an ecosystem facilitator. In this capacity, the public sector acts in a more indirect and enabling manner, aiming to stimulate innovation and collaboration within a broader ecosystem rather than engaging in specific bilateral arrangements. Typically, this involves up-front public investments with the expectation that long-term economic, technological, or societal benefits will emerge over time. Such facilitation measures often include funding mechanisms—such as knowledge-transfer programs, pilot projects, innovation clusters, or venture capital initiatives—that help reduce barriers to experimentation and collaboration among stakeholders.

Beyond financial support, governments may actively convene actors within the ecosystem through formats such as round tables, conferences, competitions, and hackathons, thereby fostering exchange and cooperation across institutional boundaries. Additionally, public authorities may establish infrastructures that enable experimentation and knowledge development, including living labs, regulatory sandboxes, collaborative networks, digital platforms, think tanks, competence centers, and observatories. These activities can serve multiple strategic objectives, such as encouraging innovation in structurally disadvantaged regions (e.g., East Germany), strengthening knowledge accumulation, promoting international cooperation, and supporting the development and diffusion of artificial intelligence technologies. Whereas partnerships are particularly widespread in our sample, the rather reserved role of the facilitator can be found especially in Germany.

Multi-level Coordination. The integration of supranational institutions and intergovernmental agreements plays a central role in achieving the stated objectives. Legal and ethical frameworks are thus being consolidated within the Dutch and German approach, particularly in the context of European institutions such as the European Commission and the Council of Europe, but also with the OECD. It is also worth mentioning here that the General Partnership for AI (GPAI), hosted by the OECD, is mentioned both in the Dutch and German documents. Capacity building, which has already been presented, is being brought into international focus within the framework of European initiatives, primarily by Germany, which also wants to establish international standards for occupational health and safety in this context. An important aspect of international coordination is the security and defense sector, which is given considerable attention in both German and Dutch strategies. While Germany often focuses on individual bilateral partnerships, e.g., within the framework of FCAS with France and Spain, the Netherlands is more oriented toward international organizations such as NATO or INTERPOL.

When building AI infrastructures, the strategies rely on international cooperation. This is emphasized in all cases, notably in the EuroHPC, ALT-EDIC, and GAIA-X projects, illustrating how AI infrastructure is embedded in European-level orchestration efforts. Worth noting in this regard is Britain's strategy, which otherwise tends to shy away from multi-level coordination, yet explicitly includes the EuroHPC project in its infrastructure efforts. Another country-specific feature is Germany's focus on development cooperation with countries in the Global South.

5 Discussion and Conclusion

5.1 Summary of Key Findings

All three national AI strategies consistently frame public administrations as AI adopters with varying emphasis on the roles of proactive adopter, regulator, and ecosystem orchestrator. Each details specific use cases (e.g., chatbots, predictive analytics) alongside capacity-building approaches, with the UK explicitly following a "scan, pilot, scale" model to drive systematic adoption. The regulator role dominates in Germany and the Netherlands: Germany aligns with national and European regulations to mitigate risk. The Netherlands utilizes partnerships and AI in accordance with the EU AI Act to unlock the innovation potential of public administration. The UK prioritizes promoting innovation as a broader economic stimulus. In the orchestrator role, the Netherlands acts most integratively (infrastructure, funding platforms), Germany emphasizes hierarchical multi-level coordination, and the UK relies on dedicated action plans.

5.2 Interpretation of Role Profiles

Our analysis reveals distinct role profiles across the three strategies.

As AI adopter, Germany and the Netherlands pursue ambitious objectives with detailed use cases (chatbots, procurement automation, climate analytics), yet both remain at preparation/piloting stages, while the UK articulates systematic "scan-to-pilot" pathways positioning government as scaling pioneer. The Netherlands mentions sophisticated implementation toolkits for innovative technologies but provides little detail on their operationalization—unlike the UK's explicit scaling responsibility.

As regulator, Germany and the Netherlands emphasize normative principles (transparency, openness, societal/labor impacts, latter a German specificity) alongside binding regulations and oversight structures, with Dutch strategies offering greater detail on European norm integration and potential legal updates. By contrast, the UK adopts a regulation-light approach favoring steering mechanisms and trust-based governance, reflecting post-Brexit competitiveness priorities over institutional mandates.

As orchestrator, differences align with administrative traditions: the Netherlands excels in public-private partnerships, Germany acts as restrained ecosystem facilitator through hubs and funding, and the UK prioritizes innovation clusters and compute infrastructure. These profiles reveal systematic trade-offs: Germany's regulatory intensity potentially constrain adoption ambition, Dutch orchestration risks accountability gaps without stronger oversight, and UK scaling focus may underemphasize multi-level coordination—each requiring institutional clarification for effective implementation.

A role-based perspective also brings into view inherent tensions and overlaps between state roles. In practice, the remit of governments as regulators may conflict with their ambitions as ecosystem orchestrators, for example when strict legal safeguards and risk controls constrain the scope for experimental partnerships or rapid scaling of AI solutions. Likewise, a strong emphasis on public administrations as proactive adopters can sit uneasily with their responsibility to enforce demanding oversight and assurance mechanisms. Viewed through a role-based lens, the strategies suggest potential tensions between different governmental functions: while the United Kingdom foregrounds innovation-friendly orchestration and scaling, the German documents place greater weight on binding regulatory alignment, and the Dutch approach seeks to reconcile both through dense multi-level coordination and partnership arrangements.

5.3 Contributions to AI Governance Debates

These findings nuance prevailing accounts of state roles in AI governance, which often treat "the state" as a unified regulator or innovation enabler rather than disaggregating the specific role conceptions attached to public administrations in strategic documents. The three strategies broadly conform to the triadic perspective of governments as adopters, regulators and orchestrators, but they also reveal that these roles are not equally specified or institutionally anchored across countries, and that they may even be implicitly sequenced (e.g., adoption before orchestration) rather than pursued in parallel. While some strands of the literature emphasize top-down regulation, our analysis suggests that strategic framings often incorporate regulatory commitments into broader agendas of administrative modernization, digital sovereignty, and ecosystem development, particularly in the Dutch and UK cases. At the same time, the comparatively strong German focus on regulatory alignment illustrates how strategies can still be primarily risk-oriented, potentially limiting the articulation of administration as an entrepreneurial orchestration hub. The emerging role profiles also resonate with debates on the "algorithmic state": the UK's explicit "scan, pilot, scale" approach and the Dutch emphasis on data infrastructures and AI ecosystems both point towards administrations that increasingly operate as platform-like, predictive service providers rather than merely digitized bureaucracies. At the same time, Germany's more cautious, regulation-centered stance suggests a more incremental trajectory towards such an algorithmic state, with stronger ex ante safeguards and multi-level coordination.

5.4 Implications for Digital Government and Policy Design

From a digital government perspective, results indicate strategies signal "algorithmic state" maturity, yet they differ in connecting objectives, capacities, and instruments. Concrete public-sector use cases (Germany/Netherlands) frame AI beyond generic modernization for domains like safety, climate, citizen services, thereby reinforcing expectations that administrations will manage complex trade-offs between efficiency, rights protection and public value creation. The limited attention to scaling logics in Germany and the Netherlands, compared to the UK's explicit adoption pathway, suggests that strategies may underestimate the organizational and political challenges of moving from pilots to system-wide transformation. In terms of policy design, our coding of instruments across the adopter, regulator and orchestrator roles highlights a tendency to combine traditional authority-based tools (e.g., legal obligations, oversight bodies) with newer, innovation-oriented instruments such as sandboxes and public-private partnerships, but often without clarifying how these tools interact over time.

Building on the cross-country patterns identified above, we derive the following policy recommendations across countries:

1. Consider more explicit scaling pathways following the UK "scan, pilot, scale" model to bridge pilot-to-production gaps.
2. Strengthen post-AI Act coordination using Dutch multi-level partnership approaches to complement German hierarchical EU alignment.
3. Develop hybrid regulator-adopter designs integrating UK innovation momentum with German/Dutch oversight traditions, for example through regulatory sandboxes with mandatory scaling requirements or certified procurement frameworks.

5.5 Methodological and Conceptual Reflections

Methodologically, the deductive-inductive qualitative content analysis and multi-step coding procedure strengthen the transparency and reliability of our reconstruction of role framings, e.g., through the explicit handling of borderline cases and analytically meaningful disagreements between coders. At the same time, this strategy-focused design also entails important limitations.

While our analysis systematically reconstructs role conceptions in national AI strategies, it remains deliberately confined to "policy-as-written" rather than implementation processes. As a result, we cannot assess to what extent the ambitious role profiles identified here have materialized in administrative practice. Furthermore, our focus on a small number of core strategy documents prioritizes analytical depth and cross-country comparability, while necessarily excluding broader digitalization and AI policy initiatives that may also shape governmental roles. Future research should therefore complement our document-based approach with process tracing of key initiatives, interviews with policy-makers and civil servants, and analyses of administrative practice. Such studies would be essential to determine whether the cross-country differences identified here translate into divergent patterns of AI adoption, regulation, and orchestration in practice. They could also offer theoretical insights into how the roles of governments are evolving and being reshaped amid changing conditions in AI governance, particularly in response to generative AI and emerging supranational regulation.

Disclosure of Interests. The authors have no competing interests to declare.

References

1. Cihon, P., Maas, M.M., Kemp, L.: Fragmentation and the future: investigating architectures for international AI governance. Global Pol. **11**(5), 545–556 (2020)
2. Department of Science, Innovation & Technology: AI Opportunities Action Plan. Government response (2025)
3. Department of Science, Innovation & Technology: AI Opportunities Action Plan. Independent report (2025)
4. Entsminger, J.: Public Sector Artificial Intelligence Strategies: Considerations for a Public Value Approach. Tech. rep., Center for the Governance of Change, IE University (2022)
5. Ferrari, F.: State roles in platform governance: AI's regulatory geographies. Competition Change **28**(2), 340–358 (2024)
6. Hjaltalin, I.T., Sigurdarson, H.T.: The strategic use of AI in the public sector: a public values analysis of national AI strategies. Gov. Inf. Q. **41**(1), 101914 (2024)
7. Howlett, M., Giest, S., Mukherjee, I., Taeihagh, A.: New policy tools and traditional policy models: better understanding behavioural, digital and collaborative instruments. Policy Des. Pract. **8**(1), 121–137 (2025)
8. Janowski, T.: Digital government evolution: from transformation to contextualization. Gov. Inf. Q. **32**(3), 221–236 (2015)
9. Jobin, A., Ienca, M., Vayena, E.: The global landscape of AI ethics guidelines. Nat. Mach. Intell. **1**(9), 389–399 (2019)
10. Latzer, M., Festic, N.: A guideline for understanding and measuring algorithmic governance in everyday life. Internet Policy Rev. **8**(2) (2019)
11. Madan, R., Ashok, M.: AI adoption and diffusion in public administration: a systematic literature review and future research agenda. Gov. Inf. Q. **40**(1), 101774 (2023)
12. Margetts, H., Dorobantu, C.: Rethink government with AI. Nature **568**(7751), 163–165 (2019)
13. Mayring, P.: Qualitative Inhaltsanalyse: Grundlagen und Techniken, 11th edn. Pädagogik, Beltz, Weinheim Basel (2010)
14. OECD: Governing with Artificial Intelligence: The State of Play and Way Forward in Core Government Functions, OECD Publishing (2025)
15. Ranchordas, S., Vinci, V.: Regulatory sandboxes and innovation-friendly regulation: between collaboration and capture. Ital. J. Publ. Law **16**(1), 107–139 (2024)
16. Taeihagh, A.: Governance of artificial intelligence. Policy Soc. **40**(2), 137–157 (2021)
17. The Federal Government: Artificial Intelligence Strategy (2018)
18. The Federal Government: Artificial Intelligence Strategy of the German Federal Government. Update (2020)
19. The Ministry of Economic Affairs and Climate Policy: Strategic Action Plan for Artificial Intelligence (2019)
20. The Ministry of the Interior and Kingdom Relations: The government-wide vision on. Generative AI of the Netherlands (2024)
21. Thierer, A.D.: Permissionless Innovation: The Continuing Case for Comprehensive Technological Freedom, revised and expanded edn. Mercatus Center, George Mason University, Arlington, Virginia (2016)

A Framework to Evaluate Government Readiness in Predictive AI Implementation

Emyana Sirait(✉), Anneke Zuiderwijk, and Marijn Janssen

Delft University of Technology, Delft, The Netherlands
{E.R.E.Sirait,A.M.G.Zuiderwijk-vanEijk, M.F.W.H.A.Janssen}@tudelft.nl

Abstract. Despite the growing use of AI in the public sector, there is limited empirical evidence on its implementation. Prior research on government readiness for AI has predominantly focused on the adoption stage from the viewpoint of IT or public managers, often yielding high-level insights and one-sided narratives. AI implementation goes beyond adoption as it involves shaping, integrating, and routinizing AI applications that are often overlooked in discussions about AI adoption. This study examines how government readiness for predictive AI implementation involving multiple stakeholders can be evaluated. It develops a framework for evaluating government readiness for predictive AI implementation by analyzing three Indonesian government use cases and one Dutch government case. The framework covers the socio-technical factors of AI and enables more nuanced analysis by involving perspectives of: 1) those of the stakeholders responsible for the development, management, and governance of AI innovation (which, in short, we call AI developers), and 2) those of the AI users. From developers' perspectives, the framework examines readiness in terms of resource availability, the organization's capability to change, and governance, while from users' perspectives, readiness is evaluated based on the perceived performance and impact of the deployed AI systems. The framework also indicates the influence of AI characteristics and individual readiness on organizational-level implementation. The scientific contribution is to extend the AI readiness model in government beyond adoption, addressing multiple stakeholders' perspectives on implementation. This framework is expected to guide public organizations in reflecting on their AI implementation and make iterative improvements to ensure responsible AI.

Keywords: Artificial Intelligence · Implementation · Government organizations · Readiness · Evaluation · Framework

1 Introduction

Many public organizations are increasingly interested in adopting, implementing, and using Artificial Intelligence (Criado & O.de Zarate-Alcarazo, 2022; Mikalef et al., 2022; Yigitcanlar et al., 2023). The reason is the various potential for AI in government,

D. Duenas-Cid et al. (Eds.): ePart 2026, LNCS 16923, pp. 382–397, 2027.
https://doi.org/10.1007/978-3-032-36676-4_25

for example, through automated systems that tackle procedural and repetitive tasks, informed decision-making that provides outcome predictions and personalized services, and chatbots and virtual assistants that enable cognitive engagement. It is expected that AI can offer benefits, including efficiency, transparency, increased citizen participation, evidence-based decisions and policies, fewer mistakes, and reduced human error (Criado & O.de Zarate-Alcarazo, 2022; Mellouli et al., 2024).

To successfully implement AI in government and maximize the benefits, public organizations should have a certain level of readiness. Previous studies on AI readiness have highlighted the factors influencing AI adoption in the public sector (Mikalef & Gupta, 2021; Mikalef et al., 2022; Wang et al., 2022), such as pressure and resource availability, that influence organizations' decisions to adopt AI. Whereas adoption is about starting to use implemented systems, the process of shaping and configuring AI in the organization has gained less attention. Implementing AI involves sociotechnical dynamics in which algorithms meet workflows, governance mechanisms, and end-user practices (Kudina & Van De Poel, 2024). The European Commission report in (Misuraca and van Noordt, 2020) exposed that the number of AI projects that proceed to the implementation phase is much smaller than the total number of proposed AI initiatives. Some determinants influence the government's readiness for implementing AI. The readiness determines how well the AI system performs after being implemented (Wang et al., 2022). A fair evaluation of AI system performance and its impacts should consider the perspectives of the users, not merely those responsible for technology development.

This study addresses the research question: *How can government readiness for the implementation of predictive AI involving multiple stakeholders be evaluated?* To answer the question, this study develops a framework that enables a thorough evaluation of government readiness from the perspectives of AI developers and AI end-users. Here, we refer to AI developers as stakeholders responsible for the development, management, and governance of AI innovation, with roles ranging from IT manager, data analyst, data scientist, digital transformation officer, and algorithm auditor. Meanwhile, AI end-users are stakeholders who directly use AI applications' predictions and recommendations to make public decisions. The scientific contribution of this research is to extend existing frameworks on AI readiness in government beyond the adoption phase and to address multiple stakeholder perspectives on AI implementation in government. Practically, this research offers insights into the factors that shape AI readiness in government decision-making.

2 Background

Among various typologies and capabilities of AI applications, one typology that is strategically used in government is AI predictive analytics. Predictive AI uses computer algorithms to learn from data, identify patterns, and make predictions, ultimately supporting informed decision-making by combining statistical modeling, data mining, and machine learning (Seifert, 2023). Data-driven technologies such as predictive AI are expected to help governments enhance their policies and services by leveraging data. As AI technology advances, it can undertake more complex tasks that require multiple cognitive capabilities within a single system. Therefore, AI is often not a standalone system but part of a larger system. It makes measuring AI complex because AI is typically bundled with other technologies (OECD, 2021). This study takes predictive AI as the object of analysis to allow deeper exploration of its distinctive characteristics.

Prior studies about organizational AI readiness have explored factors influencing the adoption and implementation of AI in the public sector. Factors influencing AI adoption generally examine why an organization intends to adopt AI and the capabilities that underlie that decision. Several pulling factors are institutional pressure (Wang et al., 2022; Rodriguez Müller et al., 2025), managerial perceptions towards AI (Criado & O.de Zarate-Alcarazo, 2022; Yigitcanlar et al., 2023; Rodriguez Müller et al., 2025), and relative advantage (Neumann et al., 2022). Whereas observed factors forming AI capabilities are tangible resources such as data and technology, intangible resources such as coordination and change capacity, and also human resources (Mikalef & Gupta, 2021; Mikalef et al., 2023; Van Noordt & Tangi, 2023).

Implementing AI involves not only resource preparation but also shaping, integrating, and routinizing AI applications. The implementation readiness should investigate how the organization can deploy the system and what factors drive successful deployment. Some studies have delved into contextual factors in AI implementation (Maragno et al., 2023). However, many contexts of AI implementation remain underexplored, including specific types of AI systems, stakeholders involved in AI implementation, and AI system users. A number of case-specific features influence AI implementation and are interconnected (Maragno et al., 2023). When AI is conceptualized as a socio-technological and adaptive system, people, policies, systems, data, AI, processes, and other elements coevolve (Janssen, 2025). Therefore, modeling readiness for AI implementation should not only include technical factors, but also human factors, ethical, and governance.

Earlier studies on AI adoption and implementation have mostly focused on perceptions from IT and managerial levels (Criado & O.de Zarate-Alcarazo, 2022; Maragno et al., 2023; Yigitcanlar et al., 2023). A few studies have taken the perspective of government employees in general (Ahn & Chen, 2022), and many claims are made by AI developers and large corporates about its substantial benefits and impact (Duan et al., 2019). As argued in (Voß & Pawlowski, 2019), not only top and middle management play a major role in successfully transforming the organization at stake. Contextual aspects, such as culture, interaction with digital change, one's mindset, and their interrelationship, are also essential. This could be explored by inviting a broader group of employees to share their perspectives on digital developments within the organization.

In practice, one organization may have initiated several AI innovations with different AI application characteristics (typologies, complexity, acquisition methods) across different departments (sizes, functions, AI use goals). Therefore, analyzing organizational AI readiness may not always reflect the full range of AI innovations. Microanalysis using a specific AI application is beneficial for providing a detailed discussion of the application's characteristics and the organization's conditions for its implementation.

3 Research Method

This section explains the method used to attain the goal of this research: to develop an evaluation framework of AI implementation in the public sector. We conducted exploratory case studies to investigate how several government organizations implement predictive AI in practice and the factors that influence its implementation. Literature mentions that a case study approach is useful for situations in which the main research questions are "how" or "why" (Yin, 2003).

In this study, we select cases in the Dutch and Indonesian governments. The cases represent variation in organizational and AI characteristics, such as administrative level, AI development stage, procurement method, and technology used. Given that two countries differ in their socio-demographic and regulatory approaches to AI, we analyze how these contexts influence the implementation of AI in each country. The Netherlands, as part of the EU, has gradually implemented the AI Act since it entered into force on 1 August 2024. This regulation establishes a foundation for Dutch AI governance, emphasizing transparency, explainability, and fairness in light of AI's potential risks. On the other hand, Indonesia has no enforceable AI regulations yet, allowing a more innovation-driven approach to AI. The existing regulation supporting AI use in Indonesia is mainly the Personal Data Protection Act, with other related rules that softly regulate the ethical aspects of AI.

For each case, we conducted interviews with relevant informants directed by the organizations, based on their suitability for the research purpose. In total, we interviewed 35 government officials from the Dutch and Indonesian governments, representing two perspectives: *AI developers* and *AI end-users*. The predictive AI applications developed and used by the studied organization vary in the techniques and algorithms they use, but they share a similar goal/function: forecast future outcomes by processing real-time historical data to support data-driven decision-making. A more detailed description of the cases included in this study is shown in Table 1 below.

We conducted semi-structured interviews using questions constructed from the lists of readiness factors identified in the literature review. The interviewees were asked to evaluate their organization's readiness for implementing a particular predictive AI application. The interviews were conducted between October 2024 and February 2025 for cases 1 and 2, and between November 2025 and February 2026 for cases 3 and 4. Each interview lasted about one hour, and all were recorded, transcribed, and thematically coded in Atlas.ti software. We applied deductive coding to the transcripts, using labels developed from the literature. We also applied inductive coding to identify unlabeled sub-factors that emerged from interviews. Through the deductive and inductive coding process, the factors influencing AI implementation in government are gathered. The labels used in deductive-inductive coding (as shown in Table 2) are then classified into two tiers of categories, which are indicated in the table by brackets. Afterward, we framed these factors in light of related IS theories to develop an evaluation framework for government readiness in predictive AI implementation. The framework takes the

Table 1. Description of AI use cases in this study

Case Characteristics	Case 1	Case 2	Case 3	Case 4
AI Applications	AI advises on the risk of citizen requests for tax reductions	AI delivers financial advisory and management to track and consolidate financial data from all local governments	AI analyzes data to support an app that serves multiple city public services to citizens	AI digitally maps the talent pool
Goals	Helps subsidy advisors make faster decisions on whether to approve or reject grant requests	Detects data anomaly, evaluates budget execution with tagging, forecasts budget realization, and analyzes spending priorities	In the form of a virtual assistant, it helps the decision-makers and related offices to seek and analyze data, for example, finances or regulations	Maps participants' competences and levels of completed digital trainings on an online training platform and predicts their completion rate on future trainings as advice for acceptance
Users	Officials responsible for making decisions on grant requests from citizens	Officials responsible for financial planning and reporting	Decision-makers or officials seeking information to inform their decisions, such as about public budgeting	Officials responsible for developing digital training programs for citizens and selecting the participants
Administrative level	Government Agency in NL	Central Government, Indonesia	City Government, Indonesia	Central Government, Indonesia
Development stages of AI	In use	In use	In use	In the pilot phase
Procurement method	Self-developed	Self-developed	Half-Procured (using RAG and Agentic AI from tech provider)	Self-developed
Technology complexity	Rule-based systems, machine learning	Rules-based system, natural language processing, machine learning	Predictive analytics combined with Generative AI	Predictive analytics, Natural language processing, Deep learning
Number of interviewees	6 AI developers, 6 AI users	8 AI developers, 5 AI users	2 AI developers, 2 AI users	3 AI developers, 3 AI users

organization, including the AI application within it, as the unit of analysis, allowing respondents to express their perceptions towards a specific AI application with which they are familiar.

Table 2. Labels used in deductive-inductive coding, and the coding categories

Deductive coding (list of readiness factors from literature, confirmed in interviews)	Inductive coding (unlisted factors or sub-factors found in interviews)
• Data availability, quality, interoperability, standardization → [Data Resource -> Resource availability] • Insufficient infrastructure, computing power, data storage, cloud-based infrastructure → [Infrastructure -> Resource availability] • Developing internal AI skills, providing training, → [AI Skills -> Resource availability] • Top management support, budget allocation → [Digital leadership -> Capability to change] • Coordination, AI awareness → [Organization culture -> Capability to change] • Regulatory support → [Regulations -> Regulations and Governance] • Fairness, accountability, transparency, explainability → [Ethical principles -> Regulations and Governance] • Benefits for efficiency, faster working, reduced cost → [Impacts -> Net benefits]	• Variation in use case characteristics: development stages, procurement method, AI technologies, function, risk → [AI characteristics] • Data-driven government, change management → [Organization culture -> Capability to change] • Human oversight, system maintenance → [Continuous learning -> Capability to change] • AI Auditing → [Ethical principle -> Regulations and Governance] • Dedicated team for AI innovation, national approach for AI, standard guidelines for AI → [Innovation governance -> Regulations and Governance] • Prediction accuracy, relevant information → [Information Quality -> System performance] • Smart features, system integration, meet users' expectations → [AI system quality -> System performance] • Ethical consideration in using AI → [Ethical compliance -> System Performance] • Actual use, user satisfaction → [Net benefits] • Leaders' mindset, attitude, and literacy gaps of users towards AI → [Individual Readiness] • Trust in the AI system → [Individual Readiness]

4 Results

Based on interview data, we recognize that AI developers and AI users highlight different factors when evaluating their organizational readiness for predictive AI implementation. From the AI developers' perspective, we identify three dimensions for evaluating implementation readiness: resource availability, the capability to change, and regulations and governance. From users' perspectives, we found that system performance and the impacts of using AI applications are the main factors they mentioned regarding AI implementation. Moreover, we identify the influence of individual readiness on organizational readiness, focusing on mindset and attitudes towards AI. Also, variations in AI characteristics affect the government's readiness for AI implementation. Findings from our case study are condensed into the four points below, which guide the development of our evaluation framework.

4.1 AI Developers' Perspective: Readiness Beyond Resource Availability

Scholars have observed challenges to the adoption of AI in public administration, including limited resources, accountability, and ethical issues (Valle-Cruz et al., 2020). Evaluating the implementation of AI in governments that have used it should cover these factors to learn how they address challenges. First, it is obvious that resource availability is the essential factor in building an AI application. At least three fundamental resources are required: data, infrastructure, and skills, as frequently mentioned in previous research on AI readiness (Jöhnk et al., 2021; Mikalef & Gupta, 2021; Pumplun et al., 2019). We learned that a system-level evaluation allows informants to be more specific in mentioning readiness determinants. For example, in our study, we found that some AI developers do not always feel they have a good grasp of the resources required for their AI implementation. In [Case 1], for example, the informant mentioned their limitations in leveraging data resources and infrastructure. "*We have got very much text data, but we cannot use it because we do not have enough processing power, and on the other hand, we cannot put the data on the cloud because it is very sensitive.*" But the organization in this use case could continue implementing because it started with a small, low-risk use case before gradually scaling up.

Beyond resource availability, we found that organizations' capability to change is also a determinant of readiness. It involves digital leadership, organizational culture, and continuous learning. In [Case 1], one interviewee said, "*We had some very strong stakeholders who committed to taking this AI model into production. I think the right top management just forced this use case. For most other projects, we do not have these kinds of stakeholders with the right mindset and priority.*" Then in [Case 2], we learned that the organization's commitment to data-driven decision-making has driven a range of data analytics and AI innovations. "*Our institution instructed all employees to explore data utilization massively because we have numerous types of financial data. Processing the data will make more impactful and data-driven decisions.*" Moreover, interviewees in cases 2 and 4 indicated that their organizations adapted their structure and operations to support ongoing AI maintenance, including data and algorithm updates.

The procurement method for AI innovation also influences how AI developers perceive the determinants of readiness. In [Case 3] of our study, for example, we see that this Indonesian local government is using modified Agentic AI services from a technology provider. The case demonstrates that procured AI can accelerate service delivery and operational efficiency when deployed appropriately. This choice allows them to benefit from AI at work, despite the lack of AI-ready data, limited infrastructure, and AI skills. "*For us, leveraging generative AI through agentic systems that are built on open LLMs is a game-changer. Such systems alleviate longstanding organizational challenges related to data management, infrastructure capacity, and specialized skills. Agentic AI can substantially reduce the technical overhead typically associated with large-scale AI deployment. Thereby enabling organizations to focus on strategic value creation, including making predictions for informed decisions, rather than operational complexity.*"

However, implementing AI should be supported by responsible governance (Janssen, 2025). For example, a standard to distinguish which use cases are sufficiently low-risk to be procured AI, and which involve sensitive data or critical processes that necessitate internal development and tightly governed AI systems. In all studied cases, there is a recurring concern about the lack of a formal national or internal regulatory framework for AI. In case 1, for instance, the lack of standardized requirements for AI innovations has hindered AI developers from deploying several AI projects. Yet, in cases 2, 3, and

4, AI developers are keen to explore AI-based innovations, despite the risks not being fully identified.

Drawing on insights from our case studies and supported by several IS theories, we conclude that, from the perspective of AI developers, evaluating government readiness for AI implementation should consider factors of resource availability, organizational capabilities to change, and governance readiness (as shown in Fig. 1).

4.2 AI Users' Perceptions: Readiness is Reflected in System Performance and Impacts

Many studies have shown the benefits of AI in increasing efficiency and driving innovation, but only a limited number of those claims come from end users. Kamal (2006) says that to achieve the anticipated benefits of technology, simply acquiring or adopting a technology is not sufficient – it must be deployed and utilized appropriately by the organization and its intended users. Earlier studies evaluating the impact of deploying AI systems on organizations, cities, and communities believed that the impacts can be personal, organizational, and societal (Collins et al., 2021). Yet, limited empirical evidence from the end-user perspective supports those statements.

Based on interviews with AI users, this study identifies several insights into how they perceived the AI implementation in the organization. First, users emphasize the importance of an AI system's performance and values in supporting their work, regardless of the underlying technology. For instance, the interviewee in [Case 2] expressed, "*The most important thing in developing AI-based solutions is that the tool answers the user's needs and makes the work easier, not merely follow the trend of technology. By that, user acceptance and engagement with the AI tool will be high.*" In line with that, users in [Case 1] expressed their expectation of better performance from the existing AI tool: *"There is still insufficient proof that the system works correctly."*

Second, users point out the need for ethical compliance of the AI system. They admit that AI has facilitated faster work, but they also expect transparency and explainability on how the system works. As we found in Case 1, users expressed a need for simple documentation that explains the AI system, as the existing documents are complex. Moreover, we observed that the studied organizations primarily focus on internal AI use, with no evaluation of its direct impact on the public. As a result, accomplishing open government goals like openness, accountability, and participation through the use of AI technologies (Mellouli et al., 2024) gets less attention.

Using these observations, we recall the DeLone and McLean (D&M) models frequently used to evaluate the IS success (Urbach & Müller, 2012). It consists of six interrelated dimensions: information, system, and service quality; (intention to) use; user satisfaction; and net benefits. In the context of an AI system, as found in our case study, we add ethical compliance as an extra dimension. Thus, we apply the adjusted D&M IS success model to evaluate government readiness in AI implementation from the perspective of AI users. In our evaluation framework (as presented in Fig. 1), we show that system performance and net benefits from using an AI system are the two main factors affecting users' perception of government readiness to implement AI.

4.3 Individual Readiness Influences Organizational Readiness in AI Implementation

Prior research indicates that individuals exhibit varying attitudes and behaviors when executing the advice provided by an AI decision-support system (Ahn & Chen, 2022; Duan et al., 2019). Human decision-makers in any position (e.g., senior managers, administrative clerks) can accept or reject the use of AI for decision-making. Hence, users' behavior in using AI should be taken into account, since personal traits, knowledge, and understanding of AI users significantly impact its use and success (Duan et al., 2019).

In [Case 1], interviews with AI developers informed us of their sense about users' resistance to using the AI system: *"Even if we have a good deep learning model, I doubt it will be used for the task because the people working on it will protest against it. They seem afraid their work will be replaced by an AI model."* Meanwhile, interviews with AI users in the same case expressed other factors, *"This AI gives me information about comparing with prior project requests, but I trust my opinion more than AI, considering my long experience. Trust and added value are something the system should work on more."*

With those insights, we recall the theoretical concept of technology readiness in people, as proposed by Parasuraman (2000). The theory refers to individuals' propensity to adopt and use new technologies to achieve goals in both personal and professional contexts. It involves the psychometric properties between drivers (optimism, innovativeness) and inhibitors (discomfort, insecurity) of technology readiness. In the context of AI, users' behavior varies. Previous studies have shown that AI has raised city managers' expectations for the future of local government services (Yigitcanlar et al., 2023), yet the main roadblocks to AI adoption in local governments are also psychometric in nature, namely trust and change management (Yigitcanlar et al., 2023). Another socio-cognitive study by O. de Zarate-Alcarazo (2022) suggested that CIOs' openness and trust towards AI technologies are positively associated with AI implementation rates in local public administration.

Drawing on insights from our case studies and supported by IS theory, we take individual readiness as one variable in our evaluation framework (as presented in Fig. 1). Individual readiness, which denotes openness to and trust in AI among AI developers and AI users, serves as an antecedent to organizational readiness for AI implementation.

4.4 AI Characteristics Influence Implementation Readiness

From the four AI use cases we studied, we identified that different AI characteristics influence the readiness factors for implementing AI applications, as shown in Table 3.

Table 3. The influence of AI characteristics on readiness factors, derived from case studies

AI Characteristics	Comparison of its readiness factors	
Method for producing the examined AI applications	*Self-developed* - the organization uses its internal infrastructure and skills to build the application [Cases 1, 2, 4] Readiness factors to be addressed: - Deal with limited infrastructure, such as computing power - Ensure continuous learning to maintain the models	*Procured AI* - utilize services from technology providers or consultants, including using Retrieval Augmented Generation or agentic AI [Case 3] Readiness factors to be addressed: - Ensure data sovereignty - Regulation and governance, for example, about data classification that is sensitive to share with the AI models - Employees' awareness of the risk of data misuse
Development stages of the examined AI applications	*In use* [Cases 1, 2, 3] Readiness factors to be addressed: - To increase the performance of the system - Evaluate the benefits for the users and the impact on public services	*In development/Pilot* [Case 4] Readiness factors to be addressed: - Coordination and clear responsibility arrangement - Users' commitment
Technology complexity	*Simple AI Techniques* - with explicit rules and limited complexity, e.g., regression analysis, time series forecasting, decision trees, rule-based systems, and other basic machine learning models [Case 1] Readiness factors to be addressed: - Limitation on the functionality of the system	*More complex AI techniques* - using models that can automatically learn patterns in large and unstructured data, such as Neural networks, Natural language processing, Deep learning, and other advanced ML algorithms [Cases 2, 3, 4] Readiness factors to be addressed: - Explainability of the black box - Fairness

(*continued*)

Table 3. (*continued*)

AI Characteristics	Comparison of its readiness factors	
National approach to AI	*Regulation-focused* [Case 1] as the impact of the AI Act that regulates AI implementation based on risk classification Readiness factors to be addressed: - Slower AI innovations - Opportunity loss of embracing the latest technology	*Innovation-focused* [Cases 2, 3, 4] as the soft regulation approach to AI. The main related regulation is the Personal Data Protection Act Readiness factors to be addressed: - Ethical use of AI - Juridical clarity on AI risks
Organizational approach to AI	*Centralized innovation* [Cases 1 and 3] Readiness factors to be addressed: - Slower AI innovations because they come only from a dedicated team - Limited AI team members to handle all the work	*Decentralized innovation* [Cases 2, 4] Readiness factors to be addressed: - Diverse AI innovations because they come from business units - Challenging for the organization to do single oversight and control

By synthesizing the insights from our case studies and integrating them into established IS theories, we propose a framework to evaluate AI implementation readiness that incorporates AI developers' and AI users' perspectives (see Fig. 1).

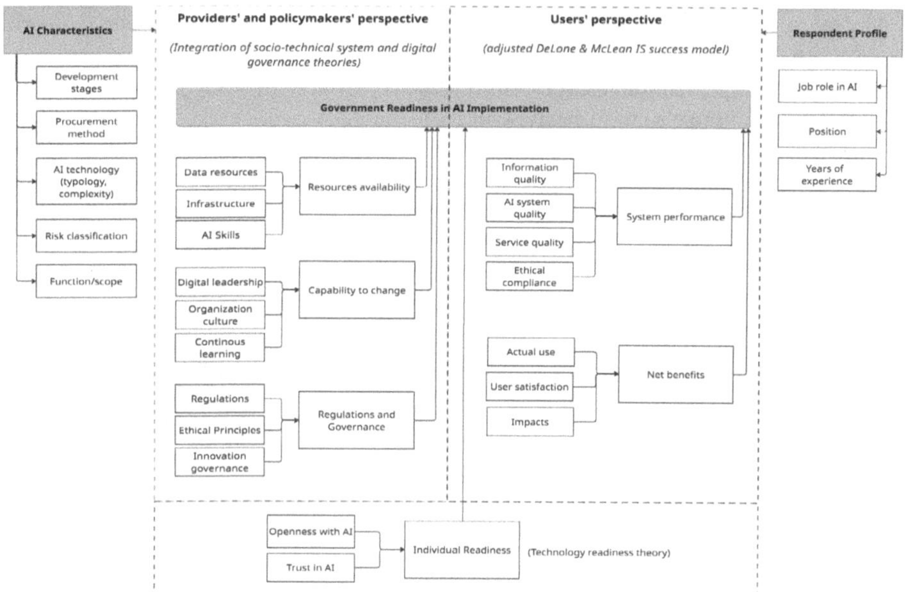

Fig. 1. A proposed dual-perspective framework to evaluate government readiness in AI implementation

5 Discussion

This study contributes to the growing discourse on AI readiness in the public sector by extending analytical attention from the adoption stage to the post-adoption stage. We suggest multiple stakeholders' perspectives for evaluating readiness during the implementation stage to cover the full process of adopting AI, from preparing resources to institutionalizing AI for daily work. A dual-perspective approach of AI developers and AI users to evaluate government readiness across four empirical cases in Indonesia and the Netherlands has identified factors influencing the implementation of predictive AI. Involving users' perceptions rather than relying solely on developers' perspectives has underscored the importance of AI system performance and its impacts on users when evaluating implementation readiness. This aligns with the success determinants found in other information systems, which consistently indicate that perceived usefulness, ease of use, and alignment with work practices significantly shape technology impacts (DeLone & McLean, 2003; Venkatesh et al., 2003).

Furthermore, integrating readiness and evaluation into a single framework appears to be an effective way to enable continuous improvement. This framework accommodates users' perceptions after deployment to ensure the ultimate purpose of technology in government is not to follow hype but to deliver tangible usefulness, public value, and improved service outcomes. This readiness evaluation framework provides a more comprehensive understanding of implementation issues, challenging the managerial, top-down interpretations that currently dominate AI research in the public sector.

In general, our evaluation framework addresses organizational readiness. However, our study also found that individuals' readiness within the organization influences implementation at the organizational level. This finding is in line with a study from Yigitcanlar et al. (2023), saying that unfamiliarity with AI technology becomes a reason some local governments are not prepared for AI. Or a study from Rodriguez Müller et al. (2025) stating that public managers' perceptions of AI, such as regarding its anticipated benefits and risks, may moderate public managers' willingness to adopt AI. It becomes obvious that individual perceptions play a role in organizational decisions, particularly in the context of innovation.

Theoretically, our proposed framework goes beyond the resource-based view theory, which is frequently used in AI readiness studies in the public sector (Mikalef & Gupta, 2021; Mikalef et al., 2023; Van Noordt & Tangi, 2023) or the Technology-Organizations-Environment (TOE) framework (Pumplun et al., 2019; Jöhnk et al., 2021; Mikalef et al., 2022; Neumann et al., 2022). The resource-based view and TOE are effective at identifying requirements for AI adoption or the willingness to adopt AI, but are limited in their ability to explore broader socio-technical aspects of AI that are significant during implementation, such as an organization's capacity for change, governance, and ethical principles. Whereas the successful deployment of AI in the public sector requires advanced technical capabilities, solid organizational capabilities, effective governance structures, and robust ethical guidelines (Ashok et al., 2022; Lanier et al., 2020).

Additionally, we discovered that the D&M success model can be applied to evaluate government readiness for AI implementation from the users' perspective. In our framework, the adjusted DeLone and McLean's IS Success Model is useful in examining the AI system performance and net benefits after implementation. The factors of information quality, system quality, and service quality, alongside actual use, user satisfaction, and impacts, indicate implementation readiness. By applying this framework to AI research in the public sector, we can examine user-centric factors that contribute to the success and failure of AI projects, confirming or challenging previous findings (Westenberger et al., 2022).

Our study also acknowledged the influence of AI characteristics on implementation readiness. This is in line with the findings in Maragno et al. (2023) that stated factors, affordances, and constraints vary due to the type of AI system, the characteristics of the service, and the relative organization. In our cases, one interesting attribute is the procurement approach used to build the AI system. As AI models and computational engines continue to advance, governments increasingly face a tactical choice between developing AI systems in-house and procuring externally. Building proprietary AI requires substantial investment in computing infrastructure, specialized expertise, and continuous maintenance, which are challenges for many governments. Some institutions, like our Case 3, choose the strategy of procuring AI solutions, as many LLM options are available in the market. This AI has become attractive because it can streamline processes like data preparation and ETL (extract, transform, load) workflows, thereby reducing operational complexity. However, this choice carries consequences for the government. In highly regulated environments such as the Netherlands, concerns about data sovereignty, algorithmic transparency, and regulatory compliance are particularly pronounced. While this cautious stance mitigates risks associated with privacy and accountability, it may also prevent public institutions from fully benefiting from emerging AI innovations. Government procurement decisions are influenced by a country's national approach to AI. As reflected in our case study, the two countries take different approaches to AI: the Netherlands tends to be regulation-focused, while Indonesia tends to be innovation-focused. But in all cases, we found that the lack of a firm regulatory framework for AI remains a concern. This implies a shared underlying principle that algorithms and AI should be adopted with caution, and selecting proper algorithms to support decision-making is essential (Janssen et al., 2022).

6 Conclusion

This study examined how government readiness for predictive AI implementation involving multiple stakeholders can be evaluated. Based on evidence from three predictive AI cases in Indonesia and one from the Netherlands, we developed a dual-perspective framework to evaluate government readiness in AI implementation that integrates AI developers' and AI users' perspectives. From developers' perspectives, the framework examines readiness in terms of resource availability, the organization's capability to change, and governance, while from users' perspectives, implementation is evaluated based on the performance and impact of deployed AI systems. This evaluation framework also indicates the influence of AI characteristics and individual readiness on the implementation of predictive AI at organization level.

The scientific contribution of this research lies in both 1) extending existing frameworks concerning the government's readiness for predictive AI beyond the adoption phase, and 2) addressing the multiple perspectives of different stakeholders on predictive AI implementation. In practice, this research encourages governments that have begun using AI to assess their readiness for more effective and impactful implementation. The evaluation should cover both sides: the AI developers and AI users, to ensure responsible use of AI in governments. This type of evaluation is still largely unexplored, but it is becoming important as governments increasingly transition from pilots to operational AI deployment.

A limitation of this study is the small number of case studies we conducted, which may introduce selection bias, and the findings may not necessarily reflect the situation in other government organizations. Future research could expand this work by testing the framework across broader institutional settings and other AI technologies (e.g., generative AI). Longitudinal studies within a particular organization would also be valuable for observing how readiness factors evolve as AI systems mature over time. Such efforts would expose the relationships among factors and further refine our understanding of how governments can be institutionally prepared to implement AI responsibly and effectively to advance public values.

References

Ahn, M.J., Chen, Y.-C.: Digital transformation toward AI-augmented public administration: the perception of government employees and the willingness to use AI in government. Gov. Inf. Q. **39**(2), 101664 (2022). https://doi.org/10.1016/j.giq.2021.101664

Ashok, M., Madan, R., Joha, A., Sivarajah, U.: Ethical framework for artificial intelligence and digital technologies. **62**, 102433 (2022). (rayyan-607430822)

Collins, C., Dennehy, D., Conboy, K., Mikalef, P.: Artificial intelligence in information systems research: a systematic literature review and research agenda. Int. J. Inf. Manage. **60**, 102383 (2021). https://doi.org/10.1016/j.ijinfomgt.2021.102383

Criado, J. I., O.de Zarate-Alcarazo, L.: Technological frames, CIOs, and Artificial Intelligence in public administration: a socio-cognitive exploratory study in Spanish local governments. Govern. Inf. Q. **39**(3), 101688 (2022). https://doi.org/10.1016/j.giq.2022.101688

Duan, Y., Edwards, J.S., Dwivedi, Y.K.: Artificial intelligence for decision making in the era of big data–evolution, challenges and research agenda. **48**, 63–71 (2019). (rayyan-607430806)

Janssen, M.: Responsible governance of generative AI: conceptualizing GenAI as complex adaptive systems. Policy Soc. **44**(1), 38–51 (2025). https://doi.org/10.1093/polsoc/puae040

Janssen, M., Hartog, M., Matheus, R., Yi Ding, A., Kuk, G.: Will algorithms blind people? The effect of explainable AI and decision-makers' experience on AI-supported decision-making in government. Soc. Sci. Comput. Rev. **40**(2), 478–493 (2022). https://doi.org/10.1177/0894439320980118

Jöhnk, J., Weißert, M., Wyrtki, K.: Ready or not, AI comes—an interview study of organizational AI readiness factors. Bus. Inf. Syst. Eng. **63**(1), 5–20 (2021). https://doi.org/10.1007/s12599-020-00676-7

Kamal, M.M.: IT innovation adoption in the government sector: identifying the critical success factors. J. Enterp. Inf. Manag. **19**(2), 192–222 (2006). https://doi.org/10.1108/17410390610645085

Kudina, O., Van De Poel, I.: A sociotechnical system perspective on AI. Minds Mach. **34**(3), 21 (2024). s11023-024-09680–09682. https://doi.org/10.1007/s11023-024-09680-2

Lanier, P., Rodriguez, M., Verbiest, S., Bryant, K., Guan, T., Zolotor, A.: Preventing infant maltreatment with predictive analytics: applying ethical principles to evidence-based child welfare policy. J. Family Violence **35**(1), 1–13 (2020). https://doi.org/10.1007/s10896-019-00074-y

Maragno, G., Tangi, L., Gastaldi, L., Benedetti, M.: Exploring the factors, affordances and constraints outlining the implementation of Artificial Intelligence in public sector organizations. Int. J. Inf. Manage. **73**, 102686 (2023). https://doi.org/10.1016/j.ijinfomgt.2023.102686

Mellouli, S., Janssen, M., Ojo, A.: Introduction to the issue on artificial intelligence in the public sector: risks and benefits of AI for governments. Digit. Govern.: Res. Pract. **5**(1), 1–6 (2024). https://doi.org/10.1145/3636550

Mikalef, P., Gupta, M.: Artificial intelligence capability: conceptualization, measurement calibration, and empirical study on its impact on organizational creativity and firm performance. Inf. Manage. **58**(3), 103434 (2021). https://doi.org/10.1016/j.im.2021.103434

Mikalef, P., et al.: Enabling AI capabilities in government agencies: a study of determinants for European municipalities. Gov. Inf. Q. **39**(4), 101596 (2022). https://doi.org/10.1016/j.giq.2021.101596

Mikalef, P., et al.: Examining how AI capabilities can foster organizational performance in public organizations. Gov. Inf. Q. **40**(2), 101797 (2023). https://doi.org/10.1016/j.giq.2022.101797

Misuraca, G., van Noordt, C.: AI Watch, artificial intelligence in public services: overview of the use and impact of AI in public services in the EU. Publications Office (2020). https://data.europa.eu/doi/10.2760/039619

Neumann, O., Guirguis, K., Steiner, R.: Exploring artificial intelligence adoption in public organizations: a comparative case study. Public Manage. Rev. 1–28 (2022). https://doi.org/10.1080/14719037.2022.2048685

Parasuraman, A.: Technology readiness index (Tri): a multiple-item scale to measure readiness to embrace new technologies. J. Serv. Res. **2**(4), 307–320 (2000). https://doi.org/10.1177/109467050024001

Pumplun, L., Tauchert, C., Heidt, M.: A new organizational chassis for artificial intelligence—exploring organizational readiness factors (2019)

Rodriguez Müller, A. P., Tangi, L., Lerusse, A.: Understanding the adoption of artificial intelligence in local government decision-making: the influence of institutional pressures and managerial perceptions. Public Administration, padm.70033 (2025). https://doi.org/10.1111/padm.70033

Urbach, N., Müller, B.: The updated DeLone and McLean model of information systems success. In: Dwivedi, Y.K., Wade, M.R., Schneberger, S.L. (eds.) Information Systems Theory, vol. 28, pp. 1–18. Springer, New York (2012). https://doi.org/10.1007/978-1-4419-6108-2_1

Van Noordt, C., Tangi, L.: The dynamics of AI capability and its influence on public value creation of AI within public administration. Gov. Inf. Q. **40**(4), 101860 (2023). https://doi.org/10.1016/j.giq.2023.101860

Voß, F.L.V., Pawlowski, J.M.: Digital readiness frameworks: current state of the art and research opportunities. In: Uden, L., Ting, I.-H., Corchado, J.M. (eds.) Knowledge Management in Organizations, vol. 1027, pp. 503–514. Springer, Cham (2019). https://doi.org/10.1007/978-3-030-21451-7_43

Wang, Y., Zhang, N., Zhao, X.: Understanding the determinants in the different government AI adoption stages: evidence of local government chatbots in China. Soc. Sci. Comput. Rev. **40**(2), 534–554 (2022). https://doi.org/10.1177/0894439320980132

Westenberger, J., Schuler, K., Schlegel, D.: Failure of AI projects: understanding the critical factors. Procedia Comput. Sci. **196**, 69–76 (2022). https://doi.org/10.1016/j.procs.2021.11.074

Yigitcanlar, T., Agdas, D., Degirmenci, K.: Artificial intelligence in local governments: perceptions of city managers on prospects, constraints and choices. AI Soc. **38**(3), 1135–1150 (2023). https://doi.org/10.1007/s00146-022-01450-x

Yin, R.K.: Case Study Research: Design and Methods,3rd edn. Sage Publications (2003)

Lost in Translation: A Typology of Multi-Level Responsible AI Governance

Dimitrios Symeonidis(✉) and Anastasija Nikiforova

Institute of Computer Science, University of Tartu, Narva Mnt 18, 51009 Tartu, Estonia
{dimitrios.symeonidis,anastasija.nikiforova}@ut.ee

Abstract. Responsible AI has become a widely adopted reference point in contemporary AI governance, reflected in a growing number of ethical guidelines, regulatory initiatives, and organizational governance frameworks. However, the meaning of "responsibility" often shifts as governance moves across institutional contexts, creating translation gaps between normative principles, regulatory frameworks, organizational practices, and citizen experiences. As a result, Responsible AI governance frequently appears coherent at the policy level while producing tensions or legitimacy concerns in practice. This study aims to explain how responsibility varies across governance levels and how such variation creates cross-level tensions in Responsible AI governance. Based on a systematic literature review and typology development, we synthesize the Responsible AI literature to develop a typology of Responsible AI governance across four governance levels - super-macro, macro, meso, and micro - each characterized by distinct institutional logics, legitimacy criteria, and modes of socio-technical enactment. The analysis shows that tensions in Responsible AI governance arise not only from implementation gaps but also from structural differences in how responsibility is defined, operationalized, and evaluated across governance levels. The study contributes to Information Systems and AI governance research by conceptualizing responsibility as a relational, multi-level governance construct, advancing understanding of persistent principle-to-practice gaps, and offering a diagnostic tool for policymakers, organizations, and researchers designing or auditing Responsible AI governance frameworks.

Keywords: AI governance · Cross-level translation · Institutional logic · Multi-level governance · Responsible AI · Typology

1 Introduction

The expanding use of artificial intelligence (AI) across public and private sectors has intensified scholarly and policy debates around how AI systems should be developed, deployed, and governed responsibly (Dignum, 2019; Mittelstadt, 2019; Nikiforova et al., 2026; Selbst et al., 2019). In response, a growing body of literature has sought to define Responsible AI (RAI), typically through sets of ethical principles such as fairness, transparency, accountability, safety, and human oversight (Floridi et al., 2018; Dignum, 2019;

Published by Springer Nature Switzerland AG 2027
D. Duenas-Cid et al. (Eds.): ePart 2026, LNCS 16923, pp. 398–413, 2027.
https://doi.org/10.1007/978-3-032-36676-4_26

Munn, 2023). These principles have been included in international guidelines, such as the OECD AI Principles (2019) and the EU AI Act (2024), national strategies, and organizational policies, such as the internal algorithmic auditing framework (Raji et al., 2020). This diffusion has produced a shared cross-jurisdictional vocabulary that structures contemporary AI governance debates. However, the apparent convergence around common principles masks substantial variation in how responsibility is interpreted, prioritized, and enacted across governance contexts. As a result, RAI remains conceptually ambiguous. While actors frequently invoke the same principles, they often attach different meanings to them and evaluate them according to different criteria (Mikalef et al., 2022; Ibitoye et al., 2025; Sadek et al., 2025). This ambiguity is not merely a problem of implementation - it raises a more fundamental question of whether responsibility itself refers to the same phenomenon across governance contexts.

A central limitation of existing RAI research is the treatment of responsibility as a conceptually uniform construct (Papagiannidis et al., 2025). Variation in RAI implementation is documented (Mittelstadt, 2019; Raji et al., 2020; Rakova et al., 2021) but attributed to execution failure, such as inadequate resources, organizational inertia, or enforcement gaps. Less attention has been paid to the possibility that responsibility itself may be constituted differently across governance levels. Actors at different governance levels invoke different governance questions, recognize different evaluative audiences, and validate responsibility claims through different legitimacy logics (Jobin et al., 2019; Shin, 2021; Ehsan et al., 2019). Frameworks that rely on a shared normative vocabulary therefore risk producing structural misalignment, because terminological consensus conceals operational divergence (Metcalf et al., 2019; Thornton et al., 2012). This conflation obscures why responsibility claims that appear coherent at one level fail at another, and why implementation gaps cannot be resolved through additional guidance without first specifying which governance level that guidance targets and under which legitimacy logic it operates.

Building on this observation, we conceptualize RAI as a multi-level construct which meaning varies systematically across governance contexts. Rather than treating responsibility as a stable concept implemented differently by different actors, we argue that responsibility is articulated, operationalized, and evaluated differently depending on the governance level at which it is enacted.

This study addresses two research questions:

RQ1: How does the operational meaning of responsibility differ across the four governance levels, namely super-macro, macro, meso and micro?

RQ2: What patterns of coordination failures and tensions arise from misalignment between responsibility meanings across governance levels?

To address these questions, we conduct a systematic literature review (SLR) combined with typology development (Kluge, 2000) integrating multi-level governance theory (Marks & Hooghe, 2004; Bache & Flinders, 2004), institutional theory (Powell & DiMaggio, 1991; Scott, 2014), and socio-technical systems perspectives (Bostrom & Heinen, 1977). Through an iterative analysis of the RAI literature, we develop a four-level typology comprising super-macro, macro, meso, and micro responsibility configurations. Each level is characterized by distinct actors, governance questions, legitimacy criteria, and failure modes. The resulting typology provides a framework for understanding how responsibility meanings shift across governance contexts and why governance coordination frequently breaks down despite apparent agreement on shared ethical principles. To demonstrate the analytical utility of the framework, we apply the typology illustratively to the Romanian AI governance context, showing how cross-level misalignment manifests within a specific national institutional setting.

The remainder of this paper proceeds as follows: Sect. 2 provides theoretical background, Sect. 3 details methodology, Sect. 4 presents findings, Sect. 5 discusses implications and limitations. Section 6 concludes the paper and defines future research directions.

2 Theoretical Background

The rapid deployment of AI systems has generated extensive scholarship on responsible AI development and governance. Early research established ethical principles, such as fairness, transparency, accountability, privacy, human oversight, as normative foundations (Floridi et al., 2018; Jobin et al., 2019; Dignum, 2019; Papagiannidis et al., 2025). Research shows convergence around these principles across international organizations, national governments, and corporations (Fjeld et al., 2020; Hagendorff, 2020; Correa et al., 2023), representing significant progress in establishing shared governance vocabulary. However, principle-based approaches face persistent implementation challenges. Mittelstadt (2019) demonstrates that ethical principles have little effect without institutional translation mechanisms. Raji et al. (2020) document the accountability gap between principle endorsement and auditing practice in technology organizations, while Rakova et al. (2021) show through interviews with practitioners at major AI companies that organizational structures systematically hinder responsible AI implementation. These failures are not confined to any single governance level or sector, but appear consistently across supranational, national, and organizational contexts (Ribeiro et al., 2025). Such systematic, cross-level breakdown cannot be fully attributed to implementation failure alone. Instead, it points to a deeper conceptual ambiguity, whereby RAI may not be sufficiently differentiated to guide actors operating under fundamentally different governance logics. Recent work reveals systematic divergences in how RAI is understood across contexts, as organizations tend to operationalize responsibility through compliance-oriented logics that are structurally misaligned with the broader normative content of RAI principles (Papagiannidis et al., 2025; Ibitoye et al., 2025; Nikiforova et al., 2026), pointing to fundamentally different definitional frames across governance actors. This divergence operates not only across actor types but across the very dimensions through which responsibility is defined. Governance frameworks that conflate these dimensions therefore address fundamentally different problems under the same label. This literature reveals the conceptual puzzle that RAI appears simultaneously as universal ethical aspiration (Floridi et al., 2018), contested political project (Metcalf et al., 2019), organizational governance challenge (Raji et al., 2020), and experiential judgment (Ehsan et al., 2019). These framings invoke different actors, which are ranging from technical developers and organizational compliance officers to regulators and affected publics (Raji et al., 2020; Rakova et al., 2021), different evaluation criteria, with organizations assessing responsibility through auditability and risk mitigation while users assess it through experiential trust and fairness outcomes (Papagiannidis et al., 2025; Shneiderman, 2020), and different legitimacy logics, where state actors ground authority in legal mandate, organizations in reputational risk management, and civil society in participatory claims (Jobin et al., 2019; Wang et al., 2025). This stands in contrast with treatments of RAI as a single phenomenon that attribute variation to implementation differences.

This treatment reflects broader tendencies in IS governance research. While IS scholars recognize that information systems operate across multiple levels (Orlikowski, 2000; Burton-Jones & Volkoff, 2017), governance concepts themselves are rarely theorized as inherently multi-level. Research examines how the *same* governance mechanism produces different outcomes across contexts (Janssen et al., 2020) but not whether governance *meanings* vary systematically by level. Literature currently lacks conceptual frameworks capable of explaining (1) what dimensions structure responsibility meanings, (2) why tensions arise when these meanings interact across governance levels, and (3) how coordination remains possible despite semantic differences. To explain this, we draw on three complementary theoretical traditions.

Multi-level governance theory (Marks & Hooghe, 2004; Bache & Flinders, 2004) provides the basis for distinguishing governance scales, highlighting how different levels involve distinct actors, authority structures, and coordination mechanisms. Institutional theory (Powell & DiMaggio, 1991; Scott, 2014) explains how responsibility claims are legitimized through different regulative, normative, and cultural-cognitive logics. Socio-technical systems theory (Bostrom & Heinen, 1977; Orlikowski, 2000; Leonardi, 2011) complements these perspectives by highlighting how responsibility is enacted through interactions among technological, organizational, and social elements. These three perspectives are complementary rather than interchangeable, where governance theory identifies where responsibility is articulated, institutional theory explains according to what legitimacy criteria it is evaluated, and socio-technical systems theory explains how it is enacted. These perspectives suggest that responsibility is not a fixed property of AI systems but a relational governance construct whose meaning depends on governance scale, legitimacy criteria, and socio-technical enactment. These dimensions function as sensitizing constructs (Kluge, 2000) that guided the initial coding process while remaining open to empirical refinement. As described in Sect. 3, they provide the analytical basis for identifying distinct responsibility logics across the literature.

3 Methodology

To construct our RAI typology, we combined a systematic literature review (SLR) with typology development (Kluge, 2000; Nickerson et al., 2013). Our objective was to identify systematic patterns in how responsibility is conceptualized, enacted, and evaluated across governance contexts. An SLR was conducted to capture contemporary RAI discourse, drawing on Scopus, Web of Science, and the AIS eLibrary to ensure both disciplinary breadth and IS-specific coverage. The search string was defined as: *("responsible artificial intelligence" OR "ethical artificial intelligence" OR "trustworthy artificial intelligence" OR "AI ethics") AND ("governance" OR "regulation" OR "policy") AND ("framework" OR "typology" OR "guideline" OR "principle" OR "legitimacy")* and it was applied to titles, abstracts, and keywords. Studies were included if they (1) addressed AI responsibility beyond technical performance, engaging governance, legitimacy, organizational, or societal implications, (2) were published as journal articles, conference papers, or book chapters in English, and (3) appeared between 2020 and 2025. After deduplication, 642 papers were identified, of which 88 could not be retrieved through available institutional access, yielding 554 papers. Full-text screening then excluded studies focused solely on technical AI performance, non-RAI governance topics, or applications without governance implications, resulting in a final corpus of 202 papers used for typology development. The selection process followed PRISMA guidelines and is presented in Fig. 1.

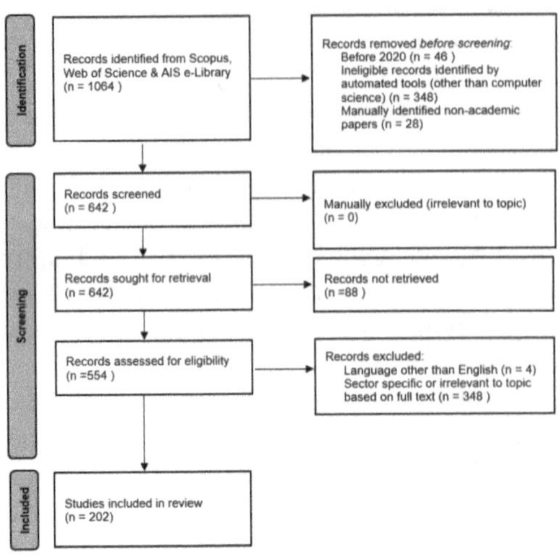

Fig. 1. Systematic Literature Review Process

To develop the typology, we followed Kluge's (2000) typology construction procedure. In the first step, we derived analytical dimensions to structure RAI meanings. Drawing on the three dimensions mentioned above, namely governance scale, institutional logic, and socio-technical configuration, we formed the initial coding framework, as summarized in Table 1, while remaining open to empirical refinement.

Table 1. Theoretical Foundations for Typology Development

Theoretical foundation	Derived dimension	Operationalization in this study
Multi-level governance theory (Marks & Hooghe, 2004; Bache & Flinders, 2004)	Governance scale	Supranational, national, organizational, individual
Institutional theory (*Powell & DiMaggio,* 1991; Scott, 2014)	Institutional logic & legitimacy criteria	Regulative, normative, and cultural-cognitive pillars
Socio-technical systems theory (Bostrom & Heinen, 1977; Orlikowski, 2000; Leonardi, 2011)	Socio-technical configuration	Responsibility enacted through discourse, regulatory processes, development practices, or user interaction

In the second step, we conducted thematic synthesis (Thomas & Harden, 2008) across the 202-paper corpus. Each study was coded according to its dominant governance level, primary actors, evaluative audience, legitimacy criteria, material artifacts, and identified failure modes. When studies addressed multiple levels, classification was based on the level most central to the research question and empirical analysis. Iterative comparison across coded cases revealed recurring configurations in which governance scale, institutional logic, and socio-technical configuration aligned systematically (Thornton et al., 2012). This process produced four responsibility types corresponding to supranational, national, organizational, and individual levels. In the third step, we examined alternative typology structures, including merging supranational and national levels and splitting the organizational level by sector, confirming that the four-level structure provided the most coherent and analytically distinct configuration (Nickerson et al., 2013). In the fourth step, we refined and characterized each type by examining its internal theoretical consistency and empirical manifestation (Kluge, 2000). We then analyzed cross-level interactions to identify recurring patterns of coordination and misalignment, which informed the development of propositions regarding cascading translation, decoupling, and conflict between governance levels (Orlikowski, 2000; Leonardi, 2011). Typology robustness was assessed through theoretical coherence, mutual exclusivity, and external plausibility checks (Kluge, 2000; Nickerson et al., 2013).

Following typology development, we apply the framework to the Romanian AI governance context as an illustrative application. This step does not constitute empirical validation of the typology but serves to demonstrate its analytical utility by mapping observable governance dynamics onto the four responsibility levels. Evidence is drawn from publicly available sources including EU regulatory instruments, Romania's National Strategy on Artificial Intelligence, corporate governance documentation, decisions and annual reports of the Romanian National Supervisory Authority for Personal Data Processing (ANSPDCP), and European Commission survey data. The Romanian context was selected because it represents a case of high institutional distance within a common supranational regulatory framework, bound by the same EU AI Act obligations as all member states but operating with comparatively lower administrative capacity, distinct national innovation priorities, and specific democratic legitimacy concerns. This makes cross-level translation failures particularly visible and analytically instructive.

4 A Typology of Multi-Level AI Governance

This study develops a four-level typology of RAI governance (Table 2) capturing systematic variations in how responsibility is articulated, institutionalized, and operationalized across governance contexts. The four levels are super-macro, macro, meso and micro and they are not mutually exclusive empirical categories but ideal-typical articulations that clarify why RAI meaning can take different form in different contexts. These levels emerged from recurring patterns in how responsibility is defined, by whom it is enacted, and according to which legitimacy criteria it is evaluated.

Table 2. Typology of Multi-Level AI Governance

Level	Super-Macro	Macro	Meso	Micro
Governance scale	Supranational	National	Organizational	Individual/Interactional
Dominant logic	Professional-Normative	State-Democratic	Market-Professional	Community-Relational
Legitimacy basis	Normative and cultural-cognitive (principal convergence, expert endorsement)	Normative-Regulative (legal authorization, cultural fit, public acceptance)	Regulative (procedural, compliance, accountability mechanisms)	Cultural-Cognitive (experiential trust, contextual appropriateness)
Socio-Technical enactment	Discourse production (principles, frameworks, guidelines)	Regulatory processes (laws, policies, enforcement)	Development practices (risk assessments, testing, documentation, deployment)	User interaction (interfaces, explanations, automated decisions)
Characteristic failure	Abstraction without translation	Legitimacy contestation despite compliance	Proceduralization without substantive impact	Experiential dissonance

At the super-macro level, RAI operates as a supranational normative project. Responsibility is articulated through global ethical principles developed by international organizations, multi-stakeholder initiatives, professional associations, and transnational expert communities (Floridi et al., 2018; Jobin et al., 2019; OECD, 2019). *Governance mechanisms* are voluntary and rely on soft coordination tools such as principle articulation, consensus-building, reputational signaling, and cross-jurisdictional diffusion (Westerstrand, 2025). *Institutional legitimacy* is grounded primarily in normative and cultural-cognitive pillars (Scott, 2014). Ethical authority emerges through professional endorsement and convergence around shared values such as fairness, transparency, accountability, safety, and human oversight (Fjeld et al., 2020; Hagendorff, 2020). The central artifacts are textual and discursive, namely guidelines, white papers, declarations, and research publications (Westerstrand, 2025; Bean et al., 2025). Responsibility exists in abstract form, detached from specific technological architectures or organizational routines. This abstraction enables broad consensus across heterogeneous political and economic systems but limits operational specificity (Beltran, 2025). As Mittelstadt (2019) argues, principles alone cannot guarantee ethical AI, because abstraction without institutional translation leaves implementation indeterminate. Failure at this level therefore manifests as abstraction without translation, as principles achieve symbolic legitimacy yet provide insufficient guidance for downstream actors. When organizations adopt such principles rhetorically without embedding them in concrete processes, the result may resemble ethics washing (Bietti, 2020).

As responsibility moves downward from supranational discourse into nation-states and societal governance systems, its meaning changes. *At the macro level*, RAI becomes embedded within sovereign authority structures and democratic institutions. Governance is enacted by governments, legislatures, regulatory agencies, courts, and civil society actors operating within specific cultural and constitutional contexts (Bache & Flinders, 2004; Janssen et al., 2020; Li and Goel, 2025). Unlike supranational bodies, these actors possess coercive authority to mandate compliance and impose sanctions. Institutional legitimacy here combines normative justification with regulative enforcement (Scott, 2014). Responsibility claims must align with constitutional values, reflect public deliberation, and withstand judicial scrutiny. Ethical principles articulated globally are translated into binding laws, regulatory frameworks, and enforcement mechanisms. For example, abstract commitments to fairness may become anti-discrimination requirements in high-risk AI applications, while transparency may be codified as documentation obligations or rights to explanation (Ananny & Crawford, 2018; Mittelstadt, 2019; Wang et al., 2025). This translation requires making explicit value trade-offs that remain implicit at the supranational level, such as balancing innovation with precaution or privacy with functionality. Cross-national research demonstrates that these trade-offs vary according to political culture and institutional context (Wong, 2020; Hagendorff, 2020; Nannini et al., 2025). Whereas the supranational level struggles with indeterminacy, the national level struggles with legitimacy contestation. Systems that comply with regulatory standards may nevertheless face public resistance if they conflict with culturally embedded expectations or societal norms (Shin, 2021). Thus, responsibility at this level is defined not only by legal compliance but also by societal acceptance and political legitimacy.

When responsibility shifts further into organizations to *meso level*, RAI becomes a matter of internal governance capability and development practice. Governance is enacted through hierarchical authority structures, including executive leadership, compliance officers, ethics boards, and technical teams (Raji et al., 2020; Rakova et al., 2021). Institutional legitimacy at this level is grounded primarily in regulative logic, often combined with market and professional considerations (Thornton et al., 2012; Scott, 2014). Organizations must demonstrate adherence to external regulations while maintaining operational efficiency and technical competence. Responsibility is evidenced through documented procedures, audit trails, impact assessments, and accountability mechanisms. The socio-technical configuration is materially intensive, as responsibility is enacted through system design, dataset curation, model training, bias testing, documentation, and monitoring (Selbst et al., 2019; Leonardi, 2011; Alshibani et al., 2025). Ethical principles and regulatory requirements are translated into concrete technical constraints under conditions of resource limitations, time pressure, and organizational complexity. Unlike national regulation, which defines external boundaries, organizational governance operationalizes these boundaries in situated contexts. However, studies repeatedly document the gap between principle endorsement and substantive implementation (Raji et al., 2020; Rakova et al., 2021). Failure at this level manifests as proceduralization without impact, where governance mechanisms become checkbox exercises disconnected from meaningful ethical reasoning. This dynamic reflects classic patterns of institutional decoupling, in which formal structures signal compliance while underlying practices remain unchanged (Meyer & Rowan, 1977; Stoyanovich et al., 2025). Thus, whereas national governance struggles with political legitimacy, organizational governance struggles with integrating ethical aspiration into operational practice.

The meaning of responsibility shifts once more when examined from the perspective of individuals interacting with AI systems. *At micro level*, RAI is evaluated through lived experience rather than through compliance documentation or regulatory authorization. Governance is enacted through relational accountability mechanisms such as user trust, feedback, complaint, and collective action rather than through formal institutional authority (Bache & Flinders, 2004; Hu, 2025). Institutional legitimacy relies primarily on cultural-cognitive alignment embedded in community norms and expectations (Scott, 2014). Responsibility is judged according to experiential criteria, namely whether systems feel fair, behave predictably, respect contextual boundaries, and provide explanations that match users' mental models (Liao et al., 2020; Ehsan et al., 2019). The socio-technical enactment is interactional. Interfaces, automated decisions, recommendations, and system responses become the sites at which responsibility is encountered. Transparency is experienced as intelligibility in context (Ananny & Crawford, 2018; Marti-Ochoa et al., 2025). Fairness is perceived through outcome equity in specific situations (Shin, 2021). Accountability is experienced as responsiveness when concerns are raised. Failure at this level manifests as experiential dissonance, where technically compliant and procedurally governed systems are perceived as opaque, unfair, or inappropriate. Such failures reveal that responsibility cannot be reduced to compliance or principle adherence; it must also be enacted convincingly in practice. Unlike organizational governance, upward influence from users to institutions is structurally weaker, making it difficult for experiential signals to systematically reshape regulatory or organizational structures (Majrashi, 2025).

5 Illustrative Vignette of the Typology: AI Governance in Romania

This section applies the typology illustratively to the Romanian AI governance context. The aim is to demonstrate analytical utility by showing how the typology structures interpretation of observable governance dynamics, not to test or confirm its claims empirically. Romania provides an instructive example operating within the European Union's supranational regulatory architecture while developing national AI governance structures. These conditions make visible the translation challenges that can emerge as RAI principles move from normative commitments to regulatory instruments, organizational implementation, and citizen-facing system interactions. In Table 3 we map governance dynamics onto the four responsibility levels, drawing on regulatory instruments, national strategy documents, corporate governance disclosures, and survey data.

Table 3. Illustrative Application of the Typology

Translation tension	Romanian evidence	Illustrated translation challenge
Tension 1: Principle → Regulation (super-macro to macro)	EU AI Act (EP & COE, 2024) and GDPR (European Union, 2016) establish abstract super-macro commitments to transparency, accountability, and risk management. The 2024 Romanian presidential election was annulled following algorithmic amplification allegations on TikTok (Stan & Zaharia, 2025; Constitutional Court of Romania, 2024); the European Commission subsequently opened formal DSA proceedings (2025)	Abstraction without translation: supranational principles did not generate anticipatory macro-level rules capable of governing recommendation algorithm interference in democratic processes and legitimacy contestation arose despite formal regulatory compliance
Tension 2: Regulation → Organizational practice (macro to meso)	Bitdefender's (2024) responsible AI commitments emphasizing privacy-preserving techniques coexisted with GDPR enforcement action by ANSPDCP (Beltran, 2025), illustrating that macro-level regulatory obligations did not automatically translate into embedded organizational change	Proceduralization without substantive impact: formal governance documentation signaled compliance while development practices remained incompletely aligned with regulatory intent, reflecting institutional decoupling (Meyer & Rowan, 1977)
Tension 3: Practice → Citizen experience (meso to micro)	Citizen complaints reported by ANSPDCP (2024) and Eurobarometer survey data (European Commission, 2023) showing widespread concerns about automated decision-making and data use, despite institutional certification of system compliance	Experiential dissonance: organizational transparency procedures did not translate into intelligibility at the point of citizen interaction; systems perceived as opaque or unfair despite procedural compliance (Ehsan et al., 2019; Shin, 2021)

These examples illustrate how governance tensions can emerge as responsibility is translated across levels. The Bitdefender case is particularly instructive at the meso level, as the company's published RAI commitments coexisted with concurrent GDPR enforcement action by ANSPDCP, a textbook instance of decoupling (Meyer & Rowan,

1977). In the Romanian context, supranational normative commitments are interpreted through national regulatory priorities, operationalized within organizational compliance practices, and ultimately evaluated through citizen experiences with AI-enabled systems. This pattern of cascading translation tensions demonstrates the typology's diagnostic value. In the Romanian case, an intervention point can be observed in the macro-to-meso translation, where regulatory obligations exist but organizational embedding remains incomplete. This suggests that enforcement mechanisms and organizational capacity-building are the priority corrective levers.

6 Discussion

The developed RAI governance typology confirms that RAI is not a single governance problem but a nested set of distinct legitimation challenges, each governed by different institutional logics. At the super-macro level, responsibility is constituted through international norm convergence and regulative legitimacy, at the macro level - through national regulatory frameworks and legal mandates, at the meso level - through organizational compliance and normative legitimacy, and at the micro level - through practitioner discretion and cultural-cognitive legitimacy. This alignment with Scott's (2014) three pillars of institutional legitimacy suggests that persistent tensions in RAI governance are not merely technical or operational, but may also reflect structural differences in how responsibility claims are legitimized across governance levels. They arise when actors operating under different legitimacy logics invoke the same normative vocabulary while attaching different meanings and evaluative criteria to responsibility. The developed typology therefore reframes RAI fragmentation as more than a coordination problem, reflecting a plurality of legitimacy logics that can generate cross-level misalignment even where actors appear to share common principles. Governance frameworks should therefore explicitly accommodate both coordination across levels and differences in how responsibility is understood and legitimized.

The typology developed in this study contributes to IS scholarship by clarifying that RAI is not a single, uniform construct but a set of analytically distinct yet interconnected responsibility configurations operating across governance scales. Much of the existing literature treats responsibility as if principles such as fairness, transparency, and accountability retain stable meaning as they move from global declarations to national regulation, organizational practice, and user experience (Floridi et al., 2018; Jobin et al., 2019). Our findings challenge this assumption. By demonstrating that responsibility is constituted differently at supranational, national, organizational, and interactional levels, we extend construct clarity research in IS and highlight the risks of cross-level conceptual conflation (Scott, 2014; Orlikowski, 2000). This distinction helps prevent the conflation of symbolic principle endorsement, regulatory compliance, procedural governance, and experiential trust, which represent analytically distinct forms of responsibility.

This perspective also advances AI governance research by highlighting and explaining persistent implementation gaps. Prior studies have documented the principle-to-practice divide and the limitations of abstract ethical frameworks (Mittelstadt, 2019; Raji et al., 2020; Rakova et al., 2021). Our analysis suggests that many of these gaps are not merely failures of execution but reflect systematic cross-level translation challenges.

Supranational norms require reinterpretation within democratic and cultural contexts (Bache & Flinders, 2004; Wong, 2020), which in turn must be operationalized within organizational constraints and development processes (Selbst et al., 2019). Finally, these processes must materialize as trustworthy experiences for individuals interacting with AI systems (Ehsan et al., 2019; Liao et al., 2020). When translation mechanisms are weak or absent, cascading misalignment occurs. For example, regulatory compliance may be achieved while experiential legitimacy remains fragile (Shin, 2021), or organizations may adopt governance structures that signal conformity without substantively altering development practices (Meyer & Rowan, 1977). By identifying cascading translation, decoupling, and conflict as recurring cross-level dynamics, this framework offers a structured explanation for governance breakdowns frequently observed in practice.

The typology further contributes to socio-technical systems theory by demonstrating that responsibility is not a fixed property of technical artifacts but emerges differently across socio-technical configurations, from impact assessments and documentation at the organizational level to interface design and decision outcomes at the interactional level (Orlikowski, 2000; Liao et al., 2020). Consequently, micro-level signals of distrust or perceived unfairness should be interpreted as indicators of cross-level misalignment (Ehsan et al., 2019).

The typology also offers operational value for three actor groups. Policymakers and regulators can use it to diagnose whether a given RAI intervention targets the appropriate governance level and invokes a matched legitimacy logic; organizations can audit whether internal policies address only meso-level concerns or adequately acknowledge supranational and macro-level constraints; and researchers gain a coding framework for comparative cross-national studies examining how the same RAI principle is operationalized differently across levels. Its value applies both at the framework design stage, where cross-level coherence can be built in proactively, and diagnostically for existing frameworks where structural misalignments need identification and correction.

Several limitations suggest directions for future research. First, while the typology identifies systematic level differences, empirical research is needed to test how cross-level coordination mechanisms influence governance outcomes. Comparative case studies could examine whether organizations with explicit translation protocols demonstrate greater alignment between regulatory compliance and user trust. Second, the literature reviewed reflects predominantly Western institutional contexts, potentially underrepresenting alternative governance logics (Wong, 2020). Cross-cultural comparative research could assess whether the four-level structure holds across diverse political systems or whether additional configurations emerge. Third, the findings are subject to limitations inherent to the SLR, as the typology is derived from academic literature and does not incorporate policy documents, practitioner perspectives, or empirical validation. As such, it should be understood as a conceptual step toward a more comprehensive understanding of RAI governance, inviting future research to test, refine, and extend the framework across empirical contexts.

7 Conclusion

This study addresses a conceptual gap in RAI research, namely the treatment of responsibility as a uniform construct despite its varying meanings across governance contexts. Through SLR combined with typology development, we decompose RAI into four analytically distinct levels, namely super-macro, macro, meso, and micro, each characterized by distinct governance questions, actors, evaluation criteria, and failure modes. Our core theoretical contribution is demonstrating that RAI operates as a multi-level construct whose meaning shifts systematically across contexts, explaining why principle-based governance encounters persistent implementation challenges, why RAI debates generate recurring conflicts, and why governance effectiveness requires explicit attention to cross-level coordination. The typology challenges governance universalism by demonstrating that the dimensions through which responsibility is defined and evaluated differ across levels. Responsibility is thereby established as a relationally constructed, context-dependent phenomenon whose surface vocabulary conceals different legitimacy logics. Future research should empirically validate the typology, examine cross-level coordination mechanisms, incorporate non-Western responsibility framings, and address normative questions about resolving cross-level conflicts.

Acknowledgments. This work was supported by the Estonian Research Council grant "Developing human-centric digital solutions" (TEM-TA120).

GenAI Use Disclosure. During the preparation of this work the authors used ChatGPT-4o in order to improve readability and language. After using this tool/service, the authors reviewed and edited the content as needed, and take full responsibility for the content of the publication.

References

Alshibani, S.M., Korayim, D., Mehrotra, A., Agarwal, V.: Fostering responsible AI adoption in startups through entrepreneurial orientation: a sustainable approach. Technol. Forecast. Soc. Chang. **219**, 124272 (2025)
Ananny, M., Crawford, K.: Seeing without knowing: limitations of the transparency ideal and its application to algorithmic accountability. New Media Soc. **20**(3), 973–989 (2018)
ANSPDCP. Annual report 2024. Autoritatea Națională de Supraveghere a Prelucrării Datelor cu Caracter Personal (2024). https://www.dataprotection.ro
Bache, I., Flinders, M. (eds.): Multi-level Governance. Oxford University Press (2004)
Bean, E., Burleigh, C., Haskell, C., Burris-Melville, T., Payne, J., Pathak, B.: Eavesdropping on UNESCO AI policy, leadership, and ethics. J. Leadersh. Stud. **18**(4), 98–110 (2025)
Beltrán, M.: AI algorithms under scrutiny: GDPR, DSA, AI Act and CRA as pillars for algorithmic security and privacy in the European Union. Comput. Secur. 104628 (2025)
Bietti, E.: From ethics washing to ethics bashing: a view on tech ethics from within moral philosophy. In: Proceedings of the 2020 Conference on Fairness, Accountability, and Transparency, pp. 210–219 (2020)
Bitdefender. Privacy and data protection (2024). https://www.bitdefender.com/consumer/support/answer/privacy-policy/
Bostrom, R.P., Heinen, J.S.: MIS problems and failures: a socio-technical perspective. MIS Q. **1**(3), 17–32 (1977)

Burton-Jones, A., Volkoff, O.: How can we develop contextualized theories of effective use? A demonstration in the context of community-care electronic health records. Inf. Syst. Res. **28**(3), 468–489 (2017)
Constitutional Court of Romania. Decision no. 32/2024 on the annulment of the first round of the presidential election (2024). https://www.ccr.ro
Corrêa, N.K., et al.: Worldwide AI ethics: a review of 200 guidelines and recommendations for AI governance. Patterns **4**(10), 100857 (2023)
Dignum, V.: Responsible Artificial Intelligence: How to Develop and Use AI in a Responsible Way. Springer (2019)
Ehsan, U., Tambwekar, P., Chan, L., Harrison, B., Riedl, M.O.: Automated rationale generation: a technique for explainable AI and its effects on human perceptions. In: Proceedings of the 24th International Conference on Intelligent User Interfaces, pp. 263–274 (2019)
European Commission. Special Eurobarometer 523: Attitudes towards artificial intelligence (2023). https://europa.eu/eurobarometer/surveys/detail/3026
European Commission. Digital Services Act: Commission opens formal proceedings against TikTok (2025). https://commission.europa.eu/strategy-and-policy/priorities-2019-2024/europe-fit-digital-age/digital-services-act_en
European Parliament & Council of the European Union (EP & COE). Regulation (EU) 2024/1689 of the European Parliament and of the Council of 13 June 2024 laying down harmonised rules on artificial intelligence (Artificial Intelligence Act). Official Journal of the European Union, L, pp. 1–144 (2024). https://eur-lex.europa.eu/legal-content/EN/TXT/?uri=OJ:L_2024_1689
European Union. Regulation (EU) 2016/679 of the European Parliament and of the Council of 27 April 2016 on the protection of natural persons with regard to the processing of personal data (General Data Protection Regulation). Official Journal of the European Union, L 119, pp. 1–88 (2016). https://eur-lex.europa.eu/legal-content/EN/TXT/?uri=CELEX:32016R0679
Fjeld, J., Achten, N., Hilligoss, H., Nagy, A., Srikumar, M.: Principled artificial intelligence: mapping consensus in ethical and rights-based approaches to principles for AI (Berkman Klein Center Research Publication No. 2020-1). Berkman Klein Center for Internet & Society (2020)
Floridi, L., et al.: AI4People—an ethical framework for a good AI society: opportunities, risks, principles, and recommendations. Mind. Mach. **28**(4), 689–707 (2018)
Hagendorff, T.: The ethics of AI ethics: an evaluation of guidelines. Mind. Mach. **30**(1), 99–120 (2020)
Hu, R.: The philosophical and axiological dimensions of AI in art: implications for creativity and cultural value in artistic production and marketing. Cult.: Int. J. Philos. Cult. Axiol. **22**(3) (2025)
Ibitoye, A.O., Nkwo, M.S., Orji, R.: Rethinking responsible AI from ethical pillars to sociotechnical practice. AI Ethics **5**(6), 6207–6223 (2025)
Institute of International and European Affairs. Romania's 2024–2025 presidential election crisis and its aftermath (2025). https://www.iiea.com/blog/romanias-20242025-presidential-election-crisis-and-its-aftermath
Janssen, M., Estevez, E., Janowski, T.: Interoperability in digital government: a systematic review. Gov. Inf. Q. **37**(1), 101472 (2020)
Jobin, A., Ienca, M., Vayena, E.: The global landscape of AI ethics guidelines. Nat. Mach. Intell. **1**(9), 389–399 (2019)
Kluge, S.: Empirically grounded construction of types and typologies in qualitative social research. Forum: Qual. Soc. Res. **1**(1), Article 14 (2000)
Leonardi, P.M.: When flexible routines meet flexible technologies: affordance, constraint, and the imbrication of human and material agencies. MIS Q. **35**(1), 147–167 (2011)
Li, Y., Goel, S.: Artificial intelligence auditability and auditor readiness for auditing artificial intelligence systems. Int. J. Account. Inf. Syst. **56**, 100739 (2025)

Liao, Q.V., Gruen, D., Miller, S.: Questioning the AI: informing design practices for explainable AI user experiences. In: Proceedings of the 2020 CHI Conference on Human Factors in Computing Systems, pp. 1–15 (2020)
Majrashi, K.: Employees' perceptions of the fairness of AI-based performance prediction features. Cogent Bus. Manage. **12**(1), 2456111 (2025)
Marks, G., Hooghe, L.: Contrasting visions of multi-level governance. In: Bache, I., Flinders, M. (eds.) Multi-level Governance, pp. 15–30. Oxford University Press, Oxford (2004)
Marti-Ochoa, J., Martin-Fuentes, E., Ferrer-Rosell, B.: AI-driven virtual travel influencers and ethical consumerism: analysing engagement with Sena Zaro's Instagram content. Young consumers **26**(5), 728–743 (2025)
Meijer, A., Bekkers, V.: A metatheory of e-government: creating some order in a fragmented research field. Gov. Inf. Q. **32**(3), 237–245 (2015)
Meyer, J.W., Rowan, B.: Institutionalized organizations: formal structure as myth and ceremony. Am. J. Sociol. **83**(2), 340–363 (1977)
Metcalf, J., Moss, E., Boyd, D.: Owning ethics: corporate logics, Silicon Valley, and the institutionalization of ethics. Soc. Res. **86**(2), 449–476 (2019)
Mikalef, P., Conboy, K., Lundström, J.E., Popovič, A.: Thinking responsibly about responsible AI: a research agenda for information systems. Inf. Syst. Front. **24**(6), 1–17 (2022). https://doi.org/10.1007/s10796-021-10175-0
Mittelstadt, B.D.: Principles alone cannot guarantee ethical AI. Nat. Mach. Intell. **1**(11), 501–507 (2019). https://doi.org/10.1038/s42256-019-0114-4
Munn, L.: The uselessness of AI ethics. AI Ethics **3**(3), 869–877 (2023). https://doi.org/10.1007/s43681-022-00209-w
Nannini, L., Huyskes, D., Panai, E., Pistilli, G., & Tartaro, A.: Nullius in Explanans: an ethical risk assessment for explainable AI: L. Nannini et al. Ethics Inf. Technol. **27**(1), 5 (2025)
Nickerson, R.C., Varshney, U., Muntermann, J.: A method for taxonomy development and its application in information systems. Eur. J. Inf. Syst. **22**(3), 336–359 (2013)
Nikiforova, A., et al.: Responsible AI adoption in the public sector: A data-centric taxonomy of AI adoption challenges. In: Proceedings of the 59th Hawaii International Conference on System Sciences (2026)
OECD. OECD principles on artificial intelligence (2019). https://oecd.ai
Orlikowski, W.J.: Using technology and constituting structures: a practice lens for studying technology in organizations. Organ. Sci. **11**(4), 404–428 (2000)
Papagiannidis, E., Mikalef, P., Conboy, K.: Responsible artificial intelligence governance: a review and research framework. J. Strat. Inf. Syst. **34**(2), 101885 (2025)
Powell, W.W., DiMaggio, P.J. (eds.): The New Institutionalism in Organizational Analysis. University of Chicago Press (1991)
Raji, I.D., et al.: Closing the AI accountability gap: defining an end-to-end framework for internal algorithmic auditing. In: Proceedings of the 2020 Conference on Fairness, Accountability, and Transparency, pp. 33–44 (2020)
Rakova, B., Yang, J., Cramer, H., Chowdhury, R.: Where responsible AI meets reality: practitioner perspectives on enablers for shifting organizational practices. In: Proceedings of the ACM on Human-Computer Interaction, vol. 5, no. CSCW1, Article 7 (2021)
Ribeiro, D., et al.: Toward effective AI governance: a review of principles. arXiv (2025)
Sadek, M., Kallina, E., Bohné, T., Mougenot, C., Calvo, R.A., Cave, S.: Challenges of responsible AI in practice: scoping review and recommended actions. AI Soc. **40**, 199–215 (2025)
Scott, W.R.: Institutions and Organizations: Ideas, Interests, and Identities, 4th edn. SAGE (2014)
Selbst, A.D., Boyd, D., Friedler, S.A., Venkatasubramanian, S., Vertesi, J.: Fairness and abstraction in sociotechnical systems. In: Proceedings of the 2019 Conference on Fairness, Accountability, and Transparency, pp. 59–68 (2019)

Shin, D.: The effects of explainability and causability on perception, trust, and acceptance: Implications for explainable AI. Int. J. Hum. Comput. Stud. **146**, 102551 (2021)

Shneiderman, B.: Bridging the gap between ethics and practice: Guidelines for reliable, safe, and trustworthy human-centered AI systems. ACM Trans. Interact. Intell. Syst. **10**(4), Article 26 (2020)

Stan, L., Zaharia, R.: Romania: political developments and data in 2024. Eur. J. Polit. Res. Polit. Data Yearb. (2025)

Stoyanovich, J., de Paula, R.K., Lewis, A., Zheng, C.: Using case studies to teach responsible AI to industry practitioners. In: Proceedings of the AAAI Conference on Artificial Intelligence, vol. 39, no. 28, pp. 29062–29069 (2025)

Thomas, J., Harden, A.: Methods for the thematic synthesis of qualitative research in systematic reviews. BMC Med. Res. Methodol. **8**, Article 45 (2008)

Thornton, P.H., Ocasio, W., Lounsbury, M.: The Institutional Logics Perspective: A New Approach to Culture, Structure, and Process. Oxford University Press, Oxford (2012)

Wang, S., Zhang, Y., Xiao, Y., Liang, Z.: Artificial intelligence policy frameworks in China, the European Union and the United States: an analysis based on structure topic model. Technol. Forecast. Soc. Chang. **212**, 123971 (2025)

Westerstrand, S.: Fairness in AI systems development: EU AI Act compliance and beyond. Inf. Softw. Technol. 107864 (2025)

Wong, P.H.: Democratizing algorithmic fairness. Philos. Technol. **33**(2), 225–244 (2020)

Sensing, Seizing, Transforming: AI-Specific Dynamic Capabilities in Public Administration

Jan Westermann(✉), Samira Krodel, Luca T. Bauer, Marcel Hofgesang, and Björn Niehaves

University of Bremen, Bremen 28359, Germany
westerma@uni-bremen.de
https://www.uni-bremen.de

Abstract. Public administrations increasingly explore the use of artificial intelligence (AI), yet little is known about the organizational capabilities required for effective and responsible AI adoption. This study examines the development of AI-specific dynamic capabilities (DCs) in nineteen German municipalities and investigates whether these organizational capability patterns are reflected in employee-level perceptions. Drawing on a mixed-methods design, we combine qualitative focus group data with an employee survey (n = 5,332). The qualitative findings show that AI-specific sensing, seizing, and transforming capabilities are not yet systematically embedded across municipalities, with notable variation in their degree of implementation. The quantitative results show that perceived organizational innovativeness is moderately associated with several AI-specific capability dimensions, whereas AI literacy shows weaker and statistically non-significant relationships. By triangulating expert-based capability assessments with employee-level indicators, the study provides a granular empirical account of the current development of AI-specific DCs in public administration and highlights implications for both research and practice.

Keywords: Dynamic Capabilities · AI Capabilities · Digital Transformation · Digital Government · Public Administration

1 Introduction

Public administrations are undergoing profound technological and organizational transformation. In particular, the rapid advancement of artificial intelligence (AI) is heightening expectations for public administrations to modernize their structures, processes, and competencies in order to meet increasing societal and organizational demands [19,38]. While AI technologies promise significant gains in efficiency and service quality, empirical evidence suggests that these benefits materialize only when organizations possess the capabilities to respond to technological change in a deliberate, coordinated, and sustainable

Published by Springer Nature Switzerland AG 2027
D. Duenas-Cid et al. (Eds.): ePart 2026, LNCS 16923, pp. 414–430, 2027.
https://doi.org/10.1007/978-3-032-36676-4_27

manner [21]. The dynamic capabilities (DCs) framework provides a lens for understanding how public administrations sense emerging technological developments, seize opportunities for innovation, and transform existing structures accordingly [31]. Given the regulatory complexity, heterogeneous stakeholder environments, and resource constraints characteristic to the public sector [17], understanding such capability-building processes is especially important in the context of AI adoption. Recent research in public administration increasingly emphasizes that effective and responsible AI use depends not only on technological assets but also on the development of corresponding organizational capabilities [21,39]. Although scholarly interest in AI-specific DCs in public administrations is growing [21,23,39], empirical evidence is still relatively rare. Therefore, the present study adopts an exploratory multi-source perspective. While RQ1 examines the existence and development of AI-specific DCs across municipalities based on expert assessments, RQ2 explores whether these organizational patterns are reflected in employee-derived indicators:

RQ1: *How are AI-specific dynamic capabilities currently developed across municipalities and which patterns characterize their maturity?*

RQ2: *To what extent do AI-specific dynamic capabilities align with employee-level indicators such as perceived organizational innovativeness and AI literacy?*

Building on dynamic capability theory [37], we derive and empirically refine an AI-specific capability framework for public administration. While prior research has largely applied generic DC categories, this study extends existing work by specifying AI-relevant sub-capabilities and validating their empirical manifestation across municipal contexts. In doing so, we integrate theory-driven conceptualization with empirically grounded refinement, resulting in a structured yet context-sensitive capability framework.

2 Conceptual Background and Related Works

2.1 Dynamic Capabilities in Public Administration

The DCs framework has become a central theoretical lens in information systems research for understanding digital transformation [42,44]. Rooted in strategic management scholarship, DCs describe an organization's ability to sense opportunities and threats, seize them through timely and informed action, and transform its resource base and processes in response to environmental change [9,37]. This perspective emphasizes intention over passive adaptation and helps explain how organizations sustain performance in evolving technological environments, extending assumptions of the resource-based view [45] by shifting attention from the possession of resources to the organizational competencies required to reconfigure them under conditions of change [13,43].

DC theory originated in the context of private firms and strategic competition [37]. Yet, its analytical value extends to organizational settings characterized by sustained environmental change and complexity. In such contexts, performance pressures arise not solely from market pressure but from evolving societal needs, political priorities, and institutional accountability requirements. Public administrations operate under these conditions of multi-stakeholder demands and shifting expectations for transparency, responsiveness, and legitimacy [5,26]. As highlighted in the public sector DC literature, these organizations must therefore continuously observe environmental signals, learn from changing requirements, and adapt structures and processes accordingly in order to generate public value under institutional constraints [25]. Accordingly, DCs are particularly relevant in this setting because they capture how organizations transform technological and other resources into adaptive responses to complex and evolving public demands. Still, applications of the DC framework to public organizations have historically been limited [31], especially compared to the private sector, where their relationship to organizational performance is well established [2,24].

However, recent research now demonstrates the relevance of the DC perspective for understanding public-sector digital transformation [25]. Recent empirical work further highlights that the effective utilization of technology in government is mediated by capability-building processes that support sensing technological opportunities, fostering cross-unit alignment, and adapting administrative structures [29]. Taken together, previous studies underscore that public organizations, like private-sector firms, must cultivate the capacity to sense environmental and technological shifts, seize opportunities for digital innovation, and transform structures and processes accordingly. DCs therefore represent a valuable theoretical foundation for understanding how public administrations navigate the challenges of contemporary digital transformation [25].

2.2 AI-Specific Dynamic Capabilities

Building on the growing recognition that DCs are essential for enabling public-sector organizations to navigate digital transformation [19], the rapid advance of AI technologies has heightened expectations for public administrations to adapt their structures, processes, and competencies [39]. Yet many public organizations struggle with the practical adoption of AI, as integrating such systems often exceeds existing skills, routines, and governance structures [1].

To address these challenges, the notion of AI capabilities has gained prominence, referring to the organizational resources and competencies required to develop, integrate, and utilize AI technologies effectively [23]. From a theoretical perspective, AI capabilities can thus be conceptualized as domain-specific manifestations of DCs. This situates AI capabilities within DCs sensing, seizing and transforming by emphasizing organization's abilities to identify AI-specific opportunities and risks; evaluate and exploit applications; and reconfigure structures and routines to integrate AI. In such environment, capability development is shaped by an organization's broader innovation orientation, which captures its readiness to pursue and integrate new (technological) opportunities [18,28]. Consistent with this view, Mikalef et al. [23] show that innovation-oriented organizations are more likely to develop AI capabilities.

In addition, prior DC research emphasizes that organizational capabilities manifest in form of underlying employee skills, cognition, and behavior, which collectively enable organizations to sense, seize, and transform in response to technological change [37]. In the context of AI, a particularly relevant microfoundation is employees' ability to understand, interpret, and meaningfully engage with AI systems [22,23,32]. AI-literacy shapes how employees recognize AI-related opportunities, evaluate their implications, and participate in AI-enabled work processes [46]. As such, it represents an individual-level competence that may both reflect and constrain the development of AI-specific DCs in organizations. Perceived Organizational Innovativeness and AI Literacy therefore provide complementary employee-derived indicators for the subsequent exploratory triangulation.

At the organizational level, we conceptualize AI-specific DCs within the established dimensions of sensing, seizing, and transforming [37]. Drawing on prior IS and public-administration research, we derived a literature-informed set of AI-specific sub-capabilities, as summarized in Table 1 and consolidated in Fig. 1. Rather than proposing an entirely novel theoretical model, the framework provides a context-sensitive specification of existing capability concepts for municipal public administration while remaining open to further refinement and extension.

Table 1. Definitions of AI-specific capability dimensions and sub-capabilities.

AI-Specific Sensing Capabilities	
Sub-Capability	Description
Strategic & Regulatory Foresight	The ability to systematically monitor technological and regulatory developments in AI [20]
AI Infrastructure Readiness	The availability of scalable, flexible, and interoperable digital infrastructures that enable AI experimentation and system deployment [4,24,30]
Data Readiness	The ability to identify, curate, govern, and maintain high-quality and interoperable data assets required for AI [7,22,39]
AI-Specific Seizing Capabilities	
AI Talent	The ability to attract, develop, and retain interdisciplinary expertise necessary to design, manage, and operationalize AI initiatives [23]
Analytical & Solution Development	The ability to translate identified AI opportunities into concrete solutions [24,36]
AI Management & Governance	The ability to coordinate, steer, and regulate AI initiatives through strategic alignment, oversight mechanisms, and clear decision-making structures [11,30,37]
Resource Mobilization	The ability to secure, allocate, and orchestrate financial and human resources to enable and scale AI initiatives [6,44]
AI-Specific Transforming Capabilities	
Organizational Transformation	The ability to redesign structures, processes, and governance arrangements so that AI outcomes become a sustained, organization-wide component rather than a pilot-level innovation [11,19,31,37,39]
Learning & Change Management	The ability to institutionalize learning processes and feedback loops that support continuous improvement in AI use [12,23,34]

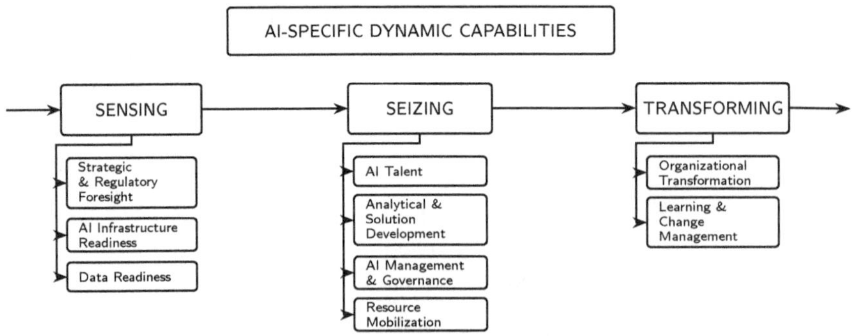

Fig. 1. AI-specific dynamic capabilities framework.

The framework served as the initial qualitative coding scheme and was subsequently refined through the analysis of the municipal cases. Its dimensions are conceptually distinct but dynamically interrelated: they may occur in parallel and mutually reinforce one another rather than follow a strict sequential order.

3 Methodology

3.1 Research Design

We employ a mixed-methods multi-case research design to examine the development of AI-specific dynamic capabilities (DCs) in public administrations [8,40]. The study combines qualitative expert-based assessments derived from structured focus group discussions across 19 German municipalities with a large-scale survey capturing perceptions of organizational innovativeness (OI) and AI literacy (AIL). The design enables exploratory triangulation between organizational capability assessments and employee-level perceptions using the municipality as a shared unit of analysis.

3.2 Data Collection

Data were collected between March and August 2025. The qualitative dataset consists of 19 focus group workshops conducted with German municipalities, involving 154 strategic decision-makers responsible for digitalization and AI-related initiatives. Discussions followed a semi-structured protocol [3], informed by prior research on digital transformation and dynamic capabilities as well as AI adoption in public sector contexts, to elicit consensual assessments of organizational capability maturity across predefined AI-specific dimensions.

To complement the qualitative insights, an employee survey was conducted in the same municipalities. OI was measured using five items adapted from Mikalef et al. [22]; AIL was measured using a three-item scale adapted from Pinski and Benlian [32]. All items were rated on a five-point Likert scale. A complete list of items is provided in the Appendix (see Table 7).

3.3 Data Analysis

Qualitative analysis followed a theory-informed coding approach based on the DCs framework of sensing, seizing, and transforming (see Fig. 1). The analysis focused on how AI-specific DCs manifest across municipalities, relying on structured, group-generated consensus statements as the primary unit of analysis. These statements were derived during the focus group discussions through structured moderation, documented as collective notes by the research team, and reflect collectively constructed meaning-making processes [14,15].

Structured group-generated records (hereafter referred to as transcripts for simplicity) were coded using MAXQDA Analytics Pro [41]. Coding proceeded iteratively, combining deductive coding based on the DC framework with inductive refinement of AI-specific sub-capabilities allowing the analytical framework to evolve in close interaction with the empirical material. This process ensured alignment between theory-driven categories and empirically grounded insights.

To ensure reliability and analytical consistency, multiple researchers independently reviewed the transcripts and reconciled coding decisions in structured sessions, following established practices in qualitative research [27,35].

To enable comparability across cases, qualitative evidence was transformed into calibrated municipality-level scores, following established calibration approaches [10,33]. For each sub-capability, coded evidence was systematically reviewed and synthesized at the case level. Final scores were assigned using a three-point ordinal scale indicating presence (+1), ambiguity (0), or absence (−1). Calibration decisions were made through structured team discussions and majority agreement based on the case-level evidence. Higher-order AI-specific DC dimensions were operationalized as the arithmetic mean of their calibrated sub-capability scores, resulting in composite municipality-level capability indices.

The quantitative analysis builds on the same municipal units. Individual survey responses were aggregated to municipality-level mean scores for OI and AIL [16]. These measures were used as complementary employee-level indicators rather than direct operationalizations of the DC dimensions. Pearson correlations were computed between the composite capability indices and the aggregated survey measures. Given the limited number of municipal cases and varying effective sample sizes, all statistical analyses are interpreted as exploratory rather than confirmatory.

4 Findings

4.1 AI-Specific DCs Across Municipalities

Qualitative analysis reveals that the development of AI-specific DCs across the nineteen municipalities remains fragmented, emergent, and only partially institutionalized. While most administrations exhibit early signs of activity across sensing, seizing, and transforming, these activities lack strategic coherence and are not yet embedded in stable organizational routines. Building on the deductively derived framework above (see Fig. 1), Fig. 2 illustrates how the conceptual framework was empirically refined based on the qualitative coding results across all nineteen municipalities. The following subsections elaborate how each capability class manifested in practice.

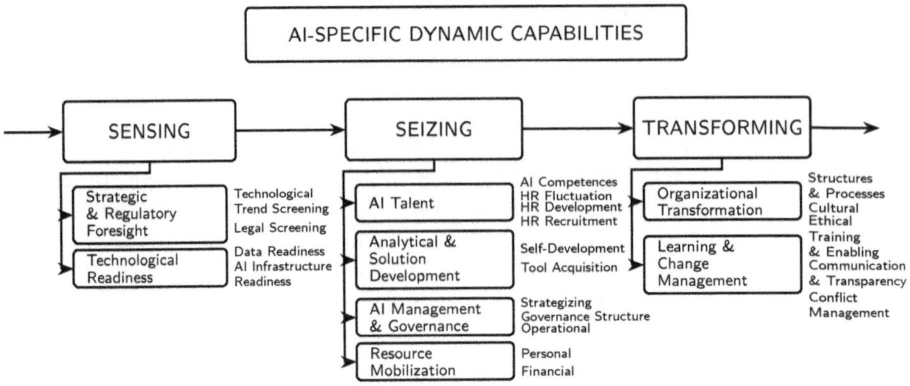

Fig. 2. Extended AI-specific dynamic capabilities framework.

AI-Specific Sensing Capabilities. Sensing capabilities emerged as the weakest and most unevenly developed domain. Although municipalities demonstrated some awareness of technological and regulatory trends, data readiness consistently represented the most pronounced gap across cases. Participants repeatedly emphasized the absence of a central data strategy, limited data governance structures, and the fragmentation of data across specialized departmental systems: "*Many data are stored in decentralized specialist procedures, partly unstructured and hybrid, for example, some parts on a network drive*" (C15) (originally documented in German and translated into English using DeepL; same procedure applies to all subsequent quotations). Another group described the situation more pointedly: "*No central overview. No data protection software in use so far*" (C13). A third expressed frustration at cultural barriers: "*There is a lack of strategy and also willingness to make one's own data accessible to others*" (C16). C refers to cases (municipalities), which are randomly numbered to ensure confidentiality.

Table 2 summarizes the occurrence of sensing-related sub-capabilities, highlighting the structural weakness of data readiness across municipalities. The reported occurrence values refer to the proportion of municipalities (n = 19) in which a given sub-capability was coded as present (+1) according to the calibration logic described in Sect. 3.3.

Table 2. Occurrence of AI-specific sensing capabilities

AI-specific Capabilities	AI-specific Sub-Capabilities	Occurrence
Strategic & Regulatory Foresight	Technological Trend Screening	63.16%
	Legal Screening	47.37%
Technological Readiness	Data Readiness	21.05%
	AI Infrastructure Readiness	52.63%

AI-Specific Seizing Capabilities. AI-specific seizing capabilities are more visible across municipalities than sensing or transforming, yet they remain at an early stage of maturity. The qualitative findings highlight that public administrations primarily pursue tool-oriented seizing, with many municipalities beginning to experiment with general-purpose AI applications such as ChatGPT or proprietary LLM platforms. This pattern is reflected in the relatively high occurrence of tool acquisition activities (78.95%), indicating that most administrations are in an exploratory phase of engaging with AI technologies.

Beyond these superficial tool-focused efforts, AI management & governance represent another set of seizing capabilities that are widely recognized but remain far from fully institutionalized. Across municipalities, participants consistently described strategic and governance frameworks as emerging but incomplete. Illustratively, some organizations reported that "*an AI competence center is under development*" (C13), while others acknowledged that "*there is no organization-wide AI strategy*" (C17). Strategic processes were frequently portrayed as only just beginning "*strategy development is now starting here*" (C14), with initial guiding principles being drafted and early alignment efforts with broader transformation agendas underway. This aligns with moderate occurrence levels for strategizing (47.37%) and governance structures (42.11%) across cases, reflecting efforts in progress but not yet institutionalized.

Despite this emerging awareness, these governance and strategic considerations have not yet translated into systematic operationalization. Participants repeatedly emphasized that AI is only loosely connected to existing digital strategies or, in some cases, not connected at all. As one group noted, "*currently no systematic operationalization in the digital strategy*" (C20).

Collectively, these accounts indicate that although governance and strategy are acknowledged as relevant, they largely remain aspirational directions rather than enacted and embedded capabilities, mirroring the overall early-stage maturity of seizing capabilities documented in Table 3.

AI-Specific Transforming Capabilities. AI-specific transforming capabilities occurred predominantly as training and communication activities, while deeper transformational practices remained sparse. For example, communication and transparency measures appeared in 78.95% of cases, and training and enabling activities in 63.16%, indicating that administrations are beginning to support employees through awareness and skill development.

Table 3. Occurrence of AI-specific seizing capabilities

AI-specific Capabilities	AI-specific Sub-Capabilities	Occurrence
AI Talent	AI Competences	52.63%
	HR Fluctuation	68.42%
	HR Development	42.11%
	HR Recruitment	36.84%
Analytical & Solution Development	Self-Development	31.58%
	Tool Acquisition	78.95%
AI Management & Governance	Strategizing	47.37%
	Governance Structure	42.11%
	Operational	26.32%
Resource Mobilization	Personal	57.89%
	Financial	47.37%

Yet more systemic elements of transformation, such as ethical governance structures, cultural adaptation, or conflict management capabilities, were far less prevalent. Participants described transformation as "*ongoing,*" (C17) or "*in early development*" (C19) with few cases reporting anchored or organization-wide transformation practices.

The distribution of occurrence rates is displayed in Table 4, where activities associated with cultural and ethical transformation appear in less than half of the municipalities.

Table 4. Occurrence of AI-specific transforming capabilities

AI-specific Capabilities	AI-specific Sub-Capabilities	Occurrence
Organizational Transformation	Structures & Processes	68.42%
	Cultural	47.37%
	Ethical	42.11%
Learning & Change Management	Training & Enabling	63.16%
	Communication & Transparency	78.95%
	Conflict Management	31.58%

4.2 Alignment with Employee-Level Indicators

Following the qualitative analysis, we examine whether the municipality-level capability assessments align with two aggregated employee-derived indicators: OI and AIL. OI captures employees' perceptions of organizational conditions, whereas AIL captures employees' self-reported competencies.

Organizational Innovativeness. Table 5 reports the Pearson correlations between the municipality-level assessments of AI-specific capability dimensions and the survey-based measure of OI. Positive coefficients indicate that municipalities with higher capability maturity tend to report higher levels of perceived OI. Several capability dimensions exhibit moderate to strong positive associations with OI. The strongest correlation is observed for Resource Mobilization (r = 0.6806**), followed by AI Talent (r = 0.5728*), AI Management & Governance (r = 0.5454*), and Strategic & Regulatory Foresight (r = 0.5114*). These correlations reach conventional levels of statistical significance. All remaining capability dimensions show weaker associations and do not reach statistical significance.

Table 5. Correlations between dynamic capability sub-dimensions and organizational innovativeness

AI-specific Sub-Capabilities	r	p-value
Strategic & Regulatory Foresight	0.5114*	0.0301
Technological Readiness	−0.0840	0.7404
AI Talent	0.5728*	0.0130
Analytical & Solution Development	0.3475	0.1577
AI Management & Governance	0.5454*	0.0192
Resource Mobilization	0.6806**	0.0019
Organizational Transformation	0.0063	0.9803
Learning & Change Management	−0.3077	0.2142

Note. Composite capability scores are calculated as the mean of the respective calibrated sub-capabilities within each DC dimension. Correlations are based on municipalities with available Organizational Innovativeness items (n = 18).

AI Literacy. Table 6 reports the correlations between the municipality-level assessments of AI-specific capability dimensions and the survey-based measure of AIL. Overall, the observed correlations between AIL and the capability dimensions are smaller in magnitude compared to those for OI. Moderate associations are observed for Resource Mobilization (r = 0.3755), Strategic & Regulatory Foresight (r = 0.3634), and AI Talent (r = 0.2841). All remaining capability dimensions show weaker correlations. None of the observed relationships reach conventional levels of statistical significance.

Table 6. Correlations between dynamic capability sub-dimensions and AI literacy

AI-specific Sub-Capabilities	r	p-value
Strategic & Regulatory Foresight	0.3634	0.1665
Technological Readiness	0.0171	0.9498
AI Talent	0.2841	0.2863
Analytical & Solution Development	0.2818	0.2903
AI Management & Governance	0.2038	0.4491
Resource Mobilization	0.3755	0.1518
Organizational Transformation	0.1317	0.6269
Learning & Change Management	0.0690	0.7997

Note: Composite capability scores are calculated as the mean of the respective calibrated sub-capabilities within each DC dimension. Correlations are based on municipalities with available AI Literacy items (n = 16).

5 Discussion

5.1 Key Findings

The qualitative analysis of the nineteen municipalities demonstrates that the concept of AI capabilities, understood here as AI-specific DCs, can be meaningfully applied as analytical framework to explain AI adoption in public administrations. Overall, the findings show that municipalities presently exhibit only limited AI-specific DCs. Across cases, the most pronounced deficiencies occur in the sensing dimension, especially regarding data readiness. Municipalities lack a coherent data (quality) management framework or a central data platform; instead, data remain dispersed across departmental applications, often with little to no awareness of pre-existing data assets. As a result, potential synergies remain unused. Seizing capabilities are more visible but remain at an early maturity stage. Currently, public administrations' seizing activities are limited to the acquisition and approval of general-purpose AI tools such as ChatGPT (or other LLMs), while other seizing capabilities are only starting to emerge: AI strategies and guiding principles are frequently under development, governance arrangements remain largely preparatory and have not yet translated into institutionalized organizational capabilities. This lack of strategic consolidation also limits progress in the transforming dimension, where current activities consist mainly of isolated training efforts but do not yet indicate systematic organizational capabilities for sustainable and profound organizational change. The quantitative analysis complements these qualitative insights by linking the expert-based capability assessments with employee-level indicators. The results show that OI exhibits moderate to strong associations with several AI-specific capability dimensions, including strategic and regulatory foresight, AI talent, AI management and governance, and resource mobilization. These correlations suggest that municipalities assessed as more mature in their AI-specific capabilities are also perceived by employees as more innovative organizations. This pattern provides initial evidence that organizational capability development becomes visible in employees' perceptions of their organization.

5.2 Implication for Research

The study provides a literature-informed and empirically refined specification of AI-specific DCs in public administration. By consolidating previously fragmented capability concepts for AI within the established sensing, seizing, and transforming dimensions, the framework offers a context-sensitive analytical structure for examining AI adoption in municipal administrations while remaining open to further refinement and extension in other administrative contexts. Furthermore, the empirical analysis identifies recurring patterns in the presence and absence of these capabilities across nineteen German municipalities. The findings provide a differentiated account of where AI-specific capabilities are already emerging and where substantial capability gaps persist. From a DCs perspective, this imbalance is noteworthy, as sensing capabilities are often considered a prerequisite for recognizing technological opportunities and initiating capability development [37]. This observation aligns with prior research suggesting that early stages of AI adoption in the public sector are frequently characterized by exploratory and technology-driven initiatives rather than strategically integrated capability development [38,39] and enriches the ongoing scholarly discourse by illustrating both the potential and the structural limitations of AI-specific DC development in public administration. A second contribution lies in the mixed-methods design that combines expert-based assessments of organizational AI-specific DCs with employee-level indicators. By linking municipality-level capability profiles with survey-based measures of OI [22] and AIL [32], the study introduces a cross-level perspective on capability development. This approach provides initial empirical evidence on whether patterns identified in expert assessments of AI-related capabilities are reflected in employee-level indicators. In doing so, the study highlights the role of employee-level competencies such as AIL as potential microfoundations of AI-specific DCs, while OI provides a complementary perception-based perspective that allows triangulating expert-based assessments of organizational capability development with employees' views of their organization.

5.3 Implication for Practice

The findings offer several actionable suggestions for public administrations. The weak development of AI-specific sensing capabilities across cases, particularly the data readiness characteristics, suggests the need to establish more robust data management foundations. Administrations will remain limited to superficial forms of AI adoption without coherent data governance, central data management and awareness of existing data assets, leaving them unable to utilize the full potential. The observed seizing capabilities are dominated by a tool-oriented focus. Instead, administrations should invest in enabling structures supporting a more strategic and sustainable approach to AI adoption. This includes designing AI governance frameworks, clarifying roles and responsibilities, and developing talents capable of evaluating, implementing, and refining AI applications in public administration. Similarly, the scarce occurrence of transforming capabilities should encourage public administrations to embed AI-specific change initiatives in more breadth and depth among organizational learning processes. Training programs, reflective learning spaces, and structured experimentation can help administrations build the organizational capabilities required to navigate the opportunities, risks and constraints of AI adoption. Overall, our insights suggest for a more coordinated and capability-oriented approach to AI adoption in public administrations. Implementing strategies that advance past isolated tool use, such as establishing structured, culture-driven, learning-oriented foundations for sustainable AI integration.

6 Limitations and Future Work

While our study provides empirically grounded insights into the current state of AI-specific DCs in public administrations, it is still beset with limitations that offer promising avenues for future work. First, case selection depended on municipalities' willingness to participate in the study, which may introduce self-selection bias. E.g., municipalities less engaged in digital transformation may be underrepresented due to capacity constraints. Second, quantifying qualitative findings simplifies complex organizational practices. Transforming coded segments into an ordinal scale (-1, 0, 1) enabled quantitative exploration but reduced the richness inherent in the underlying qualitative material. Future research may focus further by conducting within-case analyses, enabling a more granular understanding of how AI-specific capabilities unfold in individual municipalities. Third, the occurrence of correlation does not necessarily allow for conclusions about causality. While the correlation analysis linked the organizational-level capabilities with the individual-level employee-perceptions, this does not constitute causation or directionality; the identified associations should therefore be interpreted as indicators rather than explanations or predictions. Future work could more systematically investigate microfoundational mechanisms of AI capability development. In particular, longitudinal and multi-level designs could help unpack how individual perceptions, departmental structures, and organizational routines jointly shape the emergence of dynamic capabilities over time, thereby moving beyond cross-sectional correlation toward process-oriented explanations. Furthermore, because employee responses are aggregated at the municipality level and effective sample sizes vary across constructs, statistical significance should be interpreted with caution and considered exploratory.

7 Conclusion

In this paper, we examined the development of AI-specific DCs in public administrations and explored whether these organizational capability patterns are reflected in employee-level indicators. To address these objectives, we conducted a mixed method approach consisting of a qualitative analysis of nineteen municipalities across Germany and complemented these insights with an exploratory post hoc correlation analysis linking capability assessments with measures of perceived organizational innovativeness and AI literacy. Our findings show that the extent to which public administrations demonstrate AI-specific DCs remains nascent. While municipalities have begun experimenting with AI applications, the strategic development of AI-specific DCs remains at an early stage, with sensing capabilities—particularly data readiness—representing the most substantial constraint. The exploratory quantitative analysis indicates that perceived organizational innovativeness is moderately to strongly associated with several AI-specific (sub-)capability dimensions, thereby complementing prior research on AI-related capabilities [23]. In contrast, associations between AI literacy and the capability dimensions are weaker and remain statistically non-significant.

Disclosure of Interests. The authors have no competing interests to declare that are relevant to the content of this article.

A Appendix

Table 7. Survey instrument

Measure	Item
Organizational Innovativeness (Mikalef et al. 2022)	OI1. In my work unit, there is a strong openness toward digital change
	OI2. In my work unit, we actively learn from mistakes and challenges related to digital change
	OI3. In my work unit, we regularly exchange ideas to better implement digital change
	OI4. In my work unit, we actively participate in shaping digital change
	OI5. My direct supervisor actively drives digital change in our work unit
AI Literacy (Pinski & Benlian 2023)	AIL1. In general, I know the unique facets of AI and humans and their potential roles in human–AI collaboration
	AIL2. I am knowledgeable about the steps involved in AI decision-making
	AIL3. Considering all my experience, I am relatively proficient in the field of AI

References

1. Ahn, M.J., Chen, Y.C.: Digital transformation toward AI-augmented public administration: the perception of government employees and the willingness to use AI in government. Gov. Inf. Q. **39**(2), 101664 (2022)
2. Albort-Morant, G., Leal-Rodríguez, A.L., De Marchi, V.: Absorptive capacity and relationship learning mechanisms as complementary drivers of green innovation performance. J. Knowl. Manag. **22**(2), 432–452 (2018)
3. Baxter, P., Jack, S., et al.: Qualitative case study methodology: study design and implementation for novice researchers. The qualitative report **13**(4), 544–559 (2008)
4. Bharadwaj, A.S.: A resource-based perspective on information technology capability and firm performance: an empirical investigation1. MIS Q. **24**(1), 169–196 (2000)
5. Bovens, M.: Analysing and assessing accountability: a conceptual framework 1. Eur. Law J. **13**(4), 447–468 (2007)
6. Bryson, J.M., Crosby, B.C., Bloomberg, L.: Public value governance: moving beyond traditional public administration and the new public management (2014)
7. Chen, H., Chiang, R.H., Storey, V.C.: Business intelligence and analytics: from big data to big impact (2012)
8. Creswell, J.W., Clark, V.L.P.: Designing and Conducting Mixed Methods Research. Sage Publications (2017)
9. Eisenhardt, K.M., Martin, J.A.: Dynamic capabilities: what are they? In: The SMS Blackwell Handbook of Organizational Capabilities, pp. 341–363 (2017)
10. Fiss, P.C.: Building better causal theories: a fuzzy set approach to typologies in organization research. Acad. Manag. J. **54**(2), 393–420 (2011)
11. Floridi, L.: Ethics, governance, and policies in artificial intelligence, vol. 144. Springer (2021)
12. Gupta, M., George, J.F.: Toward the development of a big data analytics capability. Inf. Manag. **53**(8), 1049–1064 (2016)
13. Helfat, C.E., Peteraf, M.A.: The dynamic resource-based view: capability lifecycles. Strateg. Manag. J. **24**(10), 997–1010 (2003)
14. Kitzinger, J.: The methodology of focus groups: the importance of interaction between research participants. Sociol. Health Illness **16**(1), 103–121 (1994)
15. Krueger, R.A.: Focus Groups: A Practical Guide for Applied Research. Sage Publications (2014)
16. Kumar, N., Stern, L.W., Anderson, J.C.: Conducting interorganizational research using key informants. Acad. Manag. J. **36**(6), 1633–1651 (1993)
17. Kuziemski, M., Misuraca, G.: AI governance in the public sector: three tales from the frontiers of automated decision-making in democratic settings. Telecommun. Policy **44**(6), 101976 (2020)
18. Lai, V.S., Guynes, J.L.: A model of ISDN (integrated services digital network) adoption in u.s. corporations. Inf. Manag. **26**(2), 75–84 (1994)
19. Mergel, I., Edelmann, N., Haug, N.: Defining digital transformation: results from expert interviews. Gov. Inf. Q. **36**(4), 101385 (2019)
20. Mergel, I., Kattel, R., Lember, V., McBride, K.: Citizen-oriented digital transformation in the public sector. In: Proceedings of the 19th Annual International Conference on Digital Government Research: Governance in the Data Age, pp. 1–3. ACM (2018)

21. Mikalef, P., Gupta, M.: Artificial intelligence capability: conceptualization, measurement calibration, and empirical study on its impact on organizational creativity and firm performance. Inf. Manag. **58**(3), 103434 (2021)
22. Mikalef, P., et al.: Enabling ai capabilities in government agencies: a study of determinants for European municipalities. Gov. Inf. Q. **39**(4), 101596 (2022)
23. Mikalef, P., et al.: Examining how AI capabilities can foster organizational performance in public organizations. Gov. Inf. Q. **40**(2), 101797 (2023)
24. Mikalef, P., Pateli, A.: Information technology-enabled dynamic capabilities and their indirect effect on competitive performance: findings from PLS-SEM and fsQCA. J. Bus. Res. **70**, 1–16 (2017)
25. Miller, M., Ghaffarzadegan, N.: Dynamic capabilities in the public sector: a systematic literature review. Int. J. Public Sect. Manag. **38**(6), 717–734 (2025)
26. Moore, M.H.: Creating Public Value: Strategic Management in Government. Harvard University Press (1997)
27. Nowell, L.S., Norris, J.M., White, D.E., Moules, N.J.: Thematic analysis: striving to meet the trustworthiness criteria. Int. J. Qual. Methods **16**(1) (2017)
28. Oliveira, T., Martins, M.F.: Information technology adoption models at firm level: review of literature. In: The European Conference on Information Systems Management, p. 312. Academic Conferences International Limited (2010)
29. Panagiotopoulos, P., Protogerou, A., Caloghirou, Y.: Dynamic capabilities and ICT utilization in public organizations: an empirical testing in local government. Long Range Plan. **56**(1), 102251 (2023)
30. Pavlou, P.A., El Sawy, O.A.: Understanding the elusive black box of dynamic capabilities. Decis. Sci. **42**(1), 239–273 (2011)
31. Piening, E.P.: Dynamic capabilities in public organizations: a literature review and research agenda. Public Manag. Rev. **15**(2), 209–245 (2013)
32. Pinski, M., Benlian, A.: AI literacy - towards measuring human competency in artificial intelligence. In: Hawaii International Conference on System Sciences (2023)
33. Ragin, C.C.: Measurement versus calibration: a set-theoretic approach. In: Box-Steffensmeier, J.M., Brady, H.E., Collier, D. (eds.) The Oxford Handbook of Political Methodology, 1 edn., pp. 174–198. Oxford University Press (2009)
34. Rashman, L., Withers, E., Hartley, J.: Organizational learning and knowledge in public service organizations: a systematic review of the literature. Int. J. Manag. Rev. **11**(4), 463–494 (2009)
35. Saldaña, J.: The Coding Manual for Qualitative Researchers, 4e edn. SAGE (2021)
36. Shamim, S., Zeng, J., Shariq, S.M., Khan, Z.: Role of big data management in enhancing big data decision-making capability and quality among Chinese firms: a dynamic capabilities view. Inf. Manag. **56**(6), 103135 (2019)
37. Teece, D.J.: Explicating dynamic capabilities: the nature and microfoundations of (sustainable) enterprise performance. Strateg. Manag. J. **28**(13), 1319–1350 (2007)
38. Van Noordt, C., Misuraca, G.: Artificial intelligence for the public sector: results of landscaping the use of AI in government across the European union. Gov. Inf. Q. **39**(3), 101714 (2022)
39. Van Noordt, C., Tangi, L.: The dynamics of AI capability and its influence on public value creation of AI within public administration. Gov. Inf. Q. **40**(4), 101860 (2023)
40. Venkatesh, V., Brown, S.A., Bala, H.: Bridging the qualitative-quantitative divide: guidelines for conducting mixed methods research in information systems1. MIS Q. **37**(1), 21–54 (2013)
41. VERBI Software: MAXQDA Analytics Pro 24. VERBI Software, Berlin, Germany. https://www.maxqda.com(2024)

42. Vial, G.: Understanding digital transformation: a review and a research agenda. J. Strateg. Inf. Syst. **28**(2), 118–144 (2019)
43. Wade, H.: Review: the resource-based view and information systems research: review, extension, and suggestions for future research. MIS Q. **28**(1), 107 (2004)
44. Warner, K.S., Wäger, M.: Building dynamic capabilities for digital transformation: an ongoing process of strategic renewal. Long Range Plan. **52**(3), 326–349 (2019)
45. Wernerfelt, B.: A resource-based view of the firm. Strateg. Manag. J. **5**(2), 171–180 (1984)
46. Wyszynski, M., Westermann, J., Weber, S., Niehaves, B.: GenAI improves decision-making: revealing the debiasing capabilities of ChatGPT. In: ICIS 2025 Proceedings (2025)

Author Index

Published by Springer Nature Switzerland AG 2027
D. Duenas-Cid et al. (Eds.): ePart 2026, LNCS 16923, pp. 431–432, 2027.
https://doi.org/10.1007/978-3-032-36676-4

FSC
www.fsc.org
MIX
Papier | Fördert gute Waldnutzung
FSC® C083411

Kolibri 360 GmbH
Ferdinand-Jühlke-Straße 7
99095 Erfurt, Deutschland
produktsicherheit@kolibri360.de